AF800249

Studies in Brain and Mind

Volume 17

Series Editor
Gualtiero Piccinini, University of Missouri - St. Louis, St. Louis, MO, USA

Editorial Board
Berit Brogaard, University of Oslo, Norway, University of Miami, Coral Gables, FL, USA
Carl Craver, Washington University, St. Louis, MO, USA
Edouard Machery, University of Pittsburgh, Pittsburgh, PA, USA
Oron Shagrir, The Hebrew University of Jerusalem, Jerusalem, Israel
Mark Sprevak, University of Edinburgh, Scotland, UK

More information about this series at http://www.springer.com/series/6540

Fabrizio Calzavarini • Marco Viola
Editors

Neural Mechanisms

New Challenges in the Philosophy of Neuroscience

Springer

Editors
Fabrizio Calzavarini
Department of Letter, Philosophy, Communication
University of Bergamo LLC, Turin, Italy

Marco Viola
Department of Philosophy and Education
University of Turin
Turin, Italy

ISSN 1573-4536 ISSN 2468-399X (electronic)
Studies in Brain and Mind
ISBN 978-3-030-54091-3 ISBN 978-3-030-54092-0 (eBook)
https://doi.org/10.1007/978-3-030-54092-0

© Springer Nature Switzerland AG 2021
This work is subject to copyright. All rights are reserved by the Publisher, whether the whole or part of the material is concerned, specifically the rights of translation, reprinting, reuse of illustrations, recitation, broadcasting, reproduction on microfilms or in any other physical way, and transmission or information storage and retrieval, electronic adaptation, computer software, or by similar or dissimilar methodology now known or hereafter developed.
The use of general descriptive names, registered names, trademarks, service marks, etc. in this publication does not imply, even in the absence of a specific statement, that such names are exempt from the relevant protective laws and regulations and therefore free for general use.
The publisher, the authors, and the editors are safe to assume that the advice and information in this book are believed to be true and accurate at the date of publication. Neither the publisher nor the authors or the editors give a warranty, expressed or implied, with respect to the material contained herein or for any errors or omissions that may have been made. The publisher remains neutral with regard to jurisdictional claims in published maps and institutional affiliations.

This Springer imprint is published by the registered company Springer Nature Switzerland AG
The registered company address is: Gewerbestrasse 11, 6330 Cham, Switzerland

Contents

Chapter 1
Introduction: New Challenges in the Philosophy of Neuroscience

Fabrizio Calzavarini and Marco Viola

Abstract The present volume consists of new papers by leading philosophers of neuroscience advancing debates concerning foundational, conceptual and methodological issues in cognitive and systems neuroscience, as well as neuroscientifically inspired philosophy of mind. This introductory chapter presents the aims of the volume and provides a short overview of each contribution.

In 1978, the Sloan Foundation commissioned a State-of-the-Art report on Cognitive Science. The experts from the six disciplines involved in the research program surveyed the status of their own fields, as well as the interdisciplinary connections between them. Unlike pairings such as psychology and neuroscience, or philosophy and linguistics, each of which constituted a "well defined area of inquiry which involves the intellectual and physical tools of the two disciplines it ties together" (State of the art committee: 3), the dialogue between philosophy and neuroscience was portrayed as "a set of issues, some already familiar and important, which have not yet become the focus of formally recognized scholarly effort" (State of the art committee: 4). In short, something like a "philosophy of neuroscience" was nowhere to be found. However, this vacuum was soon to be filled. Less then a decade later, Churchland published *Neurophilosophy*, a book whose explicit aim was "to introduce philosophy and neuroscience, each to the other" (1986: 6). Since then, and partially because of this, two strands began to take shape: "philosophy of neuroscience", conceived of as a branch of philosophy of science that deals with foundational issues in neuroscience; and "neurophilosophy", referring to any application of neuroscientific concepts and evidence to traditional philosophical topics. The distinction between these two strands is somehow imperfect. Indeed, epistemological discussions about how to assign psychological labels to brain

F. Calzavarini (✉)
Department of Letter, Philosophy, Communication, University of Bergamo LLC, Turin, Italy
e-mail: fabrizio.calzavarini@unibg.it

M. Viola
Department of Philosophy and Education, University of Turin, Turin, Italy

© Springer Nature Switzerland AG 2021
F. Calzavarini, M. Viola (eds.), *Neural Mechanisms*, Studies in Brain and Mind 17,
https://doi.org/10.1007/978-3-030-54092-0_1

structures are likely to be relevant to, and to be affected by, one's metaphysical view on the mind-body problem.

Yet, notwithstanding the neurohype that surrounded the "Decade of the Brain" (1990–1999), when (in 2008) Gold and Roskies rhetorically asked "Is there a philosophy of neuroscience?" (2008: 2), their answer was still a timid "yes and no": while neurophilosophy might have had its share of attention, they claimed, "there are but a handful of philosophers of science who focus on neuroscience".

But now, 12 year later, we think that the time is ripe to answer that same question with a confident "yes". Our confidence is fueled by multiple factors. To name but a few: (a) a cursory query of "Philosophy of neuroscience" on Google Scholar finds some 2200 items, ¾ of which have been published after 2008; (b) dedicated masters programmes and summer schools can be found that bring together philosophy and neuroscience, such as the masters programme on "Philosophy of Neuroscience" at the Vrije Universiteit of Amsterdam or the *Summer Seminars in Neuroscience and Philosophy* held at Duke University since 2016; (c) the *Stanford Encyclopedia of Philosophy* hosts a dedicated section on "Philosophy of Neuroscience" since 1999, revised each 5 years; (e) in 2000, the triannual journal *Brain and Mind* was launched to collect contributions on philosophy of neuroscience and neurophilosophy. Even though by the end of 2003 the journal itself ceased to exist as such, the following year it became *de facto* a dedicated yearly section of the established philosophy journal *Synthese*, on the topic of "Neuroscience and its Philosophy"; (f) in PhilPapers, the most well-established database of philosophical writings, a specific subsection of "Philosophy of Science" is now devoted to the "Philosophy of Neuroscience".

More recently, we (i.e., the editors of this volume) have also made a modest contribution to establishing the philosophy of neuroscience by hosting *Neural Mechanisms [NM] Online*, a series of online seminars (webinars) open to anyone with an internet connection. Even before the COVID pandemic, webinars were becoming very popular in academic research due to the fact they are relatively inexpensive and can bring together experts on a given topic from all over the world. They can also be recorded so that other people can view later if they cannot attend the webinar in realtime. NM Online is the first world-wide webinar series dedicated entirely to the interaction between philosophy and neuroscience. As we write this introduction, NM Online is terminating its third year of activity[1]. Invited speakers so far have included several of the most established philosophers of neuroscience, along with some promising younger researchers. In every session, the speaker presents their paper (that has previously been shared via mailing list) and then defends it against three to five discussants (that we have previously selected on the basis of their expertise) and larger online audience (*attendees*). Right before each session, speakers and participants receive by email an invitation to join the seminar. All the sessions are available later on our YouTube channel, and reposted through the Facebook page *Neural Mechanisms Online* and Twitter account @NeuralMech.

[1]More information about the NM Online project can be found at www.neuralmechanisms.org

Overall, our mailing list includes more than 700 subscribers – and it is still growing. We have also organized a two-day online conference, the *NM Online Webconference 2018*, focusing on the topic of "New Challenges in the Philosophy of Neuroscience", i.e., the new epistemological problems and philosophical opportunities prompted by the most recent development in cognitive and systems neuroscience.

These "New Challenges in the Philosophy of Neuroscience" are also the main topic of the present volume, which builds on the experience of NM Online. The volume comprises many of the (revised and improved) articles discussed during the *NM Online 2018* series of webinars, some of the articles discussed in the *NM Online 2019* webinars, as well as contributions from some of the scholars that participated in the *NM Online* events as discussants. The contributed articles pertain to five relevant fields in current philosophy of neuroscience and neuroscientifically-inspired philosophy of mind: new forms of explanation and prediction developed in cognitive neuroscience (Sect. 1), new concepts/methods/techniques used in this field (Sect. 2), new metaphysical challenges arising from neuroscience (Sect. 3), the relation between brain sciences and mechanistic philosophy, including some issues concerning the mechanistic framework more generally (Sect. 4), and the issue of neural computations and representations (Sect. 5).

The first section opens up with Kaplan and Hewitson's article discussing the explanatory status of Bayesian modelling approaches, which are becoming increasingly popular in contemporary neuroscience, and their relation with mechanistic approaches (Chap. 2). They focus on the work of Colombo and Hatrmann (2017), one of the most developed accounts in the literature. In Chap. 3, Gessell, Stanely, Geib, and De Brigard claim that, besides the traditional focus on explanatory power, philosophers' assessment of neuroscientific frameworks should also take into account their predictive power. They argue that network neuroscience offers extremely powerful topological models for studying and predicting a number of brain-related phenomena, and that network approaches have allowed researchers to make powerful, useful predictions, regardless of whether the topological properties used in making those predictions also yield mechanistic explanations. In Chap. 4 Plebe discusses the contrast between circuital and developmental explanations in neuroscience. According to Plebe, developmental explanations provide a better explanation of an apparent tension about the cortex (i.e., the variety of its functions in the face of the uniformity of its structure), and can also be taken into account within a mechanistic framework. Section 1 of the volume closes with Chap. 5 by Weiskopf, discussing the how multivariate pattern analysis (MPVA) fares with respect of the problem of reverse inference. Weiskopf argues that MVPA faces some pervasive methodological and interpretative problems and, for this reason, it cannot provide a new solution to some of the traditional epistemic worries relating to reverse inference. He also explores a further concern, namely that the interest toward prediction that MVPA and such techniques bring about comes at the expense of explanation.

Burnston and Haueis open the second section of the volume by discussing the concept of "hierarchy" as used in systems neuroscience (Chap. 6). They explore various usages of this concept in the literature and classify them into two strands: the "representational" and the "topological" approaches. They then explore various possible relationships between representational and topological notions of hierarchy, opening a conceptual space in which further reasoning about neural hierarchy can proceed. In Chap. 7, Favela discusses another popular concept in current cognitive neuroscience, "neural reuse". He argues that neural reuse is not in itself part of a fundamental theory of brain structure and function. Instead, it is more appropriately understood as a particular mechanism of brain organization that is subsumed by a more fundamental and general theory, i.e., Neural Darwinism. In Chap. 8, Wright addresses the issue of the epistemic gap existing between experimental manipulations which produce data and their subsequent analysis. As the analysis might occur somewhere else from the production, this gap might bring about burdensome epistemic problems. Wright suggests that this problem is dealt with thanks to some 'epistemic frictions' in data manipulation, which he illustrates in relation to some new methods for detecting temporal dynamics of networks developed in Poldrack's lab. In Chap. 9 Rathkopf also discusses the concept of "neural reuse", that is, the functional reuse of a neural structure for multiple conceptually distinct tasks. He argues that, when reasoning in evolutionary terms, neural reuse must be conceptualized more abstractly than has been generally recognized and must be conceived not as a process that constrains our cognitive capacities, but as a process that liberates those capacities from evolutionary constrains. In the final chapter of the section (10), Raja and Anderson introduce the notion of an "enabling constraint" as a new conceptual tool to make sense of scalar relations in the nervous system.

At the beginning of Sect. 3 (Chap. 11), Chirimuuta discusses the explanatory value of the brain-computer comparison (e.g., circuit models of neurons and brain areas since McCulloch and Pitts [1943]) for current neuroscience. Chirimuuta argues that the relation between brain and computer should be understood as one of analogy and considers the implications of this interpretation for notions of multiple realization. Nathan (Chap. 12) offers an historical overlook of how philosophers and scientists dealt with the relation between mind and brain, identifying some shifts in what has fallen under the moniker 'the mind-body problem' over time, and trying to reframe the issue in more contemporary, neuroscientific terms. In Chap. 13, Vernazzani explores the notion of "psychoneural isomorphism", introduced by Gestalt psychologists at the beginning of the twentieth century to explain the relationship between mind and brain. The aim of his article is to provide a conceptual roadmap of psychoneural isomorphism, in order to dispel some potential misunderstanding concerning this notion and identify its precise role with reference to contemporary debates. In the final part of the article, he focuses on one example of psychoneural isomorphism from the work of Jean Petitot. In Chap. 14, Dewhurst discusses the relation between our commonsense ontology of the mental and the ontological taxonomies proposed by current cognitive neuroscience. Are they incompatible or incommensurable? He defends an 'interpretivist' approach, according to which folk psychology aims to describe coarse-grained behaviour rather than fine-grained mechanisms, and according to which the two kinds of ontology are better thought of as incommensurable rather than incompatible.

Secttion 4 deals with mechanistic explanation, arguably the most popular framework in philosophy of neuroscience. Kästner (Chap. 15) distinguishes three different kinds of explanatory projects (the "mechanistic triad"), having to do, respectively, with (i) mechanisms that explain how a given *end product* is generated; (ii) mechanisms that underlie a given *process*, and (iii) mechanisms that maintain a given *stable state* or *continuous behaviour*. Kästner critically discusses the interrelations between these three explanatory projects, providing a ground for explanatory integration within cognitive neuroscience (and within life sciences more in general). The aim of Chap. 16 (by Silberstein) is to defend a previous paper by Silberstein and Chemero (2013) from the criticisms that have been raised against it, bolstering the original claim that there are some biological and cognitive phenomena that fail in principle to be explained by localization and decomposition (i.e., the hallmarks of mechanistic explanation). Section 4 also includes also a chapter (17) by Krickel and Kohar discussing the claim that only the *relevant* details should be included in a satisfactory mechanistic explanation, and in particular Craver and Kaplan's (2020) version of this claim. Krickel and Kohar suggest some modification to Craver and Kaplan's version of the claim, as well as some potential challenges to their modified version and some replies to these challenges.

In the first chapter of the final section (Chap. 18), Colombo distinguishes different notions of *miscomputation* that are relevant for computational psychiatry. He argues that an adequate explication of miscomputation in this discipline should be explicated as "interest-relative and perspectival, although non-arbitrary, relatively clear-cut, experimentally evaluable, and instrumentally useful". Colombo then considers some implications of this claim for the adequacy of the mechanistic view in computational sciences more generally. Klein and Clutton (Chap. 19) focus on representations, rather than computation. They discuss Ramsey's "job description challenge" in the context of the cognitive science of body representations, distinguishing three possible readings of it. In their view, only one reading of this challenge is interesting, the one that integrates what they call "esoteric" and "exoteric" semantic issues. The chapters by Brozzo (Chap. 20) and Orbán and Wang (Chap. 21) also consider the case of bodily representations. In her chapter, Brozzo presents an empirical hypothesis about the way in which some bodily actions, i.e. motor actions such as grasping or reaching for something, are perceptually represented in the brain. According to this proposal, motor actions can be categorically perceived. This is compatible with a certain view about the neural mechanisms underlying the processing of motor actions: the latter could be mediated by the occurrence of motor processes in the observer. Finally, Orbán and Wang discuss De Vignemont's challenges to the internal account of bodily immunity to error through misidentification. To meet these challenges, they proposed two modified versions of the internal account: the "new internal model" and the "ecological model".

Overall, the volume provides up to date discussions of a variety of topics within philosophy of neuroscience, thus qualifying as an essential addition to several bookshelves and graduate level syllabi on philosophy of neuroscience. Needless to say, the volume does not aim at complete coverage. For instance, we have not dealt with the topic of predictive processing (an interested reader could refer to Metzinger and Wiese 2017). Nor have we been able to explore some of the exciting new techniques that are rapidly reshaping the landscape of experimental practices, such as optogenetics (see e.g., Bickle 2018). Indeed, being a vivacious and ever-growing field, philosophy of neuroscience comprises many more interesting topics than any single book can reasonably deal with. But NM Online is only in its third year of activity, and we suggest that it is not going to stop anytime soon: thus, in the following years we will do our best to cover these topics also, along with others that will surely emerge. One of the most rewarding outcomes of NM Online is that we have had the occasion to meet an astonishing number of colleagues of various ages and from various parts of the world. We dare to say that something like an international community is beginning to take shape. Without this community, neither NM Online nor this very book would have been possible. We, therefore, express our sincere gratitude to all those who contributed (and will contribute) to NM Online: to all the speakers and the discussants (see the complete list below), to all the attendees, thank you very much! We are also grateful to the Brains Blog, and especially to Nick Byrd, for helping us to disseminate the activity of NM Online; to Joe Dewhurst, who manages NM Online's Twitter account; and to the University of Turin, that made the software for NM Online available. We are also especially grateful to Brendan Ritchie for having reviewed painstakingly every single chapter and having provided useful suggestions to all authors. Last but not least, a final thanks is for Gualtiero Piccinini, not only because of all he has done for the philosophy of neuroscience, but also because, as the editor of the book series *Studies in Brain & Mind*, he encouraged us and supported the idea of this book. We hope that you will enjoy it.

1.1 Speakers of NM Online (2018, 2019, 2020 Webinars and Other Events)

Mike Anderson, Chiara Brozzo, Dan Burnston, Rosa Cao, Fausto Caruana, Mazviita Chirimuuta, David Colaco, Matteo Colombo, Carl Craver, Felipe De Brigard, Ophelia Deroy, Joe Dewhurst, Frances Egan, Luis Favela, Carrie Figdor, Bryce Gessell, Javier Gomez-Lavin, Matteo Grasso, Julia Haas, Philipp Haueis, Dan Hutto, Annelli Janssen, David M. Kaplan, Lena Kästner, Colin Klein, Matej Kohar, Beate Krickel, Edouard Machery, Manolo Martinez, Joseph McCaffrey, Marcin Miłkowski, Ruth Millikan, Marco Nathan, Luiz Pessoa, Gualtiero Piccinini, Alessio Plebe, Russ Poldrack, Vicente Raja, William Ramsey, Charles Rathkopf, Brendan Ritchie, Sarah Robins, Adina Roskies, Miguel Segundo-Ortin, Micheal Silberstein, Jackie Sullivan, Alfredo Vernazzani, Abel Wajnerman Paz, Zina Ward, Dan Weiskopf, Hong Yu Wong, Jessey Wright, and Karen Yan.

1.2 Discussants of NM Online (2018, 2019, 2020)

Mike Anderson, Alessandra Buccella, Cameron Buckner, David Barack, Dan Burnston, Wayne Christensen, Matteo Colombo, Mazviita Chirimuuta, Dimitri Coelho-Mollo, Lindley Darden, Felipe De Brigard, Joe Dewhurst, Kris Dolega, Frances Egan, Luis Favela, Carrie Figdor, Mindt Garrett, Sarah Genon, Matteo Grasso, Philipp Haueis, Casper Hesp, Eric Hochstein, Lena Kästner, Colin Klein, John Krakauer, Beate Krickel, Nikolaus Kriegeskorte, Juan Loaiza, Corey Maley, Francesco Marchi, Joseph McCaffrey, Marcin Miłkowski, Marco Nathan, David Papineau, Alfredo Paternoster, Carlotta Pavese, Gualtiero Piccinini, Alessio Plebe, Tom Polger, Emily Pritchyko, Maxwell Ramstead, Sofiia Rappe, Vicente Raja, Charles Rathkopf, Ernesto Restrepo, Brendadn Ritchie, Sarah Robins, Adina Roskies, Lauren Ross, Nicolás Serrano, Henry Shevlin, Michael Silberstein, Catherine Stinson, Marco Tamietto, Ruben Verhagen, Alfredo Vernazzani, Alberto Voltolini, Naftali Weinberger, Dan Weiskopf, Christopher Whyte, Iwan Williams, and Jessey Wright.

References

Bickle, J. (2018). From microscopes to optogenetics: Ian Hacking vindicated. *Philosophy of Science, 85*(5), 1065–1077.

Churchland, P. S. (1986). *Neurophilosophy: Toward a unified science of the mind/brain*. Cambridge, MA: MIT Press.

Colombo, M., & Hartmann, S. (2017). Bayesian cognitive science, unification, and explanation. *The British Journal for the Philosophy of Science, 68*(2), 451–484.

Craver, C., & Kaplan, D. M. (2020). Are more details better? On the norms of completeness for mechanistic explanations. *The British Journal for the Philosophy of Science, 71*(1), 287–319.

Gold, I., & Roskies, A. L. (2008). Philosophy of neuroscience. In M. Ruse (Ed.), *The Oxford handbook of philosophy of biology* (pp. 349–380). Oxford: Oxford University Press.

McCulloch, W. S., & Pitts, W. (1943). A logical Calculus of the ideas immanent in nervous activity. *Bulletin of Mathematical Biophysics, 5*, 115–133.

Metzinger, T., & Wiese, W. (Eds.). (2017). *Philosophy and predictive processing*. Frankfurt am Mein: MIND Group.

Silberstein, M., & Chemero, A. (2013). Constraints on localization and decomposition as explanatory strategies in the biological sciences. *Philosophy of Science, 80*(5), 958–970.

State of the Art Committee (1978). *Cognitive Science, 1978*. Report of the state of the art committee to the advisors of the Alfred P. Sloan foundation. Available at the URL http://www.cbi.umn.edu/hostedpublications/pdf/CognitiveScience1978_OCR.pdf

Part I
Explanation and Prediction

Chapter 2
Modelling Bayesian Computation in the Brain: Unification, Explanation, and Constraints

David M. Kaplan and Christopher L. Hewitson

Abstract Colombo and Hartmann (Br J Philos Sci 68(2):451–484. https://doi.org/10.1093/bjps/axv036, 2017) recently argued that Bayesian modelling in neuroscience can not only unify a diverse range of behavioral phenomena under a common mathematical framework, but can also place useful constraints on both mechanism discovery and confirmation among competing mechanistic models. After reviewing some reasons for decoupling unification and explanation, we raise two challenges for their view. First, although they attempt to distance themselves from the view that Bayesian models provide mechanistic explanations, to the extent that a given model successfully constrains the search space for possible mechanisms, it will convey at least some mechanistic information and therefore automatically qualify as a partial or incomplete mechanistic explanation. Second, according to their view, one widely used strategy to guide and constrain mechanism discovery involves assuming a mapping between features of a behaviorally confirmed Bayesian model and features of the neural mechanisms underlying the behavior. Using their own example of multisensory integration, we discuss how competing mechanistic models can be consistent with all available behavioral data and yet be inconsistent with each other. This tension reveals that there are exploitable degrees of freedom in the mapping relationship between models of behavioral phenomena and neural mechanisms, and points to the role that other background assumptions play including level-assumptions about the appropriate level at which the neural model should be specified (e.g., individual neuron or population level) and localization-assumptions about where in the system the underlying mechanism might occur. These considerations highlight the need for a more refined account of modelling constraints in neuroscience.

D. M. Kaplan (✉) · C. L. Hewitson
Department of Cognitive Science, Perception in Action Research Centre (PARC), Centre for Elite Performance, Expertise and Training (CEPET), Australian Hearing Hub, 16 University Drive, Macquarie University, Sydney, NSW, Australia
e-mail: david.kaplan@mq.edu.au

© Springer Nature Switzerland AG 2021

F. Calzavarini, M. Viola (eds.), *Neural Mechanisms*, Studies in Brain and Mind 17,
https://doi.org/10.1007/978-3-030-54092-0_2

2.1 Introduction

There is a major movement underway in contemporary cognitive science and neuroscience to think about the brain as a Bayesian machine that encodes information as probability distributions and performs probabilistic inference (Knill and Richards 1996; Rao et al. 2002; Clark 2013, 2015; Pouget et al. 2013; Doya 2007; Körding 2007, 2014; Ma and Jazayeri 2014). The Bayesian modelling approach combines the powerful mathematical frameworks of Bayesian statistics and statistical decision theory to formalize how new information should be combined with prior beliefs and how those updated beliefs can be used to generate optimal decisions or actions. This approach has been employed to model many different phenomena including aspects of vision (Knill and Richards 1996; Kersten et al. 2003; Stocker and Simoncelli 2006; Weiss et al. 2002), multisensory integration (van Beers et al. 1996, 1999; Ernst and Banks 2002; Fetsch et al. 2009, 2012, 2013; Alais and Burr 2004; Ernst and Bülthoff 2004; Burr and Alais 2006; Trommershauser et al. 2011), and sensorimotor control (Körding and Wolpert 2004, 2006; Tassinari et al. 2006; Orbán and Wolpert 2011; Berniker and Kording 2011; Wolpert and Landy 2012).

Despite their popularity, Bayesian models have attracted their fair share of criticism (for a review, see Hahn 2014). Challenges include the claims that they are not well supported by neural data (Jones and Love 2011; Bowers and Davis 2012a, b); are, in many circumstances, predictively equivalent to non-Bayesian models (Bowers and Davis 2012a); are unfalsifiable because of flexible model parameters including priors, likelihoods, and cost functions (Bowers and Davis 2012a, b); and are explanatorily limited (Bowers and Davis 2012a; Colombo and Hartmann 2017). It is this last issue about the explanatory power of Bayesian models that has attracted the most philosophical attention. And this issue is also the focus of the current chapter.

In this chapter, we address some open questions about Bayesian modelling, with a specific focus on the connections between explanation, unification, and modelling constraints. First, we provide a brief overview and case study of Bayesian modelling in cognitive science (Sect. 2.2). We then train our attention on Colombo and Hartmann (2017), who provide one of the most well-developed accounts of Bayesian modelling in the literature (Sect. 2.3). Colombo and Hartmann (2017) argue that Bayesian modelling in neuroscience can not only unify a diverse range of behavioral phenomena under a common mathematical framework, but can also place useful constraints on both mechanism discovery and confirmation among competing mechanistic models. After reviewing some reasons for decoupling unification and explanation (Sect. 2.4), a point of clear agreement between us and Colombo and Hartmann, we raise two challenges for their view (Sections 2.5 and 2.6).

First, although they attempt to distance themselves from the view that Bayesian models provide mechanistic explanations, we argue that to the extent a given model successfully constrains the search space for possible mechanisms, it will convey at least some mechanistic information and therefore automatically qualify as a partial or incomplete mechanistic explanation. When a model successfully reduces uncertainty about the possible mechanism (or mechanisms) underlying a target phenomenon by constraining the search space of mechanisms, it will necessarily possess some explanatory value. On our view, these are not separable or independent results. They are two sides of the same coin.

Second, according to their view, one widely used strategy to guide and constrain mechanism discovery involves assuming a mapping between features of a behaviorally confirmed Bayesian model and features of the neural mechanisms underlying the behavior. Using their own example of multisensory integration, we discuss how competing mechanistic models can be consistent with all available behavioral data and yet be inconsistent with each other. One thing this tension reveals is that often there are too many degrees of freedom in the mapping relationship between models of behavioral phenomena and neural mechanisms, and points to the role that other background assumptions play including level-assumptions about the appropriate level at which the neural model should be specified (e.g., individual neuron or population level) and localization-assumptions about where in the system the underlying mechanism might occur.[1] These considerations highlight the need for a more refined account of modelling constraints in neuroscience.

2.2 Bayesian Modelling in Cognitive Science: Overview and Case Study

Bayesian inference is a type of statistical inference where new data or information is used to update the probability that a given hypothesis is true. Bayes' theorem (eq. 2.1) specifies how to update the probability that a hypothesis H is true given some data D:

$$P(H|D) = \frac{P(D|H)\,P(H)}{P(D)} \tag{2.1}$$

Bayes' theorem states that the conditional probability of the hypothesis being true given the data, $P(H|D)$ (the *posterior probability distribution* or *posterior*), is equal to the probability of the data being true given the hypothesis, $P(D|H)$ (the *likelihood distribution* or *likelihood*), multiplied by the prior probability of the hypothesis being true, $P(H)$ (the *prior probability distribution* or *prior*), and divided by the probability of the data being true, $P(D)$. The latter ensures that the resulting probabilities sum to one. Bayes' theorem alone does not address how an agent's beliefs should be used to generate a decision or an action. A loss or utility function, specifying the expected loss for each action is also required. So-called *Bayesian Decision Theory* (BDT) puts these elements together and specifies how

[1] Although we explore these issues in the context of Bayesian models of behavior, these considerations likely have more general applicability.

to the posterior distribution can be used to generate an optimal decision or action (i.e., the one with the minimum expected loss). As indicated above, BDT is finding increasing application across the cognitive and brain sciences.

Before going any further it is important to mark a distinction between two different ways in which Bayesian models (and more specifically, BDT) can, and in fact are, being used in cognitive science.[2] Other authors have drawn similar distinctions (e.g., Jones and Love 2011; Zednik and Jäkel 2016). Bayesian modelling can be used to provide evidence for what has been termed the "Bayesian coding hypothesis", the hypothesis that "the brain represents information probabilistically, by coding and computing with probability density functions or approximations to probability density functions" (Knill and Pouget 2004, 713). Call this view *Bayesian computation.* Although obtaining clear experimental evidence is difficult, Bayesian computation lays out a relatively straightforward causal-mechanistic explanatory project like many others in computational neuroscience (Kaplan 2011). Alternatively, Bayesian modelling can be used to account for behavioral phenomena or "accommodate" behavioral data without any further commitment to the brain actually representing probability distributions or performing probabilistic inference (Maloney and Mamassian 2009; Geisler 2011). Call this view *Bayesian modelling*. According to this view, the claims about Bayesian inference should be interpreted purely instrumentally and should not be taken as assertions about how the brain actually works. Either these models are descriptive or phenomenological in so far as they provide mathematical characterizations of the phenomena for which causal-mechanistic explanations are sought (Mauk 2000; Dayan and Abbott 2001; Kaplan 2011). Or they are prescriptive in the sense that they describe optimal task performance and therefore provide an ideal or standard against which actual human performance can be compared (Maloney and Mamassian 2009; Geisler 2011). When Bayesian models are used in this way they are often called "ideal observer models". Returning to the Bayesian modelling/computation distinction, Van Gelder (1998) and Piccinini (2007) have drawn a similar — although more general — distinction among different classes of computational models that is instructive. For example, Piccinini (2007) lays out the issue as follows:

> To a first approximation, the distinction we need is that between using a computational description to model the behavior of a system—such as when meteorologists predict the weather using computers—and using it to explain the behavior of a system—such as when computer scientists explain what computers do by appealing to the programs they execute. (Piccinini 2007, 95)

[2]It should be noted that although some accounts may be more difficult to locate in terms of this binary distinction than others, our primary aim is to characterise two broad trends in the scientific literature. We do not assume that all accounts will be neatly accommodated by this distinction. For example, views that claim that probabilistic models operate at the "computational level", in Marr's sense, are difficult to place cleanly on one side of this distinction or the other because Marr's notion is itself subject to various competing interpretations (Shagrir and Bechtel 2017). Taking a stance on this debate goes well beyond the scope of the current chapter.

And he proposes the following useful way of marking this distinction terminologically:

> In *computational modelling*, the outputs of a computing system *C* are used to describe some behavior of another system *S* under some conditions. In *computational explanation*, by contrast, some behavior of a system *S* is explained by a particular kind of process internal to *S*—a computation—and by the properties of that computation. (Piccinini 2007, 96)

Similarly, it is important to distinguish Bayesian modelling from Bayesian computation for two reasons. First, without this distinction in hand it is easy to assume that the widespread success of Bayesian modelling in accommodating and/or predicting behavioral data across a wide range of domains including perception, multisensory integration, and sensorimotor learning ensures or entails that the brain performs Bayesian computations. But this is nearly as flawed an inference as is the inference from the fact that computational modelling is ubiquitous in contemporary astronomy to the conclusion that phenomena being investigated in this domain such as the solar system or planetary bodies are computing their own equations of motion (Van Gelder 1998; Piccinini 2007).[3] Second, many (arguably a majority of) researchers who employ Bayesian models are interested in testing and providing evidential support for Bayesian computation, in the sense defined above. It is this evidence we want to evaluate.

As an example of the Bayesian computation approach, consider a study by Körding and Wolpert (2004) investigating sensorimotor learning. In this experiment, subjects were asked to perform a variation on a standard visuomotor adaptation task in which visual feedback about hand position during reaching movements is manipulated so that a new sensory-motor mapping must be learned in order to perform the task correctly (Krakauer 2009). Subjects made reaches to a visual target using a mirror projection system that prevented vision of the hand. By carefully matching the distances between the projector, mirror, and table surface, all stimuli appeared to be in the same horizontal plane as the movements. On each trial, the cursor representing hand position was extinguished at movement onset and shifted laterally from the true finger position. Instead of imposing a visual shift with a fixed magnitude and direction across the entire session, as is the case in standard visuomotor adaptation studies, lateral shifts were randomly drawn on each trial from a Gaussian (normal) distribution of shifts with a fixed mean and standard deviation. This served as the imposed or true *prior* for the experiment.[4] Halfway

[3]It is only nearly as bad an inference because at least in this case we have independent reasons to believe the target system is computational. Thanks to Matteo Colombo for pointing this out.

[4]Although we are aware of the general tension between frequentist and epistemic conceptions of probability and the related debate about how to interpret priors in probabilistic models (see Feldman 2013), the interpretation of the prior in this experiment seems relatively straightforward. Because the prior distribution each subject experienced was set empirically and was the product of a random (or pseudo-random) process — each experienced shift was drawn randomly (pseudo-randomly) from a Gaussian distribution — it seems to us that the prior probabilities can be interpreted as physical probabilities and a frequentist conception is appropriate. We thank Brendan Ritchie for bringing this point to our attention.

to the target, shifted visual feedback was briefly provided (100 ms duration) with different degrees of blur or reliability: no blur (σ_0), moderate blur (σ_M), large blur (σ_L), or completely withheld (σ_∞). This served as the critical manipulation of the visual likelihood. Subjects were instructed to use whatever feedback was available on a given trial to quickly and accurately place the cursor on the target, thereby compensating for the lateral shift.

Körding and Wolpert (2004) tested three different models and found that the *Bayesian estimation model*, provided the best fit to their data. According to this model, subjects should use information about the prior distribution and current visual feedback (likelihood) to estimate the lateral shift and adjust their reaches. More specifically, they should weight their reliance on their stored prior in proportion to the reliability of the visual feedback provided on the current trial (i.e., increased reliance on the prior when visual feedback reliability is low and vice versa). This is precisely what they found.

Critically, like many other researchers who employ Bayesian models (see Sect. 2.1),[5] Körding and Wolpert (2004) interpret these behavioral results as providing evidence that the brain performs Bayesian computations. They state:

> [S]ubjects internally represent both the statistical distribution of the task and their sensory uncertainty, combining them in a manner consistent with a performance-optimizing bayesian process. The central nervous system therefore employs probabilistic models during sensorimotor learning (Körding and Wolpert 2004, 244).

Although this might seem like an overreach, we will argue that, under certain conditions, information about underlying computations and representations can be inferred reliably on the basis of behavioral evidence alone. In what follows, we argue that both behavioral and neural data can place important constraints on the search space for possible mechanisms and in doing so they provide a valuable heuristic for mechanism discovery.

2.3 Colombo and Hartmann

In a series of papers (Colombo and Seriès 2012; Colombo and Hartmann 2017), Matteo Colombo has developed a sophisticated account of Bayesian modelling in cognitive science. In an earlier paper co-authored with neuroscientist Peggy Seriès, Colombo aims to identify different uses of Bayesian models in cognitive science and then evaluate whether any such uses provide evidence that the brain implements Bayesian inference. Colombo and Seriès argue that current Bayesian models lack

[5]As another high-profile example, Ernst and Banks (2002) draw a similar conclusion about the nervous system performing Bayesian integration on the basis of behavioral performance in a cue combination task. They state: "we found that height judgements were remarkably similar to those predicted by the MLE integrator. Thus, the nervous system seems to combine visual and haptic information in fashion similar to the MLE rule." (2002, 431)

explanatory power in so far as they fail to describe underlying neural mechanisms (i.e., they do not provide causal-mechanical explanations). Despite their current explanatory shortcomings, they maintain that Bayesian models are useful "as tools for predicting, systematizing and classifying statements about people's observable performance and "should be interpreted within an instrumentalist framework." (2012, 705). Importantly, they go on to acknowledge that the predictive successes of Bayesian models often provide additional motivation to investigate and discover the neural mechanisms that underlie the observed behavioral performance. In their words, the predictive success of Bayesian modelling "motivates the Bayesian coding hypothesis at the neural level" (713). However, they do not elaborate in great detail about how this process is supposed to work. In a follow-up paper, Colombo and Hartmann (2017; hereafter, CH), provide these details and more.

A major objective of their more recent paper is to clarify how to think about the unifying power of Bayesian models. Although it is commonly acknowledged that BDT provides a powerful mathematical framework capable of unifying a diverse range of phenomena, it remains unclear whether the unification thereby achieved is explanatory in nature. To their credit, CH avoid making this mistake and do not jump on this bandwagon. Instead, they argue the connection between unification and explanation in Bayesian models is more indirect and subtle: Bayesian unification places a variety of fruitful constraints on causal-mechanical explanation. In the remainder of their paper they develop an account of the various types of constraints that flow from Bayesian unification and begin to elucidate how they work. According to CH, there are three kinds of constraint: (1) constraints on mechanism discovery; (2) constraints on the identification of causally or mechanistically relevant factors; and (3) constraints on the confirmation and selection of competing mechanistic models.

Although there are many interesting aspects of their account, in what follows we will focus on just two. First, CH assume that this shift to discussing constraints on causal-mechanical explanation allows them to sidestep the more foundational question about whether Bayesian models provide explanations in the first place. We will argue that there is no such wiggle room. Although Colombo and Hartmann attempt to distance themselves from the view that Bayesian models provide mechanistic explanations, to claim that a given model constrains the search space for possible mechanisms implies that it will convey at least some mechanistic information and will therefore qualify as a partial or incomplete mechanistic explanation.

Second, CH claim that the unifying power of the Bayesian framework can place fruitful constraints on the confirmation and selection between competing mechanistic models. Specifically, they argue (and offer a mathematical proof) that in cases where two models are equally well supported by the available data, the one that better coheres[6] with an overarching unifying model is thereby rendered more probable. While we take no issue with this particular line of reasoning, it motivates

[6]The notion of coherence is a technical one from formal epistemology. For details, see the mathematical proof supplied by Colombo and Hartmann (2017).

a slightly different challenge for their view. The challenge concerns the fact that competing mechanistic models can be (1) consistent with all available behavioral data, (2) equally well supported because they both "cohere" with a general unifying model (e.g., Eq. 2.1), and yet (3) be inconsistent with each other. This tension reveals that there are too many exploitable degrees of freedom in the mapping relationship between models of behavioral phenomena and neural mechanisms, and points to the role that other background assumptions play including level-assumptions about the appropriate level at which the neural model should be specified (e.g., individual neuron or population level) and localization-assumptions about where in the system the underlying mechanism might occur. Before laying out these two challenges in more detail, however, it is important to clarify some common ground between us and CH concerning the connection between unification and explanation. This is the task for the next section.

2.4 Unification and Explanation

Although it is commonly acknowledged that BDT provides a powerful mathematical framework to model and hence unify a diverse range of phenomena, it remains unclear whether the unification thereby achieved is explanatory in nature. Many cognitive scientists and philosophers simply assume that the unifying power of BDT must be directly related to its explanatory power. Here are a few examples[7]:

> [P]robabilistic models provide a unifying framework for explaining the inferences that people make in different settings. (Griffiths and Tenenbaum 2009, 360)

> One particularly appealing aspect of these theories is their generality: they can be used to model a wide range of tasks, from sensory processing to high-level cognition. (Pouget et al. 2013, 1170)

> Most, if not all, of the computations performed by the brain can be formalized as instances of probabilistic inference. Sensory processing, motor control, decision-making, learning and virtually all higher cognitive tasks fall into this class. By treating them as probabilistic inference problems, we may be able to derive general principles that apply to all areas of the brain (Pouget et al. 2013, 1177).

The intuitive force behind the idea that more unifying models are more explanatory is undeniable.[8] The unificationist view of scientific explanation (e.g., Kitcher 1981), which holds that explanation is a matter of providing a unified account of a range of different phenomena, derives from this basic idea.

Despite the intuitive appeal of viewing unification and explanation as inextricably linked, there are several powerful reasons to reject this idea. First, unification is

[7]Colombo and Hartmann (2017) cite some of these and many others.

[8]Similar claims about the explanatory import of mathematical unification have been made about dynamical explanation (Stepp et al. 2011), computational explanation (Chirimuuta 2014), and network explanation (Levy and Bechtel 2013; Huneman 2010; Rathkopf, C. (2018)).

not necessary for causal or mechanistic explanation. Consider first the rationale emerging from the interventionist approach to causal explanation (Hitchcock and Woodward 2003; Woodward 2004). According to the interventionist approach, unification (also known as "wide scope" or "generality") is inessential for causal explanation because explanatory power (or "depth") reflects the degree of invariance — how stable a generalization is across a set of interventions.[9] As Hitchcock and Woodward put it: "Increased scope does not always correspond to explanations that are intuitively deeper." (Hitchcock and Woodward 2003, 190). They develop their point with a simple example. They ask us to first consider a set of generalizations or models G_1-G_n that describe the functioning of a highly conserved neural circuit N_1, which is found across many different taxa. Next they ask us to consider a set of generalizations or models H_1-H_n that describe the functioning of a highly specialized neural circuit N_2, which is in one particular species of snail. According to the interventionist account, the only relevant consideration for assessing the differences in explanatory power between these two set of generalizations or models is their degree of invariance — how stable they hold across a set of interventions. If both are matched along this dimension, it is entirely inconsequential whether their scope differs. From the interventionist perspective, binding explanation and unification leads to counterintuitive results. As they put it:

> While the unificationist account seems to yield the conclusion that the generalizations governing N1 provide more unified and hence better or deeper explanations that the generalizations governing N2 simply in virtue of applying to more organisms (or more different kinds of organisms) our account avoids this unintuitive conclusion. (Hitchcock and Woodward 2003, 193)

According to the interventionist approach, then, unification is not necessary for causal explanation. It is also unnecessary for mechanistic explanation. Mechanistic models (or mechanistic explanations) vary along a number of different dimensions including completeness, detail (precision), evidential support, and scope (Craver and Darden 2013). Importantly, not all of these model dimensions bear equal explanatory weight. Completeness and empirical support are important. After all, a mechanistic model is explanatory to the extent that it completely and accurately describes the relevant parts, activities, and organization underlying the target phenomenon (Kaplan and Craver 2011).[10] Yet, scope turns out to be relatively unimportant. This is because both completeness and support can be satisfied equally well by models that apply to single cases or $n = 1$ systems (e.g., Rube-Goldberg mechanisms) or models of ubiquitous, highly conserved biological mechanisms

---

[9]Scope concerns how many systems or how many different kinds of systems there actually are to which a given model or generalization applies, and so is highly similar to (or at least strongly correlated with) unifying power.

[10]This does not entail the drive towards something like exhaustive model completeness for which the goal is that all details, relevant and irrelevant, are included. For details, see Craver and Kaplan (2018).

(e.g., DNA, voltage-gated potassium channels).[11] In this respect, the norms of interventionist and mechanist explanation are closely aligned, and they both part company with a core assumption of the unificationist approach. Neither hold that a given model explaining more things, in the sense of having wider scope, entails that it is more explanatory.

A second and closely related reason to avoid conflating unifying and explanatory power is that unification is not sufficient for causal or mechanical explanation. Although they do not dwell on the point, CH acknowledge this when they cite Margaret Morrison's influential work on unification and state that "the link between unification and explanation is far from obvious" (Colombo and Hartmann 2017, 452). What they are alluding to here is the fact that unification covers many different kinds of scientific achievement, not all of which explicitly or implicitly involve explanation (Morrison 2000). *Descriptive unification* involves the subsumption of different phenomena under a common descriptive or classificatory scheme and need not have any explanatory objective or purpose. Examples include the Linnaean taxonomic classification system. *Mathematical unification* involves subsuming different phenomena under a common mathematical framework or formalism. Examples include Lagrange-Hamilton equations (which were initially developed in mechanics but subsequently applied to domains like electromagnetism and thermodynamics), dynamical systems theory, network theory, dimensionality reduction techniques, among others. The fact that a common mathematical framework can be constructed to accommodate or deal with a range of different phenomena does not guarantee that a common set of relevant causal or mechanistic factors are responsible for those phenomena (which might in turn provide the basis for a unifying causal-mechanical explanation). And finally, *physical* (*causal or mechanistic*) *unification* involves the subsumption of different phenomena previously thought to have quite different causes or explanations are shown to be the result of a common set of causes or mechanisms. Only the latter has straightforward connections to explanation.

For these reasons, CH are well justified in rejecting the explanatory import of Bayesian unification and identifying other roles it can play in cognitive science such as providing constraints on mechanism discovery and model selection. We turn to these topics now.

[11]It is worth noting that this characterization of model scope might elide a further distinction between the ways in which scope can vary. For example, scope can refer to the same type of mechanism for the same type of phenomenon that is instantiated by many systems across different taxa. Alternatively, scope can refer to the same type of mechanism *for many different types of phenomena* that is instantiated by many systems across different species/taxa. Although these are importantly different, we would argue that wide scope in either of these senses is not required for a given model to explain. We thank Matteo Colombo for bringing this distinction to our attention.

2.5 Challenge 1: Constraints and Mechanistic Explanation

As indicated in the Introduction (Sect. 2.1), Colombo has been keenly interested in understanding the nature of Bayesian modelling in cognitive science. In his earlier paper with Seriès, they embraced an explicit position on the explanatory status of Bayesian models. In their words: "Bayesian models do not provide mechanistic explanations currently, instead they are predictive instruments." (Colombo and Seriès 2012, 719). In his more recent paper with Hartmann, they seek to sidestep the explanatory question. They state:

> Rather than addressing the issue of the conditions under which a model is explanatory, we accept that a crucial feature of many adequate explanations in the cognitive sciences is that they reveal aspects of the causal structure of the mechanism that produces the phenomenon to be explained. In light of this plausible claim, we ask a question that we consider to be more fruitful: what sorts of constraints can Bayesian unification place on causal–mechanical explanation in cognitive science? (Colombo and Hartmann 2017, 465)

Although they attempt to distance themselves from the view that Bayesian models provide mechanistic explanations, our claim is that the proposed separation does not work. To the extent that a given model successfully constrains the search space for possible mechanisms, it will convey at least some mechanistic information and therefore qualify as a partial or incomplete mechanistic explanation.[12] By defending a view about Bayesian models providing fruitful mechanistic constraints, CH implicitly endorse a (mechanistic) view about the explanatory import of these models. Or so we will argue.

Although CH identify and discuss three different types of constraints that Bayesian models can place on causal — mechanical explanation, in this section we only discuss the first concerning constraints on mechanism discovery. What then is a constraint? In the most basic sense, a constraint is simply a restriction or limitation. In its ordinary usage, the term 'constraint' often has a negative valence but the same is not true in scientific contexts. Often constraints are immensely useful in science, especially in the context of modelling. In scientific modelling contexts, a constraint can be understood as "a finding or evidence that either shapes the boundaries of the space of plausible mechanisms or changes the probability distribution over that space" (Craver 2007, 248). Constraints impose limits on the hypothesis space or the space of possible mechanisms for a given phenomenon for which an explanation is sought.

[12]There are some important parallels between the view we advocate in this chapter and the account developed by Zednik and Jäkel (2016). In that work, they offer an account of "Bayesian reverse engineering" according to which arriving at an explanatorily adequate model involves an ordered and iterative search through three different "hypothesis spaces" each of which is associated with one of Marr's levels — the computational, algorithmic, and implementational. Although there are a number of similarities between our view and theirs, there are also important differences, including different starting points: Marrian levels versus mechanistic explanations. It is, however, beyond the scope of this chapter to explore these similarities and differences and remains work for another day (for related discussion, See Bechtel and Shagrir (2015)).

Consider a simple example of a constraint at work. Suppose there are gaps in someone's understanding of the mechanism responsible for producing forward motion in a car and they discover something new about the temporal properties of the underlying mechanism such as the maximum rate at which some of the components — the pistons — can go up and down in the cylinder. This information places a hard temporal constraint (lower bound) on the rate at which other components that interact with the pistons such as the intake and exhaust valves must move. For example, this information sets their minimum rate of operation because they must be capable of opening and closing fast enough to match the speed at which the pistons can move to ensure performance is not compromised. Consequently, this information rules out many possible mechanisms with valve components that open slower than is required.

CH maintain that Bayesian models impose similar constraints on mechanism discovery. Bayesian models levy these constraints by providing a simple discovery heuristic for researchers to follow: Assume a defeasible mapping (isomorphism) between elements in behaviorally-confirmed model and neural model, which subsequently places constraints on the search space for possible mechanisms.[13] Here is how CH put the idea:

> The basic idea is to establish a mapping between psychophysics and neurophysiology that could serve as a starting point for investigating how activity in specific populations of neurons can account for the behavior observed in certain tasks. Typically, the mapping is established by using two different parameterizations of the same type of Bayesian model. (Colombo and Hartmann 2017, 467)

Rather than starting from a blank slate, Bayesian models supported by behavioral evidence can "rule in" certain hypotheses about underlying neural mechanisms and "rule out" others. CH use the work by Fetsch et al. (2012) on multisensory integration to demonstrate how Bayesian modelling of behavioral data can usefully constrain the search space for neural mechanisms.

Fetsch et al. (2012) trained monkeys to perform a heading discrimination task in which the reliability of the visual motion information was manipulated by varying the percentage of dots in the stimulus moving coherently in a single direction. During the experiment, monkeys were presented with either a single-cue (visual or vestibular) indicating heading direction, or a combined-cue (visual plus vestibular) where the two cues provided conflicting heading information. Monkeys were required to choose their current heading direction by making a saccade either to a leftward or rightward target. Behavioral thresholds from the single-cue conditions were used to estimate cue reliability (the inverse of variance), and establish the weightings that an ideal observer should apply to each cue during the combined-cue conditions. Psychometric data collected during cue conflict trials were used to compute behavioral vestibular and visual weights, which could then be compared to the optimal weights. They found that the monkeys' choices during the cue conflict trials were significantly biased towards the more reliable cue. At low visual

[13] Zednik and Jäkel (2016) use the term 'push-down heuristic' to describe something very similar.

coherence (16% coherent motion), when the vestibular cue was more reliable, the monkey made more rightward choices when the vestibular cue indicated a rightward heading, and more leftward choices when the vestibular cue indicated a leftward heading. The opposite pattern was observed when the visual cue was more reliable (60% coherent motion). These shifts in psychometric functions (and the derived vestibular and visual weights) were very close to the optimal predictions defined by the standard ideal-observer model of cue integration:

$$S_{comb} = w_{ves} S_{ves} + w_{vis} S_{vis} \tag{2.2}$$

where S_{comb} is an internal heading signal that is the weighted sum (combination) of vestibular and visual signals (S_{ves} and S_{vis}, respectively), and w_{vis} and w_{ves} are the corresponding weights ($w_{vis} = 1 - w_{ves}$). The close agreement between modeled and observed perceptual weights provides a strong indication that monkeys integrate sensory information according to its variance (i.e., in a Bayes-optimal manner).

To probe the neural mechanisms underlying task performance, Fetsch et al. recorded single unit activity in the dorsal medial superior temporal area (MSTd) — an area thought to be involved in visual and self-motion perception — while monkeys performed the heading task described above. They found that when the visual cue was more reliable and it indicated displacement in the neuron's "preferred" direction, this resulted in a higher firing rate for that neuron as compared to when the visual cue indicated no displacement or displacement in the null direction. This pattern was reversed when the visual cue was less reliable. To quantify the effect of motion coherence on individual MSTd neuron firing rates, their next step was to model the neural responses using separate neural weights for the visual and vestibular cues. This would also allow them to determine if the neural weights exhibited the same dependence on cue reliability (coherence) as the perceptual weights. Based on previous work (Morgan et al. 2008), they modeled the firing rates (tuning curves) of MSTd neurons using a simple "linear combination rule":

$$f_{comb}(\theta, c) = A_{ves}(c) f_{ves}(\theta) + A_{vis}(c) f_{vis}(\theta) \tag{2.3}$$

where f_{comb}, f_{ves}, and f_{vis} are tuning curves (firing rates) (a function of heading and/or noise) of a particular MSTd neuron for each of the combined, vestibular, and visual conditions, respectively; θ denotes heading; c denotes the coherence of the visual cue; and A_{ves} and A_{vis} are neural weights.

They found that a majority of the MSTd neurons they recorded from modulated their firing rates in a reliability-dependent way, thereby encoding information about cue reliability. More specifically, most neurons showed greater vestibular weights during trials in which visual cue reliability was low as compared to when it was high, and greater visual weights during trials in which visual cue reliability was high as compared to when it was low. In other words, the modelled weights on individual MSTd neurons varied with cue reliability on a trial-by trial basis in a manner consistent with optimal Bayesian integration.

As a final step, Fetsch et al. performed a decoding analysis to determine if behavioral performance in the heading direction task could be simulated based exclusively on MSTd population activity. For this analysis, they pooled all the individual MSTd responses for a given trial type and used maximum likelihood decoding to generate estimates of heading direction. The decoding results were highly similar to actual behavioral performance.

From all of this, CH reasonably conclude that: "Fetsch et al. (2012) provide an illustration of how Bayesian unification can constrain the search space for mechanisms" (Colombo and Hartmann 2017, 467). It is certainly true that Fetsch et al. assumed some sort of approximate mapping between elements in their behaviorally-confirmed Bayesian model and the neural model, and this imposed valuable constraints on their search for possible mechanisms. More specifically, the reliability-dependent vestibular and visual weights in the behavioral model guided their search for an analog in individual MSTd neurons, which is precisely what they found. This powerful discovery heuristic allowed them to prune the search space by ruling in some classes of possible mechanism and ruling out or excluding others. At a coarse-grained level, Fetsch et al. were able to use the behaviorally-confirmed Bayesian model to zero in on candidate mechanisms that integrate visual and vestibular information using some kind of reliability-based weighting scheme and, at the same time, set aside a range of mechanisms that integrate sensory information by doing many other things such as simply averaging visual and vestibular information without regard to reliability. At a finer-grained level, Fetsch et al. uncovered evidence that the neural weights are capable of changing very rapidly, both across and within trials, which is consistent with fast network mechanisms such as normalization (Fetsch et al. 2013; Carandini and Heeger 2012), but inconsistent with relatively slow mechanisms such as synaptic weight changes. This provides an additional constraint on the space of possible mechanisms. Given these considerations, it seems clear that Bayesian models can and do play an important role as a constraint on mechanism discovery, as CH contend.

The question we now want to consider concerns whether the constraints that flow from Bayesian models and mechanistic explanations are really as independent and separate as CH imply. Consider first the notion of a mechanism sketch. A mechanism sketch is an incomplete or partial representation of a mechanism that characterizes some but not all the parts, activities, and organizational features of the mechanism responsible for producing a given phenomenon (Craver 2007; Craver and Darden 2013). Mechanism sketches will be endowed with explanatory power to the extent that they incompletely or partially describe the mechanism, and are therefore appropriately described as incomplete or partial mechanistic explanations (Craver 2007; Craver and Darden 2013; Kaplan and Craver 2011). It is our contention that many Bayesian models including the ones discussed at length in the paper qualify as mechanism sketches. For example, the model that Fetsch et al. provide is a mechanism sketch. The model implicitly describes components — MSTd neurons — and their activities — response profiles or tuning curves exhibiting reliability-dependent modulation. Although the model is incomplete in a number of important respects including the fact that it provides no information about how this reliability-weighting is mechanistically achieved — the weights do not themselves map onto components or activities — its incompleteness does not negate its explanatory status. The Fetsch et al. neural model conveys some mechanistic information. This information is what constrains the search space of plausible mechanisms. And this is what permits the model to qualify as a (partial or incomplete) mechanistic explanation.

Going back to the initial example, we maintain that Kording and Wolpert's model also qualifies as a mechanism sketch and a partial explanation because it too serves to constrain the search space of possible mechanisms. Their findings are informative not simply because their behavioral data are merely consistent with one particular model (the Bayesian estimation model), but rather because they enable us to rule out certain possibilities (Coltheart 2006; Mole and Klein 2010). Specifically, they can rule out all schemes involving single point estimates. Although specific details about neural implementation cannot be directly inferred on the basis of behavioral evidence alone (Maloney and Mamassian 2009; Fiser et al. 2010), it remains true that the internal mechanism must at a minimum be capable of encoding uncertainty. This could be done in many ways including by representing parameters of the underlying probability distributions or through a sampling-based representational scheme (Fiser et al. 2010). Nevertheless, even in this sketchy form, the model rules out all possible mechanisms incapable of representing uncertainty. In this respect the model conveys some mechanistic information and qualifies as a partial (albeit weak) mechanistic explanation. Here is a generalized version of the basic argument:

1. If a given model M provides constraints on possible mechanisms for phenomenon P, M by definition provides evidence that rules some possible mechanisms for P in and rules some possible mechanisms for P out.
2. If M provides evidence that rules some possible mechanisms for P in and rules some possible mechanisms for P out, M must convey mechanistic information.
3. If M conveys mechanistic information (even incomplete information), M is at a minimum a mechanism sketch and provides a partial or incomplete mechanistic explanation for P.

We think this is a straightforward and palatable result that CH should readily embrace. On our view, mechanistic explanation and constraints on mechanism discovery are inextricably linked under the Bayesian computation approach. If Bayesian models do valuable work to prune the space of possible mechanisms, they are already in the explanation game. This is not to say that all Bayesian models provide equally good explanations. Far from it. Like other models, there will be a continuum of better/worse explanatory Bayesian models and distinguishing them will require assessment in terms of standard criteria including support and completeness (Craver and Darden 2013; Craver and Kaplan 2018).

2.6 Challenge 2: Degrees of Freedom

The second challenge centers on CH's claim that Bayesian unification can be useful for selecting between competing mechanistic models and ultimately confirming one model over another. Although we are sympathetic to their general point, and think that this occurs often in scientific practice, we also think the discussion masks some important complications. In particular, competing mechanistic models can be consistent with all available behavioral data, can cohere equally well with a unifying Bayesian model, and yet be inconsistent with each other in virtue of different assumptions about neural implementation. This tension reveals that there are often too many degrees of freedom in the mapping relationship between models of behavioral phenomena and neural mechanisms. Background assumptions including level-assumptions about the appropriate level at which the neural model should be specified (e.g., individual neuron or population level) and localization-assumptions about where in the system the underlying mechanism might occur often play important, under-appreciated roles.

Here is what CH say:

> Unification can also be relevant to confirmation of a mechanistic model. Specifically, Bayesian unification can be relevant to identifying which one among competing mechanistic models of a target cognitive phenomenon should be preferred. If we want to judge which one of the mechanistic models M1 and M2 is more adequate when available data, D1, confirm M1 and disconfirm M2, and D2 confirm M2 and disconfirm M1, the fact that M2 and D2 are coherent with a unifying model, U, while M1 and D1 are not provides us with evidence in favor of M2. (Colombo and Hartmann 2017, 471)

This result falls out of Bayesian confirmation theory. CH illustrate their claim by describing two competing mechanistic models of multisensory integration. One model, put forward by Stein et al. (1993), posits the non-linear combination of responses to unimodal cues and predicts superadditive multimodal responses. Importantly, none of the neural responses in this model exhibit weightings reflecting sensory uncertainty; it is a non-Bayesian model. The competing model they consider is the probabilistic population coding (PPC) model (described in more detail below), which posits a weighted linear combination of unimodal population responses and predicts additive effects in the downstream multimodal population activity. Importantly, the weights in this model do reflect sensory uncertainty; the PPC model is a Bayesian model. Consequently, in their example, only one of the two models coheres with the more abstract unifying Bayesian model. This is precisely why their argument gets a foothold.

Crucially, CH's view about the role that the Bayesian framework plays in confirmation and selection of mechanism models applies most readily to situations in which one candidate model is Bayesian and the other is not. Their account does not cope as well with situations within a "Bayesian regime" — i.e., when both candidate mechanistic models under consideration are Bayesian. In this case, both may cohere equally well with the general Bayesian framework, and yet both may be consistent with different neural data. One does not have to look far to find relevant examples. We can find precisely this tension between two models CH explicit discuss — the PPC model and the neural model proposed by Fetsch et al. To understand the tension a bit more background on the PPC model is needed.

Ma et al. (2006) proposed their probabilistic population code (PPC) model as one way the nervous system might implement Bayesian inference. The PPC model involves two basic assumptions. First, Bayesian inference is performed by relatively small neural populations rather than in individual neurons or entire brain regions. Second, the firing rates of individual neurons in the relevant population must be highly variable to the extent that they approximately obey Poisson statistics. In extreme cases where variability is essentially random and Poisson-like, the mean response of a neuron for a given condition might be equal to or even exceed its variance (Fano factor ≥ 1). Ma et al.'s critical insight is that this variability is not a nuisance or unwanted noise, but rather that neural populations automatically encode probability distributions in virtue of this Poisson-like variability. More specifically, because of the variability in individual neuron responses to a given stimulus, *s*, the overall response of the population made up of these neurons, *r*, to *s* is best described in terms of a probability distribution, p(r|s), rather than a deterministic mapping from s onto a single value of r. Importantly, p(r|s) is equivalent to the likelihood distribution from Bayes' rule. Ma et al. further assume that each distribution is represented by the activity of a distinct neural population. According to their model, the mean and variance of each distribution are encoded by population activity which can be combined and readout by a downstream population response representing the posterior.

The critical point for present purposes is that PPC model assumes fixed neural weights on individual neurons in each population that do not change with reliability. Fetsch et al. highlight this feature of the PPC model when then state: "[i]f neurons fire with Poisson statistics and tuning curves are multiplicatively scaled by coherence, then the optimal neural weights will be equal to 1 and independent of coherence" (Fetsch et al. 151). By contrast, the Fetsch et al. model makes the opposite assumption of reliability-dependent neural weights. The models of the underlying neural mechanisms supporting Bayesian multisensory integration are inconsistent with each other and yet both are consistent with the relevant behavioral data and both cohere equally well with the unifying Bayesian model (Eq. 2.1).

At least for cases like this, it is difficult to see how the Bayesian framework provides useful constraints on the confirmation and selection among competing mechanistic models. To their credit, CH do briefly address this issue. They maintain that tensions like these among competing mechanistic models when both cohere equally with a unifying general Bayesian model "provides us with a basis to figure out quantitatively what the sources of these violations might be" (Colombo and Hartmann 2017, 22). This process, they imply, will ultimately lead to revisions in models so that the violations are resolved. In the case at hand, the flagged assumption is that cue reliability has multiplicative effects on neural firing rates, which is violated by MSTd neurons (see Morgan et al. 2008, Supplementary figure

3; Heuer and Britten 2007; Angelaki et al. 2009).[14] Importantly, CH suggest that this process allows us to identify "how Ma et al.'s model should be revised so that it would predict reliability-dependent neural weights" (Colombo and Hartmann 2017, 23). However, it remains unclear how the PPC model can be modified to accommodate reliability-dependent neural weights without fundamental changing the model. As we will see below, Fetsch and colleagues seem to have something rather different in mind than using their empirical results to try to legislate changes in the PPC model. Instead, because an assumption of the PPC model is violated by features of their data, they instead want to suggest that it is simply inappropriate to apply the model in the first place (Angelaki et al. 2009; Fetsch et al. 2013). As we will see, this is a strategy that involves avoiding rather than resolving the tension between the two models.

As alluded to above, an alternative strategy is simply to deny that these are competing mechanistic models in the first place. One way of doing this involves exploiting the degrees of freedom available in these different models and highlighting how they incorporate different background assumptions about the appropriate level at which the neural model should be specified (level-assumptions) and where in the system the underlying mechanism might be located (localization-assumptions).[15] For example, Fetsch et al. provide evidence that multisensory integration is supported by reliability-dependent weighting at the level of individual MSTd neurons. Yet the fact that a neuron-level computation is performed in MSTd does not preclude the possibility of other brain regions employing a population- or network-level computation along the lines indicated by the PPC model. The different models do not necessarily compete if they apply to different levels of neural organization in different brain regions. In a review of the multisensory integration literature, Angelaki et al. (2009) adopt precisely this strategy. They maintain that:

> These results [reliability-dependent neural weights] are not necessarily in conflict with theoretical predictions [of Ma et al.'s PPC model] for two reasons. First, MSTd neurons may not adhere to the assumptions of the model (e.g. Poisson-like firing statistics and multiplicative effects of stimulus reliability). Indeed, the effect of motion coherence on visual heading tuning in MSTd does not appear to be purely multiplicative. Second, the model has not considered the effects of interactions at the network level, such as divisive normalisation. (Angelaki et al. 2009, 456)

---

[14]The details here are complex but the basic idea is that MSTd neurons exhibit lower firing rates for high coherence visual stimuli presented in the null (anti-preferred) direction than for low coherence in the null direction. This means that MSTd neuron responses are nonlinear (i.e., not multiplicative) at the flanks of their tuning curves, which violates the assumptions of the PPC model. This topic remains an active area of investigation (Fetsch, personal communication).

[15]Zednik and Jäkel (2016) introduce the useful term 'tweak' to characterize a similar practice in Bayesian ideal observer modelling of behavioral data." They maintain that tweaks reflect the available "degrees of freedom that researchers may exploit to accommodate the observed behavioral data" (Zednik and Jäkel 2016, 3959). We would argue for expanding the notion of tweaking to cover modelling practices at the neural- or implementational-level. Here we have identified several examples of this kind of model tweaking.

The apparent conflict between the two models can be resolved, according to Angelaki et al., by appealing to the different level-assumptions and localization-assumptions built into these models. But the availability of this kind of strategy raises some deeper problems for the way we have been thinking about constraints.

To connect back up with our discussion of constraints, evidence supporting the Fetsch et al. neural model may be described as constraining the space of possible mechanisms for multisensory integration (or changes the probability distribution over that space). Importantly, up until this point we have been implicitly assuming that evidence supporting the PPC model also serves to constrain the very same hypothesis space over possible mechanisms. But, if flexible background assumptions about brain levels and/or locations can be leveraged in the way just described, then it seems to follow that we are no longer dealing with the same space of possible mechanisms. Instead of both models imposing mutually reinforcing and interlocking constraints on the same space of possible mechanisms for the target phenomenon (see Craver 2007, Ch 7), each model instead levies separate constraints on different, independent (but perhaps related) spaces. This would be an extremely different — extremely local and balkanized — approach to mechanistic constraints than the more global or holistic approach that many including Craver (2007) (and perhaps CH) seem to implicitly embrace. A basic tenet of this background view of constraints seems to be that every new finding or discovery about some particular phenomenon (e.g., multisensory integration or spatial memory) imposes constraints which serve to monotonically[16] decrease the search space of possible mechanisms. But the current case suggests that this simple view might not always hold. At a minimum, what these considerations highlight is the need for a more refined account of modelling constraints in neuroscience. Although providing such an account is beyond the scope of the current chapter we have identified several features it should address. And this is a step in the right direction.

2.7 Conclusion

In this chapter we have tried to make progress on some issues concerning modelling Bayesian computation in the brain. We have focused our attention primarily on the work of Colombo and Hartmann (2017), as they provide one of the most well-developed accounts in the literature. They argue that Bayesian modelling in neuroscience can not only unify a diverse range of behavioral phenomena under a common mathematical framework, but can also place useful constraints on both mechanism discovery and confirmation among competing mechanistic models.

[16]In mathematics, a monotonic function is either entirely nonincreasing or nondecreasing. A function that increases monotonically does not exclusively have to increase, it simply must not decrease. A function that decreases monotonically does not exclusively have to decrease, it simply must not increase.

After reviewing some reasons for decoupling unification and explanation, we raised two challenges for their view. First, although they attempt to distance themselves from the view that Bayesian models provide mechanistic explanations, to the extent that a given model successfully constrains the search space for possible mechanisms, we argued that it will convey at least some mechanistic information and therefore automatically qualify as a partial or incomplete mechanistic explanation. Second, according to their view, one widely used strategy to guide and constrain mechanism discovery involves assuming a mapping between features of a behaviorally confirmed Bayesian model and features of the neural mechanisms underlying the behavior. Using their own example of multisensory integration, we discussed how competing mechanistic models can be consistent with all available behavioral data and yet be inconsistent with each other. This tension reveals that there are often too many exploitable degrees of freedom in the mapping relationship between models of behavioral phenomena and neural mechanisms, and points to the role that other background assumptions play including level-assumptions about the appropriate level at which the neural model should be specified and localization-assumptions about where in the system the underlying mechanism might occur. We ended by briefly discussing how these considerations highlight the need for a more refined account of modelling constraints in neuroscience.

Acknowledgements We would like to thank Krys Dolega, Colin Klein, Oron Shagrir, Alessio Plebe, Carlos Zednik, and especially Matteo Colombo and Brendan Ritchie for their insightful feedback.

References

Alais, D., & Burr, D. (2004). The ventriloquist effect results from near-optimal bimodal integration. *Current Biology, 14*(3), 257–262.

Angelaki, D. E., Gu, Y., & DeAngelis, G. C. (2009). Multisensory integration: Psychophysics, neurophysiology, and computation. *Current Opinion in Neurobiology, 19*(4), 452–458.

Bechtel, W., & Shagrir, O. (2015). The non-redundant contributions of Marr's three levels of analysis for explaining information-processing mechanisms. *Topics in Cognitive Science, 7*(2), 312–322.

Berniker, M., & Kording, K. P. (2011). Estimating the relevance of world disturbances to explain savings, interference and long-term motor adaptation effects. *PLoS Computational Biology, 7*(10), e1002210.

Bowers, J. S., & Davis, C. J. (2012a). Bayesian just-so stories in psychology and neuroscience. *Psychological Bulletin, 138*(3), 389.

Bowers, J. S., & Davis, C. J. (2012b). Is that what Bayesians believe? Reply to Griffiths, Chater, Norris, and Pouget (2012). Psychological Bulletin, 138 423-426

Burr, D., & Alais, D. (2006). Combining visual and auditory information. *Progress in Brain Research, 155*, 243–258.

Carandini, M., & Heeger, D. J. (2012). Normalization as a canonical neural computation. *Nature Reviews Neuroscience, 13*(1), 51.

Chirimuuta, M. (2014). Minimal models and canonical neural computations: The distinctness of computational explanation in neuroscience. *Synthese, 191*(2), 127–153.

Clark, A. (2013). Whatever next? Predictive brains, situated agents, and the future of cognitive science. *Behavioral and Brain Sciences, 36*(3), 181–204.
Clark, A. (2015). *Surfing uncertainty: Prediction, action, and the embodied mind*. Oxford: Oxford University Press.
Colombo, M., & Hartmann, S. (2017). Bayesian cognitive science, unification, and explanation. *The British Journal for the Philosophy of Science, 68*(2), 451–484. https://doi.org/10.1093/bjps/axv036.
Colombo, M., & Seriès, P. (2012). Bayes in the brain—On Bayesian modelling in neuroscience. *The British Journal for the Philosophy of Science, 63*(3), 697–723.
Coltheart, M. (2006). What has functional neuroimaging told us about the mind (so far)? *Cortex, 42*(3), 323–331.
Craver, C. F. (2007). *Explaining the brain: Mechanisms and the mosaic unity of neuroscience*. Oxford: Oxford University Press.
Craver, C. F., & Darden, L. (2013). *In search of mechanisms: Discoveries across the life sciences*. Chicago: University of Chicago Press.
Craver, C. F., & Kaplan, D. M. (2018). Are more details better? On the norms of completeness for mechanistic explanations. *British Journal for the Philosophy of Science*, axy015. https://doi.org/10.1093/bjps/axy015.
Dayan, P., & Abbott, L. F. (2001). *Theoretical neuroscience*. Cambridge, MA: MIT Press.
Doya, K. (Ed.). (2007). *Bayesian brain: Probabilistic approaches to neural coding*. Cambridge, MA: MIT press.
Ernst, M. O., & Banks, M. S. (2002). Humans integrate visual and haptic information in a statistically optimal fashion. *Nature, 415*(6870), 429–433.
Ernst, M. O., & Bülthoff, H. H. (2004). Merging the senses into a robust percept. *Trends in Cognitive Sciences, 8*(4), 162–169.
Feldman, J. (2013). Tuning your priors to the world. *Topics in Cognitive Science, 5*(1), 13–34.
Fetsch, C. R., Turner, A. H., DeAngelis, G. C., & Angelaki, D. E. (2009). Dynamic reweighting of visual and vestibular cues during self-motion perception. *The Journal of Neuroscience, 29*(49), 15601–15612.
Fetsch, C. R., Pouget, A., DeAngelis, G. C., & Angelaki, D. E. (2012). Neural correlates of reliability-based cue weighting during multisensory integration. *Nature Neuroscience, 15*(1), 146–154.
Fetsch, C. R., DeAngelis, G. C., & Angelaki, D. E. (2013). Bridging the gap between theories of sensory cue integration and the physiology of multisensory neurons. *Nature Reviews Neuroscience, 14*(6), 429–442.
Fiser, J., Berkes, P., Orbán, G., & Lengyel, M. (2010). Statistically optimal perception and learning: From behavior to neural representations. *Trends in Cognitive Sciences, 14*(3), 119–130.
Geisler, W. S. (2011). Contributions of ideal observer theory to vision research. *Vision Research, 51*(7), 771–781.
Griffiths, T. L., & Tenenbaum, J. B. (2009). Theory-based causal induction. *Psychological Review, 116*(4), 661.
Hahn, U. (2014). The Bayesian boom: Good thing or bad? *Frontiers in Psychology, 5*, 765.
Heuer, H. W., & Britten, K. H. (2007). Linear responses to stochastic motion signals in area MST. *Journal of Neurophysiology, 98*(3), 1115–1124.
Hitchcock, C., & Woodward, J. (2003). Explanatory generalizations, part II: Plumbing explanatory depth. *Nous, 37*(2), 181–199. http://www.jstor.org/stable/3506081.
Huneman, P. (2010). Topological explanations and robustness in biological sciences. *Synthese, 177*(2), 213–245.
Jones, M., & Love, B. C. (2011). Bayesian fundamentalism or enlightenment? On the explanatory status and theoretical contributions of Bayesian models of cognition. *Behavioral and Brain Sciences, 34*(04), 169–188.
Kaplan, D. M. (2011). Explanation and description in computational neuroscience. *Synthese, 183*(3), 339.

Kaplan, D. M., & Craver, C. F. (2011). The explanatory force of dynamical and mathematical models in neuroscience: A mechanistic perspective. *Philosophy of Science, 78*(4), 601–627.
Kersten, D., Mamassian, P., & Yuille, A. (2003). Object perception as Bayesian inference. *Annual Review of Psychology, 55*, 271–304.
Kitcher, P. (1981). Explanatory unification. *Philosophy of Science, 48*(4), 507–531.
Knill, D. C., & Pouget, A. (2004). The Bayesian brain: The role of uncertainty in neural coding and computation. *Trends in Neurosciences, 27*(12), 712–719.
Knill, D. C., & Richards, W. (1996). *Perception as Bayesian inference*. Cambridge: Cambridge University Press.
Körding, K. (2007). Decision theory: What "should" the nervous system do? *Science, 318*(5850), 606–610.
Kording, K. P. (2014). Bayesian statistics: Relevant for the brain? *Current Opinion in Neurobiology, 25*, 130–133.
Körding, K. P., & Wolpert, D. M. (2004). Bayesian integration in sensorimotor learning. *Nature, 427*(6971), 244–247.
Körding, K. P., & Wolpert, D. M. (2006). Bayesian decision theory in sensorimotor control. *Trends in Cognitive Sciences, 10*(7), 319–326.
Krakauer, J. W. (2009). Motor learning and consolidation: The case of visuomotor rotation. In *Progress in motor control* (pp. 405–421). Boston: Springer.
Levy, A., & Bechtel, W. (2013). Abstraction and the organization of mechanisms. *Philosophy of Science, 80*(2), 241–261.
Ma, W. J., & Jazayeri, M. (2014). Neural coding of uncertainty and probability. *Annual Review of Neuroscience, 37*, 205–220.
Ma, W. J., Beck, J. M., Latham, P. E., & Pouget, A. (2006). Bayesian inference with probabilistic population codes. *Nature Neuroscience, 9*(11), 1432–1438.
Maloney, L. T., & Mamassian, P. (2009). Bayesian decision theory as a model of human visual perception: Testing Bayesian transfer. *Visual Neuroscience, 26*(1), 147–155.
Mauk, M. D. (2000). The potential effectiveness of simulations versus phenomenological models. *Nature Neuroscience, 3*(7), 649–651.
Mole, C., & Klein, C. (2010). 9 confirmation, refutation, and the evidence of fMRI. In *Foundational issues in human brain mapping* (p. 99).
Morgan, M. L., DeAngelis, G. C., & Angelaki, D. E. (2008). Multisensory integration in macaque visual cortex depends on cue reliability. *Neuron, 59*(4), 662–673.
Morrison, M. (2000). *Unifying scientific theories: Physical concepts and mathematical structures*. Cambridge: Cambridge University Press.
Orbán, G., & Wolpert, D. M. (2011). Representations of uncertainty in sensorimotor control. *Current Opinion in Neurobiology, 21*(4), 629–635.
Piccinini, G. (2007). Computing mechanisms. *Philosophy of Science, 74*(4), 501–526.
Pouget, A., Beck, J. M., Ma, W. J., & Latham, P. E. (2013). Probabilistic brains: Knowns and unknowns. *Nature Neuroscience, 16*(9), 1170.
Rao, R. P. N., Olshausen, B. A., & Lewicki, M. S. (2002). *Probabilistic models of the brain: Perception and neural function*. Cambridge, MA: MIT Press.
Rathkopf, C. (2018). Network representation and complex systems. *Synthese, 195*(1), 55–78.
Shagrir, O., & Bechtel, W. (2017). Marr's computational level and delineating phenomena. In D. M. Kaplan (Ed.), *Explanation and integration in mind and brain science* (pp. 190–214). New York: Oxford University Press.
Stein, B. E., Meredith, M. A., & Wallace, M. T. (1993). The visually responsive neuron and beyond: Multisensory integration in cat and monkey. In *Progress in brain research* (Vol. 95, pp. 79–90). Elsevier.
Stepp, N., Chemero, A., & Turvey, M. T. (2011). Philosophy for the rest of cognitive science. *Topics in Cognitive Science, 3*(2), 425–437.
Stocker, A. A., & Simoncelli, E. P. (2006). Noise characteristics and prior expectations in human visual speed perception. *Nature Neuroscience, 9*(4), 578–585.

Tassinari, H., Hudson, T. E., & Landy, M. S. (2006). Combining priors and noisy visual cues in a rapid pointing task. *Journal of Neuroscience, 26*(40), 10154–10163.

Trommershauser, J., Kording, K., & Landy, M. S. (Eds.). (2011). *Sensory cue integration*. New York: Oxford University Press.

van Beers, R. J., Sittig, A. C., & van der Gon Denier, J. J. (1996). How humans combine simultaneous proprioceptive and visual position information. *Experimental Brain Research, 111*(2), 253–261.

van Beers, R. J., Sittig, A. C., & van Der Gon, J. J. D. (1999). Integration of proprioceptive and visual position-information: An experimentally supported model. *Journal of Neurophysiology, 81*(3), 1355–1364.

Van Gelder, T. (1998). The dynamical hypothesis in cognitive science. *Behavioral and Brain Sciences, 21*(5), 615–628.

Weiss, Y., Simoncelli, E. P., & Adelson, E. H. (2002). Motion illusions as optimal percepts. *Nature Neuroscience, 5*(6), 598–604.

Wolpert, D. M., & Landy, M. S. (2012). Motor control is decision-making. *Current Opinion in Neurobiology, 22*(6), 996–1003.

Woodward, J. (2004). Counterfactuals and causal explanation. *International Studies in the Philosophy of Science, 18*(1), 41–72.

Zednik, C., & Jäkel, F. (2016). Bayesian reverse-engineering considered as a research strategy for cognitive science. *Synthese, 193*, 3951–3985.

Chapter 3
Prediction and Topological Models in Neuroscience

Bryce Gessell, Matthew Stanley, Benjamin Geib, and Felipe De Brigard

Abstract In the last two decades, philosophy of neuroscience has predominantly focused on explanation. Indeed, it has been argued that mechanistic models are the standards of explanatory success in neuroscience over, among other things, topological models. However, explanatory power is only one virtue of a scientific model. Another is its predictive power. Unfortunately, the notion of prediction has received comparatively little attention in the philosophy of neuroscience, in part because predictions seem disconnected from interventions. In contrast, we argue that topological predictions can and do guide interventions in science, both inside and outside of neuroscience. Topological models allow researchers to predict many phenomena, including diseases, treatment outcomes, aging, and cognition, among others. Moreover, we argue that these predictions also offer strategies for useful interventions. Topology-based predictions play this role regardless of whether they do or can receive a mechanistic interpretation. We conclude by making a case for philosophers to focus on prediction in neuroscience in addition to explanation alone.

Keywords Prediction · Network science · Topology · Neuroscience

3.1 Introduction

Contemporary philosophy of neuroscience has in large part been dominated by a focus on *explanation*. This focus follows a more general trend in the philosophy of science, where causal or mechanistic explanations have overtaken law-based explanations as the preferred means for understanding much of the world. Indeed, philosophers have produced compelling arguments for mechanistic models being the standard of explanatory success in the biological sciences (Craver and Darden 2013). However, when it comes to the enterprise of science, the explanatory power

B. Gessell · M. Stanley · B. Geib · F. De Brigard (✉)
Duke University, Durham, NC, USA
e-mail: felipe.debrigard@duke.edu

© Springer Nature Switzerland AG 2021

F. Calzavarini, M. Viola (eds.), *Neural Mechanisms*, Studies in Brain and Mind 17,
https://doi.org/10.1007/978-3-030-54092-0_3

of theoretical models is only one of many virtues (Schindler 2018). Another virtue of theoretical models, which in recent years has received comparatively less attention in the philosophy of science, is *prediction*, despite it being once heralded as equally relevant as explanation among the goals of science (Hofstadter 1951; Popper 1963; Lakatos and Musgrave 1970; Salmon 1978). This absence is particularly noticeable in the *philosophy of neuroscience*, as there has been almost no discussion on the predictive power or value of theoretical models in neuroscientific research. When contrasted with the fact that contemporary neuroscience is heavily engaged in generating predictive models (e.g., Yarkoni and Westfall 2017), the dearth of discussion on prediction in the philosophy of neuroscience is even more remarkable.

Pretheoretically, many people think of prediction as synonymous with prognostication or forecasting, meaning that which is predicted has not occurred yet. This time-dependent view of prediction contrasts with a knowledge-dependent view, according to which what one predicts may or may not have already occurred, as long as it is not known. In this paper, we adopt this knowledge-dependent or epistemic reading of prediction and side with Barrett and Stanford (2006) in defining a prediction as "a claim about known matters of fact whose truth or falsity has not already been independently ascertained by some more direct method than that used to make the prediction itself" (586). Moreover, successful predictions in general enhance our epistemic standing, not by way of supplying further explanatory details, but by reducing our uncertainty as to what to expect under certain conditions, and by providing us with strategies to effectively intervene and manipulate phenomena. Of course, successful predictions often lead to improved explanations (Douglas 2009); however, even without this additional bonus, successful predictions have value in and of themselves.

Perhaps a key reason as to why there is so much emphasis on explanation (as compared to prediction) in the philosophy of science in general, and of neuroscience in particular, is that there is a clear relationship between explanation and intervention. For many scientists and philosophers, the scientific goal of unveiling the real nature of the world is at least as important as that of offering strategies to intervene and control it (e.g., Longino 2002). Given that mechanistic models provide both descriptions of natural phenomena and approaches to manipulate such phenomena, it is unsurprising that such models are taken as ideal candidates as to how to best pursue research in neuroscience (Craver 2007). The current chapter, however, puts pressure on this view by highlighting the connection between the *predictive* power of certain theoretical models in neuroscience and their value as strategies for manipulation and intervention (Douglas 2009). Importantly, the kinds of theoretical models we have in mind are *topological models*, which have recently been the subject of discussion in the philosophy of science, with some arguing that they offer an alternative kind of explanation, different from mere causal or mechanistic explanation (Huneman 2010; Lange 2016), and others arguing that they do not (Craver 2016; Povich and Craver 2018). We will largely sidestep this discussion, however, as we seek to explore the predictive rather than explanatory ambitions of topological models in neuroscience (with occasional mention of other disciplines too), and the role they can play in our capacity to intervene, manipulate, and control neural phenomena. To reiterate: our arguments seek neither to support nor to undermine the claim that topological models are explanatory, nor whether or not they are so in virtue of receiving a mechanistic interpretation. We want to argue instead for a different claim, namely that regardless of whether or not topological models receive a mechanistic interpretation, they still hold predictive value and can be reliable guides to intervention and manipulation. Moreover, we put forth the more general claim that good predictions ought to be a central goal of neuroscience, regardless of whether or not they are afforded by models that have (or even could receive) a complete mechanistic interpretation.

The chapter will proceed as follows. In Sect. 3.2, we offer a brief discussion on the relationship between prediction and explanation, and we place the role of mechanistic models in general, and in the philosophy of neuroscience in particular, within that dialectic. Next, in Sect. 3.3, we discuss the nature of topological models and their use in prediction and interventions in a number of different fields before focusing on the use of topological models in network neuroscience for prediction. We also show how these models can be useful for intervention and manipulation even absent a mechanistic understanding of their underpinnings. Finally, in Sect. 3.4, we draw some general conclusions and questions for future research.

3.2 Prediction, Explanation, and Mechanistic Models

To fully understand the relationship between intervention (or manipulability), on the one hand, and mechanistic models in neuroscience, on the other, it may be useful to begin with a brief excursus into the history of the debate on the relationship between explanation and prediction in the philosophy of science (for a recent excellent review see Douglas 2009). This will allow us to better locate the role of mechanistic models in neuroscience within this dialectic.

3.2.1 Prediction and Explanation: A Brief History

Although one can find interesting discussions about the relationship between explanation and prediction in science in the works of Hume (1748), Whewell (1840), and Mill (1843), contemporary scholarship on the subject usually starts with the deductive-nomological (DN) model proposed by Hempel and Oppenheim (1948). According to the DN model, the explanandum (i.e., the statement to be explained) must deductively follow from the explanans: a set of premises that not only should be true but also include boundary conditions and general laws. According to the DN model, in its simplest form, a scientific explanation would have the following structure:

$$\left.\begin{array}{l} C_1 \wedge C_2 \wedge C_3 \wedge \ldots \wedge C_n \\ \\ L_1 \wedge L_2 \wedge L_3 \wedge \ldots \wedge L_n \end{array}\right\} \textit{Explanans}$$

$$\therefore \textit{Explanandum}$$

Here, each C_i is a true statement of a boundary condition or particular occurrence of an event, and each L_i is a statement of a general law. Thus, a successful explanation of, say, a particular observation of a planet's location at a particular time would be given by a set of premises involving other relevant observations and empirical conditions, as well as by certain physical laws governing celestial bodies. Importantly, for Hempel and Oppenheim, both explanations and predictions shared the same logical structure, as it is possible for an explanans to state a yet unobserved event. Predictions, as it were, are explanations of future events; or, alternatively, explanations are predictions of past events (*aka* "postdictions").

This view, known since as the "symmetry thesis", was met with substantial backlash in the 1950s and 1960s. It was pointed out, for instance, that while explanations require true propositions, successful predictions need not (Scheffler 1957). Others argued that the uncertainty that applies to explanations differs from that which applies to predictions (Helmer and Rescher 1959), while still others pointed out that some theoretical models—such as in quantum mechanics (Hanson 1959) and evolution (Scriven 1959)—are good at explaining but bad at predicting. By the time Nagel's *The Structure of Science* (1961) was published, the focus in philosophy of science had almost entirely moved to explanation for, as he remarked, "the distinctive aim of the scientific enterprise is to provide systematic and responsibly supported explanations" (1961, 15). The displacement of prediction—or the "decentering of prediction" as Heather Douglas (2009) calls it—brought explanation to the forefront of philosophical scholarship in the philosophy of science.[1]

With prediction relegated to the background, most discussions focused on whether or not the DN model offered a successful analysis of scientific explanation. Philosophers quickly grew dissatisfied with the logical structure of explanations offered by the DN model. Some of the first concerns pertained to the difficulty of distinguishing statements of non-accidental generalizations from those of scientific laws (Hempel 1965). Soon after, counterexamples to the DN model started to emerge. Some pointed out explanatory asymmetries, as in the example in which the length of a flagpole is deductively derived from the length of its shadow in

[1] It is important to note that this decentering may not apply to other related areas of research, such as issues on confirmation and accommodation, both of which are related to the notion of prediction (see, for instance, Eells 2000). We thank a reviewer for inviting us to note this issue.

conjunction with relevant laws about the propagation of light (Bromberger 1966). Such a derivation, it was argued, conforms to the logical structure of the DN model, yet we feel that the explanans (i.e., the length of the flagpole) is not really explained by the explanandum (i.e., the length of the shadow plus laws pertaining to the propagation of light). Other counterexamples pointed at cases of explanatory irrelevance, as with the case of the following derivation (Salmon 1971):

(P1) All males who take birth control pills regularly fail to get pregnant
(P2) John Jones is a male who has been taking birth control pills regularly
∴ John Jones fails to get pregnant

which seems to conform to the structure of the DN model—i.e., P1 satisfies the criteria of lawfulness, and P2 states particular true observations—and yet does not constitute a successful explanation.

As a consequence, the 1970s and 1980s saw a proliferation of models of scientific explanation, including Salmon's statistical-relevance (SR) model (Salmon 1971), the causal model (Salmon 1984), and the unification model (Kitcher 1989), to name a few. Unsurprisingly, most of the scholarship on scientific explanation during those two decades boiled down to a series of exchanges between counterexamples and defenses of these various models. By the 1990s, no agreed-upon model of explanation was in the offing and, instead, philosophers of science largely moved toward some kind of explanatory pluralism. Arguments turned into discussions as to what sort of explanatory model would be more appropriate for each scientific discipline. This was the intellectual environment in which the mechanistic explanation model was fully articulated (Machamer et al. 2000), and the following years helped to strengthen it as the paramount model for scientific explanation in the life sciences, including neuroscience (Craver 2007).

3.2.2 The Mechanistic Model of Explanation in Neuroscience

Although there are several definitions of "mechanism" and "mechanistic explanation" in the philosophy of science (e.g., Machamer et al. 2000; Glennan 2002; Bechtel and Abrahamsen 2005), they all seem to agree on what Craver and Tabery (2015) call the "ecumenical" characterization of mechanism, according to which mechanisms consist of four components. First, there is the *phenomenon*, which is understood as the behavior of the system that the mechanism constitutes. Every mechanism, then, is a mechanism of some particular phenomenon—e.g., digestion, long-term potentiation, in attentional blindness—and, depending on the particular phenomenon, a mechanism can produce, underlie, or maintain it. The second component are the *parts* of the mechanism which, in turn, are linked by the third component: *causal relations*. Considerable discussion has ensued regarding the best characterization of causal relations for mechanistic explanations. For our purposes, what matters is that such causal relations are *intervenable*, that is, they can be in principle—even if not in practice—manipulated to make a difference in the phenomenon. Finally, mechanisms are also *organized* in some fashion. In the case of neuroscience, many think of mechanisms as hierarchically organized in different levels (Craver 2007), but other organizations may be possible too (Craver and Tabery 2015).

Mechanistic models in neuroscience, then, are useful for the purpose of explanation insofar as they can capture a mechanism. And they can capture a mechanism if they conform to what Kaplan and Craver (2011) call the "3 M" mapping requirement:

> (3M) A model of a target phenomenon explains that phenomenon when (a) the variables in the model correspond to identifiable components and organizational features of the target mechanism that produces, maintains, or underlies the phenomenon, and (b) the causal relations posited among these variables in the model correspond to the activities or operations among the components of the target mechanism. (Kaplan and Craver 2011, 272)

It is likely that, as of now, we do not have a single mechanistic model that fully conforms to the 3 M requirement and that provides a complete characterization of all the components. At best, we have *schematic* models: abstract or idealized descriptions of a mechanism in which many of the details are omitted and/or that include provisional place-holders for unknown components (Darden 2002). Moreover, mechanistic schematic models also vary in terms of the degree to which they capture the actual phenomenon. On one extreme, *how-possibly* models describe mechanisms in terms of how the different parts *might* be causally related and organized to produce, maintain, or support a phenomenon. On the other extreme, *how-actually* models depict how they are actually causally related, what all the parts really are, and how the parts are in reality organized to produce, maintain, or support a phenomenon. Unsurprisingly, we likely have very few—if any—how-actually models in neuroscience; these constitute a normative goal that our constantly refined how-possibly mechanistic schematic models seek to reach (Craver and Darden 2013). Much of the scientific work in contemporary neuroscience consists precisely in discovering the underlying components of a mechanistic model to provide interpretations of the filler terms that can bring a how-possibly model closer to a how-actually one. Consider our current model of long-term potentiation (LTP) in neurons in the dentate gyrus. While this may constitute one of the most thorough mechanistic models in neuroscience, researchers keep discovering new details that help to make certain assumptions and idealizations more concrete. For instance, while early models postulated that N-methyl-D-aspartate receptors (NMDAR) were necessary to trigger the induction of LTP (Collingridge et al. 1983), more recent discoveries have shown that other receptors, such as metabotropic glutamate (MGluRs) and kainate receptors can do it as well. More recently still is has been shown that even $Ca2+-$permeable α-amino-3-hydroxy-5-methyl-4-isoxazole propionic acid receptor (AMPAR) can do the trick, further inviting the revision of the actual components of our mechanistic model of LTP (Park et al. 2014). Despite being one of the most complete mechanistic models in neuroscience, our current model for LTP is not a how-actually model quite yet; at best, it is a how-nearly-actually mechanistic model (Craver and Tabery 2015).

Nevertheless, mechanistic models seem perfectly appropriate to deliver on what arguably are the two main goals of the scientific enterprise: to uncover the nature of reality, and to enable us to manipulate and control it. Mechanistic models—as opposed to the DN-, the SR-, and some variants of unificationist and mathematical models—are ideally suited to contribute toward the first goal, insofar as they care less about the logical structure of the explanation and more about its ontic commitments, that is, the kinds of actual, real structures that count as legitimately explanatory (Craver 2014).[2] The explaining is done by *real* stuff, causally related and organized in various ways in order to produce, sustain or underlie a phenomenon. Mechanistic models not only tell us *why* something happens, but also *what* makes it happen. In turn, mechanistic models contribute to the second goal thanks to their reliance on counterfactual theories of causation, particularly manipulationist views (Woodward 2003). When the causal relations are thus understood, the parts of a mechanism that constitute the *relata* can be seen as variables able to make a difference to the phenomenon—i.e., the behavior of the mechanism they are part of. In other words, mechanistic models enable us to tell what would happen to the phenomenon if one were to intervene on a particular variable (i.e., a part) at a certain level of organization. Thus, mechanistic models are ideal to tell us how the phenomenon would behave under counterfactual conditions and, consequently, they seem perfectly suited to offer predictions as well.

Given all these considerations it is hard not to think of mechanistic models as the paradigmatic model for not only scientific explanations, but also scientific *predictions* in neuroscience. In fact, some mechanists seem to suggest as much. They claim that understanding how a phenomenon works via subsuming it under a mechanistic model is perhaps the most reliable way to predict how it will behave in the future, and how it can be manipulated so that we can make it "work for us" (Kaplan and Craver 2011). A strong reading of this view would imply that models can *only* yield successful predictions if they have strong ontic commitments to the structures they represent, and if they offer, if not a how-actually, at least a how-nearly-actually or a how-plausibly mechanistic schema of the phenomenon.[3]

[2]There are some views of mechanistic models that need not have such strong ontic commitments (e.g, Bechtel 2008) and/or that need not be committed to a manipulationist/counterfactual-dependent account of causation. It is possible that some of the arguments we discuss here do not necessarily apply to these accounts. We don't discuss these accounts in depth, in part because they are not as thoroughly developed in the philosophy of neuroscience. Thanks to a reviewer for inviting us to clarify this point.

[3]We see successful predictions as those which accurately model alternative outcomes (and thus support counterfactuals to some degree), or model future states with accuracy significantly above chance. In short, good predictions estimate outcomes above randomness. Note that, on this view, how-actually and how-possibly models can both yield successful predictions; however, how-actually models may not always make predictions that are perfectly accurate, since their use is often limited to certain contexts (consider the difference between Newtonian and relativistic physics, for example).

In what follows, we argue against this strong reading according to which mechanistic models are paradigmatic models for both explanation *and* prediction, particularly as they apply to neuroscience. Instead, we argue for a weaker view, according to which, even if mechanistic models are ideally fitted for generating explanations in neuroscience, there may be some non-mechanistic models that are well suited to offer not only successful predictions but also strategies to manipulate and control certain phenomena. In particular, we defend this weaker reading in relation to topological models, which have recently been criticized by mechanicists who argue that they are not explanatory, or that, if they are, they explain precisely because they ultimately resolve into a mechanistic model (Craver 2016). The suggestion we put forth in the rest of this essay is that even if topological models only have explanatory value when translated into their mechanistic components, they still have predictive value regardless of whether or not they have clear ontic commitments and/or mechanistic interpretations.[4]

3.3 Network Science: Prediction and Interventions

Network science makes use of the mathematical tools and formalisms from graph theory to empirically investigate real-world networks. In its simplest form, a network can be thought of as a collection of differentiable elements, or *nodes*, and the pairwise relationships between them, or *edges*. Diverse real-world systems can be thought of as networks. For example, protein-protein interaction networks, structural and functional brain networks, infectious disease networks, friendship networks, and air transportation networks have all been modeled as networks for various purposes. Despite the obvious differences in the actual, real-world phenomena, all these networks can be understood as collections of nodes with certain edges between them (Butts 2009). But of course, what the nodes and edges actually represent in the world will differ across the different kinds of networks (Fig. 3.1).

Graph theoretic metrics can then be used to characterize the topological properties of these networks—regardless of how the nodes and edges are defined in practice (Watts and Strogatz 1998; Butts 2009; Huneman 2010; Sporns 2011). A simple example of a topological property is *geodesic distance*: the minimum number of edges required to transverse from one particular node i to another node j in the network. You and your Facebook friend have a geodesic distance of 1, because it only takes one edge to connect you and your friend. But the geodesic distance

[4]A clarification: we are *not* saying that Craver is necessarily committed to the strong reading. As far as we know, partisans of mechanisms have said little as to whether or not predictive models also demand the same ontic commitments that explanatory models do. Our view should rather be seen, then, as an admonition to the effect that even if one adopts a mechanistic stance vis-à-vis the way in which neuroscience ought to be pursued, then the strong ontic commitments that have been argued for explanation need not apply to prediction too.

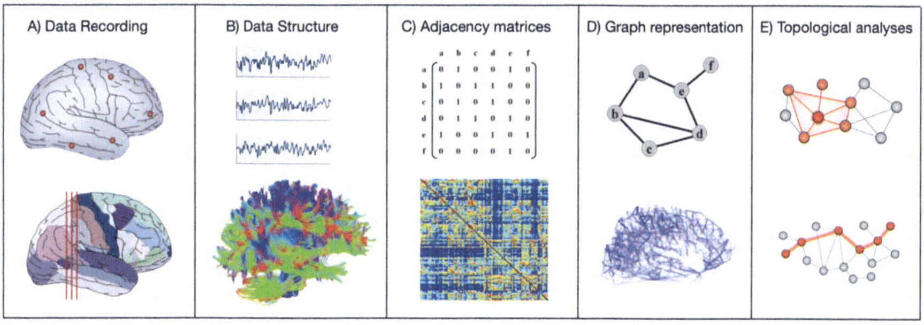

Fig. 3.1 Schematic representation of topological analyses employed in network neuroscience. (**a**) Data acquisition includes several methods, such as functional and structural MRI. (**b**) Depending on the nature of the data, their structure may vary—for example, time series in fMRI or diffusivity measures in diffusion tensor imaging (DTI). (**c**) Data is arranged in adjacency matrices, representing nodes and edges. (**d**) Data can also be represented in graphs, with lines depicting edges connecting nodes. (**e**) Topological analyses are then conducted to identify topological properties (e.g., clustering coefficient, shortest path)

between you and a friend of that friend who is not also your friend on Facebook would then be 2. Thus, geodesic distance, for instance, can help to calculate the spread of information on your Facebook wall. Relatedly, the *path length* of any node *i* in a network can be obtained by computing the average shortest number of steps necessary to get from *i* to each other node in the network (Dijkstra 1959). Path length offers an indication of how quickly or effectively information can spread throughout a network. Consider, for example, a large hierarchically structured company. The CEO likely has a relatively short path length, and information can be transmitted from the CEO to any employee in relatively few steps, whereas most low-level employees likely have a longer path length, as it takes more steps for them to communicate with members in faraway departments. A more complex graph theoretic metric is *eigenvector centrality*, a measure of the extent to which a node *i* is connected with other influential nodes in the network (nodes with lots of edges). Nodes with high eigenvector centrality are thought to be highly influential and effective in spreading information throughout a network. On social media (e.g., Twitter), for example, certain celebrities like Justin Bieber tend to have particularly high eigenvector centrality, as they tend to be connected with many other influential celebrities.

Topological and spatial scales can be changed depending on a researcher's interests. To give an example from neuroscience, the hippocampus can be studied as a single structure or unit, it can be studied as a three-part entity composed of CA1, CA3, and the dentate gyrus, or it can be studied as a more complex structure containing various cell types, layers, and their projections. Although it is often tempting to view phenomena at higher resolutions (e.g., cell types and the properties of those cells) as being the worthiest of serious investigation, it is sometimes not useful or valuable to improve the resolution with which one studies a given brain structure and its relation to cognition—especially when current computational and practical constraints are taken into account. Investigation at a more macroscopic scale often still yields useful and accurate predictions. Because it is unclear which level of granularity is the ground truth, and so unclear how best to demarcate components of the system (i.e., nodes representing functional units in the brain), topological prediction can play a central role. Comparing predictive utility at different levels of granularity can also guide future research and serve useful purposes. It is possible that different "scales of granularity" of network description will yield distinct yet complementary properties for predicting cognitive phenomena, disease states, and disease progression, among other things.

There are many other graph theoretic metrics that capture certain topological properties of networks, such as eigenvector centrality, clustering coefficients, and modularity (Newman 2010). Critically, one of the central features of network models is that topological properties can be ascertained independently of a system's physical substrates. That is, the same graph metrics can be computed on any kind of network, no matter what the nodes and edges are representing in the world; networks comprised of differently defined nodes and edges can even possess the same topological properties. To be sure, there are interesting philosophical questions about the nature of such topological properties and their relationship to the actual substrates the network models are based on. We also believe that understanding whether or not the topological properties of network models have any explanatory value above and beyond the mechanisms that underlie the system they seek to represent, is a worthwhile philosophical question (Huneman 2010; Craver 2016; Lange 2016). That being said, we also believe that the longstanding emphasis on explanation in philosophy of science, as well as the fact that network models have mainly been discussed in reference to alternative explanatory frameworks, have obscured the fact that topological properties in network models have remarkable *predictive* power. Additionally, in some instances, the predictive power of topological properties in network science has enabled us to conduct successful interventions. Let us explore some examples.

3.3.1 The Predictive Power of Topological Properties

We often obtain good predictions when causal information about the components of the system is incorporated into the model. However, in some cases, clear causal information is either unavailable, non-existent, or poorly-defined. Even in such cases, networks can still be characterized topologically, and their topological properties can produce accurate predictions. Studies of co-authorship networks, for example, capture patterns of collaboration in a given field. These networks allow us not only to identify prominent author(s) in a field, but also to successfully predict whether a publication will be well-cited in the future. For example, Sarigol et al. (2014) analyzed a dataset of over 100,000 publications from the field of computer science, and they investigated how centrality in the co-authorship network differs between authors who have highly cited papers and those who do not.

Using a machine learning classifier based only on co-authorship network centrality measures (degree centrality, eigenvector centrality, betweenness centrality, and k-core centrality), they were able to predict whether an article would belong to the 10% most cited articles in 5 years' time with a precision of 60%, well above chance. Interestingly, in order to not overemphasize one particular dimension of centrality in networks, they used several complementary measures of network centrality, and this combination of measures was crucial in adequately predicting the publication "success" of the researchers. To compute each centrality metric, however, it was first necessary to define the full set of nodes and edges in the network. By mapping out all connections in the network and computing graph metrics, they quantitatively suggested the existence of a social bias, manifesting itself in terms of visibility and attention, and influencing measurable citation "success" of researchers.

Another example pertains to traffic congestion. Consider the following question: "how can we accurately predict which roads in a city have or will have the highest occurrence of traffic jams?" A network approach might seek to predict whether a road will be congested by examining its topological properties within the larger network. This requires taking into account all other roads in the network (in this case, nodes might represent intersections, and edges might represent the road segments that link the intersections). Note that the kinds of buildings to which the roads provide access is not included in defining nodes and edges, and therefore, this information will not be included in—directly or indirectly—predicting traffic congestion.

Adopting this network approach, Wang et al. (2012) show that by incorporating information about how centrally a road is situated in graph theoretic space, they can accurately predict traffic patterns in San Francisco and Boston. The extent to which a road segment occupies a central place in the city grid is measured in terms of a mathematical property of networks known as *edge betweenness*. Edge betweenness is computed for each individual edge in a graph. To compute the edge betweenness, a search algorithm identifies the shortest possible path between each and every node in the network. It then searches the resulting data structure to determine what proportion of those paths incorporate the road segment in question. That proportion is the edge betweenness. In this particular case, edges represent road segments defined as stretches of roads between legal intersections, and nodes represent legal intersections. Wang et al. (2012) show that the traffic density on a road segment can be better predicted by modeling both the road's centrality and inherent travel demand (i.e., how often the buildings on the road are frequented) than it can by modeling inherent travel demand alone. That is, incorporating edge betweenness into the predictive model actually provides better predictions above and beyond travel demand alone. Importantly, one can only compute edge betweenness by completely searching the entire topographical structure of the system, even though the measure is computed for an individual road (i.e., edge). With this particular kind of quantitative description, one might then be able predict that in other major metropolitan cities in the United States (e.g., Houston, Phoenix, Chicago, Dallas, etc.), roads with higher edge betweenness will experience more traffic jams, on average.

Another example comes from sexual networks, whereby persons are thought of as nodes and sexual contacts as edges. Long-term and large-scale data collection has led to the production of large-scale sexual networks from Manitoba, Canada, and from Colorado Springs, USA (Woodhouse et al. 1994; Rothenberg et al. 1998; Wylie and Jolly 2001; Jolly and Wylie 2002; Potterat et al. 2002). These kinds of networks highlight the heterogeneities present in sexual networks and show the importance of core groups (i.e., highly and disproportionally interconnected subsets of people with high numbers of contacts) and 'long-distance' connections (linking otherwise distant parts of the network) in disease transmission. Note that it is only possible to uncover these core groups (i.e., network modules) and 'long-distance' connections that interconnect groups by mapping out the full structure of the sexual network. Moreover, edges are defined only by whether two individuals have sex with each other during some time period *t*. To provide a particularly salient example, Liljeros et al. (2001) showed that sexual networks, like many networks that are present in the world, have a scale-free degree-distribution (in contrast to, for example, a Gaussian distribution). This property means that the vast majority of individuals in the network have very few sexual contacts, but that there are a few individuals who have had a very large number of sexual contacts. Importantly, the fact the network has a scale-free architecture suggests that some of the individuals with a very large number of partners may bridge relatively isolated communities, i.e. they have long-distance connections in addition to many connections.[5]

On the surface, it may seem as though predictions about human sexual networks are underwritten not by the topological properties of these networks, but instead by our knowledge of the actual causal properties involved. For example, we know a lot about human sexual contact, and can give accurate microbiological explanations of how some STDs pass from one person to another. However, the force of this example is that our predictions about human sexual networks would still be accurate, *even if we had none of this causal and biological knowledge*. Suppose that we were examining sexual networks in an alien species, for example. The topological properties would still be helpful in predicting disease transmission among members of the species, even if we had no knowledge, detailed or otherwise, about the alien biology.

[5]We say that a scale-free architecture "suggests" this organization of individuals because, while not a mathematical guarantee, it appears likely to be so. In a scale-free network architecture, statically speaking, some of the high-degree nodes will be provincial hubs and some of the high-degree nodes will be connector hubs. Granted, it is not the case that networks must not follow this principle; in some scale-free networks, all the high-degree nodes might be connectors. But this seems statistically unlikely as then distinct modules are unlikely to exist. If the high-degree nodes are "randomly" arranged, then some must be connectors and some must be provincial. In other words, in scale-free networks, the nodes at the far end of the distribution have considerable influence over the other nodes in the network, more so than in other kinds of networks with other kinds of degree distributions. Some of these nodes with very many connections are likely to interconnect many different communities and be essential (in the example from the text) for diseases to propagate throughout the network.

Moreover, this example offers an interesting case in which quantitatively characterizing the topological properties of the network allows us to identify particular individuals in the network who could be targeted for a *subsequent* intervention in response to a sexual-disease outbreak, as given limited resources, it may not be practical to target all individuals in the network to promote safe-sex practices. But by specifically targeting individuals who have the greatest number of connections and those who tend to disproportionately connect otherwise distant groups or clusters in the network, it may be possible to most efficiently and effectively promote safe-sex practices to reduce the likelihood of disease transmission across the entirety of the network.

Let us summarize the three examples we have seen. The first dealt with a co-authorship networks, and focused on the predictive power of centrality measures. In this example, causal information for entities in the network is either non-existent or poorly defined; nevertheless, centrality measures are still able to help us generate successful predictions. The second example was about traffic congestion. Here, researchers used edge betweenness and other topological measures to describe global properties of the network, thereby making it possible to give accurate predictions about different cities based on the successes of a single network analysis. The third and last example, about sexual networks, illustrated the power of topological models to identify the degree distribution of a particular network—a scale-free distribution in this case, as opposed to, say, a Gaussian distribution. Most importantly, the example of sexual networks shows that causal knowledge—even when it is available and detailed—is not necessary for making predictions in virtue of topological properties.

Taken together, what these and related examples strongly suggest is that the topological properties of network models can be successfully employed to make predictions and to guide interventions on the systems they represent even when no causal or mechanistic information about the system is either known or included in the model. In fact, the examples just discussed are agnostic as to their ontic commitments, as they tend to be independent of the precise physical substrata of the modeled system. Crucially, many of these models offer clear avenues for intervention. For example, in the case of sexual contact networks, some individuals in the network have more connections than others, and certain individuals are disproportionately responsible for interconnecting relatively segregated communities in the network. By identifying and targeting those individuals, we might be more likely to stop the spread of disease. Of course, it may be possible that some of these predictions illuminate the underlying nature of the phenomenon, and as such may contribute to its explanation. But even if they don't, the topological properties of network models still hold enormous epistemic value by enabling us to make predictions and by offering the possibility of gaining some measure of control over social and natural processes (Douglas 2009; Douglas and Magnus 2013).

3.3.2 Predictions and Interventions in Network Neuroscience

The aforementioned considerations were confined to network models outside the field of neuroscience. The question now is whether or not we have evidence to the effect that network models and their topological properties can also afford predictions and strategies for manipulation *within* in neuroscience. We believe they can. Some contemporary neuroscientists treat the brain as a large-scale network. Determining what constitutes a node or an edge, however, is tricky, as it depends on the particular level of analysis, the particularities of the research questions, and the idiosyncrasies of the available technologies from which the data is acquired (Stanley et al. 2013). Regarding levels of analysis, brains can be seen as varying along at least three scales. First, there is a spatial scale that ranges from the very micro (e.g., neurons, glia) to the very macro (e.g., gross anatomical regions comprising millions of neurons and even more synapses connecting those neurons). Thus, while networks at the micro-level may include neurons as nodes and synaptic connections as edges, networks at the macro-level may include cytoarchitecturally delimited portions of brain tissue as nodes and white matter tracts as edges—in the case of structural networks—or voxels as nodes and correlations in signal over time as edges, in the case of functional connectivity networks. Second, brain networks also vary along a topological scale that goes from local (e.g., networks within a brain region) to global (e.g., networks across the whole brain). Finally, brain networks vary along a temporal scale, ranging from very fast (e.g., sub-second neural processes) to the very slow (e.g., life-span or evolutionary changes). Networks whose topological features vary along several scales are known as multi-scale. Therefore, brains can be thought of as multi-scale networks (Betzel and Bassett 2017; De Brigard 2017).

Different research questions, and their inherent practical limitations, also influence the way in which network models are constructed. We may, for instance, want to construct a brain model in which each individual neuron is represented as a node, with the edges between nodes representing synapses. Unfortunately, while this has been done successfully in the significantly less complex organism *C. Elegans* (Sporns and Kötter 2004; Towlson et al. 2013), it is not currently possible to image, record, or computationally analyze the tens of billions of neurons in the human brain, especially when neurons often have thousands of synapses (Drachman 2005). Current neuroimaging technology limits functional and structural brain network analyses to nodes above the millimeter scale, meaning that many potentially interacting neurons, synapses, and other structures will be represented as an individual node in human brain networks. The lack of a clear, obvious choice of what should represent a node in a functional brain network has resulted in the analysis of brain networks across a wide range of scales, ranging from 70-node (Wang et al. 2009) to 140,000-node whole brain networks (Eguíluz et al. 2005), using a variety of parcellation schemes dependent on wide-ranging definitional criteria (Stanley et al. 2013). The boundaries representing reasonable functional units (i.e., nodes) in the brain for investigating a particular phenomenon of interest need not line up with the surfaces of structures or other commonsense loci of demarcation, and the 'best' way to define nodes (size, brain-region, etc.) often depends on a researcher's question. Furthermore, it is possible that these different levels of granularity provide network descriptions that are distinct, yet complementary, when predicting cognitive phenomena. For example, the particular firing patterns of neurons exclusively within the hippocampus support memory encoding and retrieval (Battaglia et al. 2011), and the increased topological centrality of the hippocampus—modeled as a single node in the whole brain network—also supports memory retrieval (Geib et al. 2017a).

Finally, the data from which topological models of brain networks are built also varies as a function of the technology employed to extract them. For instance, functional brain networks have been constructed using functional MRI (fMRI) (Achard and Bullmore 2007; Achard et al. 2006; Eguíluz et al. 2005; Geib et al. 2017a, b; Salvador et al. 2005; van den Heuvel et al. 2008), electroencephalography (EEG) (Micheloyannis et al. 2006; Stam et al. 2007), magnetoencephalography (MEG) data (Bassett et al. 2006; Deuker et al. 2009; Stam 2004), and ECoG (Betzel et al. 2019). Structural brain graphs have been constructed from diffusion tensor imaging (DTI) or diffusion spectrum imaging (DSI) (Gong et al. 2008; Hagmann et al. 2008), as well as from conventional MRI data (Bassett et al. 2008; He et al. 2007).

Importantly, as in the case of the network models discussed above (Sect. 3.1), recent studies suggest that the topological properties of network models in neuroscience offer extraordinary predictive value. Consider, for instance, research on brain disease. A recent study by Khazaee et al. (2015) combined network analyses of fMRI data with advanced machine learning techniques to investigate brain network differences between patients with Alzheimer's disease (AD) and healthy, age-matched controls (see also, Khazaee et al. 2016). Alzheimer's disease is a progressive neurodegenerative disease that is accompanied by severe decline in cognitive functioning (in memory in particular; Albert et al. 2011). Graph theoretic metrics were obtained from each participant's brain network, and machine learning was used to explore the ability for graph metrics to help in the diagnosis of AD. They applied their method to resting-state fMRI data of 20 patients with AD and 20 age- and gender-matched healthy subjects. The graph measures were computed and then used as the discriminating features in the model. Extracted network-based features were fed to different feature selection algorithms to choose the most significant features. Using a set of graph metrics computed for diverse nodes (brain regions) in the network, the researchers were able to identify patients with AD relative to healthy controls with perfect accuracy (i.e., 100% correctly). So, if a new case were presented to the researchers, they would presumably be able to accurately predict whether that individual had AD based upon a set of graph theoretic metrics obtained from that individual's fMRI data. Results of this study suggest that graph theoretic metrics obtained from functional brain networks can efficiently and effectively assist in the diagnosis of AD. It may be that early diagnosis (before the onset of behavioral symptoms) is also possible by this method, regardless of whether we have a full mechanistic account explaining what occurs in the brain in AD.

A subsequent study conducted by Hojjati et al. (2017) went a step beyond Khazaee et al. (2015). Specifically, Hojjati et al. (2017) used similar graph theoretic metrics obtained from brain networks constructed from resting-state fMRI in conjunction with machine learning algorithms to predict which individuals would progress from Mild Cognitive Impairment (MCI) to AD and which individuals would not progress from MCI to AD. MCI is a transitional stage between normal age-related cognitive decline and actual AD. The researchers were able to predict with greater than 90% accuracy which individuals would progress from MCI to AD and which would not. The ability to accurately predict which individuals are likely to progress to AD offers physicians useful information to better tailor prevention and treatment programs on an individual basis. Additionally, it would be of significant use to family members of when planning for future care. (See delEtoile and Adeli (2017) for a useful recent review of similar research.)

Consider now the case of epilepsy, one of the most common neurological conditions. Epilepsy is characterized by the tendency toward recurrent, unprovoked seizures (Stam 2014). Treatment for certain severe, drug-resistant cases of epilepsy sometimes involves anterior temporal lobectomy. Using graph theoretic metrics obtained from resting-state brain networks in conjunction with machine learning algorithms, He et al. (2017) were able to predict surgical outcomes from patients who underwent anterior temporal lobectomy. More specifically, the researchers used graph theoretic measures of centrality during rest *prior* to the lobectomy to predict with high accuracy whether participants would be seizure-free a full year later. This research provides a useful potential biomarker for surgical outcomes, with the potential to usefully guide the decision-making of physicians in future cases by determining which individuals would be most likely to benefit from surgery. As in the examples above, the utility of graph theoretic metrics in predictive surgical outcomes is extremely valuable, regardless of whether or not we have a full mechanistic account explaining what occurs in the brain in epilepsy.

Cases of MCI, AD, and epilepsy offer salient examples of the predictive and intervention-guiding value of graph theoretic metrics obtained from brain networks. By "intervention-guiding," we refer primarily to the possibility of identifying sub-populations which are at greater risk for certain health problems, on which clinicians may focus their treatment (though there are other ways in which brain-based graph theoretic metrics can guide interventions as well). Similar methods have also been used to predict with incredible accuracy which individuals have common clinical disorders, such as major depressive disorder (Sacchet et al. 2015; Gong and He 2015) and attention deficit hyperactivity disorder (Colby et al. 2012). In these cases, graph theoretic metrics obtained from brain networks offer great predictive utility in diagnosis. This, in turn, has the potential to aid in optimal treatment and intervention using other established techniques. Taking all of this together, the findings just reviewed clearly indicate that the topological properties of network models in neuroscience offer extraordinary predictive value and useful information for treatment and intervention, independent of their possible mechanistic interpretations.

3.4 Conclusion

Science is undoubtedly in the business of offering explanations about natural phenomena. But it is also in the business of offering predictions and strategies to intervene and manipulate reality. Most of the research in contemporary philosophy of science has focused on explanation, and the philosophy of neuroscience has followed suit. The overarching goal of the current paper has been to shed some light on the oft-neglected issue of prediction in neuroscience. We do so through the lens of network models and their topological properties.

While current philosophers of neuroscience disagree as to whether or not network models are truly explanatory, or whether their explanatory power is based in mechanistic schemas (Klein 2012; Muldoon and Bassett 2016; Craver 2016), we focused instead on the fact that many network models have predictive value and offer strategies for manipulation and intervention even when no clear causal or mechanistic account of the phenomenon is available. As such, topological models in network neuroscience promise to enhance our epistemic status regarding the brain and its effects by way of informing many vital decisions. The ability to use topological properties from network models to make predictions also may help us to improve patient outcomes. The progression from pre-MCI to Alzheimer's disease, for example, significantly impacts an individual's life and the lives of loved ones. Predictions offer clear value for addressing these issues, even when a full-fledged mechanistic explanation of the neurological conditions isn't readily available. Cognitive neuroscience may not yet be able to give detailed instructions through mechanistic explanations for manipulating mental states; accurate predictions, on the other hand, do offer a clearer way forward for many applied problems.

References

Achard, S., & Bullmore, E. (2007). Efficiency and cost of economical brain functional networks. *PLoS Computational Biology, 3*(2), e17.

Achard, S., Salvador, R., Whitcher, B., Suckling, J., & Bullmore, E. (2006). A resilient, low-frequency, small-world human brain functional network with highly connected association cortical hubs. *Journal of Neuroscience, 26*, 63–72.

Albert, M. S., DeKosky, S. T., Dickson, D., Dubois, B., Feldman, H. H., Fox, N. C., et al. (2011). The diagnosis of mild cognitive impairment due to Alzheimer's disease: Recommendations from the National Institute on Aging-Alzheimer's association workgroups on diagnostic guidelines for Alzheimer's disease. *Alzheimer's & Dementia, 7*(3), 270–279.

Barrett, J., & Stanford, P. K. (2006). Prediction. In S. Sarkar & J. Pfeifer (Eds.), *The philosophy of science: An encyclopedia*. New York: Routledge.

Bassett, D. S., Meyer-Lindenberg, A., Achard, S., Duke, T., & Bullmore, E. (2006). Adaptive reconfiguration of fractal small-world human brain functional networks. *Proceedings of the National Academy of Sciences, 103*(51), 19518–19523.

Bassett, D. S., Bullmore, E., Verchinski, B. A., Mattay, V. S., Weinberger, D. R., & Meyer-Lindenberg, A. (2008). Hierarchical organization of human cortical networks in health and schizophrenia. *Journal of Neuroscience, 28*(37), 9239–9248.

Battaglia, F. P., Benchenane, K., Sirota, A., Pennartz, C. M., & Wiener, S. I. (2011). The hippocampus: Hub of brain network communication for memory. *Trends in Cognitive Sciences, 15*(7), 310–318.

Bechtel, W., & Abrahamsen, A. (2005). Explanation: A mechanist alternative. *Studies in History and Philosophy of Biological and Biomedical Sciences, 36*(2), 421–441.

Bechtel, William (2008). Mechanisms in cognitive psychology: What are the operations?. *Philosophy of Science, 75*(5):983–994.

Betzel, R. F., & Bassett, D. S. (2017). Multi-scale brain networks. *NeuroImage, 160*, 73–83.

Betzel, R. F., Medaglia, J. D., Kahn, A. E., Soffer, J., Schonhaut, D. R., & Bassett, D. S. (2019). Structural, geometric and genetic factors predict interregional brain connectivity patterns probed by electrocorticography. *Nature Biomedical Engineering, 1*.

Bromberger, S. (1966). Questions. *Journal of Philosophy, 63*(20), 597–606.

Butts, C. T. (2009). Revisiting the foundations of network analysis. *Science, 325*(5939), 414–416.

Colby, J. B., Rudie, J. D., Brown, J. A., Douglas, P. K., Cohen, M. S., & Shehzad, Z. (2012). Insights into multimodal imaging classification of ADHD. *Frontiers in Systems Neuroscience, 6*, 59.

Collingridge, G. L., Kehl, S. J., & McLennan, H. T. (1983). Excitatory amino acids in synaptic transmission in the Schaffer collateral-commissural pathway of the rat hippocampus. *The Journal of Physiology, 334*(1), 33–46.

Craver, C. F. (2007). *Explaining the brain: mechanisms and the mosaic unity of neuroscience*. Oxford/Ann Arbor: Oxford University Press/Clarendon Press.

Craver, C. F. (2014). The ontic account of scientific explanation. In M. I. Kaiser, O. R. Scholz, D. Plenge, & A. Hüttemann (Eds.), *Explanation in the special sciences: The case of biology and history* (pp. 27–52). Dordrecht: Springer.

Craver, C. F. (2016). The explanatory power of network models. *Philosophy of Science, 83*(5), 698–709.

Craver, C. F., & Darden, L. (2013). *In search of mechanisms: Discoveries across the life sciences*. Chicago: University of Chicago Press.

Craver, C., & Tabery, J. (2015). Mechanisms in science. In Edward N. Zalta (Ed.), *The Stanford encyclopedia of philosophy* (Summer 2019 Edition). forthcoming. https://plato.stanford.edu/archives/sum2019/entries/science-mechanisms/

Darden, L. (2002). Rethinking mechanistic explanation. *Philosophy of Science, 69*(S3), 342–353.

De Brigard, F. (2017). Cognitive systems and the changing brain. *Philosophical Explorations, 20*(2): 224–241

delEtoile, J., & Adeli, H. (2017). Graph theory and brain connectivity in Alzheimer's disease. *The Neuroscientist, 23*(6), 616–626.

Deuker, L., Bullmore, E. T., Smith, M., Christensen, S., Nathan, P. J., Rockstroh, B., & Bassett, D. S. (2009). Reproducibility of graph metrics of human brain functional networks. *NeuroImage, 47*(4), 1460–1468.

Dijkstra, E. W. (1959). A note on two problems in connexion with graphs. *Numerische Mathematik, 1*, 269 271.

Douglas, H. E. (2009). Reintroducing prediction to explanation. *Philosophy of Science, 76*(4), 444–463.

Douglas, H., & Magnus, P. D. (2013). State of the field: Why novel prediction matters. *Studies in History and Philosophy of Science Part A, 44*(4), 580–589.

Drachman, D. (2005, June). Do we have brain to spare? *Neurology, 64*(12).

Eguíluz, V. M., Chialvo, D. R., Cecchi, G. A., Baliki, M., & Apkarian, A. V. (2005). Scale-free brain functional networks. *Physical Review Letters, 94*, 018102.

Eells, E., & Fitelson, B. (2000). Measuring confirmation and evidence. *Journal of Philosophy, 97*(12), 663–672.

Geib, B. R., Stanley, M. L., Wing, E. A., Laurienti, P. J., & Cabeza, R. (2017a). Hippocampal contributions to the large-scale episodic memory network predict vivid visual memories. *Cerebral Cortex, 27*(1), 680–693.

Geib, B. R., Stanley, M. L., Dennis, N. A., Woldorff, M. G., & Cabeza, R. (2017b). From hippocampus to whole-brain: The role of integrative processing in episodic memory retrieval. *Human Brain Mapping, 38*(4), 2242–2259.

Glennan, S. (2002). Rethinking mechanistic explanation. *Philosophy of Science, 69*(S3), S342–S353.

Gong, Q., & He, Y. (2015). Depression, neuroimaging and connectomics: A selective overview. *Biological Psychiatry, 77*(3), 223–235.

Gong, G., He, Y., Concha, L., Lebel, C., Gross, D. W., Evans, A. C., & Beaulieu, C. (2008). Mapping anatomical connectivity patterns of human cerebral cortex using in vivo diffusion tensor imaging tractography. *Cerebral Cortex, 19*(3), 524–536.

Hagmann, P., Cammoun, L., Gigandet, X., Meuli, R., Honey, C. J., Wedeen, V. J., & Sporns, O. (2008). Mapping the structural core of human cerebral cortex. *PLoS Biology, 6*(7), e159.

Hanson, N. R. (1959). Copenhagen interpretation of quantum theory. *American Journal of Physics, 27*(1), 1–15.

He, Y., Chen, Z. J., & Evans, A. C. (2007). Small-world anatomical networks in the human brain revealed by cortical thickness from MRI. *Cerebral Cortex, 17*, 2407–2419.

He, X., Doucet, G. E., Pustina, D., Sperling, M. R., Sharan, A. D., & Tracy, J. I. (2017). Presurgical thalamic "hubness" predicts surgical outcome in temporal lobe epilepsy. *Neurology, 88*(24), 2285–2293.

Helmer, O., & Rescher, N. (1959). On the epistemology of the inexact sciences. *Management Science, 6*(1), 25–52.

Hempel, C. (1965). *Aspects of scientific explanation and other essays in the philosophy of science*. New York: The Free Press.

Hempel, C. G., & Oppenheim, P. (1948). Studies in the logic of explanation. *Philosophy of Science, 15*(2), 135–175.

Hofstadter, A. (1951). Explanation and necessity. *Philosophy and Phenomenological Research, 11*, 339–347.

Hojjati, S. H., Ebrahimzadeh, A., Khazaee, A., Babajani-Feremi, A., & Initiative, A.'s. D. N. (2017). Predicting conversion from MCI to AD using resting-state fMRI, graph theoretical approach and SVM. *Journal of Neuroscience Methods, 282*, 69–80.

Hume, D. (1748). *An enquiry concerning human understanding*. Glasgow.

Huneman, P. (2010). Topological explanations and robustness in biological sciences. *Synthese, 177*(2), 213–245.

Jolly, A. M., & Wylie, J. L. (2002). Gonorrhoea and chlamydia core groups and sexual networks in Manitoba. *Sexually Transmitted Infections, 78*(suppl 1), i145–i151.

Kaplan, D. M., & Craver, C. F. (2011). The explanatory force of dynamical and mathematical models in neuroscience: A mechanistic perspective. *Philosophy of Science, 78*(4), 601–627.

Khazaee, A., Ebrahimzadeh, A., & Babajani-Feremi, A. (2015). Identifying patients with Alzheimer's disease using resting-state fMRI and graph theory. *Clinical Neurophysiology, 126*(11), 2132–2141.

Khazaee, A., Ebrahimzadeh, A., & Babajani-Feremi, A. (2016). Application of advanced machine learning methods on resting-state fMRI network for identification of mild cognitive impairment and Alzheimer's disease. *Brain Imaging and Behavior, 10*(3), 799–817.

Kitcher, P. (1989). Explanatory unification and the causal structure of the world. In P. Kitcher & W. Salmon (Eds.), *Scientific explanation* (pp. 410–505). Minneapolis: University of Minnesota Press.

Klein, C. (2012). Cognitive ontology and region- versus network-oriented analyses. *Philosophy of Science, 79*(5), 952–960.

Lakatos, I., & Musgrave, A. (Eds.). (1970). *Criticism and the growth of knowledge: Volume 4: Proceedings of the international colloquium in the philosophy of science*, 1965. London:. Cambridge University Press.

Lange, M. (2016). *Because without cause: Non-causal explanations in science and mathematics*. Oxford: Oxford University Press.

Liljeros, F., Edling, C. R., Amaral, L. A. N., Stanley, H. E., & Åberg, Y. (2001). The web of human sexual contacts. *Nature, 411*(6840), 907.

Longino, H. (2002). *The fate of knowledge*. Princeton: Princeton University Press.

Machamer, P. K., Darden, L., & Craver, C. F. (2000). Thinking about mechanisms. *Philosophy of Science, 67*(1), 1–25.

Micheloyannis, S., Pachou, E., Stam, C. J., Vourkas, M., Erimaki, S., & Tsirka, V. (2006). Using graph theoretical analysis of multi channel EEG to evaluate the neural efficiency hypothesis. *Neuroscience Letters, 402*(3), 273–277.

Mill, J. (1843). *A system of logic, ratiocinative and inductive*. London.

Muldoon, S. F., & Bassett, D. S. (2016). Network and multilayer network approaches to understanding human brain dynamics. *Philosophy of Science, 83*(5), 710–720.

Nagel, E. (1961). *The structure of science*. New York: Harcourt, Brace & World.

Newman, M. (2010). *Networks: An introduction*. Oxford: Oxford University Press.

Park, P., Volianskis, A., Sanderson, T. M., Bortolotto, Z. A., Jane, D. E., Zhuo, M., et al. (2014). NMDA receptor-dependent long-term potentiation comprises a family of temporally overlapping forms of synaptic plasticity that are induced by different patterns of stimulation. *Philosophical Transactions of the Royal Society B: Biological Sciences, 369*(1633), 20130131.

Povich, Mark & Craver, Carl F. (2018). Because without Cause: Non-Causal Explanations in Science and Mathematics. *Philosophical Review, 127*(3):422–426.

Popper, K. (1963). *Conjectures and refutations: The growth of scientific knowledge*. London: Routledge & Kegan Paul.

Potterat, J. J., Muth, S. Q., Rothenberg, R. B., Zimmerman-Rogers, H., Green, D. L., Taylor, J. E., et al. (2002). Sexual network structure as an indicator of epidemic phase. *Sexually Transmitted Infections, 78*(suppl 1), i152–i158.

Rothenberg, R. B., Potterat, J. J., Woodhouse, D. E., Muth, S. Q., Darrow, W. W., & Klovdahl, A. S. (1998). Social network dynamics and HIV transmission. *AIDS, 12*(12), 1529–1536.

Sacchet, M. D., Prasad, G., Foland-Ross, L. C., Thompson, P. M., & Gotlib, I. H. (2015). Support vector machine classification of major depressive disorder using diffusion-weighted neuroimaging and graph theory. *Frontiers in Psychiatry, 6*, 21.

Salmon, W. (1971). *Statistical explanation & statistical relevance*. Pittsburgh: University of Pittsburgh Press.

Salmon, W. C. (1978). Unfinished business: The problem of induction. *Philosophical Studies, 33*(1), 1–19.

Salmon, W. (1984). *Scientific explanation and the causal structure of the world*. Princeton: Princeton University Press.

Salvador, R., Suckling, J., Schwarzbauer, C., & Bullmore, E. (2005). Undirected graphs of frequency-dependent functional connectivity in whole brain networks. *Philosophical Transactions of the Royal Society B: Biological Sciences, 360*(1457), 937–946.

Sarigöl, E., Pfitzner, R., Scholtes, I., Garas, A., & Schweitzer, F. (2014). Predicting scientific success based on coauthorship networks. *EPJ Data Science, 3*(1), 9.

Scheffler, I. (1957). Explanation, prediction, and abstraction. *The British Journal for the Philosophy of Science, 7*(28), 293–309.

Schindler, S. (2018). *Theoretical virtues in science: Uncovering reality through theory*. Cambridge: Cambridge University Press.

Scriven, M. (1959). Explanation and prediction in evolutionary theory. *Science, 130*(3374), 477–482.

Sporns, O. (2011). The human connectome: a complex network. *Annals of the New York Academy of Sciences, 1224*(1), 109–125.

Sporns, O., & Kötter, R. (2004). Motifs in brain networks. *PLoS Biology, 2*(11), e369.

Stam, C. J. (2004). Functional connectivity patterns of human magnetoencephalographic recordings: A 'small-world' network? *Neuroscience Letters, 355*(1–2), 25–28.

Stam, C. J. (2014). Modern network science of neurological disorders. *Nature Reviews Neuroscience, 15*(10), 683.

Stam, C. J., Nolte, G., & Daffertshofer, A. (2007). Phase lag index: Assessment of functional connectivity from multi channel EEG and MEG with diminished bias from common sources. *Human Brain Mapping, 28*(11), 1178–1193.

Stanley, M. L., Moussa, M. N., Paolini, B., Lyday, R. G., Burdette, J. H., & Laurienti, P. J. (2013). Defining nodes in complex brain networks. *Frontiers in Computational Neuroscience, 7*, 169.

Towlson, E. K., Vértes, P. E., Ahnert, S. E., Schafer, W. R., & Bullmore, E. T. (2013). The rich club of the C. elegans neuronal connectome. *Journal of Neuroscience, 33*(15), 6380–6387.

van den Heuvel, M. P., Stam, C. J., Boersma, M., & Pol, H. H. (2008). Small-world and scale-free organization of voxel-based resting-state functional connectivity in the human brain. *NeuroImage, 43*(3), 528–539.

Wang, J., Wang, L., Zang, Y., Yang, H., Tang, H., Gong, Q., et al. (2009). Parcellation-dependent small-world brain functional networks: A resting-state fMRI study. *Human Brain Mapping, 30*(5), 1511–1523.

Wang, P., Hunter, T., Bayen, A. M., Schechtner, K., & González, M. C. (2012). Understanding road usage patterns in urban areas. *Scientific Reports, 2*, 1001.

Watts, D. J., & Strogatz, S. H. (1998). Collective dynamics of 'small-world' networks. *Nature, 393*(6684), 440.

Whewell, W. (1840). *The philosophy of the inductive science*. London.

Woodhouse, D. E., Rothenberg, R. B., Potterat, J. J., Darrow, W. W., Muth, S. Q., Klovdahl, A. S., et al. (1994). Mapping a social network of heterosexuals at high risk for HIV infection. *AIDS, 8*(9), 1331–1336.

Woodward, J. (2003). *Making things happen: A theory of causal explanation*. Oxford: Oxford University Press.

Wylie, J. L., & Jolly, A. (2001). Patterns of chlamydia and gonorrhea infection in sexual networks in Manitoba, Canada. *Sexually Transmitted Diseases, 28*(1), 14–24.

Yarkoni, T., & Westfall, J. (2017). Choosing prediction over explanation in psychology: Lessons from machine learning. *Perspectives on Psychological Science, 12*(6), 1100–1122.

Chapter 4
Circuital and Developmental Explanations for the Cortex

Alessio Plebe

Abstract The cerebral cortex manifests a feature that puzzles researchers since early neuroscience: the functional repertoire of the cortex is incredibly vast despite its strikingly uniform structure. This work analyzes the phenomenon of the apparent clash between uniformity and variety of functions, and it pinpoints the sort of explanations that this phenomenon requests. A possible resolution of this tension has been proposed several times in terms of a basic neural circuit so successful to underlie all cortical functions. Circuital models have the virtue of belonging to the mechanistic framework of explanation, and they have greatly improved the understanding of computational properties of the cortex. However, they all lack explanations of the contrast between uniformity and multiplicity of functions in the cortex. A reason for this failure is neglecting the developmental aspect of the cortex, the most likely source of variation in functions. In biology, developmental explanations are receiving increasing attention, but they are often contrasted with the mechanistic ones. I contend that, in the case at hand, the explanandum of the development differs from the ones usually found in developmental biology, and developmental aspects in the cortex can be taken into account within a mechanistic explanation.

Keywords Cerebral cortex · Canonical circuit · Developmental explanation · Mechanistic explanation

4.1 Introduction

It is well agreed upon that the mammalian neocortex is the site of processes enabling higher cognition from consciousness to symbolic reasoning and, for humans, language (Miller et al. 2002; Fuster 2008; Noack 2012). Why the particular

A. Plebe (✉)
Department of Cognitive Science, University of Messina, Messina, Italy
e-mail: aplebe@unime.it

© Springer Nature Switzerland AG 2021

F. Calzavarini, M. Viola (eds.), *Neural Mechanisms*, Studies in Brain and Mind 17,
https://doi.org/10.1007/978-3-030-54092-0_4

arrangements of neurons in the cortex makes such a difference with respect to the rest of the brain is still, after a century of research, largely unknown. Edinger (1904) was one of the first to rank mammals as the most intelligent animals, in virtue of the brand new layered brain equipment introduced by nature. Although current comparative cognition has weakened this intellectual superiority, the cortex is still considered to be the crowning achievement of brain evolution, and the quest for understanding its computational properties is among the most prominent and yet unresolved issues in neuroscience.

This chapter addresses one of the most puzzling facts of the cortex: the clash between its strikingly uniform structure and the breadth of its functional repertoire. This issue will be spelled out precisely in Sect. 4.2 and the consistency of its premises will be assessed. The perplexing discrepancy between uniformity and variety of functions has been often cited by neuroscientists as the motivation for searching a fundamental circuit responsible for the computational power of the cortex, often called "canonical circuit" (Plebe 2018).

How these circuits are conceived, and their achievements, will be discussed in Sect. 4.3. This is currently the mainstream line of research on the cortex, extending up to the large-scale brain simulation projects (Markram et al. 2015). An epistemological virtue of the circuital direction of research is that canonical models may broadly qualify as mechanistic. However, they all fail to explain the contrast between uniformity and multiplicity of functions in the cortex. The main reason is that circuital models neglect the developmental aspect of the cortex, which is of paramount importance in the diversification of functions across cortical areas. Development is not just crucial for the early diversification of cortical functions, it is an everyday business for the cortex, as detailed in Sect. 4.2.2.

Developmental explanations in biology are anything but new (Waddington 1957; Gottlieb 1971), but they have remained marginal until recently. Today, developmentally oriented explanations and epigenetics are mainstream (Baedke 2018), stirring a growing philosophical debate on the explanatory standards in biology. Several philosophers have claimed that developmental explanations are distinct and irreducible to mechanistic explanations (Mc Manus 2012; Parkkinen 2014). However, as discussed in Sect. 4.4, the standard cases these philosophers have in mind are quite different from the case of the cortex, which seems to fit well into a multi-level mechanistic explanation.

Even if research on combining developmental mechanisms with basic cortical circuits is still marginal with respect to the mainstream research, examples of proposals in this direction are described in Sect. 4.5. These models may qualify as mechanistic sketches, even if incomplete.

4.2 Explanandum and Explanations for the Cortex

In a first approximation, much of the research on the cerebral cortex aims at progressing in some way the answer to the question "how does the cortex work?". To answer this general question, among the contributions that may broadly qualify as mechanistic, one can include the identification of the layered structure of the cortex and, within it, the structure of specific classes of neurons with their interconnections (Ramón y Cajal 1891; Lorente de Nó 1938; Braak 1980; Nieuwenhuys 1994). Since much of the activity in the cortex is electrical, a privileged path to an answer to the general question might be the search of some fundamental structure in the cortex processing electrical signals. This is the kind of explanation sought by the "canonical circuits" research effort (Plebe 2018) that will be shortly summarized in Sect. 4.3.

However, the mammalian cortex is a very special part of the brain and requires explanations to peculiar sorts of phenomena. I argue that the most compelling phenomenon is the combination of the following two facts:

$\mathcal{P}_1$ the cortex is remarkably uniform;
$\mathcal{P}_2$ the cortex is the main site of a bewildering variety of functions.

It is not possible to formulate propositions $\mathcal{P}_1$ and $\mathcal{P}_2$ precisely and deterministicaly, because of their non quantitative nature. Therefore, their combination do not lead to a paradox in a strict logical sense. Nevertheless, the clash between uniformity in structure and variability in the performed functions is a relevant issue, and is echoed by many neuroscientists when writing about the cortex, I collected here a few quotations:

> The mammalian cerebral neocortex can learn to perform a wide variety of tasks, yet its structure is strikingly uniform. It is natural to wonder whether this uniformity reflects the use of rather few underlying methods of organizing information (Marr 1970, p.163)

> The apparent uniformity of the neocortex has given rise to the speculation that [...] is designed to perform the same basic operation, or 'computation' as it is now fashionable to call it. [...] The tempting notion is then that nature's laboratory has hit on a process that enables it to use the same machinery for very different ends. If this attractive view is correct, the $64000 question is then: what is the cortex doing with its inputs? (Martin 1988, p.639–640)

> Neurobiological studies have shown that cortical circuits have a distinctive modular and laminar structure, with stereotypical connections between neurons that are repeated throughout many cortical areas. It has been conjectured that these stereotypical canonical microcircuits are [...] advantageous for generic computational operations that are carried out throughout the neocortex (Haeusler et al. 2009, p.73)

> The neocortex is the brain structure most commonly believed to give us our unique cognitive abilities. Yet the cellular organization of the neocortex is broadly similar not only between species but also between cortical areas. (Harris and Shepherd 2015, p.170)

> The cerebral cortex performs a wide range of cognitive tasks in mammals [...] Yet it processes these diverse tasks with what appears to be a remarkably uniform, primarily six-layer architecture [...] This has long suggested the idea that a piece of six-layer cortex with a surface area on the order of a square millimeter constitutes a fundamental cortical 'processing unit' (Miller 2016, p.75)

Before discussing what sort of explanation can reconcile propositions $\mathcal{P}_1$ and $\mathcal{P}_2$, let me scrutinize them one at a time. Premise $\mathcal{P}_2$ is quite straightforward since there is ample evidence of the array of heterogeneous functions engaging the cortex. One may require to provide a working definition of "function" as used in proposition $\mathcal{P}_2$, being a notoriously ambiguous notion in philosophy. Regarding the cortex, "function" can be read as: an etiological function in a realistic account (Wright 1976), as the capacities of its components (Cummins 1975), as the cognitive capacities deriving from its activities (Young et al. 2000), or as a mathematical mapping between incoming and outgoing signals (Rathkopf 2013; Burnston 2016). However, for the current purposes it is easy to verify the cortex's involvement in a multitude of functionalities under all the accounts of the term "function" just listed.

4.2.1 *Is the Cortex Uniform?*

Premise $\mathcal{P}_1$ is more controversial than $\mathcal{P}_2$. The issue of uniformity has given birth to two opposing parties in neuroscience: the "lumpers" and the "splitters" (Carlo and Stevens 2013, p.1488). The former find the idea of uniformity exciting and puzzling, whereas the latter believe that every cortical area is unique in structure. One notable radical example in the "splitters" party is found in Marcus et al. (2014, p.551–552): "What would it mean for the cortex to be diverse rather than uniform? One possibility is that neuroscience's quarry should be not a single canonical circuit". Marcus and co-workers solve the clash between $\mathcal{P}_1$ and $\mathcal{P}_2$ by denying premise $\mathcal{P}_1$: the diversity of functions in the cortex is simply explained by diverse structures.

I take that the question, as formulated in the title of this section, cannot get a sharp answer because is ill-posed. There is no suitable metrics to quantitatively assess uniformity in general. For example, the cortex is certainly not uniform down to the molecular level like a metal plate. Moreover, there are obvious diversification in the two layers of the cortex engaged in the main extracortical communication. The fourth layer is the main target of thalamocortical projections, so it is well developed in primary sensorial areas. For the opposite reason, the fifth layer is mainly populated by pyramidal cells projecting to the basal ganglia or directly to the corticospinal tract, so it is highly developed in all motor areas. The different extents of layers IV and V have been used by von Economo and Koskinas (1925) for a broad classification of: *granular* cortex, typical of sensorial areas rich of spiny stellate neurons fed by thalamic fibers; *agranular* cortex, with few spiny stellate cells, such as the motor areas. However, apart from the density of extracortical connections in layers IV and V, the laminar structure and the intracortical connectivity remain similar even between granular and agranular areas.

I think that for the purpose of the present discussion, the issue of uniformity is manageable from a relativistic perspective, confronting the available data on uniformity/disuniformity across the entire cortex with the variations in the neuroanatomical structure of the rest of the brain. By using a relativistic account of uniformity, experimental evidences seem to speak in favor of $\mathcal{P}_1$, as I will show.

The most important and investigated kind of uniformity is the regular repetition of the radial profile of the cortex, which can be grouped into six distinct layers, as first observed by Berlin (1858) and detailed by early neuroscientists such as

Ramón y Cajal (1906), Brodmann (1909), Vogt and Vogt (1919), and von Economo and Koskinas (1925). In a first attempt to assess the uniformity of the cortex on a quantitative basis, Rockel et al. (1980) counted the number of cells through the entire thickness of the cortex in most of the major cortical areas in monkeys, humans, and several other mammals. This count has been found to be surprisingly constant for the different areas and the different species, with about 110 neurons in cortical sections of 30 μm diameter. The only exception is always to be found in the primary visual cortex, with a count of about 270 neurons. Their observations have been the subject of a fierce debate for over 30 years, with doubts raised concerning whether their experimental methods were technically flawed (Rakic 2008), and other studies reporting twofold or even threefold variations in neural density across the entire cortex (Herculano-Houzel et al. 2008). Recently, Carlo and Stevens (2013) have replicated the direct count performed by Rockel and coworkers using modern stereological methods, and they confirmed the same uniformity of count.

Additional neurophysiological features of the cortex have been compared by Karbowski (2014) across species and regions. Again, he found remarkable invariance in a number of neuroanatomical measures. The postsynaptic density length (the thick part of the postsynaptic membrane hosting neurotransmitter receptors) has a mean value of 0.38 μm for the entire human cortex, with a standard deviation of only 0.04 μm. The synaptic density has a mean of $5 \times 10^{11}\,\mathrm{cm}^{-3}$ with a standard deviation as small as 0.3. The ratio of excitatory to inhibitory synapses is highly invariant even across species, with an average of 0.83 and a standard deviation of 0.03.

As mentioned above, the notion of uniformity can only be applicable to the cortex from a relativistic perspective. Therefore, it is useful to compare systematically the variation of the main stereological features within the cortex with the variation of the same features in rest of the brain. For this purpose I have adopted the most updated cell atlas for the mouse brain (Erö et al. 2018). This atlas includes densities for all cell types in 46 cortical areas and in 551 other brain structures. Table 4.1 shows the mean values and the standard deviations for neural cells, together with the results for different types of neurons. The statistics has been evaluated separately on the cortical areas, on the non-cortical brain areas, and on the whole brain. For the purpose of evaluating the relative uniformity, only the columns with standard deviations are relevant. The table reveals a striking larger variability in all cell densities in the non-cortical regions compared with the cortical areas. The standard

Table 4.1 Comparison of uniformity in neural cells density in the cortex and in the rest of the brain. (The data are from Erö et al. 2018)

	Cortex		Non-cortical regions		Whole brain	
Cell type	MEAN	STDEV	MEAN	STDEV	MEAN	STDEV
All neurons	8.59×10^4	1.75×10^4	1.02×10^5	1.74×10^5	1.01×10^5	1.67×10^5
Excitatory	7.42×10^4	1.77×10^4	8.92×10^4	1.72×10^5	8.80×10^4	1.65×10^5
Inhibitory	1.12×10^4	5.92×10^3	1.12×10^4	1.47×10^4	1.12×10^4	1.42×10^4
Modulatory	5.10×10^2	1.62×10^2	1.82×10^3	5.37×10^3	1.72×10^3	5.17×10^3

deviation is ten times greater in the non-cortical regions for all neurons and for the excitatory ones, it is more than double for inhibitory neurons, and more than thirty times greater for modulatory neurons. This relative uniformity is still evident even when comparing the cortex with the whole brain, cortex included, as visible in the rightmost column of Table 4.1.

In addition to the qualitative and statistical uniformity of the radial organization, there is a further uniformity in the cortex due to the periodical replication of a small cylindrical structure. The so-called *columnar* organization of the cortex was first suggested by von Economo and Koskinas (1925) and by Lorente de Nó (1938). It was first demonstrated by Mountcastle (1957) in the somatic sensory cortex, where vertical cylinders of neurons respond all to the same single stimulation of cutaneous receptors. A few years later, Hubel and Wiesel (1959) discovered a columnar organization in the primary visual cortex. A related concept has been introduced by Rakic (1995) – the "ontogenetic column" – a vertical stack of cells, divided by glial septa, generated during the embryonic migration of neurons into the cortical plate. This column is smaller in diameter compared to those of Mountcastle and Hubel and Wiesel. To what extent the columnar organization is ubiquitous in the cortex is an open question. For Horton and Adams (2005) there are too many and diverse concepts under the umbrella of "cortical column" to be a unifying principle of the cortical structure. Still, there is a widespread view that columnar organization is a fundamental feature of the cortex, even if not homogeneous and common to all areas (Rockland 2011; Kaas 2012; Molnár 2013; Rothschild and Mizrahi 2015; Casanova and Opris 2015)

The dimensions along which self-similarity of the cortex can be evaluated, here briefly summarized, lean toward a judgment of uniformity when compared with how the rest of the brain is organized. There are, indeed, other parts of the brain with a laminar structure, the most relevant is the cerebellum. In fact, the cerebellum is organized like a small brain, with an outer laminated "cortex" surrounding its deep non-laminated nuclei, and the cerebellar cortex is as uniform as the cerebral cortex (Ito 1984). The difference is that the cerebellar cortex has three layers, with a population of cells different from that of the cortex. Moreover, the cerebellum is much more narrow in scope than the cerebral cortex, being involved mostly in the regulation of movements and in some forms of motor learning. Most of the other parts of the brain lack any laminar structure, still there are several alternative forms of patterning of local circuits. For example, part of the ventral striatum is characterized by the alternation of *striosomes* and *matrisomes*, with the former rich of cholinergic and dopaminergic transmission, and the latter impoverished in these substances (Graybiel 1984). A second example of a typical small circuital module in the brain is the *glomerulus*, a spherical aggregation of neurons with the entire synaptic structure contained within a single glial sheath. The glomerulus is a prominent component of the olfactory bulb (Treloar et al. 2002) and is found also in the lateral geniculate nucleus of the thalamus (Sherman and Guillery 2006) and in the cerebellum (Ito 1984), but not in the cerebral cortex.

In summary, the number of neuroanatomical regularities of the cortex, unique with respect to the rest of the brain, lead to consider the cortex uniform enough to, at least, rise surprise and sense of anomaly when joining premise $\mathcal{P}_1$ and $\mathcal{P}_2$.

4.2.2 *Plasticity in the Cortex*

Having established that the premises $\mathcal{P}_1$ and $\mathcal{P}_2$ hold, the issue of the coexisting uniformity and functional diversification of the cortex still lacks an explanation. In my opinion, the weakness of the mainstream research on mechanistic cortical canonical models is in overlooking the developmental dimension of the cortical circuits. The focus is almost entirely on the mature circuits and functions, neglecting the enormous capacity of the cortex to mold its computational functions in response to patterns of input.

The developmental feature belongs to the phenomena collected under the term *neural plasticity*, which in fact comes in several different forms (Berlucchi and Buchtel 2009) and has been investigated under a variety of perspectives. A dominant perspective is the reorganization of the nervous system after injuries and strokes (Lövdén et al. 2010; Fuchs and Flügge 2014), while other streams of research focus on memory formation (Squire and Kandel 1999; Bontempi et al. 2007). A first account of plasticity as occurring in the cortex was in the landmark paper of Buonomano and Merzenich (1998), who distinguished three levels of plasticity, in relation to different methodologies of analysis. For the purpose of the current discussion, we can adopt a similar but more specific classification:

1. *synaptic plasticity*, addressing changes at single synapse level;
2. *intracortical map plasticity*, addressing internal changes at the level of a single cortical map;
3. *intercortical map plasticity*, addressing changes on a scale larger than a single cortical map.

The term "map" as used in "cortical map" is often regarded as synonymous of the more popular "area" (see for example Schüz and Miller 2002). There are, however, some methodological differences. The parcellation of the cortex into a mosaic of spatially contiguous areas is a long sought enterprise in neuroscience, which proved to be extremely challenging (Drury et al. 1996; Haueis 2012; Nieuwenhuys 2013; Glasser et al. 2016). It is not difficult to imagine that the main difficulty boils down in the uniformity of the cortex, which lacks the sharp boundaries in neurobiological properties proper to other parts of the brain. Even at the level of genetic expression, the boundaries in functional characteristics across cortical areas do not correspond to any sharp transition in the graded expressions of the transcription factors in the progenitors zones (O'Leary et al. 2007, 2013). The genetic expression across the entire cortex is highly homogeneous (with the exception of the visual area V1), in contrast to the sharp and complex differential relationships between extracortical brain areas (Hawrylycz et al. 2012).

The use of "map" instead of "area" has the advantage of implicitly adopting a parcellation policy more suited for the cortex: a lawful relation between the surface of the cortex and a relevant aspect of the representational structure. First introduced by Mountcastle (1957) for the somatosensory cortex, a cortical map is defined by the continuous map on the surface of the cortex isomorphic to the somatic sensory organ. In fact, cortical maps can be rigorously identified for all sensory and motor areas, but in higher areas the represented domain has a complicated and mostly unknown topological structure, which makes a systematic mapping on the cortical surface difficult.

Synaptic plasticity is not different from that in the rest of the brain, and it involves the known mechanisms of long-term potentiation (LTP), long-term depression (LTD) and spike-timing-dependent plasticity (STDP) (Markram et al. 1997; Feldman 2000). It is easier to observe intracortical map plasticity in maps of the sensorial cortex where it is responsible, for example, of perceptual learning (Fahle and Poggio 2002; Weinberger 2007), that is the long-term enhanced performance on a perceptual task as result of repeated experiences. While perceptual learning is an everyday business, intracortical map plasticity is responsible for the main early diversification of cortical functions driven by spontaneous neural activity (Khazipov and Buzsáki 2010; Zhang et al. 2011). Intercortical map plasticity induces modifications on a scale larger than a single map afferent. Typical case is the abnormal development in primary cortical areas: when following the loss of sensory inputs, neurons become responsive to sensory modalities different from their original one (Karlen et al. 2010).

The most striking examples of modal plasticity are the famous rewiring experiments, in which retinal axons of ferrets are connected at birth to the medial geniculate nucleus, which relay the signals to A1 instead of V1. This abnormal connectivity has induced a functional reorganization of A1, enabling visual behavior in the animals (Roe et al. 1987, 1990). A main question raised by this visual perception is how the transformation in A1 occurs. Either A1 and V1 are so similar that the change in sensory input has not been so significant, or intercortical map plasticity is powerful enough to mold the A1 small-scale circuitry to function, partially, as V1. Gao and Pallas (1999) gave a precise answer, demonstrating that A1 deeply changes its normal organization across a major tonotopic axis into a periodical, symmetrical array of orientation-tuned clusters of neurons, resembling that of V1.

It is possible now to clarify what sort of explanation we are after, when we face the clash between $\mathcal{P}_1$ and $\mathcal{P}_2$. The mainstream research on canonical circuits attempts to elaborate the following sort of explanation:

$\mathcal{P}_1$ the cortex is remarkably uniform;
$\mathcal{P}_2$ the cortex is the main site of a bewildering variety of functions;
$\mathcal{E}_C$ there must be a canonical circuit common all over the cortex, able to perform many different functions.

Explanations of the sort of $\mathcal{E}_C$ (with suffix $_C$ for "circuit") are doomed to failure, as I will argue in Sect. 4.3. If the set of premises is enforced with plasticity, a different sort of explanation can be offered:

$\mathcal{P}_1$ the cortex is remarkably uniform;
$\mathcal{P}_2$ the cortex is the main site of a bewildering variety of functions;
$\mathcal{P}_3$ the cortex is characterized by a remarkable plasticity;
$\mathcal{E}_D$ there must be a strategy common all over the cortex which enables a basic circuit to gradually change and develop a wide variety of functions, depending on the input patterns.

where in $\mathcal{E}_D$ the suffix $_D$ is for "circuital-developmental". Sketches of this sort of explanation will be discussed in Sect. 4.5.

4.3 Circuital Explanations

Most of the achievements in characterizing cortical mechanisms derive from the circuital perspective. The origin of circuital perspective is in the blending of electrical and electronic engineering with neurophysiology around half of the last century, and it represented a significant step in the epistemology of neuroscience (Brazier 1961; Rose and Abi-Rached 2013). A paradigmatic case of the impact electrical engineering has in the field of neuroscience is the cable equation, first derived by Lord Kelvin for the design of the transatlantic telegraph cable (Thomson Kelvin 1855), and later adapted by Wilfrid Rall (1957) to neural membrane potentials. The same basic equation is at the heart of the NEURON simulator (Hines and Carnevale 1997), which is currently adopted in the largest brain simulation projects.

The circuit equivalent to the cable equation is used as a model of the electrical behavior of dendrites and axons of single neurons. The circuit inherits the exact abstraction assumed in electrical engineering as a network of idealized quantized components of very few types (batteries, resistors, inductors, capacitors), connected by ideal perfect conductor lines. Moreover, the circuit behaves as node connections obeying the laws of Kirchhoff (1845). A similar set of assumptions is assumed implicitly in the microcircuits proposed to explain the cortex.

The most influential of these circuits is the "canonical microcircuit of the cortex", formulated by Douglas et al. (1989). Exactly like in the cable equation model, this cortical microcircuit inherits the main assumption of electrical circuits, approximating the electromagnetic field into a finite set of attributes that do not depend on the position of elements in physical space (Paynter and Beaman 1991). However, unlike the cable equation, the elements in the canonical microcircuit of the cortex are not standard electrical components but "neurons", abstracted in three classes. One class corresponds to the combination of superficial pyramidal neurons in layers II and III, and spiny stellate in layer IV projecting to them. The second class encompasses deep pyramidal population of layer V and VI. The third class includes generic GABA-receptor inhibitory cells.

Douglas and co-workers have implemented the circuit made of these three virtual neural units in a computational model, using rate-encoding of the outputs of the three units. The effect of outputs on connected units was computed as a change in membrane potential after a transmission delay. The output of each unit was a thresholded hyperbolic function of the average membrane potential, after a constant time relaxation. The tuning and later validation of the model was derived by intracellular recordings in the cat visual cortex (area 17), using a technique also borrowed from electrical engineering: pulse stimulation. During the stimulation, electrodes record the response to electrical pulses in range 0.2–0.4 ms, which simulate the optic radiation above the lateral geniculate nucleus. The main advantage of pulse stimulation was the availability of standard engineering system analysis tools for the evaluation of the responses. In addition, pulse stimulation is agnostic with respect to the many different natural stimuli to different cortical areas, thus making the canonical circuit general. Once tuned, the model was able to produce simulated responses to pulse signals in good agreement with the measured cortical responses. Later on, Douglas et al. (2004) confirmed the validity of their canonical model, with minor revisions to the relative strengths of the connections. The dominant excitation is now provided by intracortical connections between pyramidal neurons, so that even a relatively weak thalamic input can be greatly amplified. Even if inhibition is relatively weak, by modulating the recurrent excitation it may play an important role.

Circuits are abstractions aimed at isolating the main components of a system and their reciprocal electrical connections, providing seemingly a typical mechanistic explanation. In addition, the circuit of Douglas and Martin is complemented with a computational counterpart. Neurocomputational models, under certain conditions, are forms of mechanism with their own explanatory power (Piccinini 2015). A common criterion to ascertain which models give mechanistic explanation of the modeled system is the *model-mechanism-mapping* (3M) constraint (Kaplan 2011; Kaplan and Craver 2011):

> A model of a target phenomenon explains that phenomenon to the extent that (a) the variables in the model correspond to identifiable components, activities, and organizational features of the target mechanism that produces, maintains, or underlies the phenomenon, and (b) the (perhaps mathematical) dependencies posited among these (perhaps mathematical) variables in the model correspond to causal relations among the components of the target mechanism.

This constraint does not work as a logical binary condition. In fact, complete mechanistic models of neural behavior are unrealistic. The constraint is perfectly compatible with incomplete models, where details are omitted either for reasons of computational tractability or because these details are still unknown.

Note that the model of Douglas et al. is idealized in terms of *population*: the three elements in the model represent populations of certain categories of real neurons. Therefore, constraint (a) of 3M – "the variables in the model correspond to *identifiable* components [...] of the target mechanism" – is not met, or at least with large approximation. This approximation is different from the issue of the amount of details included in a model: it is not a matter of excluding details. In the case of the canonical circuits, units are clearly not physical single cells. Their extension in the cortex is not specified, nor the number and locations of cells on which the population of cells is averaged as a single abstract unit.

Douglas and Martin have tried to overcome this issue by constructing a more comprehensive microcircuit template of the cortex, using sophisticated statistical experimental data. Binzegger et al. (2004) have used 3-dimensional cell reconstruction on a sample of primary visual cortex, and they have analyzed the laminar pattern of synaptic boutons of 39 reconstructed neurons. The average number of synapses formed between neurons in different layers was estimated using an enhanced version of a simple rule by Peters and Payne (1993). In its simplest form, this rule states that the synapses from a given type of presynaptic neuron distribute evenly over the population of potential postsynaptic cells in the same cortical layer. In the refined version more details are taken into account, for example the fact that chandelier cells form synapses with pyramidal cells only. The final result is not a circuit anymore, rather a graph of synaptic connections between every type of cells, in five layers (layer II and III are joined together), having as edges the estimated proportion of synapses. A different way of deriving a statistical canonical circuit of the cortex is by using cellular recordings instead of cell morphology.

Thomson et al. (2002) have used paired recording – the simultaneous continuous measurement of electrical potentials from presynaptic and postsynaptic sites – obtaining about 1000 recordings on a variety of cortical neurons in several layers. Haeusler and Maass (2007) have used the data to assemble a statistical circuit made of 6 virtual cell types, corresponding to excitatory or inhibitory populations of cells distributed into layers II/III, IV, and V. This graph can include two types of edges: probability of connections, as in Binzegger et al. but also average strengths of connections. Haeusler & Maass have implemented a network of about 500 single compartment neurons, with proportion of connections matching those of the graph, and have performed a series of computational tasks, such as classifying two different sequences of spikes. The performances were compared with the same task executed by networks with the same number and type of neurons, but without the layered structure and the proportions of connections derived by real cortical data. Haeusler et al. (2009) have implemented, within the same neural simulation, the statistical graph of Binzegger et al. in order to compare the two models, with very similar performances. A different simulator was developed (Potjans and Diesmann 2014) based on the combined data of Binzegger et al. and Thomson et al. giving better accuracy in predicting certain experimental findings like spontaneous firing rates, but no performance on computational tasks was evaluated.

The most advanced cortical circuit derived from statistical cytology and connectivity data has been developed within the Human Brain Project (Markram et al. 2015). It reproduces a volume of 0.3 mm^3 of the rat somatosensory cortex with 31 thousands neurons and 37 million synapses. This microcircuitry is able to reproduce activities and several response properties recorded in vitro and in vivo experiments. However, even in the most advanced and refined form, explanations of the sort

$\mathcal{E}_C$ (see Sect. 4.2) say little, if nothing, about the paradox of the cortex expresses in the premises $\mathcal{P}_1$ and $\mathcal{P}_2$. The main reason hinges upon neglecting the premise $\mathcal{P}_3$ and addressing a static adult configuration only, discarding the development of synaptic connection in relation to the type of input patters. In the simulations of Haeusler & Maass, all synaptic strengths are necessarily equal to the statistical average derived by the data. For example, if the synaptic strengths in the circuit corresponding to one orientation-selective column in the primary cortex are all substituted with their mean value, the column would loose its selectivity, missing entirely its computational function. Considering the rewiring experiment described in Sect. 4.2, if an explanation of kind $\mathcal{E}_C$ holds, then we may expect the following two cases:

1. A1 in the rewired ferrets continues to perform its tonotopic function forever, which is useless with the new connectivity;
2. the microcircuit in A1 is versatile enough to immediately switch from the tonotopic function to orientation selectivity when the new input occurs.

Neither of these cases occur. Instead, intracortical map plasticity is powerful enough to mold A1 small-scale circuitry to function, partially, as V1. A1 deeply chang its normal organization across a major tonotopic axis into a periodical, symmetrical array of orientation-tuned clusters of neurons, resembling that of V1. Using optical imaging Sharma et al. (2000) have compared the patterns of horizontal connections in V1, normal A1 and rewired A1. While in normal A1 this pattern is elongated anteroposteriorly along the isofrequency axis, in rewired A1 the field of connections is wider, very patchy and elongated mediolaterally. This pattern is very similar to the field of horizontal connections in V1.

The explanatory limits of $\mathcal{E}_C$ can be well interpreted in the light of timescales, following Marom (2010). Canonical circuits are abstracted over a highly simplified temporal manifold, which takes care of one or just few short timescales, neglecting slower timescales at which important circuital adaptations take place.

4.4 Developmental Explanations in Biology

As mentioned in the Introduction, developmental explanations deserve their own place in biology. One of the earlier contributions to the developmental perspective in biology is found in the work of Conrad Hal Waddington (1957), who used to conceive an animal as a "developmental system". This idea blended with epigenetics, which become an empirically testable field through the innovative experimental research carried out by Gilbert Gottlieb (1971). He carefully worked in identifying developmental conditions that allow the capacity of ducklings to identify their maternal call, in particular the necessary auditory perceptual experiences during hatching. Eventually, Ford and Lerner (1992) set a systematic research agenda for developmental explanations in biology, proposing the "Developmental System Theory", in which epigenetics and biological development processes are linked to ideas coming from system theory and cybernetics. In fact, until recently concepts from developmental system theory and epigenetics were not picked up by mainstream biology, dominated by genetics. Today epigenetics and developmental system theory are among the most booming fields in biology (Griffiths and Tabery 2013; Baedke 2018). As a consequence, the relevance and validity of mechanistic explanations in developmental biological phenomena has become the topic of fervent discussions.

Mc Manus (2012) has argued that developmental phenomena cannot be accommodate within the mechanistic framework. Among the reasons, during development it seems impossible to maintain a basic principle held in mechanistic explanations, the *mutual manipulation* (Craver 2007, p.153). This principle establishes a sort of symmetry between the possibility to manipulate a part of the system and observing changes in some of its activities, and the possibility to produce globally similar changes and observe variations in one of its constitutive parts. Clearly, in a developmental phenomenon it is almost impossible to manipulate the final form of the system and observe changes in its initial constituents. For Ylikoski (2013) developmental explanations are not fully unrelated with mechanistic explanations, they combine in one some properties of causal explanations and other properties of mechanistic explanations. Causal explanations typically address changes of a system in time, seeking what triggers a specific change. Conversely, mechanistic explanations do not take time into account, and seek parts and relations between parts that empower a system with a causal capacity. A developmental explanation involves both time and changes in the causal capacity of a system. However, Parkkinen (2014) contends that in most cases the focus of development is not just in how the causal capacities of a system have changed in time, rather in the formation of novel constituents. A textbook example is the formation of a segmented body plan starting from the embryo. For this reason, Parkkinen is less compliant than Ylikoski in seeing a continuity between developmental and mechanistic explanations, and more in line with Mc Manus. The lack of time dimension in the mechanistic framework is also the topic discussed by Leuridan and Lodewyckx (2020). Specifically, they address the requirement of synchrony between the constitutive relations in multi-level mechanisms. A part in a lower-level mechanism is a constitutive relation in a multi-level mechanism when its behaviour concurs in the behaviour of the higher level mechanism, and constitutive relations are supposedly synchronic (Craver 2007). Leuridan and Lodewyckx argue for a diachronic reinterpretation of constitutive relevance, showing with logical arguments and with examples including neural plasticity, that there are cases of intralevel relations between parts that are constitutive, but operate at distinct times.

A different criticism on the possibility of developmental processes to be included in mechanistic explanations is raised by Brigandt (2015), based on the use of mathematical models. An important methodology in the study of the development of morphological structures is given by mathematical models, mostly based on reaction-diffusion equations. Brigant argues that, since mechanistic explanations are usually contrasted to mathematical explanations, the former are not appropriate for explaining biological processes such as morphological structures development.

The issue appears relevant in explaining how the cortex works, because such explanation involves mathematical models, as seen in Sect. 4.3 and as proposed in Sect. 4.5. However, the separation drawn by Brigandt between mechanistic and mathematical explanations is somehow too sharp. It is correct that for Craver (2007) certain mathematical models are just predictive and not explanatory, as reported by Brigandt, but this is the case for *certain* mathematical models. As seen in Sect. 4.3, there are criteria for discriminating between mathematical models with pure predictive scope and those that explain.

Most of the discussions of the philosophy of developmental explanations in biology targets phenomena that are different from the issue of cortical plasticity. The most common domains of development in biology focus on specific segments of ontogeny, such as the period from fertilization to birth in embryology, or from birth to the adult form of the organism (Minelli and Pradeu 2014). Developmental aspects in the cortex are not just limited to specific periods in ontogenesis, they are constitutive of the everyday working of the cortex. Development is in action, for example, every time a new mental concept is acquired, or an existing one is refined (Plebe and Mazzone 2016). Recent brain imaging techniques have demonstrated subtle changes in cortical microconnectivity in tasks such as abacus calculation training (Li et al. 2016); learning about the *Microraptor zhaoianus*[1] (Bauer and Just 2015); memorizing new names of flowers (Hofstetter et al. 2017); learning the structure of organic compounds (Just and Keller 2019).

A discussion close to the case at hand is provided by Craver and Darden (2013, pp.171–174) about LTP. First, LTP is one of the major forms of neural plasticity, therefore directly relevant to the cortex. But, most of all, Craver and Darden relate the basic mechanism that explains LTP with other higher level phenomena which depend on LTP, or depend on intermediate phenomena depending on LTP. In other words, development becomes integrated in a multi-level mechanistic explanation. In the example given by Craver and Darden, the mechanism at lower level concerns the phenomena of the activation of NMDA receptors in the postsynaptic cell and the chain of biochemical activities triggered by calcium ions that flow into the cell when NMDA receptors open. The level immediately above is the mechanism inducing the strengthening of the synaptic connection between a presynaptic and a postsynaptic cell, in which the main constituents are the phenomena and activities of the level below. A next level is the formation of place cells in the hippocampus (O'Keefe and Recce 1993), which are the basis of spatial cognition. The higher level is the exploration of a mouse in an environment (for example, a controlled Morris water maze), capturing visual cues that trigger the generation of place cells, through LTP plasticity.

This example shares aspects with the account of development needed to complete an explanation of the cortex, in particular the stratification in levels. The lowest level encompasses the same NMDA mechanism found in LTP plasticity, supplemented by other mechanisms of plasticity, as those reviewed in Sect. 4.2. The highest level

[1] A four-winged dinosaur bird species.

is one of the meaningful functions performed by a cortical area, such as visual or auditory processing, at a mature stage of its development. Possible ways of linking the lowest and the highest levels are discussed in the next section.

4.5 Developmental Explanations for the Cortex

The circuital approach for studying the cortex is dominating current mainstream computational neuroscience (Haeusler et al. 2009; Markram et al. 2015). There are, however, several strands of research that address developmental explanations. In this section I will first provide a brief historical survey of this research, followed by two examples of developmental explanation for the cortex, described in some more detail.

4.5.1 Modeling Cortical Plasticity

Several theoretical models have been proposed for cortical plasticity. One of the first, and most influential, was based on the mathematical framework of self-organization, a unified mathematical treatment of natural phenomena where a global ordering emerges from complex local interactions (Ashby 1947; Haken 1978; Kauffman 1993). The first attempts to use the mathematical framework of self-organization for describing neural phenomena are attributed to von der Malsburg (1973) and Willshaw and von der Malsburg (1976), who addressed the organization of maps in the visual cortex. There are three key mechanisms in cortical circuits that match with the premises of self-organization:

1. small signal fluctuations might be amplified, an effect highlighted in the canonical circuits described in Sect. 4.3;
2. there is cooperation between fluctuations, in that excitatory lateral connections tend to favor the firing of other connected neurons, and LTP reinforces synapses of neurons that fire frequently in synchrony;
3. there is competition as well, with the static part captured by computations like divisive normalization (Kouh and Poggio 2008), and the additional dynamics caused by synaptic homeostasis, which compensates for the gain in contribution from more active cells, by lowering the synaptic efficiency of other afferent cells.

In the cortical model devised by von der Malsburg the activity x_i of each neuron i was computed by the following system of differential equations:

$$\frac{\partial}{\partial t}x_i(t) = -\alpha_i x_i(t) + \sum_{j\in\mathcal{L}_i} w_{ij} f\left(x_j(t)\right) + \sum_{j\in\mathcal{A}_i} w_{ij} a_j(t) \tag{4.1}$$

$$f\left(x_i(t)\right) = \begin{cases} x_i(t) - \theta_i & \text{if } x_i(t) > \theta_i \\ 0 & \text{otherwise} \end{cases} \tag{4.2}$$

where $\mathcal{L}_i$ is the set of cortical neurons with lateral connections to the cell i, and $\mathcal{A}_i$ is the set of all afferent axons, each carrying a signal $a(t)$. The function $f(x)$ disables axon signal when the activation $x_i(t)$ is below a certain threshold θ_i . w_{ij} are the synaptic efficiencies between cell presynaptic j and postsynaptic i, and are modified by an amount proportional to the presynaptic and postsynaptic signals, in the case of coincidences of activity. Periodically all w_{ij} leading to the same cortical cell i are renormalized, resulting in competition, in that some synapses are increased at the expense of others. The source of afferents, in such process of self-organization, can be the external scene seen by the eyes, but also spontaneous activity generated by the brain itself (Mastronarde 1983). Equations like those in (4.1), explain different kinds of organization in the visual system ranging from retinotopy, ocular dominance, to orientation sensitivity (von der Malsburg 1995).

From then on, several further theoretical models have been proposed on how the cortex can develop functions using the basic synaptic plasticity mechanisms (Eliasmith and Anderson 2003; Deco and Rolls 2004; Ursino and La Cara 2004). Here I will give details on just one theoretical model, and show how this model succeed in explaining aspects of the functions performed in cortical areas V1 and V2, as the result of development. The model is based on a formulation of self-organization, simpler than that of von der Malsburg, called LISSOM (*Laterally Interconnected Synergetically Self-Organizing Map*) (Sirosh and Miikkulainen 1997; Miikkulainen et al. 2005) evolved in the *Topographica* neural simulator (Bednar 2009, 2014).

4.5.2 The LISSOM Architecture

I give here only the essential formulations of the LISSOM, which allow to identify the components that operate at the synchronous level of the overall mechanism, and those that belong to the diachronic level. The basic equation of the LISSOM describes the activation level x_i of a neuron i at a certain time step k:

$$x_i^{(k)} = f\left(\gamma_{\mathrm{A}}\mathbf{a}_i \cdot \mathbf{v}_i + \gamma_{\mathrm{E}}\mathbf{e}_i \cdot \mathbf{x}_i^{(k-1)} - \gamma_{\mathrm{H}}\mathbf{h}_i \cdot \mathbf{x}_i^{(k-1)}\right) \tag{4.3}$$

The vector field $\mathbf{v}_i$ is a circular area of afferents to the neuron i, and $\mathbf{x}_i$ is the circular area within the cortical map where neurons have excitatory or inhibitory connections to the neuron i. The vector $\mathbf{a}_i$ is the receptive field of the unit i. Vectors $\mathbf{e}_i$ and $\mathbf{h}_i$ are composed by all connection strengths of the excitatory or inhibitory neurons projecting to i. The scalars γ_{A}, γ_{E}, γ_{H}, are constants modulating the contribution of afferents, excitatory, inhibitory and backward projections. The function f is a non linear monotonic function, which details will be given next, k is the time step in the recursive procedure. Note that the time step k is at the very short time scale necessary for the cortical map to converge to a stable response to a stimulus, therefore it can be assumed that the neural activation deriving from equation (4.3) are synchronous.

The diachronic process, running at the time scale of cortical development, is a lower mechanism that affect all connection strengths. It is based on the combination of the general Hebbian principle, and a normalization mechanism that counterbalances the overall increase of connections of the pure Hebbian rule. All connection change in time according to the following rules:

$$\Delta \mathbf{a}_i = \frac{\mathbf{a}_i + \eta_{\mathrm{A}} x_i \mathbf{v}_i}{\|\mathbf{a}_i + \eta_{\mathrm{A}} x_i \mathbf{v}_i\|} - \mathbf{a}_i, \tag{4.4}$$

$$\Delta \mathbf{e}_i = \frac{\mathbf{e}_i + \eta_{\mathrm{E}} x_i \mathbf{x}_i}{\|\mathbf{e}_i + \eta_{\mathrm{E}} x_i \mathbf{x}_i\|} - \mathbf{e}_i, \tag{4.5}$$

$$\Delta \mathbf{i}_i = \frac{\mathbf{i}_i + \eta_{\mathrm{I}} x_i \mathbf{x}_i}{\|\mathbf{i}_i + \eta_{\mathrm{I}} x_i \mathbf{x}_i\|} - \mathbf{i}_i, \tag{4.6}$$

where $\eta_{\{\mathrm{A,E,I}\}}$ are the learning rates for the afferent, excitatory, and inhibitory weights, and $\|\cdot\|$ is the L^1-norm. All variations Δ appearing in the above equations are typically small, but their accumulation in the course of thousands of applications of the same equations on different afferents $\mathbf{v}$ will eventually form well organized topologies of the fields $\mathbf{a}$, $\mathbf{e}$, and $\mathbf{h}$. The time scale of the convergence of these fields corresponds to developmental times: – weeks or months – and is of several orders of magnitute grater than the time k – corresponding to milliseconds – that appears in equation (4.3). Note how the variables $\mathbf{a}$, $\mathbf{e}$, and $\mathbf{h}$ appear as the outcome of the lower-level mechanism described by equations (4.4), (4.5), and (4.6), while they are static components in the higher-level mechanism described by equation (4.3).

The formulation in (4.3) takes into account the following key features of cortical circuits:

1. the intercortical connections of inhibitory and excitatory types;
2. the afferent connections, of thalamic nature or incoming from lower cortical areas;
3. the organization on two dimensions of neural coding.

On the other hand, the formulations in (4.4), (4.5), and (4.6) take into account the following key principles of cortical development:

1. the reinforcement of synaptic efficiency by Hebbian learning;
2. homeostatic compensation of neural excitability.

I will discuss in 4.5.3 how far the formulations (4.3) and (4.4), (4.5), (4.6) can advance the reconciliation between propositions $\mathcal{P}_1$ and $\mathcal{P}_2$, and how they fit into the mechanistic framework. Before that, I would like to show two examples of the application of LISSOM to specific developmental phenomena.

In modeling how the cortex develops purposeful and efficient functions a crucial aspect in need of explanation is how to reconcile adaptivity on one side, and robustness and stability on the other side. Adaptivity is the key to construct connections implementing functions, driven by environment and internal experiences, but the sensitivity to changes in input patterns exposes to destabilizing forces, as seen in Sect. 4.2. Stevens et al. (2013) addressed this issue using Topographica, in the case of orientation maps development in the primary visual cortex. There is large empirical evidence for the robustness and stability of this development in several mammals and against several differences in visual experiences, see references in the paper from Stevens and co-workers. The most complete and direct evidence for robustness derives from studies on ferrets, with orientation maps recorded using chronic optical imaging at different ages (Chapman et al. 1996), showing how the earliest measurable maps are similar in form to the eventual adult map. Many computational models of orientation map development have been proposed (Goodhill 2007), but no model previous to Stevens et al. has been shown to develop with robustness and stability.

The key for developing robust and adaptive orientation maps in the model of Stevens et al. is in the nonlinear function f of equation (4.3), expressed as a piecewise linear function with threshold θ:

$$f(z) = \begin{cases} 1 & \text{when } z > 1 + \theta \\ z - \theta & \text{when } \theta < z < 1 + \theta \\ 0 & \text{otherwise} \end{cases} \tag{4.7}$$

The threshold adapts according with the neural activation level x:

$$\theta^{(k)} = \theta^{(k)} + \lambda\,(\bar{x} - \mu) \tag{4.8}$$

where $\bar{x}$ is a smoothed exponential average in time of the activation level x, and λ and μ are fixed parameters. This computation implements the biological mechanism of neural intrinsic homeostatic adaptation (Turrigiano and Nelson 2004). The model learns in a first stage from synthetic noisy disks, which simulate internal retinal waves of the prenatal phase, followed by real natural images. The results of the orientation maps in the model, taken at iterations steps between 2000 and 10000, closely resemble the orientation maps in ferrets measured at postnatal days age from P33 up to P42.

Most computational models of the cortex focus on the primary perceptual areas, especially V1, for which features of the neuron receptive fields are well documented. However, the most valuable functions of the cortex rely on a hierarchical computational strategy in which circuits at later levels in the hierarchy combine inputs from earlier levels building more complex responses. Thus, a model that nurtures canonical hopes, is required to explain how higher areas of the cortex build complexity upon lower areas. A suitable case to explore is in the secondary visual cortex, V2, as it receives most of its input from the well understood primary visual area. In addition, recently responses of neurons in V2 have been the object of several investigations (Ito and Komatsu 2004; Anzai et al. 2007; Hegdé and

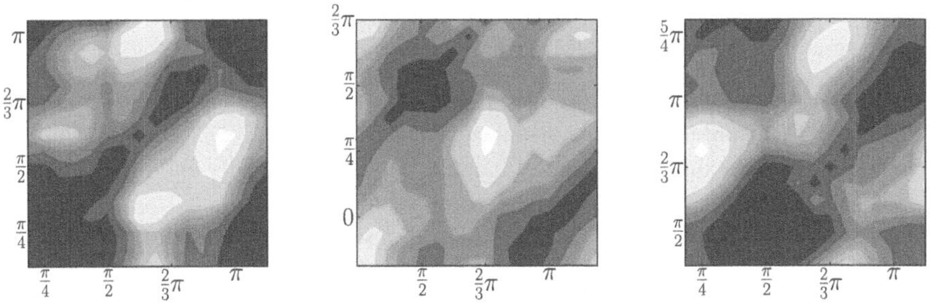

Fig. 4.1 Examples of V1 subunit interactions in the neural responses in area V2 of the model by Plebe (2012). In gray scale the level of activation of a single V2 neuron in the model, when in the receptive fields of the two V1 subunits two oriented bars are presented. The orientation of the two bars are the axes of the plots

Van Essen 2007). The model by Plebe (2012) investigated computationally the complex responses in V2 as resulting from V1 inputs. The model was based on two Topographica layers corresponding to V1 and V2, in each one the neurons are ruled by equation (4.3), with V2 receiving afferent from V1, and backprojecting to V1 as well.

The model was able to reproduce the sensitivity to angles, as measured by Ito and Komatsu (2004), and also the subtle dependencies of V2 neurons from subunits in V1 belonging to their receptive fields, found by Anzai et al. (2007). The contour plots in Fig. 4.1 reveal the mechanics of the selectivity to angles in neurons of V2, as depending from nonlinear interactions between two V1 subunits in its receptive fields. The plots are obtained by presenting simultaneously in the retina two oriented bars, centered withing the two receptive fields of the two V1 subunits, and measuring the response in the model V2 unit for every combination of the two orientations. It can be seen that there are peaks of response to specific combinations of orientation, but there are also areas in the two orientations space with an inhibitory effect, as resulting from the empirical study of Anzai et al. (2007). The complex responses in V2 were not implicit in the model definition, they result from development, achieved by a first stage of experience with simple noisy elongated patterns, followed by more complex patterns like corners and crosses. Thus, the model provides a preliminary insight of how complexity in cortical responses emerges from development (Riesenhuber 2012).

4.5.3 What the Developmental Models Have Explained

Going back to the problem of the propositions $\mathcal{P}_1$ and $\mathcal{P}_2$ stated in Sect. 4.2, shall we claim that models like Topographica give explanations of the sort of $\mathcal{E}_D$? Probably not. At least not yet, because the coverage of phenomena successfully explained by the models so far is limited, with respect to the wide range of functions in the cortex that a canonical model has the burden to explain. We can summarize the achievements of the two LISSOM-based models as follows:

- the first model is able to reproduce the orientation selectivity in V1, developed through exposure to plausible visual experiences;
- the first model exhibits the kind of balance between adaptivity and robustness in the development of orientation maps recorded in case studies;
- the second model is able to reproduce the sensitivity to angles on V2, as depending from nonlinear interactions between V1 subunits.

This is a far too limited list with respect to the breath of functions in the cortex. However, I believe this kind of models represents the most promising road toward $\mathcal{E}_D$. There are two distinct computational aspects of these models that potentially may apply all over the cortex:

1. a stereotyped essential sketch of the constituents of a cortical response, both in a initial or a mature stage, in equation (4.3)
2. a stereotyped essential sketch of the etiology of a cortical response, by its history of experiences, given by equations (4.4), (4.5), and (4.6).

The computation of point 1. describes the behavior of an abstract LISSOM unit as dependent from all its intracortical and extracortical connections. Therefore, it is not anchored to a specific circuital sketch, as in explanations of the kind of $\mathcal{E}_C$. There might be a sort of resemblance with the idea of "canonical computations", which is the mathematical operations most often carried out across different areas in the cortex. According to this idea, the general applicability of these operations makes the cortex powerful and flexible (Kouh and Poggio 2008; Carandini and Heeger 2012). It is out of the scope of this paper to discuss canonical computations, suffice it to say they identify specific operations such as divisive normalization or maximization. This is not the case of equation (4.3) of the LISSOM model.

The computation of point 2. is not anchored to a specific circuital sketch, either. A specific instance of the Topographica model, like the two here described, relies on a simulated cortical map, which circuital structure is not modified during development. The consequences of the LISSOM equations (4.4), (4.5), and (4.6) are at the level of intracortical map plasticity, which suffices to produce mature functions in the simulated experiments.

Both points 1. and 2. contribute to an explanation that might qualify as a mechanistic sketch including development effects. As seen in the previous section, several philosophers have defended, to various degrees, the autonomy of developmental explanations as distinct and irreducible to mechanistic explanations in biology (Mc Manus 2012; Ylikoski 2013; Parkkinen 2014; Brigandt 2015). However, the class of biological phenomena these philosophers take into consideration are very different from the problem of the cortex here addressed. By contrast, the phenomenon of LTP analyzed by Craver and Darden (2013), which is more relevant for the case at hand, can be well explained within a multi-level mechanistic framework.

A difficulty in regarding the LISSOM model as a two-level mechanism is that in the interlevel relations of a mechanistic framework are assumed to be synchronic (Craver 2007; Craver and Bechtel 2007). In the LISSOM model the components constitutive of the interlevel relations have **a**, **e**, and **h** as mathematical counterparts. These quantities change in developmental time under the causal effect of distal experiences, as described by equations (4.4), (4.5), and (4.6). On the other hand, the same quantities **a**, **e**, and **h** have the role of static parts in the higher-level mechanism described by equation (4.3). Even if the lower-level LISSOM model subsumes a form of LTP as analyzed by Craver and Darden (2013), the LTP of Craver & Darden does meet the synchronic requirement, while the lower-level LISSOM does not. In the LISSOM model, the relevant phenomenon is not an event of synaptic modification, rather the organization of connections over development time and its effect on the synchronic function performed by the mature cortical map.

There are, however, ways of dealing with the diachronic nature of interlevel relations in multi-level mechanisms, as mentioned in Sect. 4.4. Leuridan and Lodewyckx (2020) review a number of reasons to give up the strict requirement for interlevel constitutive relations to be synchronic, at least in biology and psychology. They also offer three scientific case studies in which constitutive relations are clearly diachronic and causally efficacious, all touching loosely upon cortical development. One of the cases is, again, the LTP phenomenon treated by Craver and Darden (2013), specifically one of the experiments reported in Kandel (2000). In this experiment, genetically modified mice have their LTP impaired. When the transgene is turned off, the LTP returns to normal and the mouse's capability for spatial memory is gradually restored. This is a striking example of diachronic process that the authors comment as follows:

> As constitutive relations supposedly are instantaneous, the question now becomes: is the causal relation in question synchronic? Interpreting this particular case as involving concrete processes, rather than abstract variables, shows that in a wet and messy biological context some time elapses before any effects actually arise. [...] The processes in question are of a complex, continuous character, and they need time to unfold and develop across the different levels of the mechanism. This seems to be the case for many, if not most, biological and neuroscientific mechanisms. (Leuridan and Lodewyckx 2020, p.12)

This seems, indeed, to be exactly the case for developmental-mechanistic models like the ones described in this section. In these models, the lower level corresponds to synaptic plasticity, in the taxonomy suggested in Sect. 4.2. This level is not the target of the explanation, so it is synthesized in equations (4.4), (4.5), and (4.6) without unfolding the details. The main level is the intracortical map plasticity described in the LISSOM equations (4.3) in conjunction with (4.4), (4.5), and (4.6). The higher level is the phenomenon of orientation selectivity in V1 in the model of Stevens et al., and the selectivity to angles in V2 by combinations of V1 subunit responses is the higher level in the model of Plebe. The models can be assumed as mechanistic sketches (Craver 2007, p.114), with most of the mathematical variables corresponding to identifiable components or activities in the cortex, even if at a coarse level. For example, the threshold θ in equation (4.8) corresponds to the regulation of firing rates in biological V1 in dependence of the average activity.

The details of how it is implemented in real neurons (by changing the number and distribution of ion channels) are omitted in the model.

4.6 Conclusions

In this chapter we have analyzed the search of an explanation of why the cortex is at the same time so uniform and so diversified in functions. This enterprise is justified only if the premise of uniformity is true, and our review of the current knowledge suggests that it is the case. Most proposals addressing this issue have followed a circuital strategy, trying to distill a fundamental circuital arrangement of cells in the cortex – often called "canonical" – at the heart of its computational power. Despite the enormous progress brought by this body of research, the answer to the paradox of the cortex is still, disappointingly, inconclusive. One reason is that all canonical solutions proposed so far have overlooked the dimension of cortical development due to plasticity, which is the main source of its computational flexibility, as supported from the reviewed evidences. Thus, a successful road towards a canonical explanation of the cortex paradox should be better construed as a mixed explanation of both the constituents essential for its computational power, and the developmental account of how cortical maps achieve their mature functions. There are several sketches of models following this direction, we provided details of two cases. Should this direction loose the epistemological advantage of a mechanistic format of explanation, that canonical circuits have to some degree? Probably not, for the development components too it would be possible to establish correspondences between mathematical elements of the models and neurophysiological correlates. Therefore, it is possible to qualify certain models of the cortex that include development as, at least, incomplete mechanistic sketches.

References

Anzai, A., Peng, X., & Van Essen, D. C. (2007). Neurons in monkey visual area V2 encode combinations of orientations. *Nature Neuroscience, 10*, 1313–1321.

Ashby, W. R. (1947). Principles of the self-organizing dynamic system. *The Journal of General Psychology, 37*, 125–128.

Baedke, J. (2018). *Above the gene, beyond biology: Toward a philosophy of epigenetics*. Pittsburgh: Pittsburgh University Press.

Bauer, A. J., & Just, M. A. (2015). Monitoring the growth of the neural representations of new animal concepts. *Human Brain Mapping, 36*, 3213–3226.

Bednar, J. A. (2009). Topographica: Building and analyzing map-level simulations from Python, C/C++, MATLAB, NEST, or NEURON components. *Frontiers in Neuroinformatics, 3*, 8.

Bednar, J. A. (2014). Topographica. In: D. Jaeger & R. Jung (Eds.), *Encyclopedia of computational neuroscience* (pp. 1–5). Berlin: Springer.

Berlin, R. (1858). Beitrag zur structurlehre der grosshirnwindungen. Ph.D. thesis, Medicinischen Fakultät zu Erlangen.

Berlucchi, G., & Buchtel, H. (2009). Neuronal plasticity: Historical roots and evolution of meaning. *Nature Reviews Neuroscience, 192*, 307–319.
Binzegger, T., Douglas, R. J., & Martin, K. A. (2004). A quantitative map of the circuit of cat primary visual cortex. *Journal of Neuroscience, 24*, 8441–8453.
Blumberg, M. S., Freeman, J. H., & Robinson, S. (Eds.). (2010). *Oxford handbook of developmental behavioral neuroscience*. Oxford: Oxford University Press.
Bontempi, B., Silva, A., & Christen, Y. (Eds.). (2007). *Memories: Molecules and circuits*. Berlin: Springer.
Braak, H. (1980). *Architectonics of the human telencephalic cortex*. Berlin: Springer.
Brazier, M. (1961). *A history of the electrical activity of the brain: The first half-century*. New York: Macmillan.
Brigandt, I. (2015). Evolutionary developmental biology and the limits of philosophical accounts of mechanistic explanation. In: P. A. Braillard & C. Malaterre (Eds.), *Explanation in biology – An enquiry into the diversity of explanatory patterns in the life sciences* (pp. 135–173). Berlin: Springer.
Brodmann, K. (1909). *Vergleichende Lokalisationslehre der Grosshirmrinde*. Leipzig: Barth.
Buonomano, D. V., & Merzenich, M. M. (1998). Cortical plasticity: From synapses to maps. *Annual Review of Neuroscience, 21*, 149–186.
Burnston, D. C. (2016). Computational neuroscience and localized neural function. *Synthese, 193*, 3741–3762.
Carandini, M., & Heeger, D. (2012). Normalization as a canonical neural computation. *Nature Reviews Neuroscience, 13*, 51–62.
Carlo, C. N., & Stevens, C. F. (2013). Structural uniformity of neocortex, revisited. *Proceedings of the Natural Academy of Science, 110*, 719–725.
Casanova, M. F., & Opris, I. (Eds.). (2015). *Recent advances on the modular organization of the cortex*. Berlin: Springer.
Chapman, B., Stryker, M. P., & Bonhoeffer, T. (1996). Development of orientation preference maps in ferret primary visual cortex. *Journal of Neuroscience, 16*, 6443–6453.
Craver, C. F. (2007). *Explaining the brain: Mechanisms and the mosaic unity of neuroscience*. Oxford: Oxford University Press.
Craver, C. F., & Bechtel, W. (2007). Top-down causation without top-down causes. *Behavioural Processes, 22*, 547–563.
Craver, C. F., & Darden, L. (2013). *In search of mechanisms: Discoveries across the life sciences*. Chicago: University of Chicago Press.
Cummins, R. (1975). Functional analysis. *Journal of Philosophy, 72*, 741–765.
Deco, G., & Rolls, E. (2004). A neurodynamical cortical model of visual attention and invariant object recognition. *Vision Research, 44*, 621–642.
Douglas, R. J., Martin, K. A., & Whitteridge, D. (1989). A canonical microcircuit for neocortex. *Neural Computation, 1*, 480–488.
Douglas, R. J., Markram, H., & Martin, K. (2004). Neocortex. In: G. M. Shepherd (Ed.), *The synaptic organization of the brain* (5th ed., pp. 499–558). Oxford: Oxford University Press.
Drury, H. A., Van Essen, D. C., Anderson, C., Lee, C., Coogan, T., Lewis, J. W. (1996). Computerized mappings of the cerebral cortex: A multiresolution flattening method and a surface-based coordinate system. *Journal of Cognitive Neuroscience, 8*, 1–28.
Edinger, L. (1904). *Vorlesungen über den Bau der nervösen Zentralorgane des Menschen und der Tiere*. Leipzig: Vogel.
Eliasmith, C., & Anderson, C. H. (2003). *Neural engineering computation, representation, and dynamics in neurobiological systems*. Cambridge, MA: MIT Press.
Erö, C., Gewaltig, M. O., Keller, D., & Markram, H. (2018). A cell atlas for the mouse brain. *Frontiers in Neuroinformatics, 12*, Article 84.
Fahle, M., & Poggio, T. (Eds.). (2002). *Perceptual learning*. Cambridge, MA: MIT Press.
Feldman, D. E. (2000). Timing-based LPT and LTD at vertical inputs to layer II/III pyramidal cells in rat barrel cortex. *Neuron, 27*, 45–56.

Ford, D. H., & Lerner, R. M. (1992). *Developmental systems theory: An integrative approach.* Newbury Park: Sage Publications.

Fuchs, E., & Flügge, G. (2014). Adult neuroplasticity: More than 40 years of research. *Neural Plasticity, 2014*, ID541870.

Fuster, J. M. (2008). *The prefrontal cortex* (4th ed.). New York: Academic.

Gao, W. J., & Pallas, S. (1999). Cross-modal reorganization of horizontal connectivity in auditory cortex without altering thalamocortical projections. *Journal of Neuroscience, 19*, 7940–7950.

Glasser, M. F., Coalson, T. S., Robinson, E. C., Hacker, C. D., Harwell, J., Yacoub, E., Ugurbil, K., Andersson, J., Beckmann, C. F., Jenkinson, M., Smith, S. M., & Essen, D. C. V. (2016). A multi-modal parcellation of human cerebral cortex. *Nature, 536*, 171–182.

Goodhill, G. J. (2007). Contributions of theoretical modeling to the understanding of neural map development. *Neuron, 56*, 301–311.

Gottlieb, G. (1971). Development of species identification in birds: An inquiry into the prenatal determinants of perception. Chicago: Chicago University Press.

Graybiel, A. M. (1984). Correspondence between the dopamine islands and striosomes of the mammalian striatum. *Neuroscience, 13*, 1157–1187.

Griffiths, P. E., & Tabery, J. (2013). Developmental systems theory: What does it explain, and how does it explain it? *Advances in Child Development and Behavior, 44*, 65–94. JAI, Berlin

Haeusler, S., & Maass, W. (2007). A statistical analysis of information-processing properties of lamina-specific cortical microcircuit models. *Cerebral Cortex, 17*, 149–162.

Haeusler, S., Schuch, K., & Maass, W. (2009). Motif distribution, dynamical properties, and computational performance of two data-based cortical microcircuit templates. *Journal of Physiology, 21*, 1229–1243.

Haken, H. (1978). *Synergetics – An introduction, nonequilibrium phase transitions and self-organization in physics, chemistry and biology* (2nd ed.). Berlin: Springer.

Harris, K. D., & Shepherd, G. M. (2015). The neocortical circuit: Themes and variations. *Nature Neuroscience, 18*, 170–181.

Haueis, P. (2012). The fuzzy brain. Vagueness and mapping connectivity of the human cerebral cortex. *Frontiers in Neuroanatomy, 6*, Article 37.

Hawrylycz, M. J., Lein, E. S., Guillozet-Bongaarts, A. L., Shen, E. H., et al. (2012). An anatomically comprehensive atlas of the adult human brain transcriptome. *Nature, 489*, 391–399.

Hegdé, J., & Van Essen, D. C. (2007). A comparative study of shape representation in macaque visual areas V2 and V4. *Cerebral Cortex, 17*, 1100–1116.

Herculano-Houzel, S., Collins, C. E., Wong, P., Kaas, J. H., & Lent, R. (2008). The basic nonuniformity of the cerebral cortex. *Proceedings of the Natural Academy of Science, 34*, 12593–12598.

Hines, M., & Carnevale, N. (1997). The NEURON simulation environment. *Neural Computation, 9*, 1179–1209.

Hofstetter, S., Friedmann, N., & Assaf, Y. (2017). Rapid language-related plasticity: Microstructural changes in the cortex after a short session of new word learning. *Brain Structure and Function, 222*, 1231–1241.

Horton, J. C., & Adams, D. L. (2005). The cortical column: A structure without a function. *Philosophical Transactions of the Royal Society B, 360*, 837–862.

Hubel, D., & Wiesel, T. (1959). Receptive fields of single neurones in the cat's striate cortex. *Journal of Physiology, 148*, 574–591.

Ito, M. (1984). *The cerebellum and neural control*. New York: Raven Press.

Ito, M., & Komatsu, H. (2004). Representation of angles embedded within contour stimuli in area V2 of macaque monkeys. *Journal of Neuroscience, 24*, 3313–3324.

Just, M. A., & Keller, T. A. (2019). Converging measures of neural change at the microstructural, informational, and cortical network levels in the hippocampus during the learning of the structure of organic compounds. *Brain Structure and Function*. https://doi.org/10.1007/s00429-019-01838-4:1-13

Kaas, J. H. (2012). Evolution of columns, modules, and domains in the neocortex of primates. *Proceedings of the Natural Academy of Science USA, 109*, 10655–10660.

Kandel, E. R. (2000). Cellular mechanisms of learning and the biological basis of individuality. In: E. R. Kandel, J. H. Schwartz, & T. M. Jessel (Eds.), *Principles of neural science* (4th ed., pp. 1247–1279). Amsterdam: Elsevier.

Kaplan, D. M. (2011). Explanation and description in computational neuroscience. *Synthese, 183*, 339–373.

Kaplan, D. M., & Craver, C. F. (2011). Towards a mechanistic philosophy of neuroscience. In: S. French & J. Saatsi (Eds.), *Continuum companion to the philosophy of science* (pp. 268–292). London: Continuum Press.

Karbowski, J. (2014). Constancy and trade-offs in the neuroanatomical and metabolic design of the cerebral cortex. *Frontiers in Neural Circuits, 8*, 9.

Karlen, S. J., Hunt, D. L., & Krubitzer, L. (2010). Cross-modal plasticity in the mammalian neocortex. In: Blumberg et al. (2010) (pp. 357–374).

Kauffman, S. A. (1993). *The origins of order – Self-organization and Selection in evolution*. Oxford: Oxford University Press.

Khazipov, R., & Buzsáki, G. (2010). Early patterns of electrical activity in the developing cortex. In: Blumberg et al. (2010) (pp. 161–177).

Kirchhoff, G. (1845). Ueber den Durchgang eines elektrischen Stromes durch eine Ebene, insbesonere durch eine kreisförmige. *Poggendorff's Annalen der Physik und Chemie, 64*, 487–514.

Kouh, M., & Poggio, T. (2008). A canonical neural circuit for cortical nonlinear operations. *Neural Computation, 20*, 1427–1451.

Leuridan, B., & Lodewyckx, T. (2020). Diachronic causal constitutive relations. *Synthese*. https://doi.org/10.1007/s11229-020-02616-0:1--31

Li, Y., Chen, F., & Huang, W. (2016). Neural plasticity following abacus training in humans: A review and future directions. *Neural Plasticity, 2016*, ID 1213723.

Lorente de Nó, R. (1938). Architectonics and structure of the cerebral cortex. In: J. Fulton (Ed.), *Physiology of the nervous system* (pp. 291–330). Oxford: Oxford University Press.

Lövdén, M., Bäckman, L., Lindenberger, U., Schaefer, S., & Schmiedek, F. (2010). A theoretical framework for the study of adult cognitive plasticity. *Psychological Bulletin, 136*, 659–676.

Marcus, G. F., Marblestone, A., & Dean, T. (2014). The atoms of neural computation. *Science, 346*, 551–552.

Markram, H., Lübke, J., Frotscher, M., & Sakmann, B. (1997). Regulation of synaptic efficacy by coincidence of postsynaptic APs and EPSPs. *Science, 275*, 213–215.

Markram, H., Muller, E., Ramaswamy, S., et al. (2015). Reconstruction and simulation of neocortical microcircuitry. *Cell, 163*, 456–492.

Marom, S. (2010). Neural timescales or lack thereof. *Progress in Neurobiology, 90*, 16–28.

Marr, D. (1970). A theory for cerebral neocortex. *Proceedings of the Royal Society of London B, 176*, 161–234.

Martin, K. A. C. (1988). The Wellcome Prize lecture – From single cells to simple circuits in the cerebral cortex. *Quarterly Journal of Experimental Physiology, 73*, 637–702.

Mastronarde, D. N. (1983). Correlated firing of retinal ganglion cells: I. Spontaneously active inputs in X- and Y-cells. *Journal of Neuroscience, 14*, 409–441.

Mc Manus, F. (2012). Development and mechanistic explanation. *Studies in History and Philosophy of Biological and Biomedical Sciences, 43*, 532–541.

Miikkulainen, R., Bednar, J., Choe, Y., & Sirosh, J. (2005). *Computational maps in the visual cortex*. New York: Springer-Science.

Miller, K. D. (2016). Canonical computations of cerebral cortex. *Current Opinion in Neurobiology, 37*, 75–84.

Miller, E. K., Freedman, D. J., & Wallis, J. D. (2002). The prefrontal cortex: Categories, concepts and cognition. *Philosophical Transactions: Biological Sciences, 357*, 1123–1136.

Minelli, A., & Pradeu, T. (2014). Theories of development in biology – Problems and perspectives. In: A. Minelli & T. Pradeu (Eds.), *Towards a theory of development* (pp. 1–14). Oxford: Oxford University Press.

Molnár, Z. (2013). Cortical columns. In: J.L.R. Rubenstein & P. Rakic (Eds.), Comprehensive developmental neuroscience: Neural circuit development and function in the healthy and diseased brain (pp. 109–129). New York: Academic.

Mountcastle, V. (1957). Modality and topographic properties of single neurons in cats somatic sensory cortex. *Journal of Neurophysiology, 20*, 408–434.

Nieuwenhuys, R. (1994). The neocortex. *Anatomy and Embryology, 190*, 307–337.

Nieuwenhuys, R. (2013). The myeloarchitectonic studies on the human cerebral cortex of the Vogt-Vogt school, and their significance for the interpretation of functional neuroimaging data. *Brain Structure and Function, 218*, 303–352.

Noack, R. A. (2012). Solving the "human problem": The frontal feedback model. *Consciousness and Cognition, 21*, 1043–1067.

O'Keefe, J., & Recce, M. (1993). Phase relationship between hippocampal place units and the EEG theta rhytm. *Hippocampus, 3*, 317–330.

O'Leary, D. D., Chou, S. J., & Sahara, S. (2007). Area patterning of the mammalian cortex. *Neuron. 56*, 252–269.

O'Leary, D. D., Stocker, A., & Zembrzycki, A. (2013). Area patterning of the mammalian cortex. In: J. L. R. Rubenstein & P. Rakic (Eds.), *Comprehensive developmental neuroscience: Patterning and cell type specification in the developing CNS and PNS* (pp. 61–85). New York: Academic.

Parkkinen, V. P. (2014). Developmental explanation. In: M. C. Galavotti, D. Dieks, W. J. Gonzalez, S. Hartmann, T. Uebel, & M. Weber (Eds.), *New directions in the philosophy of science* (pp. 157–172). Berlin: Springer.

Paynter, H., & Beaman, J. J. (1991). On the fall and rise of the circuit concept. *Journal of the Franklin Institute, 328*, 525–534.

Peters, A., & Payne, B. R. (1993). Numerical relationships between geniculocortical afferents and pyramidal cell modules in cat primary visual cortex. *Cerebral Cortex, 64*, 467–478.

Piccinini, G. (2015). *Physical computation: A mechanistic account*. Oxford: Oxford University Press.

Plebe, A. (2012). A model of the response of visual area V2 to combinations of orientations. *Network: Computation in Neural Systems, 23*, 105–122.

Plebe, A. (2018). The search of "canonical" explanations for the cerebral cortex. *History and Philosophy of the Life Sciences, 40*, 40–76.

Plebe, A., & Mazzone, M. (2016). Neural plasticity and concepts ontogeny. *Synthese, 193*, 3889–3929.

Potjans, T. C., & Diesmann, M. (2014). The cell-type specific cortical microcircuit: Relating structure and activity in a full-scale spiking network model. *Cerebral Cortex, 24*, 785–806.

Rakic, P. (1995). Radial versus tangential migration of neuronal clones in the developing cerebral cortex. *Proceedings of the Natural Academy of Science USA, 92*, 323–327.

Rakic, P. (2008). Confusing cortical columns. *Proceedings of the Natural Academy of Science USA, 34*, 12099–12100.

Rall, W. (1957). Membrane time constant of motoneurons. *Science, 126*, 454.

Ramón y Cajal, S. (1891). On the structure of the cerebral cortex in certain mammals. *La Cellule, 7*, 125–176.

Ramón y Cajal, S. (1906). In: J. DeFelipe & E. G. Jones, *Cajal on the cerebral cortex: An annotated translation of the complete writings*. Oxford: Oxford University Press. 1988.

Rathkopf, C. A. (2013). Localization and intrinsic function. *Philosophy of Science, 80*, 1–21.

Riesenhuber, M. (2012). Getting a handle on how the brain generates complexity. *Network: Computation in Neural Systems, 23*, 123–127.

Rockel, A., Hiorns, R., & Powell, T. (1980). The basic uniformity in structure of the neocortex. *Brain, 103*, 221–244.

Rockland, K. S. (2011). Five points on columns. *Frontiers in Neuroanatomy, 4*, Article 22.

Roe, A. W., Garraghty, P., & Sur, M. (1987). Retinotectal W cell plasticity: Experimentally induced retinal projections to auditory thalamus in ferrets. *Social Neuroscience Abstract, 13*, 1023.
Roe, A. W., Garraghty, P., Esguerra, M., & Sur, M. (1990). A map of visual space induced in primary auditory cortex. *Science, 250*, 818–820.
Rose, N., & Abi-Rached, J. M. (2013) *Neuro: The new brain sciences and the management of the mind*. Princeton: Princeton University Press.
Rothschild, G., & Mizrahi, A. (2015). Global order and local disorder in brain maps. *Annual Review of Neuroscience, 38*, 247–268.
Schüz, A., & Miller, R. (Eds.). (2002). *Cortical areas: Unity and diversity*. London: Taylor & Francis.
Sharma, J., Angelucci, A., & Sur, M. (2000). Induction of visual orientation modules in auditory cortex. *Nature, 404*, 841–847.
Sherman, S. M., & Guillery, R. W. (2006). *Exploring the thalamus and its role in cortical function*. Cambridge, MA: MIT Press.
Sirosh, J., & Miikkulainen, R. (1997). Topographic receptive fields and patterned lateral interaction in a self-organizing model of the primary visual cortex. *Neural Computation, 9*, 577–594.
Squire, L., & Kandel, E. R. (1999). *Memory: From mind to molecules*. New York: Scientific American Library.
Stevens, J. L. R., Law, J. S., Antolik, J., & Bednar J. A. (2013). Mechanisms for stable, robust, and adaptive development of orientation maps in the primary visual cortex. *JNS, 33*, 15747–15766.
Thomson, A. M., West, D. C., Wang, Y., & Bannister, P. (2002). Synaptic connections and small circuits involving excitatory and inhibitory neurons in layers 2-5 of adult rat and cat neocortex: Triple intracellular recordings and biocytin labelling *in vitro*. *Cerebral Cortex, 12*, 936–953.
Thomson Kelvin, W. (1855). On the theory of the electric telegraph. *Proceedings of the Royal Society of London, 7*, 382–399.
Treloar, H. B., Feinstein, P., Mombaerts, P., & Greer, C. A. (2002). Specificity of glomerular targeting by olfactory sensory axons. *Journal of Neuroscience, 22*, 2469–2477.
Turrigiano, G. G., & Nelson, S. B. (2004). Homeostatic plasticity in the developing nervous system. *Nature Reviews Neuroscience, 391*, 892–896.
Ursino, M., & La Cara, G. E. (2004). Comparison of different models of orientation selectivity based on distinct intracortical inhibition rules. *Vision Research, 44*, 1641–1658.
Vogt, C., & Vogt, O. (1919). Allgemeine Ergebnisse unserer Hirnforschung. *Journal of Psychology and Neurology, 25*, 279–461.
von der Malsburg, C. (1973). Self-organization of orientation sensitive cells in the striate cortex. *Kybernetic, 14*, 85–100.
von der Malsburg, C. (1995). Network self-organization in the ontogenesis of the mammalian visual system. In: S. F. Zornetzer, J. Davis, C. Lau, & T, McKenna (Eds.), *An introduction to neural and electronic networks* (2nd ed., pp. 447–462). New York: Academic.
von Economo, C., & Koskinas, G. N. (1925). *Die Cytoarchitektonik der Hirnrinde des erwachsenen Menschen*. Berlin: Springer.
Waddington, C. H. (1957). *The strategy of the genes: A discussion of some aspects of theoretical biology*. London: George Allen and Unwin.
Weinberger, N. M. (2007). Associative representational plasticity in the auditory cortex: A synthesis of two disciplines. *Learning and Memory, 14*, 1–16.
Willshaw, D. J., & von der Malsburg, C. (1976). How patterned neural connections can be set up by self-organization. *Proceedings of the Royal Society of London, B194*, 431–445.
Wright, L. (1976). *Teleological explanations*. Berkeley: University of California Press.
Ylikoski, P. (2013). Causal and constitutive explanation compared. *Erkenntnis, 2*, 277–297.
Young, M. P., Hilgetag, C. C., & Scannell, J. W. (2000). On imputing function to structure from the behavioural effects of brain lesions. *Philosophical Transactions of the Royal Society B, 355*, 147–161.
Zhang, J., Ackman, J., Xu, H. P., & Crair, M. C. (2011). Visual map development depends on the temporal pattern of binocular activity in mice. *Nature Neuroscience, 71*, 1141–1152.

Chapter 5
Data Mining the Brain to Decode the Mind

Daniel A. Weiskopf

Abstract In recent years, neuroscience has begun to transform itself into a "big data" enterprise with the importation of computational and statistical techniques from machine learning and informatics. In addition to their translational applications such as brain-computer interfaces and early diagnosis of neuropathology, these tools promise to advance new solutions to longstanding theoretical quandaries. Here I critically assess whether these promises will pay off, focusing on the application of multivariate pattern analysis (MVPA) to the problem of reverse inference. I argue that MVPA does not inherently provide a new answer to classical worries about reverse inference, and that the method faces pervasive interpretive problems of its own. Further, the epistemic setting of MVPA and other decoding methods contributes to a potentially worrisome shift towards prediction and away from explanation in fundamental neuroscience.

5.1 Neuroscience and the Data Revolution

From genetics to astronomy and climatology, the sciences now routinely deal with extraordinarily large quantitative datasets and deploy computational techniques to manage and extract information from them. Neuroscience is no exception to this trend. The quantity and kinds of neural data available have shifted radically in the last two decades (Van Horn and Toga 2014), a transition striking enough to prompt declarations that "massive data is the new reality in neuroscience and medicine" (Bzdok and Yeo 2017, p. 560). With this shift has come a transformation in the analytic tools used to share and process this data, as well as a new wave of optimism about the ability of these methods to overcome long-standing theoretical challenges.

The data revolution has several different fronts. Here I will focus on the impact that machine learning (ML) techniques have had on theory and practice in

D. A. Weiskopf (✉)
Department of Philosophy, Georgia State University, Atlanta, GA, USA
e-mail: dweiskopf@gsu.edu

© Springer Nature Switzerland AG 2021

F. Calzavarini, M. Viola (eds.), *Neural Mechanisms*, Studies in Brain and Mind 17,
https://doi.org/10.1007/978-3-030-54092-0_5

neuroscience. Machine learning allows us to efficiently partition complex datasets, make inferences, conduct searches, and extract hidden patterns.[1] In the following discussion I sketch one way that machine learning has transformed neuroscientific practice, namely through the application of data analytic tools to imaging studies. One such application is in the use of multivariate pattern analysis (MVPA) to uncover neural structure. MVPA-based methods have proliferated since their introduction in studies of visual processing (Haxby 2001). However, it is by no means clear how best to interpret the outputs of the increasingly complicated machine learning algorithms that lie at the heart of these methods.

My aim in this chapter is twofold. First, I argue that MVPA does not provide a new solution to the longstanding problem of reverse inference, a claim that has been advanced by Guillermo del Pinal and Marco Nathan in several papers (Del Pinal and Nathan 2017; Nathan and Del Pinal 2017), and that also comports with the interpretation of MVPA presupposed by many prominent studies. If MVPA enabled us to break new ground in overcoming the challenges of reverse inference, this would be a powerful argument in favor of multivariate studies over traditional mass-univariate approaches. I present three interpretive challenges that cast doubt on the claim that MVPA can singlehandedly resolve the reserve inference debate. These challenges center on these techniques' sensitivity and globality, the instability and interpretive opacity of their results, and their agnosticism with respect to causal structure.

Second, I want to sound a cautionary note about these new tools and statistical techniques. Such technologies are not ideologically neutral. They come with certain preferred uses, as well as a specific deployment of rhetoric which they carry over from their original computational contexts to their neuroscientific applications. With respect to machine learning, the key term often involved is *prediction*. Indeed, some neuroscientists have explicitly begun to couch their epistemic aims in terms not of explanation or understanding, but of greater predictive accuracy. While this may not yet be the prevalent view among practitioners, I suggest that in light of the ease with which machine learning tools can be turned to purely predictive ends we should be cautious about interpreting models that are based on them, and conscious about the subtle effects they may be having on the studies we design and the epistemic aims that we adopt. Prediction and explanation need not inherently be in conflict with one another, and neuroscience should develop multifaceted modeling practices that integrate these goals rather than favoring one over the other.

[1]Much recent work using machine learning in neuroscience has centered on deep convolutional neural networks (DCNNs). Classifiers such as the ones discussed here are sometimes used to assign labels to the layers of DCNNs, so the two are not entirely unrelated. Nevertheless, DCNNs are substantially different in their structure and uses from the kinds of models I focus on, so I omit further discussion of them.

5.2 Two Forms of Reverse Inference: Functional and Predictive

The ultimate aim of cognitive neuroscience and neuropsychology is to construct interfield theories bridging brain and mind. Ideally, such bridges would comprise an explanatory implementation theory that would make it comprehensible *how* and *why* specific patterns of brain activity realize the cognitive processes that they do. Explaining the neural basis of cognition requires an account of how low-level neural processes give rise to specific cognitive functions, spelled out in terms of their causal capacities and organization. Most acknowledge that this is at present a utopian prospect. A more modest goal would be to make neuroscientific data evidentially relevant to determining the structure of cognition. Debate has raged, however, over how optimistic we should be even about this goal, with skeptics such as Max Coltheart (2006, 2013) doubting whether neural evidence could *ever* be sufficient to distinguish between competing psychological models.

Building interfield theories and bringing neural evidence to bear in psychology requires a reliable inferential framework capable of crossing ontological, epistemic, and methodological boundaries. Here I focus on one facet of this framework, namely *reverse inference*.[2]Reverse inferences move from the fact that neural process N occurs to the conclusion that (with some probability) cognitive process C is engaged (Poldrack 2006). Reverse inferences hold when N's activation provides sufficient evidence for C, to the exclusion of other cognitive processes that might be taking place. For instance, suppose (1) that activity in Broca's area makes it probable that processing of sequentially structured information is taking place, (2) that this processing is unlikely to be taking place in the absence of this underlying activity, (3) that no other regional neural activity is evidence for this form of sequence processing, and (4) that activity in this area is not strong evidence for the engagement of any *other* type of cognitive process. Knowing these facts, we can use such activation to conclude that a novel experimental task that activates Broca's area involves sequential processing, which may help to decide between two different psychological models of how it is performed.

The present impasse over reverse inference centers on how to respond to the comprehensive failure of functional localization for many cognitive processes of interest. Localization is the claim that particular cognitive processes are realized by neuroanatomically circumscribed brain regions that are relatively small and functionally dedicated. It has increasingly become clear that many, perhaps most, brain regions seem to participate to some degree in several different cognitive processes (Anderson 2014; Rathkopf 2013; Weiskopf 2016). So from the fact that a pattern N occurs there is some probability that *at least one* of the processes C_1, C_2,

[2]Its other face, *forward inference*, involves moving in the opposite direction, viz. from the engagement of a cognitive process to the fact that a specific neural process is occurring (Henson 2006). For discussion of forward inferences in the context of dissociation studies rather than imaging contexts, see Davies (2010).

..., C_n is being engaged. From this probability distribution we can't conclude that any one of them, to the exclusion of all others, is localized in N. Given that some regions are involved in a wildly heterogeneous-seeming array of activities across many domains, it is hard to conclude that realizing any one of them is that region's determinate and unique function.

A number of strategies have been proposed to deal with the problem of reverse inference under conditions of functional heterogeneity (Burnston 2016a; Glymour and Hanson 2016; Hutzler 2014; Klein 2012; Machery 2014; McCaffrey 2015; Roskies 2009). Rather than survey all of these here, I will consider a recent proposal to solve the problem by applying machine learning techniques to imaging data.

First, though, we should distinguish two purposes for which one may seek out reverse inferences. Call these *functional* reverse inference and *predictive* reverse inference. A functional reverse inference involves two claims: that activity in N indicates engagement of C, *and* that this relationship holds because the function of N is to realize C.[3]Functional RI is distinguished by the fact that it incorporates a justification for why the inferential relationship is reliable. It is not merely an accidental-but-reliable co-occurrence: the neural process or region that is active is one that has a certain assigned cognitive function. Having such a function imposes highly specific requirements on the causal organization of the underlying region, namely that it be capable of underwriting the pattern of effects that characterize the target cognitive process. This constraint in turn secures an *explanatory* connection between what is happening in N and C's engagement. Seeking out functional RIs such as these is essential to fleshing out the sort of interfield theory sketched earlier.

Predictive RI, by contrast, merely says that activity in N indicates a certain probability of engagement of C. Its focus is on finding reliable indicators of cognition, no matter what the function of those markers is within the mind/brain system. These might be thought of as "cognitive biomarkers". In medicine, a biomarker is any detectable biological signature that is correlated either with the presence or progression of a disorder. Examples include hemoglobin A1C levels for diabetes or the BRCA1 gene for breast cancer. Biomarkers are sometimes linked directly with the underlying causal factors that drive a disorder, but often may reflect effects or other secondary processes that are restricted to their clinical or prognostic utility.

By analogy, neural activity in a region understood as a cognitive biomarker can serve predictive RI perfectly well, despite not being apt for functional RI. The reason is that this activity can be exploited to predict cognitive processing even when it is not what *realizes* that processing. For instance, the "ground truth" might be that N_1 realizes C, but N_1 might also reliably co-occur with N_2—either because N_1 and N_2 are directly causally related (e.g., N_1 causes N_2), or because they are common effects of a distinct cause. Here both N_1 and N_2 would be equally suited

[3]Alternately, the second claim can be formulated in mechanistic terms: the neural mechanism involved in N has the function of realizing or implementing cognitive process C. I won't make any assumptions here about whether all realizing structures for cognitive processes are mechanistic.

for predictive RI, but not equally good for functional RI. The grounds for predicting cognitive processing differ from those that explain it.

Functional and predictive RI are distinguished in terms of the purposes or goals that lie behind them. This is not to deny that they may work together in many contexts. There is no contradiction between gathering information about brain-mind correlations for the purpose of finding realizers and seeking such correlations for the aim of finding strongly predictive neural signatures. Nevertheless, they can also be pursued exclusively, and prescribe different programs of experimental interventions, interpretation of evidence, and statistical analysis. A neural signature of deception, for instance, might be highly predictive and legally probative without tracking the neural implementation of the intent to deceive. Theorists have not always been explicit on which conception of reverse inference is at issue, although most of the debate over bringing neuroscientific evidence to bear on cognitive theories has tacitly assumed a functional conception of RI. Carefully observing this distinction becomes especially important with the recent turn to machine learning methods, because the rhetoric of decoding, and the striking success of ML classifiers on prediction tasks, has begun to drive some neuroscientists towards abandoning explanation in favor of prediction. It is not an accident that the rise of decoding methods in neuroscience has coincided with the more general adoption of predictive machine learning tools in science, medicine, industry, and marketing (see, e.g., Agrawal et al. 2018).

Proponents of this "predictive turn" argue that it injects much needed rigor into neuroscience and psychology. They correctly point out that these fields have disappointing track records of real-world prediction. The traditional significance tests they frequently use are hard to interpret in predictive terms, and merely fitting statistical models to existing datasets often leaves us unable generate any useful forecasts. These shortcomings have also been obscured to some degree by the focus of recent philosophy of science on questions concerning explanation, to the exclusion of prediction.[4]

The extent to which the predictive turn is becoming more prominent in neuroscience at large is hard to measure given the size and diversity of the field. Nevertheless, passages such as the following, drawn from position papers by major participants in the debate, represent a few straws in the wind:

> "Perhaps the biggest benefits of a prediction oriented within psychology are likely to be realized when psychologists start asking research questions that are naturally amenable to predictive analysis. Doing so requires setting aside, at least some of the time, deeply ingrained preoccupations with identifying the underlying causal mechanisms that are mostly likely to have given rise to some data." (Yarkoni and Westfall 2017, p. 18).

> "Isolating components of mental processing leads to studying them only via oppositions, and this reductionism prevents the building of broad theories of the mind. We believe that

[4]There are some notable exceptions to this. For instance, Douglas (2009) argues that despite the philosophical neglect of prediction, it remains central to defining the scientific enterprise, and Northcott (2017) points out that in many domains such as political polling, prediction is often a more desirable epistemic trait than understanding.

> predictive modeling provides new tools to tackle this formidable task" (Varoquaux and Poldrack 2019, p. 1).

> "the main goal of the prediction enterprise is to put the built model, with already estimated model parameters, to the test against some independent data... she [the investigator] is not necessarily worrying about how the model works or whether its fitted parameters carry biological insight" (Bzdok and Ioannidis 2019, p. 3).

The thrust of these passages is clear: prediction should be given *at least* equal (if not greater) epistemic weight as explanation in modeling cognitive and neural phenomena.

Of course, these are merely three papers that stake out their high-level methodological claims relatively quickly. For another indicator of prediction's rise, consider the rapidly growing field of neuroforecasting, the explicit goal of which is to find neural signals that predict individual, group, or society-wide behaviors, attitudes, and trends (Berkman and Falk 2013). In some representative studies, activity in medial prefrontal regions of individual smokers exposed to antismoking public health messages has been said to predict the population-level success of those campaigns (Falk et al. 2012), and nucleus accumbens activation has been singled out as a predictor of aggregate success of crowdfunded projects on the Internet (Genevsky et al. 2017). Often these neural predictors outperform behaviors or expressed attitudes, which makes them especially attractive targets for marketing purposes.

To the extent that there is a move towards predictively oriented studies taking place, this may in part be an effect of the new tools that neuroscientists have at their disposal. The predictive turn is a concomitant of the adoption of techniques from machine learning. Since these tools have a natural epistemic habitat in data science tasks where computationally efficient prediction is the goal, they tend to carry aspects of this habitat with them when they take root in new domains.

5.3 Decoding the Mind with Multivariate Pattern Analysis

Much of the excitement surrounding the use of machine learning in neuroscience is that it offers the possibility of decoding brain activity, a process that its advocates often colorfully refer to as "reading" the mind off of the brain (Norman et al. 2006; Poldrack 2018).[5] In a typical decoding experiment, participants perform a set of

[5]The mindreading rhetoric is handled cagily in the literature. For instance, despite his book's title, Poldrack hedges on the aptness of the "reading" metaphor, referring to it as "audacious" at one point (p. 2). Others have been less cautious: Haynes et al. (2007) explicitly refer to "reading intentions" out from brain activity, and in a review essay Haynes (2012) remarks that thanks to "combining fMRI with pattern recognition" it "has been possible to read increasingly detailed contents of a person's thoughts" (p. 30). He later comments that in practice this form of mindreading will likely be most useful with respect to broad categories of mental states such as the intent to deceive. Finally, Tong and Pratte (2012) helpfully distinguish between "brain reading"

tasks during a data collection phase. In principle any sort of data can serve as input to a decoding process (EEG, MEG, direct electrode recordings, etc.), but I will focus on functional MRI studies. Participants are scanned while performing tasks that are typically selected for their differences in the information and the processes that they draw on.[6]

The data from these tasks consists of a vector of numbers measuring the change in the BOLD signal at each voxel at each time step of the scanning sequence. In a procedure known as cross-classification validation, each input sequence is labeled according to the task or stimulus condition that it was gathered in (with labels just being binary features), and the data is separated into two piles: a training set and a test set. Typically, data from a certain number of subjects is reserved for testing. The labeled training sequences are then fed into a supervised machine learning classifier until it reaches criterion performance. Testing is then carried out on the remaining reserved data. This process is iterated across different training subsets, and the classifier's overall performance is reported as the average of its performance on each run.[7]

There are many possible classifiers to use in MVPA studies. To streamline discussion, I will focus on a single commonly used example, namely support vector machines (SVM). SVMs efficiently learn to assign each voxel a weight according to how well its activity can help to predict the target category. In linear SVMs, each voxel is assigned a positive or negative weight according to its contribution to correct labeling. The SVM's goal is to draw an optimal hyperplane in voxel (feature) space partitioning the space of possible activity patterns into regions corresponding to each label. There are usually many linear partitions available, but optimality means that the hyperplane maximizes the margin from itself to the nearest members of each category. Data sets that cannot be linearly partitioned in their raw form can be transformed using kernel methods into spaces where such partitioning is possible.[8]Once an SVM learns to achieve an optimal degree of separation with the training set, its weights are frozen and its performance is judged by averaging over repeated folds of out-of-sample transfer (i.e., how well it classifies members of the unseen test set).

---

and "mind reading", where the former refers to predicting overt or observable behaviors from brain activity, while the latter refers to predicting subjective cognitive states. They regard MVPA methods as having contributed to progress in both (pp. 485–6).

[6]Many studies also use naturalistic tasks (e.g., movie watching) that engage more widespread cognitive processes. For more details on experimental design, see Tong and Pratte (2012), Haxby et al. (2014), and Haynes (2015).

[7]There is reason to think that these prevalent leave-*k*-out training regimes aren't adequately variance-minimizing, however; see Varoquaux et al. (2017), who recommend leaving out 10–20% of the data and using repeated random splits. Because of the relative youth of these paradigms, best experimental practices are still stabilizing.

[8]Most neuroimaging studies use the standard linear kernel. Higher-order relationships among voxels are considered only in nonlinear classifiers, including so-called "deep" neural networks. Since almost everyone considers these too powerful and unconstrained for use with imaging data, I continue to omit them here.

Decoding, then, is defined as a classifier's performing adequately well at inferring from neural data to a category label standing for something extra-neural, e.g., a perceptual stimulus, a behavioral response, a task condition, or a cognitive process.[9] This decoding paradigm can be illustrated by Kamitani and Tong's (2005) landmark study of visual attention. Participants were initially scanned while viewing gratings oriented at either 45° or 135°, and the resulting images were used to train a classifier on voxels selected from regions V1-V4. They were then shown a grating that superimposed both of the previous ones and asked to direct their attention selectively to one or another of the orientations. The data from the second phase was fed into the classifier trained on the first phase, which was able to discriminate between the two attention conditions with nearly 80% accuracy. They concluded that information about a participant's attentional state can be decoded from activity in visual cortex.

As Varoquaux and Poldrack (2019) emphasize, classifiers' "validity is established by successful predictions from new data, and not by isolating significant differences across observations" (p. 2). In this sense the statistical regime that underlies MVPA is fundamentally different from that of mass univariate analysis. It focuses not primarily on detection of univariate statistical differences in activation patterns, but on extracting predictive information—in any form whatsoever—from distributed neural activity (Hebart and Baker 2018). The epistemic regime of prediction is therefore entwined with MVPA at a fundamental level.

Machine learning applied to neural data has proven fruitful across many practical domains. Examples include classifying patients into neuropsychiatric groups on the basis of resting scans, diagnosis of neuropathological conditions by biomarkers rather than symptoms, creating brain-computer interfaces and other neuroprosthetics, extracting the contents of ongoing visual perception, and detection of consciousness in unresponsive patients. The success of these clinical and translational applications is more than enough to justify the interest in solving more fundamental theoretical problems using the same analytic toolkit.

5.4 Decoding as a Solution to Reverse Inference

In experimental setups where what is being decoded is the occurrence of a cognitive process (rather than, say, the presence of a disorder), decoding can be interpreted

[9] *Encoding*, by contrast, involves the reverse operation: training classifiers to predict measurements of neural activation given an experimental task, condition, or stimulus input. Note that the encoding/decoding distinction has to do with the *direction of inference* relative to available neural data. In either direction, it is couched in terms of the measured information made available. Further inferences are required to move from this data to conclusions about content or actual neural ground truths. The encoding/decoding distinction also shouldn't be confused with direction of *causality*. Both decoding and encoding are predictive modeling techniques that can be applied to experimental setups in which neural activity is either the cause or effect of the state being predicted.

as the use of classifiers to perform reverse inference tasks. It is a very short step from (1) MVPA reveals that information about mental states can be extracted from measured brain activity to (2) MVPA can be used to infer the occurrence of mental states on the basis of measured brain activity.[10] In several papers, Guillermo del Pinal and Marco Nathan have taken this step. They argue that MVPA provides a new solution to the problem of reverse inference (Del Pinal and Nathan 2017; Nathan and Del Pinal 2017). They call this pattern-based reverse inference, by contrast with classical location-based reverse inference.

Their central argument for preferring MVPA to location-based approaches rests on the fact that classifier-based studies satisfy what they call the linking condition (Del Pinal and Nathan 2017, p. 129). Suppose we want to know whether a task-evoked pattern of neural activity N engages cognitive processes C_1 or C_2. To do so requires *independent evidence* that N is positively linked with, say, C_1 (rather than C_2). In traditional univariate analysis this evidence is precisely what is missing, thanks to the multifunctionality of regions across studies (see Sect. 5.2). However, MVPA involves training classifiers on data gathered within phases of the same experiment, rather than making comparisons across experiments. It therefore circumvents the problem by directly comparing activation patterns, where the reliability with which these patterns are distinguishable is determined within the experiment (pp. 135–6). Moreover, MVPA does this without importing any problematic assumptions either about the localization of cognitive processes in brain regions, or about the previously established cognitive functions of those regions.

From these points we can extract the following methodological prescription concerning the utility of *decoding for cognitive difference*:

> (DCD): If a decoder can be trained to distinguish neural patterns elicited by two tasks, then the tasks involve different cognitive processes.

DCD relies on the principle that any differences in cognitive processing will be reflected in their underlying neural realization, so no two processes can have (within an individual performing a specific task) the same realization.

Appeal to the DCD principle is implicit in Del Pinal and Nathan's arguments. They propose that multivariate imaging analysis can "overcome the challenge of determining the reliability of bridge laws and, as a result, promise to be a more useful technique for discriminating among competing cognitive-level hypotheses" (Nathan and Del Pinal 2017, p. 5). Suppose that we begin with a classifier trained to decode cognitive processes C_1 and C_2 from distinct equivalence classes of neural patterns. Then we have the leverage needed to decide whether an arbitrary novel task taps that one or the other of these processes by seeing how that classifier performs on data collected from imaging that task (pp. 5–7). Successful decoding here is presented as sufficiently strong evidence to license functional reverse inferences.

---

[10] A closely related inference concerning the decoding of representational content from MVPA classification studies has been challenged by Ritchie et al. (2019). See especially pp. 11–13 for a detailed unpacking of the premises that these inferences rely on.

In a related vein, Ritchie et al. (2019) articulate a principle they call the "decoder's dictum" that they argue persuasively drives the interpretation of many MVPA studies. According to the dictum, "If information can be decoded from patterns of neural activity, then this provides strong evidence about what information those patterns represent" (p. 2). DCD as presented here can be viewed as complementary to the decoder's dictum: the latter focuses on the decodability of information, while the former concerns the use of decoding to discover cognitive processes. Information and processing are tightly related but nevertheless distinct. Cognitive processes may differ in the informational or representational content that they manipulate, but they may also make distinct uses of the same body of information (if, for instance, the goal of the information processing is different in each case). Ritchie, Kaplan, & Klein's arguments against the decoder's dictum thus dovetail with the ones presented here against the DCD principle. Each attempts to separate and target one strand in the familiarly entwined notion of "information processing."

DCD can also be seen as tacitly driving the interpretation of a number of imaging studies. Varoquaux and Thirion (2014), for instance, propose that decoding provides a "principled methodological framework for reverse inferences" (p. 4), where the latter are understood in the functional sense. Moreover, DCD-like principles aren't confined to the pages of theoretical papers. Consider studies of visual perception such as Haynes and Rees (2005), in which participants simultaneously viewed two stimuli designed to induce binocular rivalry while indicating via button-pressing which of the two they were experiencing at a particular moment. A pattern classifier was trained on activity in 50 voxels of V1 and used to predict the timing with which one or the other visual stimulus became conscious, achieving an 80% success rate. In a separate condition, a classifier trained to distinguish presentations of monocular non-rivalrous stimuli could predict binocular switching similarly well. Haynes & Rees conclude that "[their] data could be taken to represent a simple form of 'mind reading,' in which brain responses were sufficient to predict dynamic changes in conscious perception in the absence of any behavioral clues" (p. 1302). That is, they interpret this study's methods as licensing an inference from accurate machine classification of neural patterns to changes in people's perceptual states.

Similar inferences crop up in studies of pain perception. In one widely cited study, Wager et al. (2013) subjected participants to thermal stimuli varying from warm to painful. These stimuli were both classified and rated according to intensity on a 100-point scale. A sparse pattern classifier (see Sect. 5.2 below) was trained on a map of anatomical regions preselected for their known involvement in pain processing, and this classifier was tested on scans of neural activity during the stimulation period. The classifier was used to generate predictions of how the stimulus was experienced, and to predict its intensity.[11] It was able to discriminate

[11] These predictions were calculated in terms of a "signature response", here defined as the dot product of the trained classifier weights and the activation map for each temperature within participants (see p. 1391 and the Supplementary Materials). Signature response was used in two

painful from nonpainful conditions with 93% specificity and sensitivity, and to predict pain intensity well (although warmth intensity was less successfully captured). These results, among others, lead them to conclude that the regions of interest (ROIs) driving classifier performance constitute a "neurologic signature" (p. 1396) or biomarker of subjective pain experience. This again is consistent with DCD, since biomarker regions (as determined by classifier weight assignments) are singled out for their role in predicting participants' experiential reports, which are assumed to reflect their phenomenal state. The logic of this study is representative of that presented in a recent survey and critique of the pain prediction literature by Hu and Iannetti (2016).[12]

Finally, moving from experiential states to cognitive ones, DCD also drives studies aimed at predicting intentions to act. Soon et al. (2013) trained classifiers to find regions that are predictive of conscious decisions to carry out abstract actions (in this case, adding or subtracting single digit numbers). Participants viewed a sequence of slides containing a matrix of numbers plus a single letter cue, and were free to choose at any time to either add or subtract the numbers. After indicating readiness and carrying out the arithmetic operation, they reported the result along with which letter was present when they became aware of their decision. Classifiers were trained on scans from the 8–18 s preceding their awareness, with the aim of distinguishing between the operations that later they carried out. At 4 s prior to awareness of the intention, two regions were able to successfully decode (with 59% accuracy) which type of mental arithmetic the participants carried out. This decoding success was interpreted as evidence for the presence of an unconscious intention to execute a mental action. In their discussion section, they say: "Our results show that regions of medial frontopolar cortex and posterior cingulate/precuneus encode freely chosen abstract intentions before the decisions have been consciously made" (p. 6219). An additional explicit invocation of a DCD-like principle occurs in their methods section, where they note that "[g]ood classification implied that the local cluster of voxels spatially encoded information about the participant's specific current intention" (p. 6221).

These examples suggest that DCD-style inferences of the kind recommended by Del Pinal and Nathan are employed across a number of domains in contemporary imaging studies. Nevertheless, I argue we should reject the claim that decodability of differences between tasks is generally sufficient to reveal cognitive differences. Classifiers are powerful tools, but they often achieve their results for reasons that are opaque or flat out in conflict with the wider epistemic purposes that drive the debate over reverse inference. In the following sections I survey three problems that plague

ways: to directly predict rated intensity of a stimulus, and with an imposed threshold to predict pain/no pain.

[12]This review also distinguishes between two objectives in decoding: discovering a pain-specific *neural signature* and discovering a reliable pain *predictor*. This approximately corresponds to the distinction drawn here between functional and predictive RI. As the authors note, these two goals prescribe distinct experimental and statistical logics and should be more cleanly separated in practice.

the interpretation of decoding results. The picture that emerges is one on which even when they can attain a high degree of predictive success, we may not be able to confidently infer from this fact to either ground truths about neural functioning or to facts about cognitive processing.

5.4.1 *The Problem of Sensitivity and Globality*

Two core traits for which classifiers are touted are their high degree of sensitivity to variations in neural activity and their globality, meaning that in making predictions they inherently take into account spatially distributed voxel patterns. Del Pinal and Nathan specifically cite globality as a virtue when they note that MVPA does not rely on assumptions about localization of cognitive functions in the brain. They remark that "classifiers can employ multi-voxel patterns, which are distributed across traditional brain regions of interest. Hence, the use of [pattern-based reverse inference] is compatible with the possibility that the sources from which to decode cognitive processes are widely distributed patterns" (Nathan and Del Pinal 2017, p. 7). And this sensitivity to distributed or global patterns in turn means that MVPA methods can be used to detect cognitive processes whose realization spans several multifunctional local regions. This emphasis on the ability of MVPA to track global patterns of interest is often couched in terms of evidence for a highly distributed neural code, with task-relevant information being encoded by subtle activation differences within and across regions (Kragel et al. 2018).[13]

From this perspective the globality of classifiers is a virtue, since it meshes appropriately with the structure of the underlying neural realizers. Both sensitivity and globality, however, can lead to scenarios in which labeled patterns are distinguished with high accuracy without this necessarily being a sign that different cognitive processes are engaged. In short, classifiers can be *oversensitive* relative to our interest in reverse inference.

To see this, consider that classifiers may succeed for reasons that do not seem related to the functions of the underlying regions or the task being carried out. For example, regions of motor cortex frequently show distinctive activity across task contexts, due to the demands of the specific responses each task requires. A classifier might assign these some predictive value, without their being relevant to the "core" cognitive processes of interest (Jimura and Poldrack 2012, p. 550). Indeed, in one

[13]However, despite the fact that it remains common to see successful applications of MVPA described in terms of distributed neural representations, it has been shown that we cannot infer from the dimensionality of the measurements to that of the underlying neural code itself. Linear classifiers will use any number of voxel features that they are trained on, but this does not establish that the brain itself encodes this information in this way (Davis et al. 2014). For a real-world example, single electrode studies can recover information about face identity in macaque visual cortex, but this information cannot be decoded with MVPA, plausibly because of weak clustering of similarly-responding neurons (Dubois et al. 2015).

often-cited study dozens of cortical regions could support successful classification between 30% and 50% of the time (Poldrack et al. 2009). Cases like these show that a neural pattern can be useful for distinguishing the engagement of two processes without being the realizer of either.

One response to this problem is to be more selective about the regions that are used to train and test classifiers. If motor processes are not thought to be functionally relevant, voxels in motor cortex should be stripped out by deleting them from the input vectors prior to classifier training. But while *a priori* selection of ROIs can remove regions that are believed to be irrelevant to the cognitive processing that we are interested in, this solution doesn't generalize. Sometimes the information that classifiers exploit is present in regions that we *are* interested in, and so it can't be successfully stripped out. For example, regions of primary visual cortex contribute to discrimination of high level visual features despite the fact that we don't have strong reasons to think that they actually compute using the information that can be decoded from them (Cox and Savoy 2003). Decoders' sensitivity to available information *within* ROIs can easily outstrip the ground truths about whether and how that information is causally used (de-Wit et al. 2016; Ritchie et al. 2019).

This interpretive problem arises even in the most methodologically sophisticated of studies. Searchlight analysis is a widely used exploratory technique that avoids presupposing anything about specific assignments of functions to local regions (Etzel et al. 2013; Kriegeskorte and Bandettini 2007a, b; Kriegeskorte et al. 2006). Briefly, it involves dividing the brain (or some region thereof) up into three-dimensional volumes each of which centers on a voxel. A new classifier is then trained on the signals within each such volume to see whether its activity patterns can discriminate among conditions of interest. The metaphorical "searchlight" can be visualized as the iteration of an MVPA detector systematically through these relatively small brain regions. In principle this gives an unbiased procedure for sorting brain volumes by how predictive their activity patterns are. The output of searchlight analysis is typically a map of those regions that can decode the target condition with greater than chance accuracy.

As an illustration, consider Vickery et al.'s (2011) study of reward processing. In one of their experiments, they asked participants to play a penny-matching game against a computer, in which the players won on average 48% of the time. They then examined trials on which players won to see whether there were regions from which signs of reward or reinforcement (operationalized as wins vs. losses) could be decoded using both an ROI-based and a searchlight analysis. In the former, they found reinforcement signals decodable in 37 of 43 prechosen bilateral regions, while in the latter ~30% of all voxels elicited significant decoding. In the authors' words, "[v]irtually every major cortical and subcortical division contained a significant cluster in one or both hemispheres" (p. 169). Neural signals linked with reinforcement, then, are far from localized. They can be decoded from an extremely widespread set of brain regions. Moreover, these globally distributed patterns are also more sensitive than paired univariate analyses: only between 9 and 7 of the 43 ROIs were significant when a standard general linear model is applied to the same data.

However, the ability to sensitively decode winning trials from globally distributed patterns does not inherently support the claim that these regions realize or have the function of tracking wins. There may be some very general cognitive process labeled "reinforcement" that is involved in these regions' activity—although whether it is precisely the same process in each case or not would require much more precise specification. But there are many forms that this involvement may take. Detecting wins may modulate other processes carried out within those regions without those regions being in any sense *for* detecting wins. Vickery, Chun, & Lee themselves are cautious on this point, saying that "the functional neuroanatomy exists for positive and negative outcomes to directly influence neural processing throughout nearly the entire brain" (p. 175). A region's processing being influenced by the valence of an outcome does not require that the region has the function of processing that valence, nor that there be any single cognitive process that those regions share. It is compatible with any form of influence strong enough to make the region a good predictor.

It is certainly defensible for some translational purposes to focus just on decoding success. Perhaps engineering brain-computer interfaces or clinical diagnosis are examples. However, doing so involves privileging predictive RI over functional RI. This carries the risk that our models are ignoring potentially explanatory ground truths. Insofar as a model is insensitive to such truths, we should not treat it as directly illuminating cognitive processing.

5.4.2 The Problem of Tradeoffs and Interpretability

A second problem facing MVPA methods is that even when classifiers can distinguish between task states, increased prediction accuracy *per se* does not guarantee other epistemically desirable properties. Here the problem lies in the fact that what is decoded depends in part on the specific modeling choices made by experimenters. Because classifier performance turns on model selection and tuning of parameters, it embodies certain familiar trade-offs. In particular there is a tension between the *stability* of the weights and the performance of the classifier (Baldassarre et al. 2017; Rasmussen et al. 2012; Varoquaux et al. 2017). Stability is a measure of how reliably the same weight pattern will be reproduced by different classifiers, or by different runs of the same classifier. Machine learning research has increasingly focused on the quantifying these tradeoffs, and one consistent result that emerges from these studies is that if we choose parameter assignments that maximize the predictive success of a classifier, we are necessarily sacrificing other potentially important properties.

A typical linear classifier like SVM has a soft margin parameter that determines how much misclassifications are counted against a weight assignment.[14] Sparse classifiers include various regularization terms, which impose parsimony constraints (degree of fit to the data, contiguity, smoothness, etc.) on the resulting weights. These classifiers are used to select only some of the possible input features to drive the weight vector, but a great deal turns on exactly how these parameters are tuned. In one study, Rasmussen et al. (2012) found that as the regularization parameter is varied, predictive accuracy decreases (from ~72% to 50% correct) while pattern reproducibility as measured by Pearson's correlation increases (from 0.0 to 0.5). More accurate prediction, in other words, is purchased at the cost of high variability in the spatial weight map. This implies that credit assigned to one region could be revoked if the same classifier were retrained without alteration.

The tradeoff for a model's high degree of success, then, is a lack of reliable informativeness about what regions are most responsible for that success. This has obvious consequences for the interpretation of classifier performance: we may know that a certain region is predictive without having generalizable insight into why this is the case. These types of tradeoffs apply even within the domain of sparse classifiers, which attempt to group weights into relatively few internally homogeneous or structurally adjacent clusters. In a comparison across six sparse models trained on fMRI datasets, systematic accuracy-stability tradeoffs arise for each one (Baldassarre et al. 2017). A typical sparse classifier such as LASSO can achieve high accuracy (85%) at a corrected overlap score of just under 0.6, while a higher overlap score (around 0.7) returns much worse accuracy (~65%).

If predictive accuracy is all that we care about, it is clear which parameter tuning we should prefer. But in practice, modelers often prefer sparse solutions. What sparseness costs in predictive accuracy it purportedly gains in making models more interpretable and biologically plausible. A non-sparse model can assign decoding importance to a scattered, buckshot-like distribution of regions that lacks any neurophysiological sense. Even sparse models are not interpretively transparent, though. While the best-performing sparse classifiers converged in assigning the same five regions the highest weight (although not in the same order), they still varied widely in how many regions they included overall (from 10 to 106 total). Human-legible interpretation remains challenging with dozens of small, anatomically insignificant regions participating.

The situation with respect to tradeoffs among classifier performance, stability, and interpretability is strongly akin to what Gelman and Loken (2014) famously refer to as the "garden of forking paths" in statistical analysis. The number of available off-the-shelf classifiers plus the number of tunable parameters for each gives rise to potentially quite distinct assignments to each of these three valuable properties. The choice of any particular model-parameter pairing in imaging studies can be epistemically consequential, and can even shape whether a result

[14] The choice of kernel is also significant, but many neuroimaging applications use a linear kernel, so I ignore this complication here.

is considered significant. But such choices are often undermotivated. We should be cautious about interpreting results where the choice of data analysis methods is largely unconstrained except by custom and experimenter preference, and where these choices can make a difference to the outcome of the analysis.

Classifier interpretability is further complicated by the fact that weights may be assigned to voxels that are not the origin of the underlying neural signal, and that low (or even negative) weights may be assigned to voxels where the signal is located.[15] To take one example of this phenomenon, suppose that we have BOLD measurements from two regions, and that the ground truth is that one of these regions contains information that can discriminate moderately well between two labeled conditions, while the other contains no such information. Nevertheless, the weight vector to achieve optimal discrimination can (under the right circumstances) be one that assigns double the weight to the latter region than to the former—despite the fact that the latter region is by hypothesis one that is informationally empty (Haufe et al. 2014).[16]

This idealized case illustrates two broader points about MVPA: first, it runs together signal and noise, treating both as potential information; and second, it assigns weights based not on individual voxel importance but on how well overall classification performance is affected. Classifiers are holistic and opportunistic (see also Ritchie et al. 2019, p. 14). Two voxels that contain no genuine information about which condition obtains can nevertheless be used for discrimination if they have different noise variances in each condition (Hebart and Baker 2018). So weight assignments are at least sometimes performed on grounds other than the causal-explanatory significance of voxel activity, further complicating interpretability.

Practitioners may object that typical studies look mainly at overall classifier accuracy as their measure of interest. So it may seem unclear why these issues about their precise internal structure may matter. With respect to the reverse inference debate, however, the concern is that we cannot easily analyze classifier successes in terms of the underlying neural ground truths. One can't, for example. conclude from the fact that a successful classifier assigns a certain weight to a voxel that the voxel's activity contains a signal whose function is realizing the cognitive operation that is being decoded. Weights are assigned for the purpose of maximizing overall success using any available cues. As Haufe et al. note: "*A widespread misconception about multivariate classifier weights is that (the brain regions corresponding to)*

[15] A related warning is that positive weights on a voxel can reflect *decreases* in its activation, since if these decreases are reliable they may convey information about certain stimulus conditions.

[16] This artificial example has been criticized by Schrouff & Mourão-Miranda (Schrouff and Mourao-Miranda 2018), who argue that it holds only for low signal-to-noise ratio cases. However, given that it is often unclear what the SNR is for particular ROIs, it is fair to say we cannot across the board rule out the presence of "false positive" voxel weights. Moreover, the type of noise matters. As Haufe et al. point out, it is sometimes possible to correct for the presence of Gaussian noise to recover underlying signal, but this doesn't hold for noise induced by scanner drift, head motion, and periodic noise (P. K. Douglas and Anderson 2017), all of which are present in imaging data.

measurement channels with large weights are strongly related to the experimental condition" (Haufe et al. 2014, p. 97). If this assumption doesn't hold in general, the undeniable success of classifiers may end up being causally opaque.

Even so, one may wonder why issues such as the interpretability of models should matter from a perspective such as that of DCD, where the express goal of decoding is simply to find evidence that decides between two possible cognitive models. Given del Pinal and Nathan's emphasis on the fact that MVPA does not depend on any specific localizationist assignment of functions to regions, prioritizing sparseness at all might seem beside the point. DCD as a criterion of reverse inference cares only about predictive success, not other epistemic traits of models. Once we no longer seek to map cognitive functions onto regions in a way that respects their underlying causal organization, there is no added evidential value in the mere fact that a weight map is sparsely interpretable, let alone stable.

For these purposes, decoding that is based on an unstable weight map or one that is hard to interpret may indeed be adequate. A more traditional concern for functional RI might lead us to have a different set of goals in mind, however, including the desire to *explain* how neural patterns realize cognitive processes. For these goals, interpretability and plausibility matter. Focusing attention on a sparse subset of regions is best understood as motivated by a search for neural structures that play the appropriate causal and explanatory roles. As we will see, though, even this goal often proves elusive.

5.4.3 The Problem of Causality

Suppose a classifier achieves what we regard as a good balance of accuracy and production of a reproducible and plausibly interpretable weight map. It is tempting to infer from such success to claims about causality and processing. Kriegeskorte and Douglas (2019), for instance, propose that classifiers can perform double duty as causal models: "if a decoder is used to predict behavioral responses, for example judgments of categorization or continuous stimulus variables . . . then the decoder can be interpreted as a model (at a high level of abstraction) of the brain computations generating the behavioral responses from the encoding of the stimuli in the decoded brain region" (p. 171).

However, decoders do not give us enough evidence to conclude that the predictively weighted regions *cause* behavioral effects. There are several reasons for this. One is that decoders have no inherent causal directionality built into them. Procedures to find the best boundary to enable pattern-to-label associations are agnostic on whether there are any causal relations between the two. This is easy to see once we step outside the domain of neural data, since classifiers are frequently used on datasets that have no such causal relations among features and labels. SVMs can be used to parse handwritten ZIP codes on envelopes, or for image analysis and facial recognition. Success in these contexts implies nothing about causal structure in the target materials. Even within neuroscience, it is common to train classifiers on multimodal data sets (combining imaging, MEG/EEG, and other physiological or clinical biomarkers) that do not have a clear joint causal interpretation of their features (Meng et al. 2017; Woo et al. 2017).

Moreover, good predictors in machine learning tasks do not always overlap well with good targets of intervention (Athey 2017). Consider a non-neural example. Marketing firms use machine learning tools to discover "high churn" customers—those that are likely to stop using a company's products or services. But the population of customers who respond well to interventions such as marketing appeals only overlaps by 50% with those who are in the predictively isolated high-churn group. So we can often know that churn will take place in a certain population without being able to use that information to intervene causally on it. The same holds for many neurodiagnostic classifications. A classifier might use anatomical features such as hippocampal volume or the presence of amyloid plaques to diagnose Alzheimer's disease, but neither of these is a *cause* of the disorder. Drugs targeting amyloid, in particular, have regularly failed to produce clinical improvements in DAT patients. Decoders in this case are not reliably tracking causes. This is a recurring problem across fields using similar classifier-based analyses, such as genome-wide association studies: significant variables are often not predictive, and vice-versa (Lo et al. 2015).

Indeed, decoders can just as easily operate in an *anti*-causal direction (Weichwald et al. 2015). Consider two experimental designs, one in which a stimulus is presented followed by BOLD imaging, and another in which BOLD imaging occurs prior to production of a behavioral or cognitive response. In a stimulus-first design, decoding the category of the stimulus operates *against* the direction of causation in the experiment. Clearly in this design we can't treat decoders as causal models. But we cannot do so even in cases where the direction of decoding and the direction of causation are consistent. The reason is that decoding weights are not designed to be measures of causal contribution. As noted above, some factors may result in weights being assigned to features that are not causes of a phenomenon, such as incidentally correlated noise within voxels. Some actual causes may even receive low or zero weights simply because they are not most useful for decoding purposes *in the context* of the other voxel weight assignments.

Neither can we generally regard classifiers as processing models. Decoding is typically presented as a way of extracting the information present in regional activation patterns. But within cognitive modeling, there is an important distinction between information and processes, as evinced by the fact that such models posit representation-process pairs that work together to execute cognitive functions. This is a fundamental commitment of cognitive modeling within the broadly Marrian tradition (Barsalou 2017). Algorithmic cognitive models describe how representations are constructed, stored, and transformed in carrying out a cognitive operation. Task performance is a product of the joint operation of both factors (along with architectural facts such as resource constraints). The neural decodability of a distinction between two task conditions does not tell us whether this stems from representational differences, processing differences, demand characteristics and resource usage, or some combination of these. From the point of view of causal modeling, this simply amounts to conflating potentially separate contributions. The sort of information that decoding provides, then, does not inherently tell us about causal-explanatory structure, particularly as it relates to cognitive processing.

This point can be illustrated by studies that explicitly attempt to use classifiers to derive causal structure by using their performance as predictive of later events such as behavior. Consider a nicely designed set of studies by Grootswagers et al. (2018). They asked participants to view images and make binary categorization decisions about them, e.g., judging whether a banana was animate or inanimate. In the first analysis phase they used a SVM-based searchlight procedure to generate a map of regions whose activation predicted correct category decisions. The crucial step lies in the second analysis phase, which involved running another searchlight analysis in which a new SVM classifier was trained for each region and the distance of each presented exemplar from the classifier's decision hypersurface was computed. The logic behind this analysis stems from signal detection theory, which holds that an option's distance to a decision boundary is determined by the evidence; i.e., stronger evidence places items farther away in space. This, in turn, generates the prediction that items located close to the decision boundary should be ones for which there is relatively little evidence, or evidence of ambiguous quality. For items such as these, choice is more difficult. Finally, there is the assumption that choice difficulty is reflected linearly in decision time. The second map created depicts the averaged degree of negative correlation between distance to the classifier's decision hyperplane and RT.

Their key finding is that the two maps overlap somewhat, but not entirely: while animacy can be successfully decoded throughout the ventral visual processing stream, RT is predicted by only a subset of those regions, predominantly ones located in anterior ventral temporal cortex. This suggests that mere decodability does not imply that the information present in regional patterns is formatted correctly to be "read out" in behavior (see also Williams et al. 2007). To bridge this gap successfully requires analysis that systematically links the properties of classifiers with behavioral variables. Specifically, it depends on interpreting the formal structure of classifiers in terms of established computational theories (Ritchie and Carlson 2016). Here the relevant theory is a distance-to-bound model of choice transposed to the neural domain. Applying this model turns essentially on giving explanatory significance to the distances defined by classifier hypersurfaces, since these are treated as both reflecting the processing of evidence and as affecting behavioral responses.

While the approach taken by Grootswagers et al. is promising, some caveats remain. Most importantly for present purposes, the success of this computational framework only drives home further the need to take seriously the model choice considerations raised in Sect. 5.2, particularly since different classifier structures may give rise to different RT predictions. If classifiers are to be treated as causal models of the computational processes that mediate behavior, they need to be both stable and interpretable. This, again, is just to emphasize that they need to be chosen with an eye towards facilitating functional RI.

5.5 Decoding as Data Exploration

The problems surveyed here converge on the following conclusions. In terms of our original distinction, classifiers can be extraordinarily useful tools for *predictive* reverse inference. For *functional* reverse inference—the discovery not only of neural activity that is indicative of cognitive processing, but also of a prospective implementation theory for that processing—their utility is significantly less clear. The reason is that they are driven, in an unknown proportion of cases, by factors besides the ground truth concerning what patterns of neural activation are causally and explanatorily responsible for the cognitive processing we are investigating. Disentangling genuinely explanatory factors from the rest is difficult given that classifiers inherently conflate them. Decoding, in short, allows reverse inference of an often opaque kind that does not suit all of our investigative ends equally well.[17]

In fact, MVPA itself is demonstrably not a panacea for the ills of localization-based reverse inference, since the same problem of multiple functional assignments can arise just as readily within it as in univariate analysis. To see this, consider a widely discussed study by Knops et al. (2009). In the first phase of their study they scanned participants during a random left/right eye movement task and trained a classifier on a group of six pre-selected cortical ROIs. This classifier could decode direction of motion with ~70% accuracy across all participants. They then had the same participants perform a simple arithmetic task: either add or subtract two displayed numbers and choose the closest correct answer (out of seven choices). The classifier trained on activation patterns from the bilateral posterior superior parietal lobule (PSPL) was then applied, without alteration, to activation patterns in that region from the arithmetic task. The classifier succeeded ~55% of the time with addition mapped onto rightward eye motion and subtraction onto leftward motion (breakdown by condition was 61% correct for addition and 49% for subtraction).

Knops et al. concluded from the fact that the same classifier achieved predictive success on both datasets that that the PSPL is involved in computations underlying both L/R eye motion and addition/subtraction. But now we face exactly the problem of multifunctional regions again. The information present in PSPL might indicate rightward eye motion *or* (some unknown cognitive component of) mental addition—or, for that matter, some more abstract but unknown operation that is implicated in both of them. We are not appreciably closer to understanding what "the" function of PSLP is, except to say that it contains information that can contribute to this pattern of discriminative success across tasks. The classifier transfer paradigm, then, is not *in itself* an advance in understanding the cognitive processing that goes on in particular brain regions.

I should stress that this conclusion is one that del Pinal and Nathan might not object to. At one point they seem to reject the search for an explanatory

[17]To be clear, the preceding arguments are obviously not meant as blanket condemnations of the use of MVPA and machine learning in neuroscience. The issue concerns only whether the successful use of ML-based decoding methods is sufficient for making reverse inferences.

implementation theory of the sort that functional RI is concerned with, arguing that rather than focusing on "how cognitive algorithms are neurally implemented", reverse inference should address only the question of "which cognitive processes are more or less likely to be engaged in certain tasks whose nature is under dispute" (Nathan and Del Pinal 2017, p. 9). This approach is quite consistent with the predictive turn, although they don't couch their claim in those terms. As Bzdok and Ionnadis note, "predictive approaches put less emphasis on mechanistic insight into the biological underpinnings of the coherent behavioral phenotype" (2019, p. 3). Shifting focus away from discovering the cognitive function of brain regions (or even distributed brain networks) is of a piece with this move from explanatory understanding towards successful prediction.

Supposing, however, that neuroscientists wish to retain explanation as an epistemic goal, how can we reconceive the role of experimental practices such as MVPA, given that decoding models are not *themselves* explanatory? The rhetoric surrounding prediction depicts it as competing with explanation, or at least on the opposite side of a continuum from it (Bzdok and Yeo 2017). I suggest, to the contrary, that we view decoding models not as competing for epistemic real estate with causal-explanatory ones, but as cooperating as part of a modeling pipeline. Decoding results constitute both *heuristic input* to explanatory models as well as *constraints* on them.

Decoding is a useful heuristic insofar as it suggests a menu of possible sites for further investigation and intervention. Regions whose activity can be decoded to distinguish between presented visual objects can also be scrutinized for whether that activity predicts behavioral outcomes. There is no guarantee that it will, as shown by the Grootswagers et al. study discussed above, but this needs to be investigated using methods that are geared towards generating and testing potential causal explanations, not just predictive adequacy. Such regions can also be probed using other methods to uncover their operations. For example, regions that support decoding of information might exhibit repetition suppression for that same information (though see Ward et al. 2013 for some doubts about this).

In a case where we know that decodable information can be correlated with behavioral or cognitive outcomes, we have the following constraint: for any region from which information can be read out, any account of that region's function should explain how the region can contribute towards producing the behavior in question. That is, decoding results help to establish and clarify the explanandum phenomena that characterize the regions targeted by explanatory modeling. Something like this provides a useful way to understand Representational Similarity Analysis (RSA), a procedure of correlating the geometry of the space of stimuli (such as pictures of artifacts or faces) with that of the activation space of a collection of voxels (Kriegeskorte 2011; Kriegeskorte and Kievit 2013). RSA returns a numerical measure (using, e.g., rank correlation) of the extent to which stimuli that are close together in visual similarity space remain so in activation space. As its name suggests, RSA is sometimes described as characterizing what a region represents. But as normally practiced it does not offer hypotheses about algorithmic-level vehicles or processes. Rather, it articulates abstract structures of correspondence between regions and stimuli or behaviors. These correspondences, often discovered through decoding studies, can be regarded as tentative functional assignments. In this way they become part of the phenomena to be fed into the explanatory modeling pipeline.

In short, the place of decoding models is not in competition with explanatory modeling, but prior to and in concert with it. This aligns with the guiding and constraining role that data modeling has often been assigned within a mechanistic framework (Bechtel and Abrahamsen 2005; but see Burnston 2016b for an alternate interpretation). Imaging data is noisy and complex. Machine learning tools provide one way of extracting useful patterns from this data, which can help to stabilize new phenomena and discover new explanatory targets. This dovetails nicely with one of the original functions for which linear classifiers were developed, namely the partitioning of large datasets according to the varieties of hidden information that they contain. As tools for simplifying and exploring neuroscientific datasets, they can contribute to explanatory modeling without displacing it.

Finally, with respect to the question of reverse inference, the mere existence of decoding differences between task conditions does not establish differences in the underlying cognitive processes. What it does, however, is provide a set of phenomena to be investigated further; specifically, it suggests a plausible hypothesis about the structured information that is robustly detectable (in the brain at large, in an ROI, or within a cluster of searchlights) and that can be connected with measurable outcomes. If decoding results are stable across a wide range of models, parameter settings, and training regimes, and if they are systematically connectable with cognitive or behavioral outcomes, then the most predictive interpretable regions these converging models pick out are plausible targets for explanatory modeling. MVPA achieves the role of potential evidence for or against cognitive hypotheses by playing a supporting (though not individually sufficient) role in this sort of data modeling pipeline.[18]

5.6 Conclusion

I've argued that MVPA's ability to make predictive inferences from activation patterns does not offer us a transparent interpretive window onto the ground truths that drive this success. This form of predictive modeling is useful not because it can

[18]This point is similar to Kriegeskorte and Douglas's (2019) warning against committing the *single-model-significance* fallacy: that is, assuming that because a model explains some significant variance that it *thereby* captures facts about processing or causal structure. To reach such conclusions we need to integrate information from many models operating over a wide range of training data and parameter settings. This many-model integration process is what I have referred to here as a modeling pipeline. This notion is also discussed at length by Wright (2018), who emphasizes that in practice multiple analyses of data make distinct contributions to the characterization of phenomena in neuroimaging.

serve as a replacement for explanatory modeling, but because, seen in the proper perspective, it is an essential complement to it. Techniques from data science have their natural home in the analysis and modeling of data, even when deployed within neuroscience. To the extent that neuroscience continues to import and adapt machine learning tools, with their associated epistemic focus on prediction over explanation, there may be strong temptations to focus on the success of these tools without inquiring into the underlying causal-explanatory facts that enable them to succeed or fail. This temptation is understandable, given their striking translational successes, but I've argued that giving in to it would be a mistake. We should welcome the return of prediction as an important scientific desideratum without granting it dominance over our epistemic regime.

Acknowledgments Thanks to Nikolaus Kriegeskorte, Marco Nathan, J. Brendan Ritchie, and Jessey Wright for their thoughtful comments on a previous draft of this paper, which was presented as part of the Neural Mechanisms webinar in June, 2019. I'm grateful to Fabrizio Calzavarini and Marco Viola for generously inviting me to participate in this series.

References

Agrawal, A., Gans, J., & Goldfarb, A. (2018). *Prediction machines: The simple economics of artificial intelligence*. Boston: Harvard Business School Publishing.

Anderson, M. L. (2014). *After phrenology*. Cambridge, MA: MIT Press.

Athey, S. (2017). Beyond prediction: Using big data for policy problems. *Science, 355*, 483–485.

Baldassarre, L., Pontil, M., & Mourão-Miranda, J. (2017). Sparsity is better with stability: Combining accuracy and stability for model selection in brain decoding. *Frontiers in Neuroscience, 11*, 62. https://doi.org/10.3389/fnins.2017.00062.

Barsalou, L. W. (2017). What does semantic tiling of the cortex tell us about semantics? *Neuropsychologia, 105*, 18–38.

Bechtel, W., & Abrahamsen, A. (2005). Explanation: A mechanist alternative. *Studies in History and Philosophy of Science Part C: Studies in History and Philosophy of Biological and Biomedical Sciences, 36*, 421–441.

Berkman, E. T., & Falk, E. B. (2013). Beyond brain mapping: Using neural measures to predict real-world outcomes. *Current Directions in Psychological Science, 22*, 45–50.

Burnston, D. C. (2016a). A contextualist approach to functional localization in the brain. *Biology and Philosophy, 31*, 527–550.

Burnston, D. C. (2016b). Data graphs and mechanistic explanation. *Studies in History and Philosophy of Science Part C: Studies in History and Philosophy of Biological and Biomedical Sciences, 57*, 1–12.

Bzdok, D., & Ioannidis, J. P. A. (2019). Exploration, inference, and prediction in neuroscience and biomedicine. *Trends in Neurosciences, 42*, 251–262.

Bzdok, D., & Yeo, B. T. T. (2017). Inference in the age of big data: Future perspectives on neuroscience. *NeuroImage, 155*, 549–564.

Coltheart, M. (2006). What has functional neuroimaging told us about the mind (so far)? *Cortex, 42*, 323–331.

Coltheart, M. (2013). How can functional neuroimaging inform cognitive theories? *Perspectives on Psychological Science, 8*, 98–103.

Cox, D. D., & Savoy, R. L. (2003). Functional magnetic resonance imaging (fMRI) "brain reading": Detecting and classifying distributed patterns of fMRI activity in human visual cortex. *NeuroImage, 19*, 261–270.
Davies, M. (2010). Double dissociation: Understanding its role in cognitive neuropsychology. *Mind & Language, 25*, 500–540.
Davis, T., LaRocque, K. F., Mumford, J. A., Norman, K. A., Wagner, A. D., & Poldrack, R. A. (2014). What do differences between multi-voxel and univariate analysis mean? How subject-, voxel-, and trial-level variance impact fMRI analysis. *NeuroImage, 97*, 271–283.
de -Wit, L., Alexander, D., Ekroll, V., & Wagemans, J. (2016). Is neuroimaging measuring information in the brain? *Psychonomic Bulletin & Review, 23*, 1415–1428.
Del Pinal, G., & Nathan, M. J. (2017). Two kinds of reverse inference in cognitive neuroscience. In J. Leefman & E. Hildt (Eds.), *The human sciences after the decade of the brain* (pp. 121–139). London: Academic Press.
Douglas, H. E. (2009). Reintroducing prediction to explanation. *Philosophy of Science, 76*, 444–463.
Douglas, P. K., & Anderson, A. (2017). Interpreting fMRI decoding weights: Additional considerations. In *31st conference on Neural Information Processing Systems (NIPS 2017)* (pp. 1–7).
Dubois, J., de Berker, A. O., & Tsao, D. Y. (2015). Single-unit recordings in the macaque face patch system reveal limitations of fMRI MVPA. *Journal of Neuroscience, 35*, 2791–2802.
Etzel, J. A., Zacks, J. M., & Braver, T. S. (2013). Searchlight analysis: Promise, pitfalls, and potential. *NeuroImage, 78*, 261–269.
Falk, E. B., Berkman, E. T., & Lieberman, M. D. (2012). From neural responses to population behavior: Neural focus group predicts population-level media effects. *Psychological Science, 23*, 439–445.
Gelman, A., & Loken, E. (2014). The statistical crisis in science. *American Scientist, 102*, 460–465.
Genevsky, A., Yoon, C., & Knutson, B. (2017). When brain beats behavior: Neuroforecasting crowdfunding outcomes. *The Journal of Neuroscience, 37*, 8625–8634.
Glymour, C., & Hanson, C. (2016). Reverse inference in neuropsychology. *The British Journal for the Philosophy of Science, 67*, 1139–1153.
Grootswagers, T., Cichy, R. M., & Carlson, T. A. (2018). Finding decodable information that can be read out in behaviour. *NeuroImage, 179*, 252–262.
Haufe, S., Meinecke, F., Görgen, K., Dähne, S., Haynes, J.-D., Blankertz, B., & Bießmann, F. (2014). On the interpretation of weight vectors of linear models in multivariate neuroimaging. *NeuroImage, 87*, 96–110.
Haxby, J. V. (2001). Distributed and overlapping representations of faces and objects in ventral temporal cortex. *Science, 293*, 2425–2430.
Haxby, J. V., Connolly, A. C., & Guntupalli, J. S. (2014). Decoding neural representational spaces using multivariate pattern analysis. *Annual Review of Neuroscience, 37*, 435–456.
Haynes, J.-D. (2012). Brain reading. In S. Richmond, G. Rees, & S. Edwards (Eds.), *I know what you're thinking: Brain imaging and mental privacy* (pp. 29–40). Oxford: Oxford University Press.
Haynes, J.-D. (2015). A primer on pattern-based approaches to fMRI: Principles, pitfalls, and perspectives. *Neuron, 87*, 257–270.
Haynes, J.-D., & Rees, G. (2005). Predicting the stream of consciousness from activity in human visual cortex. *Current Biology, 15*, 1301–1307.
Haynes, J.-D., Sakai, K., Rees, G., Gilbert, S., Frith, C., & Passingham, R. E. (2007). Reading hidden intentions in the human brain. *Current Biology, 17*, 323–328.
Hebart, M. N., & Baker, C. I. (2018). Deconstructing multivariate decoding for the study of brain function. *NeuroImage, 180*, 4–18.
Henson, R. (2006). Forward inference using functional neuroimaging: Dissociations versus associations. *Trends in Cognitive Sciences, 10*, 64–69.
Hu, L., & Iannetti, G. D. (2016). Painful issues in pain prediction. *Trends in Neurosciences, 39*, 212–220.

Hutzler, F. (2014). Reverse inference is not a fallacy per se: Cognitive processes can be inferred from functional imaging data. *NeuroImage, 84*, 1061–1069.
Jimura, K., & Poldrack, R. A. (2012). Analyses of regional-average activation and multivoxel pattern information tell complementary stories. *Neuropsychologia, 50*, 544–552.
Kamitani, Y., & Tong, F. (2005). Decoding the visual and subjective contents of the human brain. *Nature Neuroscience, 8*, 679–685.
Klein, C. (2012). Cognitive ontology and region- versus network-oriented analyses. *Philosophy of Science, 79*, 952–960.
Knops, A., Thirion, B., Hubbard, E. M., Michel, V., & Dehaene, S. (2009). Recruitment of an area involved in eye movements during mental arithmetic. *Science, 324*, 1583–1585.
Kragel, P. A., Koban, L., Barrett, L. F., & Wager, T. D. (2018). Representation, pattern information, and brain signatures: From neurons to neuroimaging. *Neuron, 99*, 257–273.
Kriegeskorte, N. (2011). Pattern-information analysis: From stimulus decoding to computational-model testing. *NeuroImage, 56*, 411–421.
Kriegeskorte, N., & Bandettini, P. (2007a). Analyzing for information, not activation, to exploit high-resolution fMRI. *NeuroImage, 38*, 649–662.
Kriegeskorte, N., & Bandettini, P. (2007b). Combining the tools: Activation- and information-based fMRI analysis. *NeuroImage, 38*, 666–668.
Kriegeskorte, N., & Douglas, P. K. (2019). Interpreting encoding and decoding models. *Current Opinion in Neurobiology, 55*, 167–179.
Kriegeskorte, N., & Kievit, R. A. (2013). Representational geometry: Integrating cognition, computation, and the brain. *Trends in Cognitive Sciences, 17*, 401–412.
Kriegeskorte, N., Goebel, R., & Bandettini, P. (2006). Information-based functional brain mapping. *Proceedings of the National Academy of Sciences, 103*, 3863–3868.
Lo, A., Chernoff, H., Zheng, T., & Lo, S.-H. (2015). Why significant variables aren't automatically good predictors. *Proceedings of the National Academy of Sciences, 112*, 13892–13897.
Machery, E. (2014). In defense of reverse inference. *The British Journal for the Philosophy of Science, 65*, 251–267.
McCaffrey, J. B. (2015). The brain's heterogeneous functional landscape. *Philosophy of Science, 82*, 1010–1022.
Meng, X., Jiang, R., Lin, D., Bustillo, J., Jones, T., Chen, J., et al. (2017). Predicting individualized clinical measures by a generalized prediction framework and multimodal fusion of MRI data. *NeuroImage, 145*, 218–229.
Nathan, M. J., & Del Pinal, G. (2017). The future of cognitive neuroscience? Reverse inference in focus. *Philosophy Compass, 12*, 1–11.
Norman, K. A., Polyn, S. M., Detre, G. J., & Haxby, J. V. (2006). Beyond mind-reading: Multi-voxel pattern analysis of fMRI data. *Trends in Cognitive Sciences, 10*, 424–430.
Northcott, R. (2017). When are purely predictive models best? *Disputatio, 9*, 631–656.
Poldrack, R. A. (2006). Can cognitive processes be inferred from neuroimaging data? *Trends in Cognitive Sciences, 10*, 59–63.
Poldrack, R. A. (2018). *The new mind readers: What neuroimaging can and cannot reveal about our thoughts*. Princeton: Princeton University Press.
Poldrack, R. A., Halchenko, Y. O., & Hanson, S. J. (2009). Decoding the large-scale structure of brain function by classifying mental states across individuals. *Psychological Science, 20*, 1364–1372.
Rasmussen, P. M., Hansen, L. K., Madsen, K. H., Churchill, N. W., & Strother, S. C. (2012). Model sparsity and brain pattern interpretation of classification models in neuroimaging. *Pattern Recognition, 45*, 2085–2100.
Rathkopf, C. A. (2013). Localization and intrinsic function. *Philosophy of Science, 80*, 1–21.
Ritchie, J. B., & Carlson, T. A. (2016). Neural decoding and "inner" psychophysics: A distance-to-bound approach for linking mind, brain, and behavior. *Frontiers in Neuroscience, 10*, 190. https://doi.org/10.3389/fnins.2016.00190.

Ritchie, J. B., Kaplan, D. M., & Klein, C. (2019). Decoding the brain: Neural representation and the limits of multivariate pattern analysis in cognitive neuroscience. *British Journal for the Philosophy of Science, 70*, 581–607.

Roskies, A. (2009). Brain-mind and structure-function relationships: A methodological response to Coltheart. *Philosophy of Science, 76*, 1–14.

Schrouff, J., & Mourao-Miranda, J. (2018). Interpreting weight maps in terms of cognitive or clinical neuroscience: Nonsense? In *2018 international workshop on Pattern Recognition in Neuroimaging (PRNI)* (pp. 1–4). Singapore: IEEE.

Soon, C. S., He, A. H., Bode, S., & Haynes, J.-D. (2013). Predicting free choices for abstract intentions. *Proceedings of the National Academy of Sciences, 110*, 6217–6222.

Tong, F., & Pratte, M. S. (2012). Decoding patterns of human brain activity. *Annual Review of Psychology, 63*, 483–509.

Van Horn, J. D., & Toga, A. W. (2014). Human neuroimaging as a "Big Data" science. *Brain Imaging and Behavior, 8*, 323–331.

Varoquaux, G., & Poldrack, R. A. (2019). Predictive models avoid excessive reductionism in cognitive neuroimaging. *Current Opinion in Neurobiology, 55*, 1–6.

Varoquaux, G., & Thirion, B. (2014). How machine learning is shaping cognitive neuroimaging. *GigaScience, 3*, 1–7.

Varoquaux, G., Raamana, P. R., Engemann, D. A., Hoyos-Idrobo, A., Schwartz, Y., & Thirion, B. (2017). Assessing and tuning brain decoders: Cross-validation, caveats, and guidelines. *NeuroImage, 145*, 166–179.

Vickery, T. J., Chun, M. M., & Lee, D. (2011). Ubiquity and specificity of reinforcement signals throughout the human brain. *Neuron, 72*, 166–177.

Wager, T. D., Atlas, L. Y., Lindquist, M. A., Roy, M., Woo, C.-W., & Kross, E. (2013). An fMRI-based neurologic signature of physical pain. *New England Journal of Medicine, 368*, 1388–1397.

Ward, E. J., Chun, M. M., & Kuhl, B. A. (2013). Repetition suppression and multi-voxel pattern similarity differentially track implicit and explicit visual memory. *Journal of Neuroscience, 33*, 14749–14757.

Weichwald, S., Meyer, T., Özdenizci, O., Schölkopf, B., Ball, T., & Grosse-Wentrup, M. (2015). Causal interpretation rules for encoding and decoding models in neuroimaging. *NeuroImage, 110*, 48–59.

Weiskopf, D. A. (2016). Integrative modeling and the role of neural constraints. *Philosophy of Science, 83*, 674–685.

Williams, M. A., Dang, S., & Kanwisher, N. G. (2007). Only some spatial patterns of fMRI response are read out in task performance. *Nature Neuroscience, 10*, 685–686.

Woo, C.-W., Chang, L. J., Lindquist, M. A., & Wager, T. D. (2017). Building better biomarkers: Brain models in translational neuroimaging. *Nature Neuroscience, 20*, 365–377.

Wright, J. (2018). The analysis of data and the evidential scope of neuroimaging results. *The British Journal for the Philosophy of Science, 69*, 1179–1203.

Yarkoni, T., & Westfall, J. (2017). Choosing prediction over explanation in psychology: Lessons from machine learning. *Perspectives on Psychological Science, 12*, 1100–1122.

Part II
Concepts and Tools

Chapter 6
Evolving Concepts of "Hierarchy" in Systems Neuroscience

Daniel C. Burnston and Philipp Haueis

Abstract The notion of "hierarchy" is one of the most commonly posited organizational principles in systems neuroscience. To this date, however, it has received little philosophical analysis. This is unfortunate, because the general concept of hierarchy ranges over two approaches with distinct empirical commitments, and whose conceptual relations remain unclear. We call the first approach the "representational hierarchy" view, which posits that an anatomical hierarchy of feed forward, feedback, and lateral connections underlies a signal processing hierarchy of input-output relations. Because the representational hierarchy view holds that unimodal sensory representations are subsequently elaborated into more categorical and rule-based ones, it is committed to an increasing degree of abstraction along the hierarchy. The second view, which we call "topological hierarchy," is not committed to different representational functions or degrees of abstraction at different levels. Topological approaches instead posit that the hierarchical level of a part of the brain depends on how central it is to the pattern of connections in the system. Based on the current evidence, we argue that three conceptual relations between the two approaches are possible: topological hierarchies could *substantiate* the traditional representational hierarchy, *conflict* with it, or contribute to a *plurality* of approaches needed to understand the organization of the brain. By articulating each of these possibilities, our analysis attempts to open a conceptual space in which further neuroscientific and philosophical reasoning about neural hierarchy can proceed.

Authors appear in alphabetical order and contributed equally to this article.

D. C. Burnston (✉)
Department of Philosophy, Tulane Brain Institute, Tulane University, New Orleans, LA, USA
e-mail: dburnsto@tulane.edu

P. Haueis
Department of Philosophy, Bielefeld University, Bielefeld, Germany
e-mail: philipp.haueis@uni-bielefeld.de

© Springer Nature Switzerland AG 2021

F. Calzavarini, M. Viola (eds.), *Neural Mechanisms*, Studies in Brain and Mind 17,
https://doi.org/10.1007/978-3-030-54092-0_6

Keywords Hierarchy · Systems neuroscience · Representation · Topology · Abstraction

6.1 Introduction

Scientific concepts evolve over time. As researchers generate new data and explore an increasing number of related yet subtly different phenomena, concepts frequently acquire novel connotations and expand their reference to novel properties. What is often left is a patchwork of multiple meanings and uses operating under the guise of a univocal concept. Because they result from the exploration of related phenomena, patchwork concepts are polysemous, i.e. they have multiple related meanings (as opposed to ambiguous words, whose distinct meanings are unrelated, cf. Sennet 2016). Recent case studies in the physical and life sciences suggest that such polysemous patchwork concepts help researchers to describe distinct but related phenomena efficiently (Wilson 2006), classify properties at different scales (Bursten 2016), or integrate seemingly incompatible uses of a concept in theoretically fruitful ways (Novick 2018; Haueis 2018). Scholars within this literature have primarily focused on patchworks as a descriptive claim about concept development within science, and on the positive contributions of patchwork concepts to the projects researchers pursue.

We agree on the descriptive claim that polysemous patchwork concepts are a pervasive feature of scientific language. We suggest, however, that the normative status of concepts with multiple related meanings is a genuinely open issue. Why *should* patchwork concepts be developed during investigation? We suggest that although patchwork concepts allow the investigation of phenomena that are closely related, they do not determine the exact relationship between them. Thus, how any two meanings of a conceptual patchwork are properly related depends on the exact relationship between the phenomena they describe. The meanings may overlap if the phenomena they describe are identical. Or the meanings may diverge if the phenomena they describe are distinct. Or one meaning may be an accurate description of some phenomenon, while another is not.

So, developing a patchwork concept allows for investigation of closely related phenomena to proceed without proscribing the relationship between them. But there is a downside to this process – concepts often change "silently," with new connotations emerging in the course of investigation, and without those differences explicitly acknowledged. The appropriate normative attitude to patchworks involves a commitment to explicitly cashing out the distinct aspects of the patchwork, so that the relationships between the phenomena they describe can be investigated empirically.

We explore these issues by analyzing the concept of "hierarchy" in systems neuroscience. As we will outline, the idea that the brain is hierarchically organized has had a long and influential history in the field. Neuroscientists have just begun to recognize, however, that the concept comprises multiple distinct connotations that are often not distinguished (Hilgetag and Goulas 2020). We analyze (i) why the patchwork has developed, (ii) the different connotations it currently comprises, and (iii) the different possible relationships between connotations within the patchwork. Our analysis thus advances both the descriptive and normative aspects of the patchwork approach, and provides clarity on a conceptually difficult issue within the neurosciences.

Posits of hierarchical organization are practically ubiquitous in systems neuroscience, but we contend that the concept currently ranges over two broadly distinct approaches with different core commitments. The first, which we call the "representational hierarchy" view, is extremely influential in the field. The representational hierarchy view posits an anatomical hierarchy of feed forward, feed back, and lateral connections which underlies a sequence of input-output relations between brain areas. During this process, simple, unimodal sensory representations are subsequently elaborated into categorical, multimodal, and rule-based ones. The second, much newer view we call the "topological" approach, which is primarily based on the notion of *centrality*. A brain area is at a higher hierarchical level if it has more widespread influence on the network of brain areas. The topological approach primarily employs tools from graph theory and also focuses on an area's temporal contribution to evolving brain dynamics.

Although the two views are deeply intertwined in current systems neuroscience, we suggest that they have distinct central commitments.[1] The representational hierarchy view is committed to specific hypotheses about the representational roles of brain parts at distinct hierarchical levels. The topological view has no such commitments. Establishing the distinction between the views allows us to ask about the relationship between them. We consider three possibilities. First, the *substantiation* view suggests that the topological hierarchies provide a more detailed view of the anatomical underpinnings of representational hierarchies. Second, the *conflict* view states that the topological approach is a potential *replacement* for the representational view. Finally, there are several possible varieties of *pluralism*, which hold that the representational and topological approaches are mutually compatible depictions of distinct aspects of brain organization.

Our discussions will be internal to the neuroscience literature, but we hasten to add that frameworks in cognitive science and philosophy of mind often employ the representational hierarchy view. Consider debates about cognitive penetration and higher-level content, which implicitly presume that "lower-level" perception involves representation of simpler perceptual features. The question is whether perception can represent more abstract categories at a "higher" level of processing

[1]Hilgetag and Goulas (2020) distinguish four instead of two senses of hierarchy. Although a detailed comparison of both taxonomies is beyond the scope of this chapter, we think that their definitions of hierarchy as laminar projection patterns and as spatial gradients of structural features share the commitments of what we call the "representational" notion of hierarchy (Sect. 6.4.2, Fig. 6.3). Similarly, we think that their definitions of hierarchy in terms of topological projection sequences and as multilevel modular networks share the commitments of what we call the "topological approach" (Sect. 6.3., Fig. 6.4).

(Orlandi 2010), and whether this is due to "top-down" influence from brain parts that represent concepts (Vetter and Newen 2014). Or consider predictive coding models, which often cite hierarchical processing in the brain to argue that feedback connections deliver predictions based on higher-level generalizations to sensory areas (Bastos et al. 2012; Hohwy 2013). Each of these positions is broadly committed to the representational hierarchy view, and thus entails either the substantiation view or some variety of pluralism. Given that the conflict view is also possible, this cannot simply be assumed.

We proceed as follows. In Sect. 6.2, we introduce the representational hierarchy approach, and in Sect. 6.3 the topological approach. Section 6.4 then articulates the substantiation, conflict, and pluralist views. In Sect. 6.5, we consider studies of the rich club phenomenon within the topological approach as a test case for the different views of the relationship. Section 6.6 concludes.

6.2 The Representational Approach to Hierarchy

The traditional – and, by far, the most common – approach takes anatomical connections in the cortex to reveal a hierarchical organization in patterns of feedforward and feedback connections. The locus classicus of this approach is Felleman and Van Essen (1991). Drawing on histological data, they posited definitions of hierarchical "level" as depicted in Fig. 6.1.

The first row of Fig. 6.1 shows two connection patterns that, according to Felleman and Van Essen's framework, count as *ascending* or feed forward: either the connection begins in "supragranular" layers (layers 1–3 of cortex, left panel) and terminate in layer 4 (middle panel). Or it originates in both supra- and "infragranular" layers (5–6, right panel) and terminates in layer 4 (middle panel). The second row of Fig. 6.1 shows that lateral connections begin in both supra- and infragranular layers and terminate in all layers. The third row shows that descending/feedback connections also begin at supra- and infragranular layers but terminate in all but layer 4. This scheme can be used to classify different parts of the brain into hierarchical levels, based purely on anatomical connectivity. A given area A is at a higher hierarchical level than another area B if A receives only feedforward connections from B, and B receives only feedback connections from A. Two areas are on the same hierarchical level if (i) they share only lateral connections, or (ii) they have similar patterns of feed forward and feedback connections to already established levels. Based on this scheme, Felleman and Van Essen constructed a hierarchical description of the visual cortex comprising ten levels. The overall picture, as shown in Fig. 6.2, has been extraordinarily influential, and is often taken as an exemplar for describing organization in the brain (Bechtel 2008, ch. 3).

While their analysis was based on anatomy, Felleman and Van Essen did not shy away from applying a functional and representational interpretation of their framework:

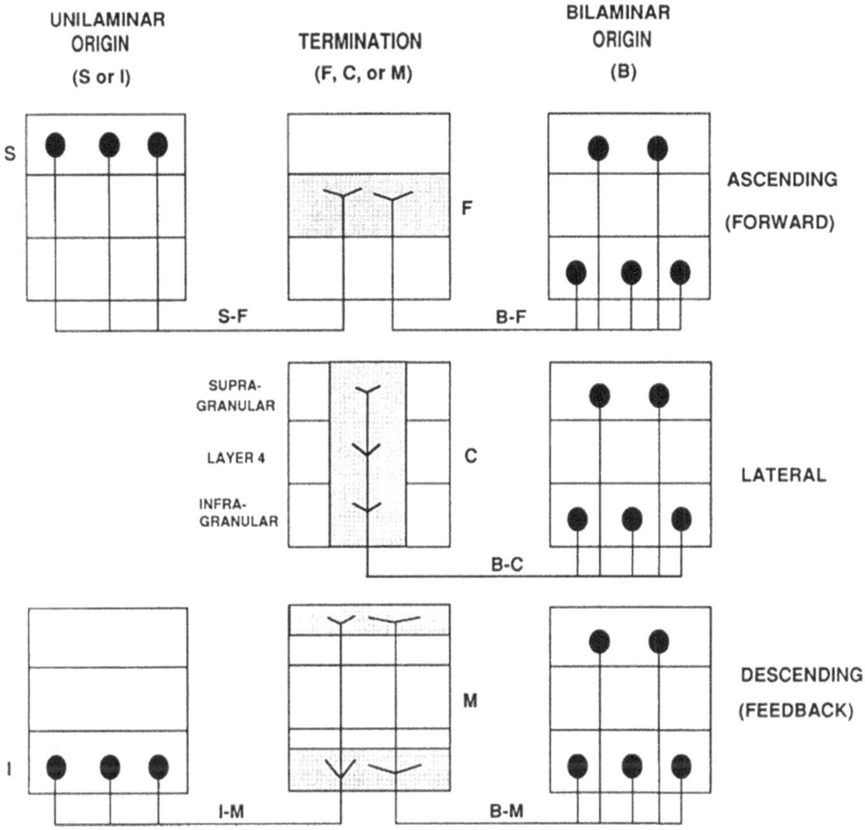

Fig. 6.1 Definitions of hierarchical relationships. From Felleman and Van Essen (1991)

> The physiological properties of any given cortical neuron will, in general, reflect many descending as well as ascending influences. Nevertheless, the cell may represent a well-defined hierarchical position in terms of the types of information it represents explicitly and the way in which that information is used. (Felleman and Van Essen 1991, p. 32).

On this view, the hierarchical position of a brain area connotes a functional and representational specificity: occupying a specific place in the hierarchy involves representing certain types of information and representing that information for further use elsewhere in the system. This approach is generally seen as a way of extending Hubel and Wiesel (1962), who showed how patterns of anatomical connectivity can combine to produce new functional representations. In Fig. 6.3, three "simple cells" (upper right) represent the orientation of an edge at a particular place in the visual field (small triangles and crosses on the left). The simple cells then forward these representations to a single "complex" cell (lower right). The complex cell will then represent the orientation *wherever* it occurs across the receptive fields of the simple cells (dotted rectangle, right).

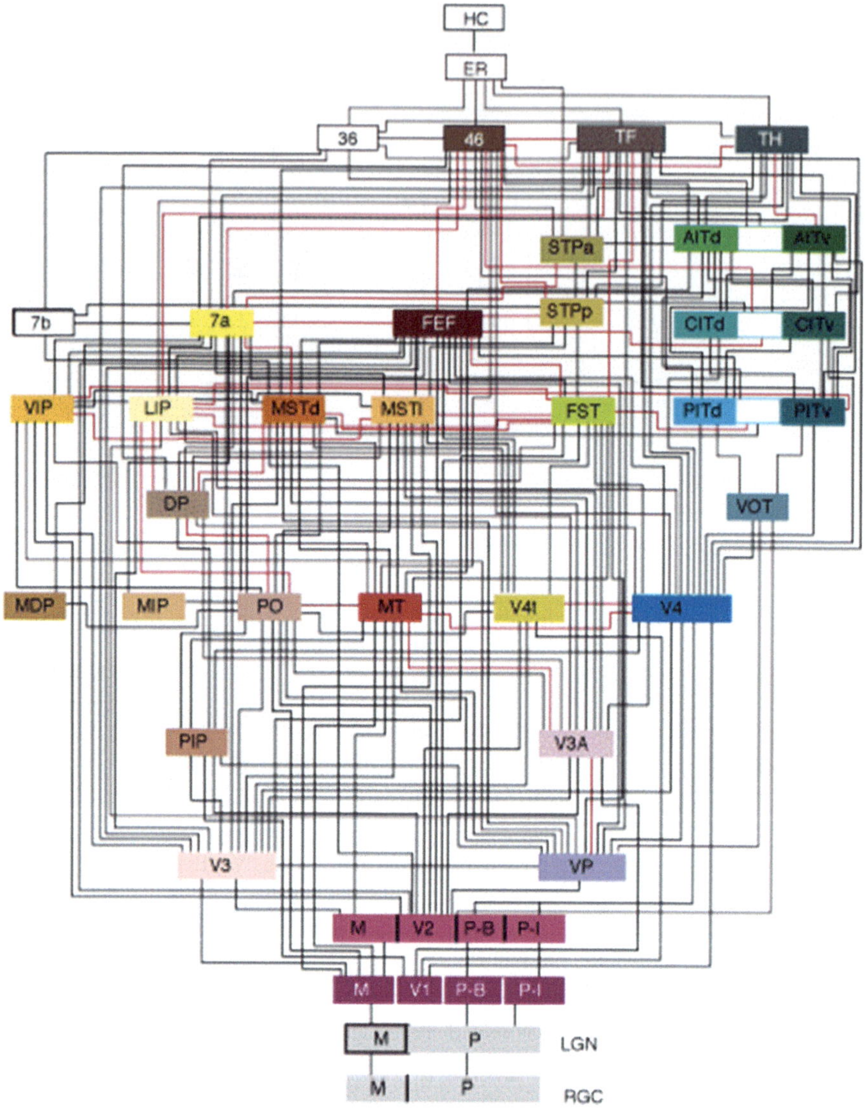

Fig. 6.2 The hierarchical wiring diagram of the macaque visual cortex. From Felleman and Van Essen (1991)

Figure 6.3 points to principles of processing within the hierarchy – *specific* information is passed along feedforward pathways, and then is represented *more abstractly* by higher levels in the hierarchy. The representational hierarchy view extends this logic to the rest of the visual system: lower levels of the hierarchy (including V1, V2, and V3) represent extremely simple features (such as orientation, wavelength, and displacement) at specific places in the visual field. At higher levels

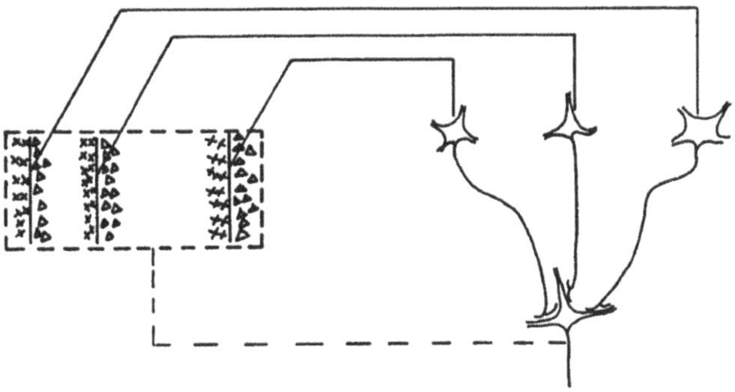

Fig. 6.3 The hierarchical logic explaining complex receptive field properties of V1 neurons in cat cortex. From Hubel and Wiesel (1962)

of the hierarchy more abstract information is represented. Within the dorsal stream for instance, MT represents general patterns of motion whereas V1 represents only local displacement. Within the ventral stream, a dedicated part of V4 represents categories of color whereas V1 represents only wavelength. A different part of V4 represents complex shapes rather than V1's representation of local orientation. Higher-level areas such as the inferotemporal cortex represent objects when they belong to a *category*, such as faces or hands, despite variation in their specific lower-level feature values (Gross et al. 1972). Due to its view of functional and representational organization, Burnston (2016a, b) has dubbed this view the "modular functional hierarchy" (MFH) picture of visual cortex organization.

Early on, it was noted that there were serious empirical shortcomings with Felleman and Van Essen's approach. In particular, many different possible attributions of hierarchical levels were compatible with the known data (Hilgetag et al. 1996). Still, the MFH view in general has had an astounding effect on the field of systems neuroscience and has extended well beyond the visual system. Here is a small set of examples.

First, the MFH view has intersected with computer vision to produce a picture of how categorical perception comes about. Influential approaches by Poggio (e.g., Riesenhuber and Poggio 1999) and Ullman (2007) have implemented feedforward networks that begin with representations of simple features and subsequently represent more abstract categories. Ullman's hierarchy is based explicitly on representing fragments of lesser complexity at lower levels, and then, on the basis of these, representing the category of the object at a subsequent stage of processing. These feedforward approaches, however, are increasingly being replaced by recurrent deep neural network architectures in computational approaches to visual object recognition.

Second, the MFH view has been used to analyze *other* sensory systems. The idea is that analogues to the simple features of the visual system can be found, and that these will be represented at lower levels of an anatomical hierarchy that works similarly to the one in the visual system. Such views have been proposed for both the olfactory and the auditory system (Savic et al. 2000; Wessinger et al. 2001).

Third, the MFH view is taken to describe motor systems. Interestingly, however, in these systems the primary direction of influence is taken to be the reverse of sensory systems. Abstract goal representations are encoded at the top of the hierarchy, localized to areas such as the premotor cortex and the inferior parietal lobule (Grafton and Hamilton 2007; for further discussion see Uithol et al. 2014), and these are subsequently expanded into a representation of the detailed object properties and motor kinematics needed to attain the outcome. Grafton and Hamilton (2007) explicitly analogize this to the kind of sequential representation in the visual system (cf. Haggard 2005).

Finally, a hierarchy of abstraction for action control is often posited to explain the organization of the dorsolateral prefrontal cortex. In a classic fMRI study, Koechlin et al. (2003) had subjects perform a series of successively more complex actions. In the simplest case, subjects had to perform a motor action in response to a visual cue. In the harder case, the stimulus-response associations shifted, depending on a second cue. In the hardest case, the overall pattern of associations between cues and sensorimotor associations changed depending on still another cue. The structure of this task is hierarchical, with sensorimotor associations nested under conditions, and conditions nested under episodes. More anterior areas of the dlPFC were activated with increasing hierarchical nesting of the needed cognitive control. Badre et al. (2010) take these and similar results to show that anterior areas are involved in the employment of abstract rules.

The representational hierarchy approach thus supports an overall view of brain function. On this picture, unimodal and motor cortices each embody a representational hierarchy. The outputs of perceptual systems are brought together in "association" cortices, including frontal and parietal areas (Mesulam 1998). Multimodal information is processed according to rules in executive control areas such as the dlPFC, and motor systems implement goals via specific representations of motor kinematics. Thus, the representational hierarchy view posits principles based on increasing abstraction for both unimodal and association cortices and for the overall functional architecture of the brain. In the next section, we discuss topological hierarchies, before moving on to discuss potential relationships between the two views.

6.3 Topological Approaches to Hierarchy

Topological approaches to hierarchy use the mathematical tools of graph theory to describe the brain as a network comprising nodes (e.g., brain areas or individual neurons) and edges (e.g., axonal connections, fiber pathways). Topological approaches are distinct from the representational hierarchy view because topological hierarchies quantitatively describe the potential influence of a given node on the system, rather than positing a specific type or degree of abstraction of the information it processes. This focus on potential influence makes topological approaches neutral to the representational functions of given parts of the brain. After introducing the graph theoretical concepts used to describe brains as hierarchical, we describe two ways to specify topological hierarchy and argue that neither of them is committed to representational functions. This neutrality prepares our argument that different relations between representational and topological approaches are possible (Sect. 6.4).

The representational neutrality we emphasize here intersects with, but is distinct from, several recent discussions in the philosophy of science literature, which attempt to address the relationships between network-based explanations and mechanistic explanations. Several authors have stressed the *distinctness* of these forms of explanation and debated the relationship between them. In particular, those who think that topological explanations are entirely distinct from mechanistic ones tend to stress their abstraction (Huneman 2010), or the fact that they describe *global* properties of systems, rather than local causal interactions (Kostić 2016; Rathkopf 2018).

We also rely on the abstractness of graph-theoretic explanation in articulating the difference between distinct conceptions of hierarchy. Graph-theoretical descriptions are neutral with respect to the representational role of different hierarchical levels because they are not committed to a particular way of functionally typing the causal interactions in the brain. However, we do not take this itself to show that topological explanation is always global, or is in conflict with mechanistic explanation in general. This is compatible with the *explanation* of particular phenomena invoking local causal interactions as well as global organizational properties. We take no particular stand on the issue here (but see Burnston 2019).

The graph-theoretical notion of hierarchy is based on the concept of *centrality* (see van den Heuvel and Sporns 2013 for review and further references). A node is at a higher hierarchical level if it is more central to the overall connectivity of the network, and at a lower level if it is more peripheral. Centrality can be analyzed in different ways. One notion is simply *degree* – a node with a large number of connections (measured as percentage of actual out of possible connections) will have a large influence in the network. An example of a degree measurement is given in the left panel of Fig. 6.4 below. A second notion of centrality is *betweenness centrality*, i.e. how many shortest paths between any two nodes pass through the node of interest. Nodes with high betweenness centrality are crucial for mediating interactions across the entire network. Finally, the *clustering coefficient* of a node measures the degree to which the node's connections are themselves connected. It is measured as the proportion of actual out of possible edges between nodes that are connected to the node of interest.

Centrality measures can be used to describe the overall properties of the network as well as particular nodes, particularly in how network organization is distributed amongst *modules* and *hubs*. Nodes in a module are more connected amongst each other than to nodes outside the module (Sporns and Betzel 2016). Consequently, these within-module nodes will influence each other more directly than other nodes. Hubs are nodes (or groups of nodes) which score high on one or multiple

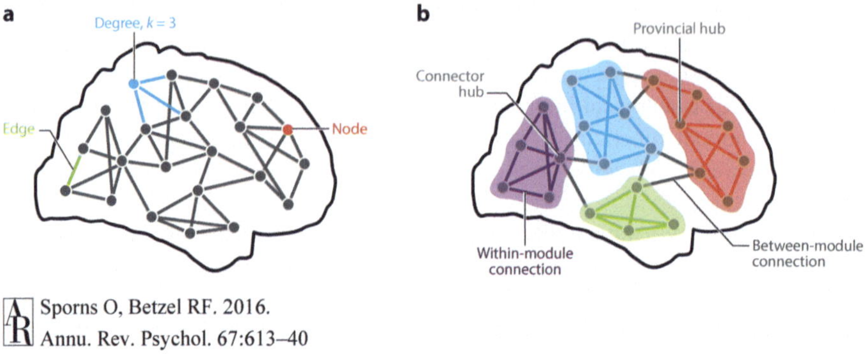

Fig. 6.4 Hierarchical measurements in the topological approach (from Sporns and Betzel 2016). Part (**a**) conveys basic network concepts, and part (**b**) a stylized module- and hub-based architecture

centrality measures, which are usually correlated (van den Heuvel and Sporns 2013). A hub with a high clustering coefficient is likely to connect several modules, and thus provide information transfer across otherwise segregated subsystems ("connector hubs"; Fig. 6.4 above). A node can also serve as a hub primarily within, rather than between modules, by mostly connecting to other nodes in the same module ("provincial hubs"; Fig. 6.4 above). The extent to which networks exhibit modularity and contain hubs gives a helpful characterization of their overall capacity to process information. When a network contains primarily modules with a smaller number of hubs, it can maximize both localized information processing through within-module connections, and information integration across the network through hub-mediated connections (Sporns 2011).

From these definitions one can already see why "influence on the network" is the primary notion for any topological approach to neural hierarchies.[2] If nodes are defined as brain areas, then activity in a highly central area will influence activity in many other areas, and thus shape the global behavior of the network. A topological hierarchy description of the brain is generated by applying the aforementioned centrality measures to anatomical or functional connectivity data. Some of these datasets include the kind of histological data cited in the discussion of Felleman and Van Essen (e.g., the CoCoMac database), but have been updated to include more complete data about neural connections. Functional connectivity is, basically, a measure of the statistical correlation in activity between brain areas over time (it

[2]While we focus on the influence notion of hierarchy, other network investigations employ a more compositional notion of hierarchy as well. For instance, researchers also talk of "hierarchy" if network structure is self-similar, e.g. when smaller modules are nested within larger modules (Hilgetag and Goulas 2020). While it may be interesting to analyze how such "encapsulation hierarchies" relate to compositional hierarchies in the mechanistic literature (Craver 2007, ch. 5), in the following we assume that systems neuroscientists studying encapsulation hierarchies are usually interested in its implications for neural signaling, i.e. on how influential a brain part is within the network (Müller-Linow et al. 2008; Sporns and Betzel 2016).

can be measured in different ways, and we won't go into the details here; see Haueis 2012 for discussion). Here we give some specific examples where researchers have employed the topological approach to hierarchy to make sense of brain organization.

An early example of the topological approach, as applied to anatomical connectivity, is from da Costa and Sporns (2005), who used degree and clustering coefficient to study the hierarchical organization of the macaque visual system. Their analysis was based on how closely a starting brain area (a "reference node") was connected to the rest of the system. They thus analyzed each area in terms of *degree* distance. From a given reference node, for instance, they asked how many other nodes it connected to with only one synaptic connection, how many at two synapses distant, etc. They defined "levels" as degree measures at distinct synaptic distances, and showed that six areas in the visual system, predominantly in the dorsal stream, connect to more than half of the rest of the visual system at the first hierarchical level. These areas thus have the most direct influence on many other areas of the visual network. Ventral stream areas predominantly connect to other nodes at the second and third hierarchical level, which means that their influence is less central. An exception was area V4, which is in the ventral stream, but had similarly high degree measures at a degree distance of one. (We will discuss their analysis of clustering coefficients in Sect. 6.4.)

Centrality-based analyses of structural connectivity have also been used to study the entire brain. For instance, Zamora-Lopéz et al. (2010) used degree and betweenness centrality to determine the distribution of hubs in the cat cortex. Their analysis revealed that most nodes with high betweenness centrality lie in frontal and limbic cortex, and only few in sensory cortices. In addition to purely structural connectivity in cats and primates, centrality measures have been applied to functional connectivity in humans. Meunier et al. (2009) used degree and modularity measures to describe functional connectivity data recorded with fMRI during the experimental resting state. They showed that only 5% of the nodes qualify as hubs that connect several modules, suggesting that these areas of the brain are particularly central. In particular, they showed that the areas of the "default-mode" network (DMN), which have been shown to be highly active during rest, are themselves both highly interconnected (thus forming a module) *and* highly connected to the rest of the brain (thus forming a hub). We discuss the DMN more thoroughly in subsequent sections.[3]

Both anatomical and functional connectivity measures are importantly *static* – they describe the state of the brain as a constant within a period of time (e.g., during rest). But network measures can also be used to describe dynamics. In the temporal

[3]Note that there are methodological issues with identifying functional hubs based on degree alone. In Pearson correlation networks, degree is partially driven by the size and not only the amount of influence a subnetwork has. Thus, nodes in larger brain areas tend to be identified as hubs in because they are part of large physical entities (Power et al. 2013). Yet some areas consistently come out as hubs in functional connectivity studies using different measures, such as anterior and posterior cingulate gyrus of the DMN (van den Heuvel and Sporns 2013).

domain, topological hierarchies posit that nodes with activity at shorter timescales have less influence on the network than nodes with activity at longer timescales. There are two ways in which this has been measured, one comparing temporal activity between areas in response to a given event, and another focusing on the oscillatory properties of brain areas.

In a measure of the first type, Deco and Kringelbach (2017) determined the *integration value* of a node's activity in response to an event – for instance the presentation of a stimulus. A node's integration value is given by the number of other nodes to which it is functionally connected after the event. The higher the integration value, the higher is its influence on the network during the time period in question. This can be extended to changes in overall functional states of the brain, such as the change from wakefulness to sleep, or the induction of a coma.

Deco and Kringelbach's computational modeling of the distribution of integration values suggests that the brain is organized into a graded, non-uniform hierarchy. There exists a continuum between nodes with a small and local influence and nodes with a large and global influence on the network. Only few nodes are situated at the top of this hierarchy, because they have large integration values and respond flexibly to neural events. Although Deco and Kringelbach do not report where these nodes are located in the brain, their modeling results mirror other functional connectivity studies which report a graded hierarchy (Margulies et al. 2016), with few hub nodes at the top (Meunier et al. 2009).

The second way of applying the topological approach to the temporal domain involves *oscillatory* hierarchies (Lakatos et al. 2005). Background activity within a brain area, often known as a *local field potential*, oscillates at characteristic frequencies. It is a widespread finding that lower-frequency oscillations constrain or modulate activity at higher frequencies and spiking behavior, either via phase coupling or phase-amplitude coupling (Canolty and Knight 2010). Moreover, *synchrony* in oscillatory phase between distinct brain areas, especially at lower frequencies, is often posited to be a key principle underlying neuronal communication and functional cooperation, and these principles have been posited to underlie recruitment of task-specific networks (Canolty et al. 2010). Intriguingly, different oscillatory frequencies have different distributions in the brain, and low-frequency oscillations are highly exhibited in hubs which overlap with the DMN (De Domenico et al. 2016). Thus, oscillatory hierarchies are one way in which network centrality can integrate information across the brain (cf. Burnston 2019).

The above examples show that researchers using a topological approach understand hierarchical position as the amount of influence a node has on the network, either by anatomically connecting many other nodes in space (centrality) or by functionally connecting them in time (integration value or phase synchrony). This focus on network influence makes topological approaches *neutral* with regard to the representational architecture of the brain. Although many studies we describe in this section do interpret their results functionally, the assumptions from which these interpretations are derived are not part of the graph-theoretic measures themselves (see Sect. 6.4.2 below). A graph-theoretic description of a node simply characterizes and quantifies its relationships to other nodes. It does not determine what information is exchanged via these connections or how.

Some researchers make this neutrality explicit: "our goal was not to identify unique hierarchical arrangements of brain regions, in terms of representational stages of streams, an approach taken in earlier work" (da Costa and Sporns 2005, p. 573; "earlier work" refers to studies following the representational approach). Instead of determining which perceptual features are represented at each level of the visual representational hierarchy, da Costa and Sporns analyzed how each node spreads its outgoing connections throughout the network hierarchy, defined in terms of degree distance. Similarly, topological methods can detect modules in a "purely data-driven way" (Sporns and Betzel 2016, p. 19.3), without using prior knowledge about the representational function of brain systems to detect modular community boundaries. Because they are neutral about representational function, topological approaches are also not committed to the claim that more abstract representations are processed at higher "levels" of the hierarchy. A high-degree node can be central regardless of whether it spreads modality-specific or multimodal information throughout the network. Hubs can be detected by their centrality measurements without assigning degrees of abstraction to what they may represent.

Dynamic measurements of topological hierarchy are similarly neutral about representational architecture. For example: intrinsic ignition capability is defined by a node's integration value, i.e. the degree of broadcasting information in the network, not the type of information a node represents (Deco and Kringelbach 2017). In sum, novel topological approaches to hierarchy focus on the influence and the spatiotemporal propagation structure of signals and are neutral with regard to the representational function at different levels of neural hierarchy. This very neutrality is what allows for the variety of possible relationships one might posit between the representational and topological hierarchy. We move to discuss those relationships in the next section.

6.4 The Relationship Between the Representational and Topological Views

6.4.1 Stage Setting

Neuroscientists using graph-theory are often unclear about the precise relationship between representational and topological approaches. Sporns (2011) sometimes seems to suggest that both approaches can be combined. He claims that network structure in the brain reveals that neural function is both "integrated" and "segregated". Segregation involves the separation of the network into distinct functional units, and integration involves the exchange of information between those units. However, Sporns also writes that the topological hierarchy presents a challenge to the representational view: "Even cursory examination of structural brain connectivity reveals that the basic plan is incompatible with a model based on predominantly feedforward processing within a uniquely specified serial hierarchy" (Sporns 2011, p. 150). How should we interpret these opposing tendencies?

We suggest construing the situation as follows. The concept of hierarchy is currently a patchwork, consisting of two approaches to hierarchical relations between brain parts. The representational approach provides researchers with particular explanatory schemas, which interpret hierarchical levels based on how abstract the representations they process are, and the input-output relations between them. The topological approach provides researchers with graph theoretical concepts like topological centrality or temporal integration to infer hierarchical levels based on a node's influence on the network. It is, however, currently an open question how these different connotations of "hierarchy" are *related* to one another.

In the following we discuss three possible relationships. On the one hand the fact that network models could explain how functions can be differentiated and how information can flow between them might suggest that "presumed aspects of the sequential organization of brain networks can be confirmed and clarified through formal topological analysis" (Hilgetag and Goulas 2020, 5). We call this the *substantiation view*. On the other hand, the high degree of interactivity in networks suggests that clear hierarchical orderings in the processing of information may not be feasible. If this is the case, then network models may offer up *alternative* organizing principles for the brain, based around the topological notion of hierarchy, which will displace the more traditional representational view. We call this the *conflict view* of the relationship. Finally, a *pluralist view* would take both motivations into account and state that there are multiple distinct hierarchical organizations instantiated in the brain. Some situations may involve modeling it as a representational hierarchy, and some a topological one, where these neither conflict nor entirely overlap.

In what follows, we discuss the commitments of each view of the relationship, and the evidential standing of those commitments. Importantly, we note examples of individual scientists who adopt, without conceptual argument, one kind of view or another. This shows that scientists themselves are being guided by particular semantic intuitions about the notion of hierarchy. The analysis thus exposes both the current state of the concept of hierarchy and articulates the argumentative burden of different approaches to its patchwork structure.

6.4.2 The Substantiation View

The substantiation view holds that the representational and topological approaches, despite using different methods, measure the *same* hierarchical organization in the brain, although the latter perhaps with a more detailed understanding of connectivity. The perspectives, after all, draw from overlapping datasets. The CocoMac database, for instance, is a database of anatomical connections based on histological data. It is frequently used for analyses within the topological approach, but includes the data that Felleman and Van Essen used to model the representational hierarchy. Two further motivations for the substantiation view are (i) that the modularity of networks can be interpreted as underlying distinct functions of the type posited in the classical hierarchy, and (ii) that the topological divisions revealed through network analysis often match functional divisions posited by the representational approach. We will discuss these briefly in turn.

First, point (i). Recall that, on the representational view, each neural system (visual, motor, frontal, etc.) exhibits significant functional autonomy from other systems. Further, within each system, the distinct areas play different functional roles in performing the system's overall function. One possible way of reading the modular architecture of topological hierarchies is as implementing functionally specified subsystems, whose integration then proceeds in, at least roughly, the way described by the representational view. Modules, recall, are characterized as parts of the network with primarily intra-module connections, thus supporting the notion that they are computational units dedicated to specific kinds of problems. Indeed, Meunier et al. (2010) suggest that a hierarchy of modules allows for each module to "specialize in sub-problems." Breakspear and Stam (2005) argue that lower levels of the topological hierarchy "represent specific features." (To be fair, both papers note that integrating information from distinct modules may be a global process.) The conceptual possibility of topological modules underlying the specific functions and interactions posited in the representational view is alluring to those friendly to the representational approach.

The support for point (ii) is empirical. It turns out that, in fact, many divisions made within the topological approach correspond to divisions made within the representational approach. This is especially true for large-scale divisions (but see Zerilli 2017). For instance, modularity analyses at the level of the whole brain reveal that visual cortex is more tightly interconnected than it is connected to other large-scale networks. In cats and macaques visual cortex is much more tightly interconnected than it is connected to somatosensory cortex, and vice versa (Sporns et al. 2007). This is true for both structural and functional connections (Honey et al. 2007). Even within these parts of the cortex, functional divisions can be made that match the representational view – for instance, structural connectivity in humans shows a distinction between the dorsal and ventral streams of the visual cortex (Hagmann et al. 2008), which are standardly taken to perform very different functions in vision (Mishkin et al. 1983).

Moreover, areas of cortex that have traditionally been called "association areas," including areas in the parietal and prefrontal cortices, standardly come out as hubs in graph-theoretic network analyses (Sporns et al. 2007; van den Heuvel and Sporns 2013). If their role is to associate (and perhaps abstract from) multiple kinds of information from unimodal cortices, then one would expect them to have a wide range of connections to those areas. Sporns (2011) himself cites approvingly the unimodal-to-association area progression posited by Mesulam and others (cf. Meyer and Damasio 2009). Passingham et al. (2002), in an influential analysis, proposed that areas such as premotor and frontal cortices will differ in the amount of different

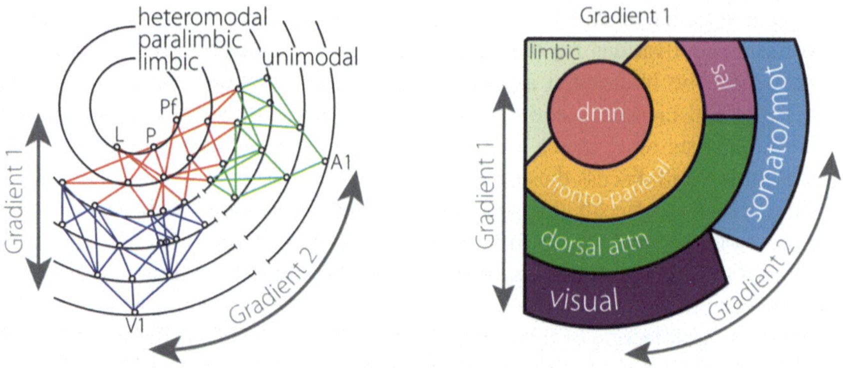

Fig. 6.5 The Mesulam model (left) and the Margulies model (right) of the cortical abstraction hierarchy. Adapted from Margulies et al. (2016)

information they will respond to from sensory cortices, and that these differences are due to differences in the patterns of connections exhibited by different areas.

Researchers using resting state functional connectivity studies have also embraced the substantiation view. Margulies et al. (2016) used diffusion map embedding, a variety of dimensionality reduction technique, on human resting state functional connectivity data. This technique involved constructing dimensions along which connected areas could be grouped, with closely connected areas close together along each dimension. The sum of all dimensions forms a so-called *embedding space*, which positions nodes according to the similarity of their functional connectivity profiles. In Fig. 6.5. Margulies et al. use two of these dimensions to describe the greatest and second greatest amount of variance in functional connectivity between areas, which they call the first and second *gradient* of connectivity.

Figure 6.5 shows that Margulies et al. interpret the two gradients of functional connectivity as revealing a *hierarchical gradient of abstraction* which runs from primary sensory areas to regions of the default mode network (DMN). According to this interpretation, default mode regions are involved in cognitive functions such semantic memory or reward-guided decision making because default mode activity processes abstract informational content, largely independent of transient environmental stimuli processed by sensory systems.

This interpretation substantiates Mesulam's representational hierarchy model (see Sect. 6.2) because it situates the DMN at the top of a known representational hierarchy that proceeds from unimodal sensory to transmodal association areas. Note, however, that this substantiation interpretation is not necessary to apply the diffusion map embedding algorithm to resting state fMRI data. This procedure places nodes closer in embedding space if they are more strongly functionally connected, or as we put it, if they influence each other more strongly than other nodes. Additional assumptions about functional connectivity directly reflecting information representation (Schölvinck et al. 2013) and topographical structure constraining cognitive processes are required (Margulies et al. 2016). To arrive at the substantiation view, these assumptions need to be combined with the supposition that the representational approach is a correct approximation of the brain's hierarchical organization.

6.4.3 *The Conflict View*

There are two primary motivations for the conflict view: (i) graph-theoretical results that conflict with the representational hierarchy; and (ii) independent evidence that speaks against the representational but not the topological approach. We take these motivations in turn.

There are individual cases in which the consistency between topological and representational approaches to hierarchy breaks down. Let us consider one case – V4 – in detail. V4 is, according to the representational approach, a "mid-level" visual area (level 5 of Felleman and Van Essen's hierarchy), which comprises two sub-areas in charge of representing color and complex shape. This clear place in the representational hierarchy is questioned by graph theoretic analyses of anatomical connectivity, which reveal that V4 scores extremely highly in measures of degree and centrality. This is shown in Fig. 6.6 below.

Figure 6.6 shows that V4 scores very highly, relative to the whole-brain network, on degree and betweenness centrality. It also ranks high on closeness centrality, which is a related measure of the average path length between the node and all other nodes in the network (shown in the inverse here for comparative ranking). V4 *also* has connections to other high centrality nodes, such as area 46 in the frontal cortex. Similarly, nodes that are directly connected with V4 (da Costa and Sporns' hierarchical level 1), have a low clustering coefficient, but nodes that are connected to *those* nodes (da Costa and Sporns' hierarchical level 2) have a very high clustering coefficient. This suggests that V4 connects, with a small number of synaptic steps, to multiple modular areas (da Costa and Sporns 2005). For areas in the dorsal stream such as MT and MST, by contrast, clustering is greater at nodes only one edge away. The way to interpret this is that most connections for dorsal stream areas are intra-modular, whereas connections for V4 are widely spread across modules. Thus, V4 is potentially a more integrative area than areas that are traditionally posited to be at the same or higher levels of the representational hierarchy.

This result suggests that, in terms of topological centrality, V4 is at the *highest* levels of the *overall* brain hierarchy, in extreme contradistinction to the low level posited for it in the representational hierarchy. Hence, there is a direct conflict between the results within the two different perspectives. Does the centrality of V4 make a functional difference? As Sporns notes, hubs are well-situated to play multiple diverse functional roles, and this is in fact borne out by the data – V4 has a much more complex functional profile than the representational hierarchy posits (Burnston 2016b; Roe et al. 2012), and lesions to V4 cause a diverse range of effects

(Schiller 1993). This puts pressure on the representational view in two ways. First, V4 may not have a well-defined place in a representational hierarchy, such that it sends a specific signal onwards to subsequent areas of the hierarchy. Second, it pressures the idea that sensory representation occurs *first,* prior to the integration of multimodal information by association areas.

The second motivation for the conflict view is independent anatomical and physiological data that conflict with the functional posits of the representational hierarchy. We can only summarize this data here, but it will suffice to get the picture across. First, both direct and subcortically mediated connections exist between primary sensory cortices in different modalities, and these are posited to underlie a variety of cross-modal effects (Driver and Spence 2000; Ghazanfar and Schroeder 2006). Second, the representational approach suggests a preferred pathway for signals in sensory cortices, such that information is represented first at lower levels, then only subsequently at higher levels (Lamme and Roelfsema 2000). However, both anatomical and time course data question the existence of such a pathway. Parts of V4 have both bidirectional and direct connections to higher visual areas which bypass the putative central ventral pathway, "violating a strict serial hierarchy at even the earliest stages of visual processing" (Kravitz et al. 2013). Temporal data show V4 in fact is *slower* to represent information than areas traditionally seen as "above" it in the hierarchy such as MST and the FEF, whereas MT is roughly tied

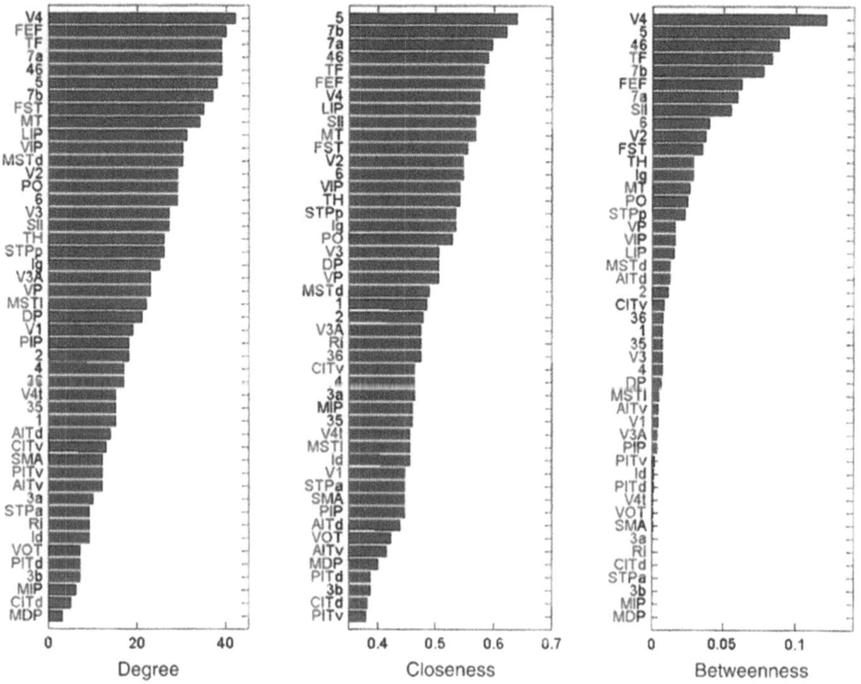

Fig. 6.6 Centrality measurements of V4. From Sporns (2011)

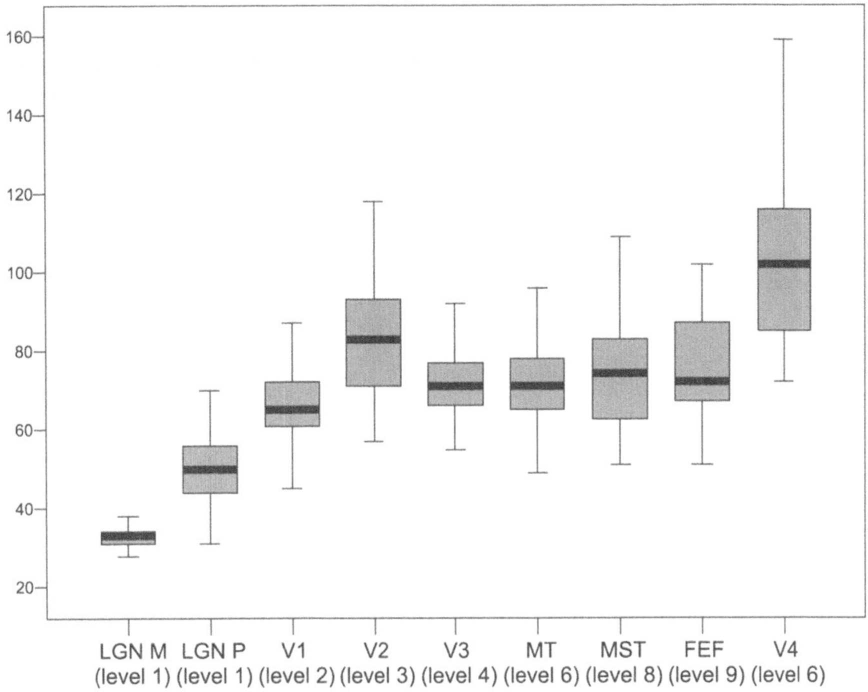

Fig. 6.7 Time-from-stimulus onset measurements for physiological activation of visual cortical areas. From Capalbo et al. (2008). "Level" refers to hierarchical level, in the sense of Felleman and Van Essen (1991), except Capalbo et al. begin counting from the LGN, rather than V1. Hence, e.g., MT and V4 are labelled as "level 6" here, but they are level 5 in Felleman and Van Essen

with these areas in terms of response latency. This result is summarized in Fig. 6.7 below.

Third, physiological results question the idea that increasingly abstract representation occurs at higher levels. Hegdé and Van Essen (2007) measured physiological responses in V1, V2, and V4 to a wide range of shapes. Examples are shown in Fig. 6.8 below.

According to the representational hierarchy, more complex shapes should be represented in higher areas of the hierarchy – in this example, simple sinusoidal gratings should be represented at V1 and V2, while increasingly complex hyperbolic and polar/radial shapes should be represented at V4. But this is not what Hegde and Van Essen found. Instead, they showed that different populations of cells in each area had greater responses to shapes *across the categories*, without one type of shape being privileged at any area. Strikingly, the authors – including Van Essen, one of the key progenitors of the representational hierarchy view – argue that their data undermines any strict division between what is represented at distinct representational stages in the visual cortex.

These results generalize both to relationships "higher up" in the purported processing hierarchy, and to the motor domain. For instance, Meyers et al. (2008)

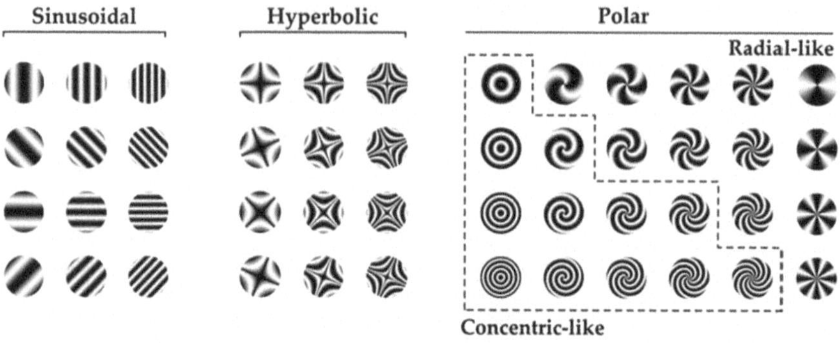

Fig. 6.8 Shapes of increasing complexity. From Hegdé and Van Essen (2007)

compared how much category information about a stimulus is extractable from populations in the inferotemporal and prefrontal cortices. There was no difference in the degree of abstraction of information that can be discerned from these populations (using decoding methods). What differed is what information co-existed with abstract category information in each population. IT tended to retain more visual detail, whereas PFC tended to combine stimulus category information with task variables. Similarly, Murray et al. (2017) modeled the circuit between prefrontal and posterior parietal cortex involved in working memory. They showed that the difference between the PFC and PPC is not how abstractly they represent information, but instead in terms of whether they *also* represent distractors – PPC does whereas PFC does not. These results suggest that areas at different levels of the traditional hierarchy are not distinguished by how abstractly they represent information, but in terms of how they represent different combinations of information that are relevant for a task.

As a final example, consider the notion of rule-representation in the frontal cortex. Rules are often construed as related to either conditional stimulus-response associations or as generalizations of those associations. So, in a same-different task, one might have neurons that respond both to the stimulus and to its repetition, or one might have cells that signal when the task is a same-different task, *regardless* of the stimulus. The latter is generally construed as more abstract, but cells with significant responses for rules do not distribute hierarchically in the cortex. In fact, rule selectivity has been shown to occur *more strongly* and *earlier* in a task in the premotor cortex than the prefrontal cortex (Wallis and Miller 2003). Moreover, there are not individual areas that represent rules at the expense of stimuli – in fact, the much more common finding is that cells even in areas traditionally construed as higher-level exhibit *mixed selectivity*, with responses mediated by combinations of stimulus, response, and rule (for review, see Rigotti et al. 2013).

In general, the results here, along with the results in visual cortex discussed above, do not support the notion of a clear hierarchy of representational abstraction either in visual or "association areas" – instead, what differentiates areas is how they combine different information in different ways.

One worry about the conflict view is that because topological approaches are neutral with regard to representational architecture, there is no inherent reason to align them with independent evidence against the plausibility of the representational view. The fact that topological approaches are compatible with that evidence does not entail that they positively support it. Our reply is that at least in some cases, graph-theoretical analyses *do* support evidence against the representational view, despite their neutrality towards representational function. Consider, for instance Goulas et al. (2014), who tested predictions about anatomical connectivity entailed by the anterior-posterior gradient of abstraction in the prefrontal cortex. They reasoned that, if more anterior areas of the prefrontal cortex were in charge of more abstract control functions, then they should send more efferent connections to areas lower in the purported hierarchy than they receive. Goulas et al. (2014) could not confirm this prediction of the abstraction gradient model, however. More posterior prefrontal regions, Brodmann areas 45 and 46, consistently sent more efferent connections than the most anterior region, area 10. Therefore, the anatomical connectivity of these regions conflicts with the anterior-posterior model.

We have shown that, despite the consistencies between the representational and topological approaches, there is also data that the topological approach can accommodate, that the representational one cannot, or at least not easily. Hence, the two views are empirically distinguishable. If one finds the data reviewed in this section compelling, one is likely to adopt the conflict view and suggest displacement of the representational approach by the topological one.

6.4.4 The Pluralist View

Both the substantiation and the conflict view seek to resolve the patchwork structure in favor of a univocal meaning of the concept of "hierarchy", referring to a distinctive organizational property. Substantiation implies that distinct hierarchical levels must always correspond to degrees of representational abstraction, and are individuated in terms of representational function. Conflict implies that hierarchical distinctions are always specified in terms of amount of influence, and are individuated with no representational commitments.

One might reasonably suspect, however, that any attempt to build a universal conceptual structure of "hierarchy" is mistaken, given the piecemeal data upon which the substantiation and conflict views are founded. Instead, one could propose a *pluralist* view about the relation between representational and topological approaches: they represent multiple, equally legitimate meanings of "hierarchy" in neuroscience which overlap in some domains and diverge in others. Pluralism suggests that both representational hierarchy and topological approaches, while having distinct constitutive commitments, are explanatorily important for understanding neural organization. Pluralists hold that the extant patchwork structure of scientific concepts is epistemically useful and – to a certain extent – reflects the structure of the underlying phenomena (Wilson 2006; Bursten 2016; Novick 2018; Haueis

2018). Below we highlight three pluralist options and discuss their advantages and drawbacks.

The first option is that there are different *processes* in the brain which will be best explained by the representational and topological approaches. On this view, there is a large amount that is correct to the representational approach – the basically serial and abstractive nature of processing, for instance – but this process breaks down at some point and gives way to a different form of organization that relies more on global interactivity. This form of pluralism is suggested by some of the comments from theorists discussed in Sect. 6.4.1. The basic problem with this form of pluralism is that it does little to answer any of the data that speaks against the representational hierarchy, since it basically accepts the traditional picture and views the topological hierarchy as a kind of integrative add-on.

The second form of pluralism is a *modelling*-based pluralism, which treats the representational and topological approaches as ways of *representing* the brain. On this view, both the representational and topological approaches can be seen as strategies for understanding neural organization, where the reason for adopting one over another depends on the explanandum. Network representations can be used to think about, for instance, efficiency of communication given constraints such as minimizing wiring length (Meunier et al. 2010; van den Heuvel and Sporns 2011). This might be contrasted with the representational hierarchy, which is meant to explain how signals are in fact processed in the brain. While this view has some advantages, and connects up with larger debates about the role of different forms of models in explanation in biology (Green et al. 2017), an explanation will have to be given about the situations in which these models conflict, such as in the case of V4 discussed above.

The other way to accommodate conflicting data is *organizational pluralism*, which suggests that the brain can in fact instantiate many different *forms* of organization, and that the representational hierarchy is one but not the only one. For instance, in many studies that inspire the representational approach, animals are studied in very limited behavioral circumstances, having to make specific perceptual judgments on the basis of presented stimuli (in the perceptual case), or having a well-defined task set that they must learn (in the prefrontal case). Perhaps, however, perception *in the context of action* requires more dynamic interaction with wider brain networks, or action in the case of *deliberation* requires broader access to, e.g., motivational and evaluative influences. On this view, there is a simple hierarchical organization for simple behavioral contexts, but this organization might be replaced by more complicated forms of signal processing, which might also be mediated by the topological hierarchy (cf. Silberstein and Chemero 2013).

We think the last view is in many ways the most promising, although not without limitations. One advantage of organizational pluralism is that it comports with a wide range of data suggesting that the network organization of the brain is not *constant* (Honey et al. 2007). When analyzing functional connectivity, different nodes attain different degrees of centrality in different contexts, and different networks are enlisted that are relevant to the task (Burnston 2019; Stanley et al. 2019). Organizational pluralism accounts for this possibility while making room for the traditional representational picture as one kind of organization that the network can adopt. Another advantage is that organizational pluralism can in principle account for *both* the data in favor of, and the data against, the representational hierarchy view. If the organization of the brain changes dynamically, then in some cases it might instantiate a representational hierarchy, while in some cases it may not – hence the traditional data in favor of, as well as the newer data against, the representational view.

The main worry about this last view is that it may be too permissive. For instance, the latency data from Capalbo et al., as well as the physiological data from Hegde and Van Essen, seem to cause problems for the representational view *even in the kind of contexts* for which it was originally proposed. Organizational pluralists must be able to account for data in the same contexts via proposed changes in organization.

In sum, we suggest that the substantiation, conflict and pluralist views are all both independently motivated (to some degree) and at work in the current literature. Given that they are all *distinct* views, however, they need to be articulated, and their commitments understood, in order for conceptual progress to be made. We have offered a preliminary version of such a framework above. In the next section, we showcase the utility of this framework by applying it to recent research on brain dynamics and rich club topology.

6.5 A Test Case: Rich Club Organization

As research into brain networks has progressed, attention has turned heavily towards brain dynamics, and how they are shaped by network features, including hierarchical centrality. Earlier, we discussed how the hub-and-module organization of the brain is often seen as a way of implementing the balance between segregation and integration of function. A dynamical corollary to this view is that highly central nodes allow for a balance of diffusion and efficiency – diffusion means that information can be broadcast widely in the network, while efficiency means that it can be routed to where it is needed (Avena-Koenigsberger et al. 2017). Whole-brain dynamics shift between rest and task, and between tasks (Shine and Poldrack 2017), and are mediated by widespread oscillatory synchronization (Deco and Kringelbach 2016).

In this section, we briefly discuss the role of the "rich-club" architecture in the brain for mediating dynamics. A network contains a rich club if its highest-degree nodes are also highly connected to each other. A rich club measurement begins with a degree threshold, k, and then asks what proportion of possible connections between nodes with degree > k obtains in the network. Rich club architectures occur in many networks, including the human brain. Simulations have shown that brain networks with a rich-club architecture have a greater range of dynamic attractors than networks without one (Senden et al. 2014).

Rich-club architecture provides an interesting test case for the different positions relating representational and topological hierarchies. First, rich club areas are at the highest levels of network-based centrality, judged by degree and influence on cortical dynamics. It also operates at particularly slow oscillatory frequencies (Senden et al. 2017a), placing it at a higher level of the oscillatory hierarchy. Finally, the rich club also significantly overlaps with the DMN which, as we saw above, is posited within the substantiation view to be a transmodal network with the highest degree of abstraction in the brain (Margulies et al. 2016). So, the rich club is a particularly rich network structure with which to analyze concepts of hierarchy.

The representational hierarchy view posits that the rich club should be involved in processing abstract representations. However, an alternative hypothesis has emerged from within the network literature. Senden et al. (2017b) studied functional connectivity between the rich club and other brain areas as subjects switched between rest and four different kinds of tasks, including working memory (n-back), response inhibition, mental rotation, and verbal reasoning. They showed, intriguingly, that during rest the rich club network had greater in-degree, meaning it received more input from other brain areas, but that this switched during tasks, with the rich club providing more output than receiving input. Moreover, rich club outputs targeted a similar set of brain areas across tasks, but the network relationship between these target areas, as well as which brain areas *they* interacted with, changed depending on the task.

The hypothesis constructed by Senden et al. is that the rich club serves as a *gate* that *mediates competition* between networks elsewhere in the brain that control the specific tasks. Note that, as befitting the different core commitments of the topological view, the gating hypothesis contains no commitments about whether the rich club does this by conveying abstract representations about task context to the rest of the network, or even whether it represents anything at all. Hence, all of the positions with regards to the relationship between the two views of hierarchy are on the table. We will not attempt to adjudicate which is correct here, but we will close by listing the explanatory obligations that each view of the relationship takes on.

The substantiation view suggests that the rich club's influence on the rest of the network is dependent on the rich club processing particularly abstract representations. A proponent of the substantiation view must then define *what* those representations are, how they are propagated to the non-rich-club nodes that receive input from the rich club, and how these are used to guide behavior. A proponent of the conflict view must argue that the competition-process is mediated primarily by the slow-oscillation and conflicting inputs about task settings coming into the rich club, and that no abstract representations are required to subsequently re-organize its output targets in the appropriate configuration for the task.

Each variety of pluralist view is also possible. *Process* pluralists would suggest that input about the task settings, and perhaps motor actions involved in implementing particular tasks, follow a representational hierarchy, but that coordinating the different subnetworks is itself a topological, and not a representational hierarchy-based, process. Model-based pluralism will suggest that the roles of the rich club in mediating diffuse yet efficient communication, as well as providing robust communication (van den Heuvel and Sporns 2011) are best described from the topological perspective, but that this is compatible with abstract representations being what is communicated diffusely and efficiently. Finally, organizational pluralism states that rich club organization, which is topological, co-exists with representational hierarchies in the brain, perhaps explaining why in-degree is significantly higher to the rich-club between tasks, but out-degree higher when task-related representations are occurring.

Each of these views in turn takes on commitments, particularly with regards to how the other areas with connections to the rich club operate. The point is that *none* of these moves is trivial, and hence whatever position one takes requires extensive justification. So, our approach to the patchwork concept helps clarify the state of the hierarchy concept with regards to extant research strategies and the available empirical data.

6.6 Conclusion

In this paper we have argued that there are two distinct approaches to the concept of hierarchy in neuroscience, whose relations have not been sufficiently scrutinized in the previous literature. While the representational approach takes progressively more abstract information processing and representational function as the core property which sorts anatomical areas hierarchically (Sect. 6.2), topological approaches take influence on the network and propagation structure to be central and are neutral with regard to abstraction and representational function (Sect. 6.3).

Our analysis of these two approaches supports the descriptive claim that many scientific concepts develop into a patchwork when researchers use them to pursue various descriptive and explanatory projects (Wilson 2006; Bursten 2016; Novick 2018; Haueis 2018). Our central contribution is the point that such conceptual patchworks leave researchers with multiple options of how to relate different uses of a concept to each other. We argued that current evidence suggests three possible conceptual relations between the two approaches to "hierarchy" (Sect. 6.4): topological hierarchies could *substantiate* the traditional representational hierarchy, *conflict* with it, or contribute to a *plurality* of approaches needed to understand the hierarchical organization of the brain. We do not wish to argue which of these relations is the correct one. We take the foregoing to have shown, however, that the conceptual landscape surrounding the notion of "hierarchy" in systems neuroscience is *extremely* complicated. Without explicating its different connotations and their relations, "use of the term 'hierarchy' can become meaningless, or worse, misleading" (Hilgetag and Goulas 2020, 8). There are no obvious answers, and there is especially no justification to presuming one view of the relationships between different notions of hierarchy over another.

Because hierarchical thinking is deeply engrained in neuroscience and is also used to defend computational (Pylyshyn 2007) and evolutionary (Barrett 2014) accounts of the mind, theorizing about relationship between the representational and topological views is of no small consequence for cognitive science. A substantiation view allows for standard conceptions of the general architecture of the brain and mind to be kept in place, with perhaps some network concepts used to fill in details or account for information integration in a more perspicuous way. The conflict view, however, promotes – and we want to stress this – a *radical* revision to our general conception of neural and mental organization, for which there are not well-articulated alternatives. Thinking about the representational and functional organization of the brain if the conflict view is true is a major conceptual project. Finally, if one pursues a pluralist option then examining the nature of the interaction between different notions of hierarchy will generate insight about functional architecture and the roles of distinct concepts in neuroscience. By articulating different possibilities of answering that question, we hope to have opened up a conceptual space in which further neuroscientific and philosophical reasoning about neural hierarchy can proceed.

Acknowledgments We thank audiences at the AISC Midterm Conference 2018 (University of Genoa) and at the Neural Mechanisms web conference 2018, as well as four anonymous reviewers for helpful feedback on earlier versions of this manuscript.

References

Avena-Koenigsberger, A., Misic, B., & Sporns, O. (2017). Communication dynamics in complex brain networks. *Nature Reviews Neuroscience, 19*(1), 17–33. https://doi.org/10.1038/nrn.2017.149.

Badre, D., Kayser, A. S., & D'Esposito, M. (2010). Frontal cortex and the discovery of abstract action rules. *Neuron, 66*(2), 315–326.

Barrett, H. C. (2014). *The shape of thought: How mental adaptations evolve*. Oxford: Oxford University Press.

Bastos, A. M., Usrey, W. M., Adams, R. A., Mangun, G. R., Fries, P., & Friston, K. J. (2012). Canonical microcircuits for predictive coding. *Neuron, 76*, 695–671.

Bechtel, W. (2008). *Mental mechanisms: Philosophical perspectives on cognitive neuroscience*. New York: Routledge.

Breakspear, M., & Stam, C. J. (2005). Dynamics of a neural system with a multiscale architecture. *Philosophical Transactions of the Royal Society of London. Series B, Biological Sciences, 360*(1457), 1051–1074.

Burnston, D. C. (2016a). Computational neuroscience and localized neural function. *Synthese, 193*(12), 3741–3762.

Burnston, D. C. (2016b). A contextualist approach to functional localization in the brain. *Biology and Philosophy, 31*(4), 527–550.

Burnston, D. C. (2019). Getting over atomism: Functional decomposition in complex neural systems. *British Journal for the Philosophy of Science*. https://doi.org/10.1093/bjps/axz039.

Bursten, J. (2016). Smaller than a breadbox: Scale and natural kinds. *British Journal for the Philosophy of Science, 69*(1), 1–23.

Canolty, R. T., & Knight, R. T. (2010). The functional role of cross-frequency coupling. *Trends in Cognitive Sciences, 14*(11), 506–515.

Canolty, R. T., Ganguly, K., Kennerley, S. W., Cadieu, C. F., Koepsell, K., Wallis, J. D., & Carmena, J. M. (2010). Oscillatory phase coupling coordinates anatomically dispersed functional cell assemblies. *Proceedings of the National Academy of Sciences, 107*(40), 17356–17361.

Capalbo, M., Postma, E., & Goebel, R. (2008). Combining structural connectivity and response latencies to model the structure of the visual system. *PLoS Computational Biology, 4*(8), e1000159.

Craver, C. F. (2007). *Explaining the brain. Mechanistic explanation and the mosaic unity of neuroscience*. Oxford: Oxford University Press.

da Costa, F. L., & Sporns, O. (2005). Hierarchical features of large-scale cortical connectivity. *The European Physical Journal B, 48*(4), 567–573.

De Domenico, M., Sasai, S., & Arenas, A. (2016). Mapping multiplex hubs in human functional brain networks. *Frontiers in Neuroscience, 10*, 326. https://doi.org/10.3389/fnins.2016.00326.

Deco, G., & Kringelbach, M. L. (2016). Metastability and coherence: Extending the communication through coherence hypothesis using a whole-brain computational perspective. *Trends in Neurosciences, 39*(3), 125–135. https://doi.org/10.1016/j.tins.2016.01.001.

Deco, G., & Kringelbach, M. L. (2017). Hierarchy of information representational in the brain: A novel 'intrinsic ignition' framework. *Neuron, 94*, 961–968.

Driver, J., & Spence, C. (2000). Multisensory perception: Beyond modularity and convergence. *Current Biology, 10*(20), R731–R735.

Felleman, D. J., & Van Essen, D. C. (1991). Distributed hierarchical representational in the primate cerebral cortex. *Cerebral Cortex, 1*(1), 1–47.

Ghazanfar, A. A., & Schroeder, C. E. (2006). Is neocortex essentially multisensory? *Trends in Cognitive Science, 10*(6), 278–285.

Goulas, A., Uylings, H. B. M., & Stiers, P. (2014). Mapping the hierarchical layout of the structural network of the macaque prefrontal cortex. *Cerebral Cortex, 24*, 1178–1194.

Grafton, S. T., & de Hamilton, A. F. C. (2007). Evidence for a distributed hierarchy of action representation in the brain. *Human Movement Science, 26*(4), 590–616.

Green, S., Şerban, M., Scholl, R., Jones, N., Brigandt, I., & Bechtel, W. (2017). Network analyses in systems biology: New strategies for dealing with biological complexity. *Synthese, 195*(4), 1751–1777.

Gross, C. G., Rocha-Miranda, C., & Bender, D. (1972). Visual properties of neurons in inferotemporal cortex of the Macaque. *Journal of Neurophysiology, 35*(1), 96–111.

Haggard, P. (2005). Conscious intention and motor cognition. *Trends in Cognitive Sciences, 9*(6), 290–295.

Hagmann, P., Cammoun, L., Gigandet, X., Meuli, R., Honey, C. J., Wedeen, V. J., & Sporns, O. (2008). Mapping the structural core of human cerebral cortex. *PLoS Biology, 6*(7), e159–e159. https://doi.org/10.1371/journal.pbio.0060159.

Haueis, P. (2012). The fuzzy brain: Vagueness and mapping connectivity in the human cerebral cortex. *Frontiers in Neuroanatomy, 6*(37). https://doi.org/10.3389/fnana.2012.00037.

Haueis, P. (2018). Beyond cognitive myopia: A patchwork approach to the concept of neural function. *Synthese, 195*(12), 5373–5402. https://doi.org/10.1007/s11229-018-01991-z.

Hegdé, J., & Van Essen, D. C. (2007). A comparative study of shape representation in macaque visual areas V2 and V4. *Cerebral Cortex, 17*(5), 1100–1116. https://doi.org/10.1093/cercor/bhl020.

Hilgetag, C. C., & Goulas, A. (2020). 'Hierarchy' in the organization of brain networks. *Philosophical Transactions of the Royal Society B, 375*, 20190319. https://doi.org/10.1098/rstb.2019.0319.

Hilgetag, C. C., O'Neill, M., & Young, M. P. (1996). Indeterminate organization of the visual system. *Science, 271*(5250), 776–777.

Hohwy, J. (2013). *The predictive mind*. Oxford: Oxford University Press.

Honey, C. J., Kötter, R., Breakspear, M., & Sporns, O. (2007). Network structure of cerebral cortex shapes functional connectivity on multiple time scales. *Proceedings of the National Academy of Sciences of the United States of America, 104*(24), 10240–10245.

Hubel, D. H., & Wiesel, T. N. (1962). Receptive fields, binocular interaction and functional architecture in the cat's visual cortex. *The Journal of Physiology, 160*(1), 106.

Huneman, P. (2010). Topological explanations and robustness in biological sciences. *Synthese, 177*, 213–245.

Koechlin, E., Ody, C., & Kouneiher, F. (2003). The architecture of cognitive control in the human prefrontal cortex. *Science, 302*(5648), 1181–1185.

Kostić, D. (2016). The topological realization. *Synthese, 195*(1), 79–98. https://doi.org/10.1007/s11229-016-1248-0.

Kravitz, D. J., Saleem, K. S., Baker, C. I., Ungerleider, L. G., & Mishkin, M. (2013). The ventral visual pathway: An expanded neural framework for the processing of object quality. *Trends in Cognitive Sciences, 17*(1), 26–49.

Lakatos, P., Shah, A. S., Knuth, K. H., Ulbert, I., Karmos, G., & Schroeder, C. E. (2005). An oscillatory hierarchy controlling neuronal excitability and stimulus representational in the auditory cortex. *Journal of Neurophysiology, 94*(3), 1904–1911.

Lamme, V. A., & Roelfsema, P. R. (2000). The distinct modes of vision offered by feedforward and recurrent representational. *Trends in Neurosciences, 23*(11), 571–579.

Margulies, D. S., Ghosh, S. S., Goulas, A., Falkiewiz, M., Huntenburg, J. M., Langs, M., Bezgin, G., Eickhoff, S. B., Castellanos, F. X., Petrides, M., Jefferies, E., & Smallwood, J. (2016). Situating the default mode network along a gradient of macroscale cortical organization. *PNAS, 113*(44), 12574–12579.

Mesulam, M. (1998). From sensation to cognition. *Brain, 121*(6), 1013–1052.

Meunier, D., Lambiotte, R., & Bullmore, E. T. (2009). Hierarchical modularity in human brain functional networks. *Frontiers in Neuroinformatics, 3*(37). https://doi.org/10.3389/neuro.11.037.2009.

Meunier, D., Lambiotte, R., & Bullmore, E. T. (2010). Modular and hierarchically modular organization of brain networks. *Frontiers in Neuroscience, 4*, 200–200.

Meyer, K., & Damasio, A. (2009). Convergence and divergence in a neural architecture for recognition and memory. *Trends in Neurosciences, 32*(7), 376–382.

Meyers, E. M., Freedman, D. J., Kreiman, G., Miller, E. K., & Poggio, T. (2008). Dynamic population coding of category information in inferior temporal and prefrontal cortex. *Journal of Neurophysiology, 100*(3), 1407–1419.

Mishkin, M., Ungerleider, L. G., & Macko, K. A. (1983). Object vision and spatial vision: Two cortical pathways. *Trends in Neurosciences, 6*, 414–417.

Müller-Linow, M., Hilgetag, C. C., & Hütt, M.-T. (2008). Organization of excitable dynamics in hierarchical biological networks. *PLoS One, 4*(9), e1000190. https://doi.org/10.1371/journal.pcbi.1000190.

Murray, J. D., Jaramillo, J., & Wang, X. J. (2017). Working memory and decision-making in a frontoparietal circuit model. *The Journal of Neuroscience, 37*(50), 12167–12186.

Novick, A. (2018). The fine structure of 'homology'. *Biology and Philosophy, 33*(6). https://doi.org/10.1007/s10539-018-9617-3.

Orlandi, N. (2010). Are sensory properties represented in perceptual experience? *Philosophical Psychology, 23*(6), 721–740.

Passingham, R. E., Stephan, K. E., & Kötter, R. (2002). The anatomical basis of functional localization in the cortex. *Nature Reviews Neuroscience, 3*(8), 606–616.

Power, J., Schlaggar, B. L., Lessov-Shlaggar, C. N., & Petersen, S. E. (2013). Evidence for hubs in human functional brain networks. *Neuron, 79*(4), 798–813.

Pylyshyn, Z. W. (2007). *Things and places: How the mind connects with the world*. Cambridge, MA: MIT Press.

Rathkopf, C. (2018). Network representation and complex systems. *Synthese, 195*(1), 55–78. https://doi.org/10.1007/s11229-015-0726-0.

Riesenhuber, M., & Poggio, T. (1999). Hierarchical models of object recognition in cortex. *Nature Neuroscience, 2*(11), 1019–1025.

Rigotti, M., Barak, O., Warden, M. R., Wang, X.-J., Daw, N. D., Miller, E. K., & Fusi, S. (2013). The importance of mixed selectivity in complex cognitive tasks. *Nature, 497*(7451), 585–590.

Roe, A. W., Chelazzi, L., Connor, C. E., Conway, B. R., Fujita, I., Gallant, J. L., et al. (2012). Toward a unified theory of visual area V4. *Neuron, 74*(1), 12–29.

Savic, I., Gulyas, B., Larsson, M., & Roland, P. (2000). Olfactory functions are mediated by parallel and hierarchical representational. *Neuron, 26*(3), 735–745.

Schiller, P. (1993). The effects of V4 and middle temporal (MT) area lesions on visual performance in the rhesus monkey. *Visual Neuroscience, 10*(4), 717–746.
Schölvinck, M. L., Leopold, D. A., Brookes, M. J., & Khader, P. H. (2013). The contribution of electrophysiology to functional connectivity mapping. *NeuroImage, 80*, 297–306.
Senden, M., Deco, G., de Reus, M. A., Goebel, R., & van den Heuvel, M. P. (2014). Rich club organization supports a diverse set of functional network configurations. *NeuroImage, 96*, 174–182.
Senden, M., Reuter, M., van den Heuvel, M. P., Goebel, R., & Deco, G. (2017a). Rich club regions can organize state-dependent functional network organization by engaging in oscillatory behavior. *NeuroImage, 146*, 561–574.
Senden, M., Reuter, M., van den Heuvel, M. P., Goebel, R., Deco, G., & Gilson, M. (2017b). Task-related effective connectivity reveals that the cortical rich club gates cortex-wide communication. *Human Brain Mapping., 39*(3), 1246–1262.
Sennet, A. (2016). *Polysemy*. Oxford Handbooks Online. https://doi.org/10.1093/oxfordhb/9780199935314.013.3
Shine, J. M., & Poldrack, R. A. (2017). Principles of dynamic network reconfiguration across diverse brain states. *NeuroImage, 180*(B), 396–405. https://doi.org/10.1016/j.neuroimage.2017.08.010.
Silberstein, M., & Chemero, A. (2013). Constraints on localization and decomposition as explanatory strategies in the biological sciences. *Philosophy of Science, 80*(5), 958–970.
Sporns, O. (2011). *Networks of the brain*. Cambridge, MA: MIT Press.
Sporns, O., & Betzel, R. (2016). Modular brain networks. *Annual Review of Psychology, 4*(67), 613–640.
Sporns, O., Honey, C. J., & Kötter, R. (2007). Identification and classification of hubs in brain networks. *PLoS One, 2*(10), e1049–e1049.
Stanley, M. L., Gessell, B., & De Brigard, F. (2019). Network modularity as a foundation for neural reuse. *Philosophy of Science, 86*(1), 23–46.
Uithol, S., Burnston, D. C., & Haselager, P. (2014). Why we may not find intentions in the brain. *Neuropsychologia, 56*, 129–139. https://doi.org/10.1016/j.neuropsychologia.2014.01.010.
Ullman, S. (2007). Object recognition and segmentation by a fragment-based hierarchy. *Trends in Cognitive Science, 11*(2), 58–64. https://doi.org/10.1016/j.tics.2006.11.009.
van den Heuvel, M. P., & Sporns, O. (2011). Rich-club organization of the human connectome. *Journal of Neuroscience, 31*(44), 15775–15786. https://doi.org/10.1523/JNEUROSCI.3539-11.2011.
van den Heuvel, M. P., & Sporns, O. (2013). Network hubs in the human brain. *Trends in Cogntive Science, 17*(12), 683–696.
Vetter, P., & Newen, A. (2014). Varieties of cognitive penetration in visual perception. *Consciousness and Cognition, 27*, 62–75.
Wallis, J. D., & Miller, E. K. (2003). From rule to response: Neuronal processes in the premotor and prefrontal cortex. *Journal of Neurophysiology, 90*(3), 1790–1806.
Wessinger, C., VanMeter, J., Tian, B., Van Lare, J., Pekar, J., & Rauschecker, J. (2001). Hierarchical organization of the human auditory cortex revealed by functional magnetic resonance imaging. *Journal of Cognitive Neuroscience, 13*(1), 1–7.
Wilson, M. (2006). *Wandering significance. An essay in conceptual behavior*. Oxford: Clarendon Press.
Zamora-López, G., Zhou, C., & Kurths, J. (2010). Cortical hubs form a module for multisensory integration on top of the hierarchy of cortical networks. *Frontiers in Neuroinformatics, 4*(March), 1–1. https://doi.org/10.3389/neuro.11.001.2010.
Zerilli, J. (2017). Against the "system" module. *Philosophical Psychology, 30*(3), 235–250.

Chapter 7
Fundamental Theories in Neuroscience: Why Neural Darwinism Encompasses Neural Reuse

Luis H. Favela

Abstract Various theories have been put forward to provide theoretical unification in neuroscience. The "data rich and theory poor" state of neuroscience makes such theories worth pursuing. An overarching theory can facilitate data interpretation and provide a general framework for explanation and understanding across the various subfields of neuroscience. Neural reuse is a recent and increasingly popular attempt at such a unifying theory. At its core, neural reuse is a claim about the brain's architecture that centers on the idea that brain regions are used for multiple tasks across multiple domains. Here, I claim that although neural reuse has many merits, it does not provide a fundamental theory of brain structure and function. Neural reuse is appropriately understood as a general organizational principle that is encompassed by a more fundamental theory. That theory is Neural Darwinism, which applies broadly Darwinian selectionist principles across scales of investigation to explain and understand brain structure and function.

Keywords Neural Darwinism · Neural reuse · Plasticity · Selectionism · Theory

7.1 Introduction

The neurosciences are often described as "data rich and theory poor" (e.g., Ascoli 2002; Churchland and Sejnowski 2016; Favela 2014; Hawkins et al. 2019; Woodward 2011; Zimmerman 2008). The "data rich" state of neuroscience is not surprising given the rapid development of technologies with evermore spatial and temporal resolution. For example, a complete electron microscopy volume of an adult fruit fly brain is approximately 106 terabytes (Zheng et al. 2018). It is estimated that an ultrahigh-resolution 3-D model of a single human brain could

L. H. Favela (✉)
Department of Philosophy and Cognitive Sciences Program, University of Central Florida, Orlando, FL, USA
e-mail: luis.favela@ucf.edu

© Springer Nature Switzerland AG 2021

F. Calzavarini, M. Viola (eds.), *Neural Mechanisms*, Studies in Brain and Mind 17,
https://doi.org/10.1007/978-3-030-54092-0_7

result in 21,000 terabytes of data (Amunts et al. 2013). In addition to technology-related reasons, such enormous amounts of data should not be surprising given that the brain is a prototypical complex system. Even the simplest of brain processes involve numerous variables. Single-neuron activity (Izhikevich 2006, 2007), for example, displays such complex systems features as nonlinear interactions, self-organization, and strong feedback among parts (Érdi 2008; Freeman 2005). Due to those reasons—as well as pragmatic ones—it is common that laboratories conduct research primarily on single brain areas, processes, or pathologies, such as Alzheimer's disease, declarative memory, visual perception, etc. Though there is a vast amount of data about the brain, there is minimal theoretical integration across explanations. This may be due to the piecemeal nature of the neuroscientific enterprise just described. Such is the state of affairs in neuroscience that has driven a number of key figures and institutions in the field to identify the "urgent need" (Sporns 2011) for a solid theoretical foundation to meaningfully integrate and interpret discovered mechanisms and data (He et al. 2013).

Various theories have been put forward to provide theoretical unification in neuroscience, for example, Bayesian brain (Doya et al. 2007), coordination dynamics (Bressler and Kelso 2016), free-energy principle (Friston 2010), network theory (McIntosh 2000), and neural masses (Freeman 1975). Neural reuse is a recent and increasingly popular attempt at a unifying brain theory (Anderson 2010). At its core, neural reuse is a claim about the brain's architecture that centers on the idea that brain regions are used for multiple tasks across multiple domains. Here, I claim that although neural reuse has many merits, it does not provide a fundamental theory of brain structure and function. Neural reuse is appropriately understood as a general organizational principle that is encompassed by a more fundamental theory. That theory is Neural Darwinism, which applies broadly Darwinian selectionist principles across scales of investigation to explain and understand brain structure and function (Edelman 1987). In the next two sections, I provide overviews of neural reuse and Neural Darwinism. With an understanding of those two frameworks, I will then be able to explain why Neural Darwinism encompasses neural reuse and why the former meets the criteria for a fundamental theory of neuroscience.

7.2 Neural Reuse

At its most basic, neural reuse is a claim about the brain's architecture. It states that local brain regions are used for multiple tasks across multiple domains (Anderson 2014, p. 4). What counts as a "brain region" can vary among tasks, ranging from single neurons to small networks (e.g., Anderson 2014, p. 30). For example, though commonly referred to as the "part of the brain for syntactic processing," activity in Broca's area has been experimentally associated with various action-related tasks (Anderson 2014, p. 4; Nishitani et al. 2005). Accordingly, neural reuse is a kind of neuroplasticity (Anderson 2016, p. 1), or, vice versa, neural plasticity may be a form of neural reuse (Anderson 2010, p. 245). Different cognitive and behavioral

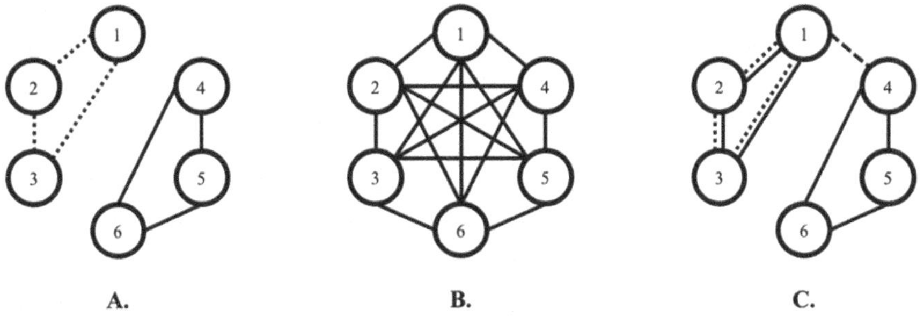

Fig. 7.1 Comparing modular, holism, and reuse conceptions of the nature of neuronal functional connections underlying cognitive and behavioral capabilities. (**a**) In modular conceptions, capabilities are underlaid by distinct neuronal networks, for example, capability X occurs via neurons 1, 2, and 3, whereas Y occurs via 4, 5, and 6. (**b**) In holistic conceptions, capabilities are underlaid by fully connected neuronal networks, whereby capability X occurs via changes in weights and order of connections, much like connectionist networks, and capability Y occurs via those same neurons but with connections of different weight and order. (C) In reuse conceptions, the same neuronal network can underlie various capabilities, for example, neurons 1, 2, and 3 can underlie both X and Y depending on various bodily and environmental conditions, but they are not involved in all capabilities. (Figure inspired by figure 1.1 in Anderson 2014, p. 8)

activities are achieved via neural coalitions, where neurons may also participate in other coalitions for other cognitive and behavioral activities.

In the current discussion, I refer to Michael Anderson's specific kind of "neural reuse," which centers on the "massive redeployment hypothesis" (MRH; Anderson 2007). According to MRH, brain areas are specialized in that they have the same activity, but that same brain activity does not underlie specific cognitive functions (Anderson 2007, p. 330). Due to the fact that brains are embodied in organisms that exist in environments, those same brain areas can be redeployed along with various body and environment conditions in order to underlie various functions (Fig. 7.1). Anderson's MRH is one of several types that are labeled "neural reuse" (Anderson 2010, p. 246). Others include the neural exploitation hypothesis (Gallese 2008), neuronal recycling theory (Dehaene 2005), and the shared circuits model (Hurley 2008). For current purposes, I use "neural reuse" as specifically referring to Anderson's MRH.

Anderson (e.g., 2010, 2014, 2016) claims that the evidence for neural reuse pushes neuroscience to rethink common assumptions concerning modularity and evolutionary psychology. The brain is understood as not composed of domain-specific modules (e.g., Broca's area as being for syntactic processing). Instead, cognitive and behavioral abilities occur as a function of "neural, behavioral, and environmental resources ... reused and redeployed in support of any newly emerging ... capacities" (Anderson 2014, p. 7). Three primary implications result from that core claim (Anderson 2016, pp. 1–2): First, new capacities are supported by mixing neural elements. Second, neural reuse supports both procedural and behavioral reuse, that is, it has both biological and behavioral implications.

Third, higher-order cognitive capacities do not have their own unique and specific neural architecture, but instead are built from existing neural structures. Notably, neural reuse is treated as a theory of neural architecture underlying cognitive and behavioral capacities that fits with embodied and ecological approaches to cognition (e.g., Anderson 2014, pp. 170–174). As Anderson puts it, "Thinking, calculating and speaking are adaptive behaviors and, as such, involve the whole organism acting in and with its environment" (Anderson 2016, p. 2).

There is compelling empirical support for neural reuse (e.g., Anderson 2008; Anderson et al. 2013; Anderson and Pessoa 2011; Pulvermüller 2018; Ziegler et al. 2018), suggesting that there is at least some evidence for its being an accurate description of brain organization, namely, that much of the brain is put to work for various ends (Anderson 2015; cf. McCaffrey and Machery 2016; Poldrack and Yarkoni 2016). As mentioned above, Broca's area is not the "syntactic processing part of the brain;" that area supports a variety of other capacities as well, such as those involving bodily action. Nevertheless, is neural reuse a *fundamental theory* that can serve as a unifying framework for brain structure and function? Before answering that question, I will first introduce Neural Darwinism in the next section. Then, I will be positioned to defend the claim that although neural reuse is appropriately understood as a general organizational principle, it is subsumed by the fundamental theory of Neural Darwinism.

7.3 Neural Darwinism

Neural Darwinism—also known as the "theory of neuronal group selection"—is the brain child of Gerald Edelman (1987). It is a theory to explain brain structure and function that is guided by the Darwinian principle of population thinking, or selectionism. Population thinking is the idea that variance within biological populations is necessary for the process of evolution, such that those individuals who can best cope with their environment will reproduce more successfully (Edelman 1988). Edelman had great success in applying the principles of population thinking to the immune system. He demonstrated that organisms do not inherit immune systems with preprogrammed antibodies ready to handle bacteria and other unwelcomed microorganisms. Instead, via Darwinian selectionist processes, antigens facilitate and constrain the scope of selected antibodies from the countless variations the immune system could randomly produces (Edelman and Gally 1967; Grumet and Edelman 1988). In an effort to continue the line of success that began with his explanation of immune systems as "immune selective systems" (1988, p. 190), Edelman attempted to bootstrap those same principles onto the brain's functions (1987). The following are the foundational points that define Neural Darwinism (Edelman 1987; Edelman et al. 2011):

1. Developmental selection leads to *primary repertoire*.
2. Experiential selection to yield *secondary repertoire*.
3. *Reentry*.

These three points explain the formation of the brain and development of cognitive and behavioral capabilities from an organism's embryonic and postnatal stages through maturation.

The first step is the development of *primary repertoires*. Edelman starts at the beginning of an organism's development in order to bolster his account of how organisms learn and cope with their environmental niche. The primary repertoires are those morphological features that develop early in an organism's life, such as the general layout of the body and early outgrowths of neural networks. Genetic constraints are at their most potent at this stage. However, even then epigenetic influences are present. He goes into great detail to explain the molecular effects of cell adhesion molecules (CAM) and substrate adhesion molecules (SAM) in the regulation and expression of cell development (1988, pp. 86–115). The vital message to receive from Edelman's weighty discussion is that cells of all kinds form groups based upon genetic information that is expressed in a controlled manner based upon the effects of CAM and SAM regulation (1989, pp. 44–46). This step is important because it demonstrates that the notions of environmental influence and selectionism-guided development express influences at the very beginning of an organism's development and at the molecular level. As Edelman notes, "The internal environment during development can exert as great a selective force ... as the external environment" (1988, p. 52). Whether due to genetic coding or CAM and SAM effects, cell division and death is affected by both the forces of selectionism and the environment inhabited by the cells. Gravity, toxins, and temperature are a few examples of the environmental conditions that affect the development of cell groups. If the environment were not novel, then selectionism would not be necessary. The fact is that the environment is novel and occurrences such as temperature and toxicity must be accounted for even at the early stages of development and at the molecular level. Despite such variations as temperature and toxicity, the environment of the womb or egg of an organism in early development is about as predictable and controlled as it will get in that organism's life. Moreover, it is at these earliest of developmental stages that the most minimal degrees of variation are demonstrated among species. Increased novelty in the environment of the organism occurs after the earlier stages of development and once the organism leaves the controlled environment of the womb or egg. Accordingly, with the ability to successfully cope with environments of increased complexity come decreases in inherited morphology. This is especially true for neurons, for the more inherited a capacity is, the less that capacity can cope with a novel factor.

Once the primary repertoire is in place, that is, once the basic genotype has been expressed in a particular environment, the *secondary repertoire* goes into effect. This is an important step in the overall theory of Neural Darwinism because it purports to overcome the shortcomings of domain-specific modular architecture, namely, the idea that brains are collections of modules for specific purposes, such as a module for visual perception and a module for fast reasoning. At the same time that Neural Darwinism pushes back against modular conceptions of mind, it also pushes back against those evolutionary psychology approaches that defend similar understandings of the mind as comprised of collections of modules selected for specific purposes over the course of a species' evolutionary history (e.g., Carruthers 2006; Cosmides and Tooby 1987; Sperber 1994; Tooby and Cosmides 1992). The secondary repertoire accounts for experiential selection via changes in synaptic strength and network organization (i.e., neural plasticity). Based upon morphologically constrained behavioral experiences, the corresponding neural activity will be strengthened or weakened. Once an organism's primary repertoires are in the process of expression in an environment, epigenetic development and alterations take place as a result of the experiences had by the organism. A human who plays the piano for many years, for example, will strengthen connectivity in neuronal groups associated with finger dexterity (Gaser and Schlaug 2003).

The behavior resulting from interactions with the environment induces effects upon neuronal coordination and organization. Interacting with the environment does not cause changes at the macroscale anatomical structures of the brain, but they do cause changes of varying strengths and weaknesses at meso- and microscale. These connections begin to develop into neuronal groups called *maps*. These maps are groupings of populations whose signals have been strengthened by environment-influenced behaviors (Edelman 1989, p. 45). This development is not confined to preestablished, domain-specific modules, such as those entailed by evolutionary psychology. In comparing primary and secondary repertoires, it helps to think of the primary repertoires as the product of genotype, weakly influenced by the environment, and pre-experiential morphological development. Secondary repertoires are the epigenetic, strongly environment influenced, experience-based selection and alteration of the fine structures of morphology such as the synaptic connectivity of brains.

Reentry, or reentrant signaling, is the core of Neural Darwinism. Due to the fact that the concept went through a number of revisions and was refined over the course of Edelman's work (e.g., Edelman 1989, p. 49; 2003, p. 5521; Edelman and Tononi 2000, pp. 114–120), I provide the following synthesized definition in an attempt to capture what is common and central to the various definitions found in the literature:

> Reentry is the dynamic process whereby an organism's cognitive and behavioral capacities (resulting from the primary and secondary repertoires) are supported by anatomically distant maps in the brain, which are linked by reciprocal signals that coordinate (via synchronization and integration) with each other *and* the physical dimensions of the body and world with a high degree of spatiotemporal accuracy.

A number of key features of reentry are worth highlighting. First, reentry is not feedback (Edelman and Gally 2013). As a term from control theory, feedback is a process that requires correction and control of signals based on prespecified paths and desired outputs, or, prescribed relationships among variables (Mayr 1970). Although the primary repertoire is genetically inherited and can be thought of as "prespecified," its expression and the secondary repertoire are not. Organisms are selectionist systems that develop via experience. For that reason, reentry is a process that synchronizes and integrates signals simultaneously from multiple neuronal populations, which are themselves receiving signals from the body and world. Second, reentry is not a form of neural plasticity, but is the process that enables plasticity. Neural plasticity refers to the ability of the nervous system to modify and reorganize its connections, function, and structure due to experience (von Bernhardi et al. 2017). Such modifications can be explained via reentry: neuronal connections can have their structure and function modified due to the nature and strength of the reciprocal connections they have among other maps and the body and world. Without those connections to synchronize and integrate signals, plasticity could be said to not have valuable alterations; where "valuable" refers to those changes that allow for useful cognitive and behavioral responses (Edelman and Tononi 2000, p. 88). Third, and most important for the current topic, reentry contributes to the degenerate nature of the components and activities that underlay behavior and cognition.

In regard to the brain, *degeneracy* refers to the ability of structurally different neuronal circuits and maps to give rise to the same function or output (Edelman 2003; Edelman and Gally 2001). Since experiential selection (i.e., secondary repertoires) is ongoing throughout an organism's life, degeneracy is also ongoing. What that means is that over the course of an organism's lifetime, various structures will give rise to similar capacities, for example, consciousness (Edelman 2003), motor movements (Sporns and Edelman 1993), and visual perception (Sporns et al. 2000). As is made evident by the preceding examples, 'structures' is used broadly to include neuronal as well as behavioral and bodily configurations. A consequence of treating all of those structures as degenerate is that their various combinations can underlay the same or similar capacities. For example, different neuronal structures in the same environment could give rise to the same capacity. This is a desirable capability for an organism to have because it means that those capacities that facilitate success in various environments can be achieved by a range of neuronal configurations. Consequently, domain-specific modules (e.g., such as those posited by evolutionary psychology) are unlikely to play major roles in defining behavior and cognition, especially past the developmental stage of primary repertoire expression, and especially in complex organisms such as mammals.

Neural Darwinism is not merely a set of assertions. There is also empirical evidence in its favor. From neuronal group selection to reentry, Neural Darwinism has both provided the theory to interpret experimental findings as well as inform hypotheses and experimental design. The following is a sample of such empirical support for Neural Darwinism:

- Binocular rivalry (Srinivasan et al. 1999)
- Brain-based robots (Krichmar and Edelman 2002)
- Consciousness (Seth and Baars 2005)
- Figure-ground segregation (Sporns et al. 1991)
- Immune system (Edelman and Tononi 2000)
- Neural network connectivity (Sporns et al. 2000)
- Object awareness (Edelman 2006)
- Schizophrenia (Tononi and Edelman 2000)
- Sensorimotor development and motor synergies (Sporns and Edelman 1993)
- Synaptic plasticity (Seth and Edelman 2007)

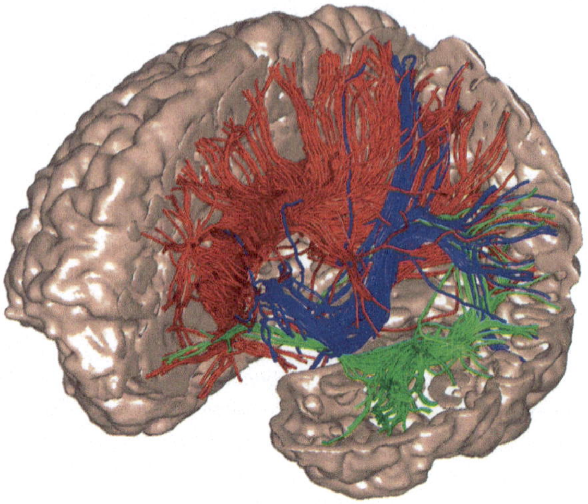

Fig. 7.2 Large-scale model of a mammalian thalamocortical system. This biologically-realistic model has 22 different neuron types contributing to the one million neuron simulation. (Figure 1 from Izhikevich and Edelman 2008, p. 3353, CC BY-NC-ND)

- Visual system (Tononi et al. 1996)

Additionally, Neural Darwinism has provided the foundation for one of the first, and most impressive, large-scale models of the brain (Fig. 7.2). The Izhikevich and Edelman model of a mammalian thalamocortical system is biologically realistic at one million neurons that simultaneously span multiple anatomical scales: global white matter, multilayered cortical microcircuitry, and 22 types of neurons with their dendritic branching (Izhikevich and Edelman 2008).

Functional integration among neuronal systems provides a vivid example of a feature of brains discussed in terms of neural reuse that is subsumed by Neural Darwinism. In the current context, "functional integration" refers to the idea that neuronal processes, or functions—i.e., cognitive (e.g., decision making) and perceptual-motor (e.g., catching a flyball)—involve contributions from various areas across the brain. Anderson has stated over a number of works that the MRH can account for such functional integration (e.g., Anderson 2007, 2016). In addition, he has conducted empirical research supporting hypotheses concerning integration, such as the claim that more evolutionarily-recent cognitive capacities are distributed across more brain regions than older ones (e.g., Anderson 2008), which is consistent with the core principles of neural reuse. However, though functional integration has been demonstrated empirically—as Anderson has done—and though that characteristic is consistent with the core claims of neural reuse, that does not mean that neural reuse has *explained how* such integration occurs. In fact, in at least some works, Anderson himself has not explicitly stated that the MRH actually explains functional integration, but that it provides a way to "understand" it, for example, by providing a "vocabulary for characterizing" it (2007, p. 343). Neural Darwinism, however, both provides an account for *why* there would be such distributed connections and *how* they integrate. Edelman and colleagues have repeatedly demonstrated that not only are evolutionarily-recent functions highly distributed (e.g., language; Edelman 2003), but that older functions (e.g., vision; Tononi et al. 1998) are highly distributed as well. Thus, like neural reuse, research from within a Neural Darwinism investigative framework has provided empirical evidence that there are such distributions. In addition, it provides a framework to understand *why*, namely, primary and secondary repertoire development via the theory of neuronal group selection. Moreover, it provides an empirically-supported account for *how* such integration occurs, namely, reentry, or reentrant dynamics (e.g., Tononi et al. 1998; Tononi et al. 1992). The entire framework has also been successfully implemented in artificial systems, or, what Edelman calls, "brain-based devices" (e.g., Krichmar and Edelman 2005). It is important to make clear that claiming that neural reuse is subsumed by Neural Darwinism is not to say that the former should be rejected. In fact, the MRH may be the better way to frame—or "characterize"—specific questions concerning aspects of the brain, such as network connectivity. In this manner, neural reuse could be understood as part of Neural Darwinism. Nevertheless, Neural Darwinism is broader in scope by providing accounts of what neural reuse does and more.

In summary, Neural Darwinism provides a theory of brain, behavioral, and cognitive structure and function. Neural Darwinism accounts for developmental stages (primary repertoires and genetics), the role of experience (secondary repertoires and epigenetics), and the processes of spatiotemporal coordination among neuronal circuits and maps (reentry). Supporting these processes, and central to Neural Darwinism, is the degenerate nature of behavioral and cognitive functions and outputs. That is to say, what matters from a Darwinian perspective is not the specific material constitution of an organism, but the ability of its brain-body-environment organization to coordinate so as to enable functions that facilitate, among other things, "the four F's: feeding, fleeing, fighting, and reproduction" (Churchland 1994, p. 31). With these introductions to neural reuse and Neural Darwinism concluded, in the next sect. I provide a sketch of what can be expected from a fundamental theory in neuroscience. After, I explain why Neural Darwinism encompasses neural reuse and why the former meets the criteria for a fundamental theory of neuroscience.

7.4 Fundamental Theories

My aim in this work is to demonstrate that Neural Darwinism encompasses neural reuse. The primary reason it does so is because Neural Darwinism does the explanatory work and provides theoretical understanding that neural reuse does and more. In that way, Neural Darwinism is more of a fundamental theory than neural reuse. Moreover, the two are not equal contenders, for the relationship between the two is asymmetrical. If Neural Darwinism were not true, then it is likely that neither would neural reuse; but the same consequence does not hold the other direction. In order to defend this position, I provide a sketch of how to understand what a "fundamental theory" in neuroscience could be like. My aim is not to provide the necessary and sufficient conditions for what a "fundamental theory" is. With that said, I must provide at least an approximation of what I mean by "fundamental theory" in the current context. As a starting point, and at its most general, a scientific theory is,

> a plausible or scientifically acceptable, well-substantiated explanation of some aspect of the natural world; an organized system of accepted knowledge that applies in a variety of circumstances to explain a specific set of phenomena and predict the characteristics of as yet unobserved phenomena. (U.S. National Academy of Sciences 2018)

Three parts of that definition are noteworthy (Bordens and Abbott 2014, pp. 33–34). First, it claims that theories provide *explanations*. That is, theories answer "How?" and "Why?" questions such as, "How does vision work?" and "Why are certain memories faster to recall?" Second, it must be *plausible*. That is, the explanation must reasonably follow from acceptable commitments, for example, produce data that facilitate explanations consistent with those produced by experimental work involving auxiliary hypotheses. Third, it must be *predictable*. That is, it must lead to the generation of testable hypotheses with expected outcomes.

To limit talk of scientific theories to the mind sciences, those three parts are also identified by Allen Newell in his discussion of features of unified theories of cognition. Newell claims that a cognitive theory is not just a collection of facts, but provides explanations, answers to questions, predictions, and prescriptions for control, among other things (Newell 1990, pp. 13–15). Also limited to the mind sciences, William Uttal defines scientific theories as,

> an integrated interpretation of a body of related empirical evidence. As such, a theory incorporates or summarizes a body of observations by extracting general principles, rules, and laws implied by the data. Theories come in many types—some mathematically formal and some ambiguously verbal—but all of which transcend the particular to illuminate the general. (Uttal 2016, p. 8)

Uttal's definition differs from the previous two in its focus on generalization and integration. Theories are not just descriptions, facts, or laws (Uttal 2005, pp. 9, 15). They must provide ideas that generalize results into comprehensive and unifying statements that transcend individual or few observations. A theory encompasses a wide range of observations into a unified set of principles (Uttal 2005, p. 23). I understand Uttal's points about generalization and integration as connecting with the "data rich" state of neuroscience: Neuroscience has plenty of descriptions and facts, but it must integrate those facts in order to develop generalizations about the phenomena of interest. I take the point about unification and principles as connecting with the "theory poor" state of neuroscience: Generalizations and integration of data are necessary, but that is not enough; neuroscience needs to employ that data in the task of developing theories that can provide explanatory unification and understanding. For that, descriptions of mechanisms and evermore data will not be sufficient to serve that end. That is what, I think, Olaf Sporns is stressing when he says that, "The point of building brain models … is to advance *understanding* of brain function, not creating in silico replicas that are as complex and *incomprehensible* as the real thing" (Sporns 2012, p. 169; italics added). Unfortunately, it seems that neuroscience continues to emphasize big data over theory (Frégnac 2017; Landhuis 2017; Sejnowski et al. 2014). Consequently, those more theory-minded neuroscientists see an "urgent need" for a "theoretical foundation for understanding the brain" (Sporns 2011, pp. 77, 130). What could neuroscience look for in a theoretical foundation, or, a fundamental theory?

A fundamental theory for neuroscience ought to keep in mind that its claims must be open to empirical evaluation and the pragmatic features of actual neuroscientific research, such as limits on the scope of experiments (e.g., budgets, computational resources, time, and personnel). Thus, the theory will depend on the particular problems that a researcher is investigating (Brigandt 2010). Along these lines, a fundamental theory for neuroscience should not be evaluated for its ability to provide an ontological base at the bottom of a hierarchy of less fundamental ontological entities (cf. Cat 2017); nor should it be concerned with epistemic values defined a priori such as simplicity (cf. Scorzato 2013). Fundamental theories hold asymmetrical relationships with secondary theories. It should be the case that the truths of the fundamental theory entail the truths of the secondary theories. This has the consequence that if the fundamental theory is demonstrated to be false, then the secondary theories would be false as well; but, if the secondary theories are demonstrated to be false, the fundamental theory can remain true.

In line with claims made by Newell, Sporns, and Uttal, when considering what a fundamental theory for neuroscience should do, the following criteria ought to be met: First, it should be able to provide explanations of structure and function. Second, it should be plausible, that is, it should produce data that facilitate explanations consistent with those produced by experimental work involving auxiliary hypotheses. Third, it should facilitate the generation of hypotheses and predictions. Fourth, it should be able to encompass, subsume, and unify other less fundamental theories. So, back to the issue of neural reuse: Is it a fundamental theory for neuroscience?

7.5 Why Neural Darwinism Encompasses Neural Reuse

There is no doubt that in the general sense of what a scientific theory is, neural reuse is certainly that: it is plausible (e.g., it informs experiments that generate data consistent with those produced by experimental work involving auxiliary hypotheses, such as embodied cognition and network science), it explains a specific set of phenomena (e.g., synesthesia and cross-modal plasticity; Anderson 2014, pp. 54–57; D'Souza and Karmiloff-Smith 2016, p. 12), and it allows for the generation of predictions (Anderson 2008; Anderson et al. 2013; Anderson and Pessoa 2011). In spite of those merits, neural reuse is not a fundamental theory of neuroscience for four, interrelated reasons.

First, although it provides an integrated interpretation of a body of related empirical evidence, it does not then necessarily lead to the extracting of general principles, rules, or laws. That is, neural reuse does not "transcend the particular to illuminate the general" in a way necessary of fundamental theories (Uttal 2016, p. 8). Other processes crucial to the brain's structure and function are likely not accounted for via reuse, for example, action potentials, ephaptic coupling, genetic expression, neurotransmitter systems, and synaptic transmission. The reason it does not is because neural reuse is primarily descriptive in nature. That is the second reason why it is not a fundamental theory: Neural reuse highlights a particular organizational feature of the brain, namely, plasticity, but it does not answer "Why?" questions concerning plasticity. Neural reuse centers on the claim that "circuits can continue to acquire new uses after an initial or original function is established" (Anderson 2010, p. 245). In other words, neural reuse tells us that brains use the same circuits and maps to facilitate various processes and outcomes. In that way, it "is a form of neuroplasticity whereby neural elements originally developed for one purpose are put to multiple uses" (Anderson 2016, p. 1). Plasticity may be a general feature of brains, but neuroplasticity is not a fundamental theory of the brain, let alone of cognitive and behavioral capacities. The various types of neuroplasticity—such as Hebbian, massive redeployment, and shared circuits—are descriptions or mechanisms hypothesized to underlie certain phenomena such as learning and memory (Berlucchi and Buchtel 2009; Fuchs and Flugge 2014). But plasticity does not serve as something from which to extract general principles about the brain from. Ephaptic coupling, for example, is likely not explained as a form of plasticity. If neural reuse is indeed a description or mechanism, then it would be inappropriate to attempt to extract general principles or rules that apply across various brain-related phenomena. Moreover, if neural reuse is just a description or mechanism, then that would explain why it is not a theory: descriptions and mechanisms alone are not theories. This leads us to the third reason: neural reuse is atheoretical.

Neural reuse, like many—perhaps most or all—mechanistic explanations, can be employed in an atheoretical manner. That is, neural reuse describes the ways in which the brain is organized, but such descriptions are theoretically underdetermined, that is to say, the description and mechanisms can be part of various theoretical frameworks. To state that reuse occurs in the brain is to just describe that a phenomenon is so (n.b. Anderson does attempt to make the case that neural reuse can answer the "Why?" question as to why the brain is built that way; e.g., Anderson 2014, pp. 5–43). It is possible to present neural reuse and other descriptions of brain function and structure (e.g., action potentials) within various overarching theories. Neural reuse could conceivably fit in various unifying theories, such as coordination dynamics, free-energy principle, and network theory. Fourth, as a description or mechanism that is atheoretical and does not generalize, neural reuse does not encompass, subsume or unify other theories. While much of the brain involves plasticity in some form or another, plasticity is surely not the brain's only process. As mentioned above, it seems unlikely that other common brain processes—such as action potentials, ephaptic coupling, genetic expression, neurotransmitter systems, and synaptic transmission—are properly understood as forms of reuse. Neural Darwinism, however, can provide a broad theoretical base for a range of neural phenomena, including reuse.

Neural Darwinism and neural reuse both attempt to do at least some of the same work: they both purport to explain brain organization and how that structure facilitates behavioral and cognitive capabilities. However, for the reasons mentioned above, though neural reuse has many descriptive virtues, it does not have strong theoretical ones. The principle reason is that neural reuse does not provide a generalizable answer to the "Why?" questions concerning a range of neural phenomena. Neural Darwinism, on the other hand, does. In response to the questions, "Why is the brain structed that way and why does it function that way?" Neural Darwinism states that from the earliest stages of an organism's development, through experiences over the course of a lifetime, from the spatial and temporal scales of molecules to overt behavior, the brain follows selectionist principles that facilitate natural selection (Edelman 1987, 1988, 1989).

In regard to general scientific theories, Neural Darwinism meets those requirements as well. First, the primary repertoire part of Neural Darwinism explains the beginnings of how brain structure and function develop. From the scale of molecular effects of CAM and SAM in the regulation and expression of cell development, to the significance of fetal environmental conditions, it is clear how selectionist principles are operating on the embodied and situated organism. Next, the secondary repertoire accounts for the role of experience on an organism's development, both neural and bodily. Then, reentry provides the process by which various neuronal maps synchronize and integrate with each other and the body and world to give rise to various cognitive and behavioral capabilities.

Neural Darwinism is also plausible in that it is consistent with other experimental and theoretical commitments. Centered on the roles of selectionism and experience, the framework Neural Darwinism provides fits with empirical evidence from the basics of cell development to the most sophisticated of cognitive and behavioral capacities such as consciousness (Edelman 1989, 2003; Edelman and Tononi 2000). Neural Darwinism also facilitates the generation of hypotheses and predictions. In addition to the above list of empirical support, which are primarily from research by Edelman and colleagues, others outside of Edelman's circle have utilized Neural Darwinism to generate hypotheses and predictions for a range of phenomena, for example, functional map plasticity (Chervyakov et al. 2016), lifespan motor development (Leversen et al. 2012), neural networks for financial data forecasting (Reid et al. 2014), and neuronal topology mapping (Fernando et al. 2008), just to name a few.

Finally, Neural Darwinism is able to encompass, subsume, and unify other theories. As long as other theories do not hold contrary commitments, then it is plausible for them to be unified under Neural Darwinism. Examples of contrary commitments include the type of modularity adhered to by evolutionary psychology (Tooby and Cosmides 1992), computational theories of mind (e.g., Fodor 1998), and non-embodied conceptions of cognition (Goldinger et al. 2016). With that said, if there is no conflict with selectionism, primary and secondary repertoire development, and reentry, then it is possible that the Bayesian brain, coordination dynamics, free-energy principle, network theory, and neural masses to be subsumed and unified under Neural Darwinism. This is true of neural reuse as well. In fact, it is especially true of neural reuse. The reuse of neural networks can be straightaway explained via selectionist processes under the pressures of natural selection. Though he does not refer to selectionism (at least from what I have read), Anderson makes clear his commitment to neural reuse being consistent with evolution (e.g., Anderson 2007, 2010, 2014). A key difference between neural reuse and Neural Darwinism in regard to evolution is that while reuse occurs *during* evolution (Anderson 2010, p. 244) and has its *origins* in evolution (Anderson 2016, p. 8), neural reuse is not provided with a theoretical underpinning in terms of natural selection or otherwise; specifically, it is not understood as enabling natural selection. Neural Darwinism, on the other hand, is explicitly given a theoretical underpinning that drives primary and secondary repertoire development and reentry: selectionism. Specifically, selectionism across spatial and temporal scales—from the molecular to overt behavior—for the purpose of the four F's. Whereas neural reuse states that there is plasticity, Neural Darwinism explains such plasticity as occurring via selectionist pressures for the purpose of feeding, fleeing, fighting, and reproduction. Since neural reuse does not adhere to commitments that are contrary to Neural Darwinism, and since Neural Darwinism accounts for phenomena that Neural reuse does and more, Neural Darwinism encompasses neural reuse and gives it a solid theoretical underpinning.

If neural reuse is encompassed by Neural Darwinism, does that mean it has nothing unique to add to our understanding of brain structure and function? No. Although, as I have argued, neural reuse is encompassed by Neural Darwinism because the latter accounts for phenomena that the former does not, the former adds to the latter in a very important way. Central to Neural Darwinism is degeneracy, or the idea that structurally different neuronal circuits and maps can give rise to the same function. For example, areas of the brain typically associated with vision can process language (Bedny et al. 2011), and those associated with the tongue can process visual perception (Sampaio et al. 2001). A related feature of brain structure and function not highlighted in the Neural Darwinism literature is pluripotency, that is, when structurally similar neuronal circuits and maps can give rise to different functions. Although Anderson does not describe reuse in terms of pluripotency, it seems clear that the former is a type of the latter—Anderson does mention pluripotency at least one time (Anderson 2015, p. 76) and Klein has discussed it in those terms (Klein 2010, p. 281). Since, as stated above, there is no clear conflict between neural reuse and Neural Darwinism, and since the former explicitly accounts for pluripotency whereas the latter does not, then it seems that by encompassing neural reuse Neural Darwinism would gain the ability to account for an additional brain phenomenon. Along those lines, and consistent with the expectations laid out above for a fundamental theory of neuroscience, just because Neural Darwinism does not currently account for all brain structure and function, it does not mean that the framework cannot be supplemented by other theories that hold consistent overall commitments.

The claim that Neural Darwinism is currently the best—that is, most encompassing and unifying—fundamental theory of brain structure and function does not necessitate the further claim that neural reuse is useless or false. As stated a number of times above, neural reuse has many virtues and is supported by much experimental evidence. Though it is not the fundamental theory of neuroscience, it is certainly a smaller-scale theory of certain aspects of brain architecture, such as its pluripotent features. With that said, even if neural reuse explains the mechanisms/processes of pluripotency, it does not provide a theory of behavioral and cognitive structure and function in toto. The selectionism at the heart of Neural Darwinism does provide such an encompassing and unifying theory. Furthermore, Neural Darwinism holds an asymmetrical position to that of neural reuse. The truths of Neural Darwinism (e.g., selectionism) entail the truths of neural reuse (e.g., plasticity). However, the converse does not hold: neural reuse does not entail Neural Darwinism. Neural reuse could be demonstrated to be false and Neural Darwinism would still be true. But if Neural Darwinism was demonstrated to be false—for example, if selectionism is not occurring in brains— then neural reuse would likely be false as well. Consequently, neural reuse is a secondary theory that is subsumed by the fundamental theory of Neural Darwinism (Fig. 7.3).

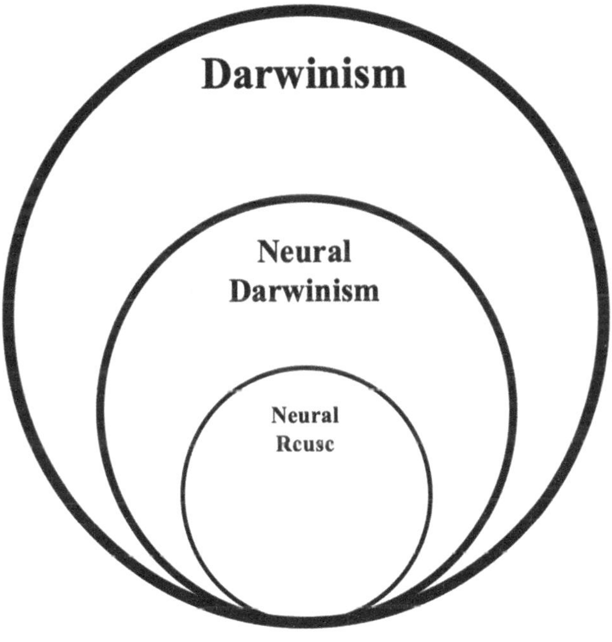

Fig. 7.3 The relationship of Darwinism, Neural Darwinism, and neural reuse. Neural reuse is encompassed by Neural Darwinism. Neural Darwinism may be the fundamental theory of neuroscience, but it is not the fundamental theory across the life sciences; that is Darwinism. It is a further question if Darwinism provides a fundamental theory outside the life sciences

7.6 Conclusion

While neuroscience has never produced more data about the brain, it is currently a piecemeal enterprise that lacks theoretical unification. Although various researchers and laboratories share common experimental practices (e.g., searching for mechanisms), there is no fundamental theory to interpret, explain, and understand the products of those practices. Such a theory could remove the "data rich and theory poor" description of neuroscience. Neural reuse is one such contender. Neural reuse is a kind of neuroplasticity that aims to account for the brain's architecture. Contrary to conceptions of the brain as a collection of domain-specific modules, neural reuse centers on the idea that local brain regions are used for multiple tasks across multiple domains. Neural reuse has many merits, for example, it is consistent with embodied cognition and network science, and it has compelling empirical support. With that said, neural reuse does not provide the kind of fundamental theory needed to explain and understand the wide range of phenomena investigated across the neurosciences.

Alternatively, I have argued that Neural Darwinism is an appropriate fundamental theory of brain structure and function that can provide theoretical unification across neuroscience. Centering on selectionist principles, Neural Darwinism offers an account of development (primary repertoire), experiential selection (secondary repertoire), and neuronal coordination (reentry). In so doing, it provides a generalizable and unified framework for explaining and understanding cognitive and behavioral capabilities, and the contributions made to those phenomena across spatial and temporal scales from molecular activity to embodied behavior. I argued that accepting Neural Darwinism as the fundamental theory of neuroscience does not necessitate dispensing with neural reuse. On the contrary, neural reuse is bolstered by being encompassed by Neural Darwinism and thereby obtaining a strong theoretical underpinning. So too does Neural Darwinism gain as neural reuse fills a gap by accounting for neural pluripotency. Thus, although neural reuse does not meet the criteria for a fundamental theory of neuroscience, it serves as a useful secondary theory within the broader framework of Neural Darwinism.

Acknowledgements The author thanks audiences at the *Neural Mechanisms Online Webconference 2018 New Challenges in the Philosophy of Neuroscience* and the meeting of the *Southern Society for Philosophy and Psychology 2019* for helpful comments and questions. The author is very thankful for constructive feedback and suggestions from the editors and reviewers. This work is partially based on material from Favela (2009).

References

Amunts, K., Lepage, C., Borgeat, L., Mohlberg, H., Dickscheid, T., Rousseau, M.-E., et al. (2013). BigBrain: An ultrahigh-resolution 3D human brain model. *Science, 340*, 1472–1475.

Anderson, M. L. (2007). Massive redeployment, exaptation, and the functional integration of cognitive operations. *Synthese, 159*(3), 329–345.

Anderson, M. L. (2008). Circuit sharing and the implementation of intelligent systems. *Connection Science, 20*, 239–251.
Anderson, M. L. (2010). Neural reuse: A fundamental organizational principle of the brain. *Behavioral and Brain Sciences, 33*, 245–313.
Anderson, M. L. (2014). *After phrenology: Neural reuse and the interactive brain*. Cambridge, MA: MIT Press.
Anderson, M. L. (2015). Mining the brain for a new taxonomy of the mind. *Philosophy Compass, 10*, 68–77.
Anderson, M. L. (2016). Précis of after phrenology: Neural reuse and the interactive brain. *Behavioral and Brain Sciences, 39*, 1–45. https://doi.org/10.1017/S0140525X15000631.
Anderson, M. L., & Pessoa, L. (2011). Quantifying the diversity of neural activations in individual brain regions. In L. Carlson, C. Hölscher, & T. Shipley (Eds.), *Proceedings of the 33rd annual conference of the cognitive science society* (pp. 2421–2426). Austin, TX: Cognitive Science Society.
Anderson, M. L., Kinnison, J., & Pessoa, L. (2013). Describing functional diversity of brain regions and brain networks. *NeuroImage, 73*, 50–58.
Ascoli, G. A. (2002). Computing the brain and the computing brain. In G. A. Ascoli (Ed.), *Computational neuroanatomy: Principles and methods* (pp. 3–23). Totowa, NJ: Humana Press.
Bedny, M., Pascual-Leone, A., Dodell-Feder, D., Fedorenko, E., & Saxe, R. (2011). Language processing in the occipital cortex of congenitally blind adults. *Proceedings of the National Academy of Sciences, 108*(11), 4429–4434.
Berlucchi, G., & Buchtel, H. A. (2009). Neuronal plasticity: Historical roots and evolution meaning. *Experimental Brain Research, 192*, 307–319.
Bordens, K. S., & Abbott, B. B. (2014). *Research design and methods: A process approach* (9th ed.). New York: McGraw-Hill Education.
Bressler, S. L., & Kelso, J. A. S. (2016). Coordination dynamics cognitive neuroscience. *Frontiers in Neuroscience, 10*(397), 1–7. https://doi.org/10.3389/fnins.2016.00397.
Brigandt, I. (2010). Beyond reduction and pluralism: Toward an epistemology of explanatory integration in biology. *Erkenntnis, 73*(3), 295–311.
Carruthers, P. (2006). *The architecture of the mind: Massive modularity and the flexibility of thought*. Oxford: Oxford University Press.
Cat, J. (2017). The unity of science. In E. N. Zalta (Ed.), *The Stanford encyclopedia of philosophy* (fall 2017 edition). Retrieved May 15, 2019 from https://plato.stanford.edu/archives/fall2017/entries/scientific-unity/
Chervyakov, A. V., Sinitsyn, D. O., & Piradov, M. A. (2016). Variability of neuronal responses: Types and functional significance in neuroplasticity and neural Darwinism. *Frontiers in Human Neuroscience, 10*, 603. https://doi.org/10.3389/fnhum.2016.00603.
Churchland, P. S. (1994). Can neurobiology teach us anything about consciousness? *Proceedings and Addresses of the American Philosophical Association, 67*, 23–40.
Churchland, P. S., & Sejnowski, T. J. (2016). Blending computational and experimental neuroscience. *Nature Reviews: Neuroscience, 17*, 667–668.
Cosmides, L., & Tooby. (1987). From evolution to behavior: Evolutionary psychology as the missing link. In J. Dupre (Ed.), *The latest on the best: Essays on evolution and optimality* (pp. 277–306). Cambridge, MA: MIT Press.
D'Souza, D., & Karmiloff-Smith, A. (2016). Why a developmental perspective is critical for understanding human cognition. *Behavioral and Brain Sciences, 39*, 11–13.
Dehaene, S. (2005). Evolution of human cortical circuits for reading and arithmetic: The "neuronal recycling" hypothesis. In S. Dehaene, J.-R. Duhamel, M. D. Hauser, & G. Rizzolatti (Eds.), *From monkey brain to human brain: A Fyssen Foundation symposium* (pp. 133–157). Cambridge, MA: MIT Press.
Doya, K., Ishii, S., Pouget, A., & Rao, R. P. (Eds.). (2007). *Bayesian brain: Probabilistic approaches to neural coding*. Cambridge, MA: MIT press.
Edelman, G. M. (1987). *Neural Darwinism: The theory of neuronal group selection*. New York: Basic Books.

Edelman, G. M. (1988). *Topobiology: An introduction to molecular embryology*. New York: Basic Books.
Edelman, G. M. (1989). *The remembered present: A biological theory of consciousness*. New York: Basic Books.
Edelman, G. M. (2003). Naturalizing consciousness: A theoretical framework. *Proceedings of the National Academy of Sciences, 100*, 5520–5524.
Edelman, G. M. (2006). The embodiment of mind. *Daedalus, Summer*, 23–32.
Edelman, G. M., & Gally, J. A. (1967). Somatic recombination of duplicated genes: An hypothesis on the origin of antibody diversity. *Proceedings of the National Academy of Sciences, 57*, 353–358.
Edelman, G. M., & Gally, J. A. (2001). Degeneracy and complexity in biological systems. *Proceedings of the National Academy of Sciences, 98*, 13763–13768. https://doi.org/10.1073/pnas.231499798.
Edelman, G. M., & Gally, J. A. (2013). Reentry: A key mechanism for integration of brain function. *Frontiers in Integrative Neuroscience, 7*, 63. https://doi.org/10.3389/fnint.2013.00063.
Edelman, G. M., & Tononi, G. (2000). *A universe of consciousness*. New York: Basic Books.
Edelman, G. M., Gally, J. A., & Baars, B. J. (2011). Biology of consciousness. *Frontiers in Psychology, 2*(4), 1–7. https://doi.org/10.3389/fpsyg.2011.00004.
Érdi, P. (2008). *Complexity explained*. Berlin: Springer.
Favela, L. H. (2009). *Biological theories of consciousness: The search for experience*. (Thesis). San Diego State University.
Favela, L. H. (2014). Radical embodied cognitive neuroscience: Addressing "grand challenges" of the mind sciences. *Frontiers in Human Neuroscience, 8*, 796. https://doi.org/10.3389/fnhum.2014.00796.
Fernando, C., Karishma, K. K., & Szathmáry, E. (2008). Copying and evolution of neuronal topology. *PLoS One, 3*(11), e3775.
Fodor, J. A. (1998). *Concepts: Where cognitive science went wrong*. New York: Oxford University Press.
Freeman, W. J. (1975). *Mass action in the nervous system*. New York: Academic.
Freeman, W. J. (2005). A field-theoretic approach to understanding scale-free neocortical dynamics. *Biological Cybernetics, 92*(6), 350–359.
Frégnac, Y. (2017). Big data and the industrialization of neuroscience: A safe roadmap for understanding the brain? *Science, 358*(6362), 470–477.
Friston, K. (2010). The free-energy principle: A unified brain theory? *Nature Reviews Neuroscience, 11*(2), 127–138.
Fuchs, E., & Flugge, G. (2014). Adult neuroplasticity: More than 40 years of research. *Neural Plasticity, 2014*, 541870, 1–10.
Gallese, V. (2008). Mirror neurons and the social nature of language: The neural exploitation hypothesis. *Social Neuroscience, 3*(3–4), 317–333.
Gaser, C., & Schlaug, G. (2003). Brain structures differ between musicians and non-musicians. *Journal of Neuroscience, 23*, 9240–9245.
Goldinger, S. D., Papesh, M. H., Barnhart, A. S., Hansen, W. A., & Hout, M. C. (2016). The poverty of embodied cognition. *Psychonomic Bulletin & Review, 23*(4), 959–978.
Grumet, M., & Edelman, G. M. (1988). Neuron-glia cell adhesion molecule interacts with neurons and astroglia via different binding mechanism. *The Journal of Cell Biology, 106*, 487–503.
Hawkins, J., Lewis, M., Klukas, M., Purdy, S., & Ahmad, S. (2019). A framework for intelligence and cortical function based on grid cells in the neocortex. *Frontiers in Neural Circuits, 12*, 121. https://doi.org/10.3389/fncir.2018.00121.
He, B., Coleman, T., Genin, G. M., Glover, G., Hu, X., Johnson, N., et al. (2013). Grand challenges in mapping the human brain: NSF workshop report. *IEEE Transactions on Biomedical Engineering, 60*(11), 2983–2992.
Hurley, S. L. (2008). The shared circuits model (SCM): How control, mirroring, and simulation can enable imitation, deliberation, and mindreading. *Behavioral and Brain Sciences, 31*(1), 1–22.

Izhikevich, E. M. (2006). Polychronization: Computation with spikes. *Neural Computation, 18*(2), 245–282.
Izhikevich, E. M. (2007). *Dynamical systems in neuroscience: The geometry of excitability and bursting*. Cambridge, MA: MIT Press.
Izhikevich, E. M., & Edelman, G. M. (2008). Large-scale model of mammalian thalamocortical systems. *Proceedings of the National Academy of Sciences, 105*(9), 3593–3598.
Klein, C. (2010). Redeployed functions versus spreading activation: A potential confound. *Behavioral and Brain Sciences, 33*(4), 280–281.
Krichmar, J. L., & Edelman, G. M. (2002). Machine psychology: Autonomous behavior, perceptual categorization and conditioning in a brain-based device. *Cerebral Cortex, 12*(8), 818–830.
Krichmar, J. L., & Edelman, G. M. (2005). Brain-based devices for the study of nervous systems and the development of intelligent machines. *Artificial Life, 11*(1–2), 63–77.
Landhuis, E. (2017). Neuroscience: Big brain, big data. *Nature, 541*, 559–561.
Leversen, J. S., Haga, M., & Sigmundsson, H. (2012). From children to adults: Motor performance across the life-span. *PLoS One, 7*(6), e38830.
Mayr, O. (1970). *The origins of feedback control*. Cambridge, MA: MIT Press.
McCaffrey, J. B., & Machery, E. (2016). The reification objection to bottom-up cognitive ontology revision. *Behavioral and Brain Sciences, 39*, 16–18. https://doi.org/10.1017/S0140525X15001594.
McIntosh, A. R. (2000). Towards a network theory of cognition. *Neural Networks, 13*, 861–870.
Newell, A. (1990). *Unified theories of cognition*. Cambridge, MA: Harvard University Press.
Nishitani, N., Schürmann, M., Amunts, K., & Hari, R. (2005). Broca's region: From action to language. *Physiology, 20*, 60–69.
Poldrack, R. A., & Yarkoni, T. (2016). From brain maps to cognitive ontologies: Informatics and the search for mental structure. *Annual Review of Psychology, 67*, 587–612.
Pulvermüller, F. (2018). Neural reuse of action perception circuits for language, concepts and communication. *Progress in Neurobiology, 160*, 1–44.
Reid, D., Hussain, A. J., & Tawfik, H. (2014). Financial time series prediction using spiking neural networks. *PLoS One, 9*(8), e103656.
Sampaio, E., Maris, S., & Bach-y-Rita, P. (2001). Brain plasticity: 'Visual' acuity of blind persons via the tongue. *Brain Research, 908*(2), 204–207.
Scorzato, L. (2013). On the role of simplicity in science. *Synthese, 190*(14), 2867–2895.
Sejnowski, T. J., Churchland, P. S., & Movshon, J. A. (2014). Putting big data to good use in neuroscience. *Nature Neuroscience, 17*(11), 1440.
Seth, A. K., & Baars, B. J. (2005). Neural Darwinism and consciousness. *Consciousness and Cognition, 14*(1), 140–168.
Seth, A. K., & Edelman, G. M. (2007). Distinguishing causal interactions in neural populations. *Neural Computation, 19*(4), 910–933.
Sperber, D. (1994). The modularity of thought and the epidemiology of representations. In L. Hirschfeld & S. Gelman (Eds.), *Mapping the mind* (pp. 39–67). Cambridge, MA: Cambridge University Press.
Sporns, O. (2011). *Networks of the brain*. Cambridge, MA: MIT Press.
Sporns, O. (2012). *Discovering the human connectome*. Cambridge, MA: MIT Press.
Sporns, O., & Edelman, G. M. (1993). Solving Bernstein's problem: A proposal for the development of coordinated movement by selection. *Child Development, 64*(4), 960–981.
Sporns, O., Tononi, G., & Edelman, G. M. (1991). Modeling perceptual grouping and figure-ground segregation by means of active reentrant connections. *Proceedings of the National Academy of Sciences, 88*(1), 129–133.
Sporns, O., Tononi, G., & Edelman, G. M. (2000). Connectivity and complexity: The relationship between neuroanatomy and brain dynamics. *Neural Networks, 13*(8–9), 909–922.
Srinivasan, R., Russell, D. P., Edelman, G. M., & Tononi, G. (1999). Increased synchronization of neuromagnetic responses during conscious perception. *Journal of Neuroscience, 19*(13), 5435–5448.

Tononi, G., & Edelman, G. M. (2000). Schizophrenia and the mechanisms of conscious integration. *Brain Research Reviews, 31*(2–3), 391–400.
Tononi, G., Sporns, O., & Edelman, G. M. (1992). Reentry and the problem of integrating multiple cortical areas: Simulation of dynamic integration in the visual system. *Cerebral Cortex, 2*(4), 310–335.
Tononi, G., Sporns, O., & Edelman, G. M. (1996). A complexity measure for selective matching of signals by the brain. *Proceedings of the National Academy of Sciences, 93*(8), 3422–3427.
Tononi, G., Edelman, G. M., & Sporns, O. (1998). Complexity and coherency: Integrating information in the brain. *Trends in Cognitive Sciences, 2*(12), 474–484.
Tooby, J., & Cosmides, L. (1992). The psychological foundations of culture. In J. Barkow, L. Cosmides, & L. Tooby (Eds.), *The adapted mind: Evolutionary psychology and the generation of culture* (pp. 19–136). New York: Oxford University Press.
U.S. National Academy of Sciences. (2018). Definitions of evolutionary terms. *The National Academies of Sciences, Engineering, Medicine: Evolution resources*. Washington, DC. Retrieved July 19, 2018 from http://nationalacademies.org/evolution/Definitions.html
Uttal, W. R. (2005). *Neural theories of mind: Why the mind-brain problem may never be solved*. Mahwah, NJ: Lawrence Erlbaum Associates, Inc.
Uttal, W. R. (2016). *Macroneural theories in cognitive neuroscience*. New York: Psychology Press.
von Bernhardi, R., Eugenín-von Bernhardi, L., & Eugenín, J. (2017). What is neural plasticity? In R. von Bernhardi, L. Eugenín-von Bernhardi, & J. Eugenín (Eds.), *The plastic brain* (pp. 1–15). Cham: Springer.
Woodward, J. F. (2011). Data and phenomena: A restatement and defense. *Synthese, 182*(1), 165–179.
Zheng, Z., Lauritzen, J. S., Perlman, E., Robinson, C. G., Nichols, M., Milkie, D., et al. (2018). A complete electron microscopy volume of the brain of adult Drosophila melanogaster. *Cell, 174*(3), 730–743.
Ziegler, J. C., Montant, M., Briesemeister, B. B., Brink, T. T., Wicker, B., Ponz, A., et al. (2018). Do words stink? Neural reuse as a principle for understanding emotions in reading. *Journal of Cognitive Neuroscience, 30*(7), 1023–1032.
Zimmerman, A. W. (2008). Preface. In A. W. Zimmerman (Ed.), *Autism: Current theories and evidence* (pp. v–ix). Totowa, NJ: Humana Press.

Chapter 8
Saving Data Analysis: Epistemic Friction and Progress in Neuroimaging Research

Jessey Wright

Abstract Data must be manipulated for their evidential import to be assessed. However, data analysis is regarded as a source of inferential errors by scientists and critics of neuroscience alike. In this chapter I argue that of data analysis is epistemically challenged in part because data are causally separated from the events that they are intended to provide evidence for claims about. Experimental manipulations place researchers in epistemically advantageous positions by making contact with the objects and phenomena of interest. Data manipulations, on the other hand, are applied to material objects that are not in causal contact with the events they are used to learn about. I then propose that some of the inferential liabilities that go along with data manipulation are partly overcome through the occurrence of epistemic friction. I consider two forthcoming contributions to network neuroscience to illustrate the benefits, and risks, of the data analyst's reliance on epistemic friction.

8.1 Introduction

Debates about progress in the sciences of the mind and brain tend to be organized around technological innovations that have changed the epistemic landscape of neuroscience. This includes debates about the promise and prospects of neuroimaging technologies (Roskies 2010a; Klein 2010), the empirical potential of brain computer interfaces and other neural augmentations (Datteri 2009; Chirimuuta 2013), and the revolutionary status of optogenetic interventions (Bickle 2016; Sullivan 2018). These technologies are rightly recognized as significant. Afterall, they push neuroscience forward by providing researchers with the ability to create new kinds of data, to perform new interventions, and to test new hypotheses and theories. Attending primarily to measurement technologies, however, has led to

J. Wright (✉)
Department of Psychology, Jordan Hall, Stanford University, Stanford, CA, USA
e-mail: jessey.wright@gmail.com

© Springer Nature Switzerland AG 2021

F. Calzavarini, M. Viola (eds.), *Neural Mechanisms*, Studies in Brain and Mind 17,
https://doi.org/10.1007/978-3-030-54092-0_8

philosophers to overlook more common but less obviously impactful innovations in research methods. The particular innovations I have in mind are the development of tools and techniques for handling, manipulating, and analyzing data.

The importance of data analysis is hard to overstate. Consider functional magnetic resonance imaging (fMRI) research. In its early days, critics of the technology focused their attention on the subtractive method that was used to analyze fMRI data. This line of critique explicitly drew a continuity between the scientific utility of a measurement tool and the methods used to analyze data it produces (e.g., van Orden and Paap 1997; Uttal 2001). While historically used to pair cognitive processes with discrete parts of the brain, now neuroimaging technologies like fMRI are, for example, used to identify the information represented in patterns of brain activity and to articulate the relationship between network dynamics and cognitive capacities. The technology itself improved in this time, most notably with the recent approval of more powerful 7-Tesla scanners for human use. However, higher resolution measurements were not sufficient to lead neuroscientists to treat fMRI data as evidence for claims about, for instance, the representational content of brain activity. It was the development of multi-voxel pattern analysis methods, which preceded the human use of 7T scanners, that catalyzed efforts to search for neural representations with fMRI data (Haxby 2012; Horikawa and Kamitani 2017). Similarly, research into brain networks has been supported by the translation of network theory methods and principles into the context of neuroimaging research (Pessoa 2014; Thompson et al. 2017).

Another reason for philosophers of neuroscience to put more effort into identifying epistemically relevant aspects of data analysis is that data sharing has made it possible for scientists to have productive careers that do not involve the design of experiments or collection of data. Data sharing mandates and initiatives are motivated by the recognition that data are important for advancing our understanding of the world, and that they provide greater benefits to science and society when shared and reused (Leonelli 2016). Through data repositories such as OpenNeuro (Poldrack and Gorgolewski 2015) and large scale data acquisition initiatives such as the Human Connectome Project (Van Essen et al. 2012), the accessibility of imaging data has created an environment in which data are being analyzed by scientists who played no role in acquiring them. Accounting for how this division of cognitive labor and skills is influencing the trajectory of the neurosciences requires explicating the epistemic characteristics of data analysis.

The primary aim of this chapter is to advocate for data analysis by explaining how, in light of inferential risks inherent to the practice of data manipulation, data analysis protocols could possibly advance knowledge. To do this I propose that epistemic friction occurring when an analyst interacts with data helps to overcome some of the epistemic obstacles associated with data manipulation.

I proceed as follows: In the next section I provide an overview of neuroimaging experiments. Along the way I outline epistemic challenges that are associated with data analysis. In Sect. 8.3 I argue that data analysis techniques assist researchers in making judgements about the evidential significance of data. In Sect. 8.4, I locate the epistemic obstacles associated with data analysis in the separation of data from the causal forces that played a role in its production. Then, I draw on Jose Medina's epistemology of resistance (2012) to ground the search for 'epistemically frictional forces' that may be operative in the context of data analysis and interpretation. In Sect. 8.5 I examine two forthcoming contributions to methods in network neuroscience. I identify frictional interactions that occurred during the development, testing and validation of these contributions. The first is a critique of the way the participation coefficient, a derivable network measure, is applied in temporal network analyses (Thompson et al. 2020). The second is a new method for deriving network-level communities from time-series data (Thompson et al. 2019).[1] Finally, I conclude with a forward-looking reflection on epistemic friction.

8.2 Epistemic Gaps in Neuroimaging Research

Functional magnetic resonance imaging (fMRI) is one of the most widely used measurement technologies in human neuroscience. It is popular because it is non-invasive and can be used to investigate the relationship between brain activity and cognition in healthy human subjects. Broadly speaking, MRI scans measure magnetic properties of chemicals in the brain which are represented as volumetric pixels, or voxels. To do this, the scanner creates a uniform magnetic field within its bore. Then, radio pulses with specific frequencies are sent into the bore at regular intervals. By leveraging the magnetic properties of different chemicals in the body, scanning protocols can be created that are able to detect the location of different tissues within the bore. For instance, fMRI scans measure the blood oxygenation level dependent (BOLD) signal by leveraging the magnetic properties of hydrogen atoms. As neurons (and other cells in the brain) become more active, they need more energy. This causes oxygenated blood to flow in greater volume to the area and provide the cells with oxygen. This creates a local change in the ratio of oxygenated to deoxygenated blood. This change is what the BOLD signal is sensitive to (see Huettel et al. 2008 for an introduction to resonance imaging).

Neuroimaging data analysis pipelines are themselves quite intricate, consisting of dozens of distinct data transformations each aimed at addressing different epistemic gaps between the data and the claims scientists hope it will help them to evaluate.

The first stage of analysis is pre-processing. It is during this step that the most the artifacts and confounds which can be detected and corrected-for are eliminated. These include detecting and correcting for head motion, magnetic field drift, and, depending on the target of research, the small inhomogeneities in the magnetic field caused by differences in tissues. In addition to cleaning the data, pre-processing

[1] At the time this chapter was written the cases examined were pre-prints. Pre-print material was chosen because I had the ability to observe as these contributions were conceived, developed, and written up. It was through observing and collaborating on these projects that the philosophical perspectives presented in this chapter were developed.

also includes procedures that prepare the data for subsequent analyses. Functional scans, which capture the BOLD signal, need to be aligned with structural scans that represent the subject's brain in finer grained detail, and, data from each participant may need to be projected onto a common brain atlas so that data may be more easily compared between subjects.

Once the data are pre-processed, it is typical to model the hemodynamic response. This step is important because the BOLD signal, which is fundamentally a measure of how blood is flowing through the brain, is causally influenced by more than just neural activity that is relevant to the cognitive process researchers are interested in. Modelling a hemodynamic response is a procedure aimed at extracting the portion of the BOLD signal that corresponds to 'signals of interest'. For now, it is sufficient to note that, just as there are a variety of methods and parameter settings that can be used to pre-process data, there is no universally appropriate way to model the hemodynamic response.

It is at this stage that the more intricate statistical analysis methods are applied, including but not limited to, subtractive analyses, multi-voxel pattern analyses, or functional connectivity analyses. It is the results of these methods that are often presented as part of the evidential support for claims about how the brain and cognition are related. Like with each prior step in data analysis, each distinct research project and study is likely to have a uniquely tuned advanced analysis protocol.

Enthusiasm for fMRI is tempered by skepticism about the utility of the BOLD signal in the study of human cognition (e.g., van Orden and Paap 1997; Uttal 2001; Roskies 2010a; Aktunc 2014). Skeptics typically draw attention to two related challenges for fMRI research. The first is that neuroimaging technologies measure phenomena that are causally distant from the targets of research, such as the use of blood oxygenation to investigate neural activity. The second is that data manipulations, which seem to be used to overcome that distance, are themselves a source of inferential error.

The indirectness of measurements certainly presents investigators with a challenge, but it is not one that stops fMRI research from making progress. Afterall, measurements that are indirectly related to the targets of investigation are typical of neuroimaging. Diffusion tensor imaging (DTI) is another example. A DTI scan uses a dual echo protocol to 'tag' and 'track' the motion of water molecules. The first echo sends a magnetic pulse that adds spin to a subset of water molecules in the brain. Then, a second echo leverages the added spin factor to identify which of the tagged water molecules have remained stationary since the first echo. Information about water diffusion is not what researchers are interested in. Instead, it is used to infer the presence of long-distance neural projections in the brain. Long distance connections between neurons are made via axons sheathed in a fatty tissue called myelin. The inference from DTI scans to claims about bundles of axonal projections is based, in part, on the fact that it is easier for water to flow down the length of a myelinated axon than it is for the water to flow across the myelin (Assaf and Pasternak 2008).

Experimental design plays an important role in improving the epistemic circumstances of imaging research. For example, carefully designed and monitored tasks are used to ensure, as much as possible, that the cognitive and neural processes of interest are part of the nexus of causal factors that gave rise to the acquired data. Scientists interested in memory may have participants perform tasks in the scanner that require them to identify images as familiar or novel (Martin et al. 2018), while researchers interested in our ability to exert control over our actions might have subject's perform a variant of the stop signal task (Bissett and Logan 2011). Controlling the circumstances of data production and using indirect measures with known causal links to the targets of inference furnishes data with the potential to be used as evidence. However, even the best designed experiment in cognitive neuroscience doesn't result in data that can be immediately situated as evidence against or in support of theory. The realization of data's epistemic potential requires researchers to extract information relevant to their research questions from data. This is what data analysis is used to do, and why perceived flaws with analysis protocols are at the heart of most skeptical attacks on neuroimaging research (see Wright 2017 for a discussion).

Determining the extent to which information about a particular event or phenomena can be obtained from the available evidence is to evaluate what Currie calls the 'epistemic retrievability' of that event (2018). The more difficult it is to identify and exploit causal links between an event of interest and available data the lower its retrievability is (p. 125). Currie focuses on the historical sciences, drawing attention to challenges associated with linking 'traces' of the past, such as fossils, with the events from which such artifacts originate. The biggest challenging historical scientists face is that the causal factors responsible for the formation of a fossil, and the causal forces that acted on the specimen over the millions of years between its creation and the present, are not fully known. This is where explanations of and theories about the causal processes that lead to the formation of specific artifacts enter the picture. The richer and more sophisticated these theories are the better scientists are able to (a) identify causal links between the events of interest and data at hand, and (b) exploit those links to extract from information about those events from data. Using data to learn about the events involved in its production requires, at least to some degree, exploiting known and hypothesized causal links between events of interest and data.

The causal chain of events linking observations of blood oxygenation to underlying neural activity creates a dependency relationship between the intensity of the BOLD signal in a particular part of the brain and the intensity of the cellular activity that is a partial cause of the changes in local hemodynamics. This gives the BOLD signal the potential to be used as evidence for claims about neural activity.

What is known about the dependency relationship linking changes in neural activity to changes in the BOLD signal informs what can and cannot be inferred about neural activity from imaging data. For example, the BOLD signal is known to show the same change in magnitude if neural activity increases or decreases, making it unable to "...easily differentiate between function-specific processing and neuromodulation, between bottom-up and top-down signals, and it may potentially confuse excitation with inhibition" (Logothetis 2008, p. 877). Even though neural excitation may have been causally involved in an observed BOLD signal, that the BOLD signal is insensitive to the difference between excitation and inhibition renders it unable to provide evidence for claims about those kinds of fine-grained neural actions. In addition to placing limits on the claims that data can be used as evidence for, what is and is not known about the causal connections between the target of research and data at hand informs how data are analyzed.

Data analysis techniques are used to exploit the known and hypothesized features of the relationship between the BOLD signal and underlying neural activity to bring fMRI experiments to bear as evidence on claims about cognitive and neural processes (Buckner 2003). To transform the BOLD signal into patterns that reflect neural activity, a hemodynamic response function (HRF) is constructed. An HRF is a mathematical formula, or model, that relates observed changes in hemodynamics—the BOLD signal—to underlying metabolic activity associated with changes in neural activity. Decisions about the parameters and shape of the HRF are partly informed by the current state of knowledge about the causal relations linking the BOLD signal to neural activity (see Poldrack et al. 2011 for an introduction to fMRI data analysis). The full causal story that links the BOLD signal to neural activity is not known, and sometimes known details are not relevant to a particular analysis. Additional considerations including the specific hypotheses under test, how conservative researchers want their analysis results to be, and the quality of the data that they are analyzing play a role in HRF modelling (i.e., Lindquist et al. 2009).

On one hand, data analysis techniques, like modelling the HRF, are flexible and require researchers to make a substantial number of decisions to implement them. This flexibility is valuable because it allows analysis methods to be adapted to different experimental circumstances, allows models to be fit to different kinds of data, allows researchers to soften assumptions about the causal forces and entities of interest in the face of uncertainty about them, and enables exploratory research to be conducted efficiently. On the other hand, that there are many valid and defensible choices that can be made is one reason that analysis is regarded as a source of epistemic liabilities.

The rapid increase in the sophistication and variety of analysis methods available to researchers has prompted critical reflections on the negative impact of 'analytic flexibility', or the freedom to make choices during data analysis. As a matter of research pragmatics, researchers can't apply every possible method and parameter setting to their data and report all findings. They must make choices about what methods to use, and when to stop analyzing their data and write up a paper. As the number of decisions researchers make in the course of analysis goes up the probability that they will find a data pattern that supports a given hypothesis also goes up (Carp 2012). This implies that increases in the degrees of freedom researchers have during analysis correspond with increases to the frequency of false positive findings appearing in the literature. This concern has been reinforced by recent work showing that using different software tools to conduct the same fMRI data analysis procedures produce significantly different results (Bowring et al. 2019; see Taylor et al. 2018 for a response).

The epistemic status of data analysis is further complicated by the line of attack often adopted by skeptics about neuroimaging. Uttal was concerned that positive results in neuroimaging may primarily rest on decisions made about signal thresholding (2001), van Orden and Paap famously critiqued the logic of subtractive analysis (1997), and more recently Ritchie, Kaplan and Klein have challenged assumptions implicit in common uses of pattern classification analysis (2019). Each of these critiques identifies an inferentially undermining assumption that goes hand in hand with a specific approach to data analysis.

For example, pattern classification analyses are a now-popular method for bringing neuroimaging data to bear on hypotheses about the content of neural representation, or, more loosely, information available from patterns of brain activity. It involves training a machine learning classifier to assign cognitive labels to BOLD signal data, and then testing that classifier on novel data. If the classifier's accuracy is above chance this is often regarded as evidence that the labelled patterns of brain activity represent, or otherwise contain, information relevant to the cognitive categories labelled. Ritchie, Kaplan and Klein (2019) rightly criticize inferences that leap to claims about information represented in the brain from high classification accuracies. They offer a number of counterpoints, including the observation that this inference rests upon the assumption that the classifier is successful because it is leveraging information about cognitive processing that is latent in the signal. If this assumption doesn't hold, and there are good reasons to think that it often does not, then classification results cannot substantiate claims about information represented in neural signals.

Even optimists about the potential of neuroimaging research recognize that epistemic obstacles are integrated into the processes of data manipulation. Consider Roskies discussion of inferential distance in neuroimaging research (2010a). She characterizes the inferential distance between evidence and claims that it purports to be about as the number and certainty of the inferential steps one needs to take in order to move from evidence to claim. As the numbers of steps increase, or their relativity certainty decreases, the inference becomes less reliable. She argues that data manipulations increase inferential distance because of the assumptions that go along with choices made about which methods to use and how to implement them. As Roskies puts it, the problem is that "... the same raw data can produce different results depending on reasonable choices about data processing ..." (2010a, p. 203).

Data patterns may not only fail to be sensitive to the causal factors they are used to make inferences about, but they may even falsely appear to be explanatorily relevant because of a difficult to detect sensitivity to decisions made during the analysis process. Put succinctly: there is no guarantee that a difference between data patterns corresponds with differences in the causal factors that played a role in creating the data. This leaves us with a philosophical puzzle: how can data analysis play an essential role in neuroimaging research without corrupting the quality of the resulting inferences?

In the next section I take the first step towards addressing this puzzle: examining the epistemic status of data patterns. I propose that their primary role is to assist in evaluating the evidential import of data.

8.3 The Status of Data Patterns

The hemodynamic response function is an estimate of the neural activity underlying the observed BOLD signal. To create an HRF model from fMRI data is to transform it into, or extract from it, data patterns – HRF parameter values – that correspond with the underlying neural activity. Thus, one possible explanation for the utility of data manipulations is that data analysis facilitates the use of data as evidence by transforming it into an approximation of the results of an ideal experiment. That is, the greater the similarity between the product of a data analysis procedure and what the data would have looked like had it been acquired in an ideal experiment, the more information it provides about the target of investigation. Philosophical treatments of data analysis that have this flavor can be found in Mayo's error statistical philosophy (1996), Woodward's account of data interpretation (2000), and McAllister's treatment of data patterns as phenomena (1997).

If the epistemic significance of data patterns, which are the product of data analysis, are evaluated by their similarity to the hypothetical products of ideal experiments, then the sources of error outlined in the previous section are serious threats to the utility of neuroimaging research. Indeed, arguments that infer from epistemic problems with data analysis to skepticism about neuroimaging technology tend to rely on the assumption that the proper role of data manipulations is estimating the products of an ideal experiment (e.g., van Orden and Paap 1997; Aktunc 2014). Counter-arguments notice that, by looking at how data are evaluated in practice, this treatment of data analysis (Wright 2017), and style of criticism (Roskies 2010b), artificially restricts the epistemic roles that data and data analysis techniques can play. While some data manipulations are directed towards correcting for measurement error, or otherwise overcoming inferential distance by approximating the results of ideal experiments, this account does not generally capture what most analysis techniques do.

Modelling the HRF as well as many of the transformations automatically performed by an MRI scanner at the time of data acquisition (see Israel-Jost 2016) are examples of data transformations that are used to isolate, or extract, approximations of an ideal experiment within data. Techniques that eliminate artifacts arising from head motion and inhomogeneities in the scanner's magnetic field, as well data manipulations that conform data from different subjects onto the same 'brain template' so that different brains can be compared, are done to emphasize the signal in data. Most of these data transformations are classified as pre-processing. The aim of automatic processing done by the scanner, modelling steps like constructing an HRF, and pre-processing in general is to transform data such that they become similar to what the data would have looked like had the experimental circumstances been more ideal. That is, had the subject not moved, had the scanner's field been homogenous, were it the case that all brains have the same shape, or had it been possible to directly measure neural activity.

However, pre-processing data does not mark the end of data analysis. Most of the manipulations applied to data after preprocessing are not intended to create something that could have been obtained in an experimental setting or would even be created were ideal experiments possible. Instead, analysis methods are used to isolate patterns that are informative about the evidential import of data with respect to the targets of research. Patterns that are useful for assessing data's evidential significance need not, and often do not correspond with what would be measured if researchers had better experimental tools. As an example, consider functional connectivity analysis.

To show that two regions of the brain are functionally connected is to show that the time course of neural activity in those regions covaries (Friston 1994, p. 57). A functional connectivity analysis involves three steps. In the first, the brain is divided into 'parcels', or regions of interest. Parcellating the brain involves drawing lines along the cortical surface that mark the boundary between two distinct regions, or parcels. Once the brain is parcellated, researchers compute an activation time-series for each parcel. One way to do so would be to take the BOLD signal in each voxel within a parcel and average them into a composite BOLD measurement for each parcel. The BOLD signal time series from each parcel can then be compared. Parcels that have strongly correlated average BOLD signals are said to be 'functionally connected' or 'co-activated'.

Functional connectivity does not provide evidence of actual interactions occurring between the functionally connected parts of the brain. It only shows that activity in spatially distinct parts of the brain are, in some way, synchronized. To show that two parts of the brain are interacting is to identify an instance of effective connectivity (Friston 1994, p. 57). Effective connectivity is very difficult to establish in neuroimaging research. Under ideal measurement circumstances, such as if investigators had access to real-time information about when and how different parts of the brain were communicating with each other, there would be no need to calculate functional connectivity. Effective connectivity would be directly, or more directly, observable.

Not only does it not correspond with effective connectivity, which most cognitive neuroscientists would prefer to gather evidence about, but like many widely used analysis methods it is not fully understood what causal factors it is actually sensitive to. In particular, it is unclear what links there are, if any, between functional connectivity and the neural substrates that underlie the isolated data patterns (Horowitz 2003). Complicating the picture is suggestive evidence linking functional connectivity analysis to movement (Van Dijk et al. 2012), to noise in the global signal (Murphy and Fox 2017), and even research showing that subject's with split brains in which no physical connection exists between their two hemispheres display strong correlations in activity between the segregated regions (Uddin et al. 2008). These challenges to functional connectivity analyses have spurred investigations into the underspecified links between the isolated data patterns and neural functions.

It has since been shown that functional connectivity is sensitive to some changes in neural responses (Schölvinck et al. 2010; Chang et al. 2013). Even with all of this uncertainty, it's still a popular method for analyzing neuroimaging data.

The uncertainties inherent in data patterns, such as incomplete information about what causal factors an analysis method is sensitive to, is not a unique problem for interpreting the outputs of data analysis. Uncertainties are present in all stages of research, and accounting for that is an important task for the philosopher of science. Feest, for instance, construes the process of research as "... one of simultaneously exploring a specific subject domain and of applying, revising, and extending existing concepts" (2017, p. 1168). She locates uncertainty in research by arguing that the explanatory targets of psychology, such as 'working memory' or 'response inhibition', are best understood as 'epistemically blurry' insofar as "... the very question that empirical data are even descriptively relevant to the object in question is part of the investigative project" (p. 1167). Data analysis techniques are also epistemically blurry as the significance of a derived data pattern is itself something that must be determined in the course of research. The realities of day to day neuroscientific research are that investigators are applying epistemically blurry tools to advance their understanding of epistemically blurry targets of research.

The interpretation of a data pattern is not only determined by facts about how that pattern was arrived at, but also by auxiliary facts about data acquisition, and by comparison with other patterns derived via other methods. For instance, confidence that functional connectivity analysis may be indicative of coordination in information processing or some form of communication between spatially separated parts of the brain is partly based on graph-theoretic analyses. The relevant results show that networks identified with functional connectivity have an efficient 'small world topology' that allows for the effective integration and processing of information across distinct sub-systems of a network (see van den Heuval and Hulshoff Pol 2010). Consistent with Roskies' response to criticisms of subtraction, multiple data patterns, often derived from multiple data sets are used to triangulate on explanations (Roskies 2010b). Together, multiple patterns provide researchers with a more complete picture of the causal forces involved in data's production than a single pattern could.

Data patterns are the result of processes that selectively distort data, exemplifying some features and suppressing others. Theses transformations can introduce assumptions, suppress information relevant for evaluating the claims of interest, and may not even be the result of reliable processes since analysis outcomes can depend significantly on decisions made during their implementation (see Wright 2018). While the results of functional connectivity analysis may be epistemically blurry, clarity is not established prior to the application of the method, but instead is achieved as a consequence of its use. Interpretations are challenged and the analysis method itself is refined in subsequent empirical research. The epistemic drawbacks of analysis methods that stem from the uncertainties inherent in their application are, at least to some degree, addressed over time through community-level interactions with patterns the method isolates.

Like manipulations of experimental systems, data analysis involves intervening on an object of interest with the aim of revealing features that are meaningful to the analyst (Boem and Ratti 2016). At the same time, decisions involved in the selection and implementation of a data manipulation provide an opportunity for biases, implicit assumptions, and other source of inferential error to enter into the research process. In the next section I argue that data manipulations have inferior epistemic status when compared to experimental manipulations because of the separation between data and the causal factors that created them. Thus, the problem with data analysis isn't that it's flexible, it's that the manipulations are not appropriately constrained by the causal factors of interest. I then propose that data analysis may be better off than it appears if other epistemically frictional forces are at work when analysts are interpreting data.

8.4 Introducing Epistemic Friction

Controlled experimental manipulations are effective tools for confirming hypotheses and theories (Currie and Levy 2019). Data manipulations—as discussed at length above—are constitutive of the error-prone process of data analysis. The differences between the targets of these manipulations forms the basis for differences in their epistemic status.[2] Experimental manipulations alter the circumstances of data production, while data manipulations alter data.

The causal proximity of the targets of experimental interventions to the events they are used to learn about is appealed to in arguments that experiments have epistemic priority over computer simulations (Guala 2002; Roush 2018). It is this material correspondence between the objects and the targets of investigation that is often recognized as "... responsible for experiments' advantage over simulations in terms of inferential power" (Parke 2014, p. 519). Currie and Levy expand on this view, arguing that control and correspondence work together to grant experiments their confirmatory power. Control is important because it provides insight into the materials under investigation (forthcoming, p. 5). On their view, the correspondence between the objects manipulated and the target of inference does not have to be material. Instead, the manipulated objects just need to be a representative member of a broader class by virtue of "... sharing focal properties with the target" (p. 7). The sharing of focal properties, they argue, is what enables researchers to generalize

[2]Those familiar with neuroimaging research may notice that experiments, and so experimental manipulations, are often designed with the data manipulations that will be carried out downstream in mind. In this way, experimental manipulations are methodologically beholden to data manipulations and so it may seem odd to classify one as epistemically inferior to the other. It is important here to note that the use of shared and otherwise open access data has begun to decouple experimental design from analysis design. As it becomes more common for researchers to analyze and interpret data that they did not produce it is important to consider the data and experimental manipulations as disentangled processes. I thank a reviewer for pressing this point.

what is learned about the targets of investigation in the lab to similar phenomena that occur under less controlled circumstances.

Data manipulations are applied to the products of experimental manipulations. If data manipulations are epistemically inferior, then something beneficial must be lost in the transition from experimentation to data interpretation. Morgan's distinction between surprising and confounding results is useful here (2005). Surprising observations are unexpected. Confounding observations are unexplainable with the theoretical and conceptual resources available to investigators, and so provoke further inquiry.

Unexpected experimental observations can be confounding because explaining them may require discovering new causal aspects of the experiment. By confounding researchers, experiments lead to learning more about the circumstances of the experiment. Unexpected data patterns typically lead to learning more about the data manipulation that produced them because the unexpected results can often be explained by appealing to the decisions that went into creating those patterns.

When experimental manipulations are found to be flawed, such as when an experiment fails to replicate or when observations are incongruent with theoretical predictions, divergent results can often be explained by appealing to the circumstances of data production. This is one reason offered for publishing and investigating replication failures. Failures, if explanations for them are sought, can lead to discoveries (Crandall and Sherman 2016, p. 98). Discovering flaws in an experimental manipulation often reveal something about the causal factors involved in producing the data or the backgrounding theory that was used to devise an experiment. Alternatively, in domains like psychology, where individuating the phenomena of interest is a primary task of research, incongruent experimental findings are useful for exploring the boundaries of conceptual and theoretical constructs (Feest forthcoming).

Differences between data patterns, on the other hand, can be explained by a variety of factors that have nothing to do with the circumstances of data production, such as method selection, model parameter choices, programming errors, and oversensitivity to noise in data. Discovering that a manipulation of fMRI data is sensitive to head motion advances knowledge about the method itself but does not provide additional insight into the neural or cognitive phenomena it is being used to study.

Experimental manipulations are valued for their potential to confirm hypotheses and uncover new facts about the world. They are able to play both of these roles because experimental interventions change objects of interest or change the conditions under which those objects are observed. That is, good experimental manipulations interfere with causal processes in a detectable way (Woodward 2003). Data manipulations, on the other hand, interfere with the data produced by an experiment. Data are, once produced, separated from the causal forces that generated them. This causal separation underwrites the epistemic inferiority of data manipulations. When researchers experimentally intervene on a system there is, to speak metaphorically, friction created between the target of investigation and the means of measurement.

Friction is important for knowledge production in general. Sher proposes that 'epistemic friction', which reflects the constraints on how we obtain and formulate knowledge claims, is one of two necessary principles for knowledge. When a knowledge generation process is constrained by the world it has epistemic friction. This is important because which helps to avoid developing 'idly hovering theories' that do not accurately describe the world (2010). Sher recognizes that friction, constraints, and resistance to the formation of knowledge claims can arise from a variety of places. Sources of epistemic friction include our standards for theorizing, rational or pragmatic desiderata, and from physical constraints set by the world itself.

Sher's wide view of friction provides some hope for the data analyst. Data manipulations do not make contact with neural and cognitive causal forces they are used to learn about, and so those causal factors cannot directly impose constrains on analysis results. The lack of direct friction between the world and the analysis, however, does not undermine the epistemic utility of data analysis as it is not the only relevant source of epistemic friction. During the process of data analysis and interpretation friction may arise from community-set standards and best practices for analysis, from the anticipation of negative reviews, or from critical and constructive interactions with lab mates and collaborators. If this is the case, then concerns about the efficacy of data analysis could be addressed by fine tuning research practices to ensure that appropriately resistive frictional forces are present in all stages of the data interpretation process.

Concerns about the epistemic obstacles inherent in data analysis such as analytic flexibility are not directed at community level practices. They are concerns about how data analysis is implemented and executed by individuals. Thus, relevant sources of resistance that might improve the circumstances of data analysis are those that affect scientists as they make decisions about parameters, models, software and ultimately determine the significance of the data patterns they discover. These decisions are made by scientists as they work write and test their code or software, and so are influenced by scientists own internal beliefs and perspectives. These decisions are also made through community interactions, be those conversations with colleagues in the halls, lab mates in the office, or peers during conference presentations or lab.

Medina argues that epistemic resistance, or friction, is valuable because it cultivates epistemic virtues of openness, curiosity/diligence, and humility (Medina 2012, chapter 1). An agent exhibits openness when they are attentive to and take seriously the perspectives of others, curiosity/diligence when they meet epistemic challenges head on and actively seek new information, and humility when they recognize that their own knowledge has gaps and limitations. When an agent encounters too much or too little friction epistemic vices such as closed mindedness, laziness, and overconfidence are cultivated instead. It is noteworthy that these virtues involve engaging with beliefs and points of view that are external to the agent in question. The vices push agents to disregard, via overconfident, or avoid through laziness, external sources that conflict with their beliefs. The epistemic value of diversity is a common feature of socio-epistemic accounts of progress in science (e.g., Borgerson 2011; Longino 2012), and is visible in frictional interactions at the community-level.

Debates about the efficacy of a method like functional connectivity advance knowledge because they involve different researchers with different points of view. Criticisms of functional connectivity cast doubt on its capacity to provide evidence of communication between regions. This lead to investigations into the causal dependencies between neural actions and functional connectivity. The participants in this debate have different stakes and interests in functional connectivity, and so are able to productively resist each other's points of view. This suggests that ameliorating the obstacles inherent to data analysis may require conflicts with the analyst's internal beliefs, desires, or expectations to arise during the interpretation of data.

In the next section I elaborate upon this notion of epistemic friction by considering examples of friction that arise during analysis method development and critique.

8.5 Friction in Network Neuroscience

The most recent trend in network neuroscience has been to use tools concurrently developed in temporal network theory (Holme and Saramäki 2012) to examine how brain networks change from moment to moment (Lurie et al. pre-print).

A network consists of nodes related to one another via edges. Creating a network requires dividing data into nodes, determining which nodes should be connected by edges, and quantifying the strength of each connection. Nodes are the members of a network, such as people or organizations in a social network. Edges represent relationships between nodes. In a friendship network, for example, edges may connect nodes if the people they represent are friends (Fig. 8.1).

A temporal network often consists of a collection of sequentially ordered static networks. The static networks that make up a temporal network are snapshots. Since snapshots are static researchers have to make all of the decisions and perform all of the transformations necessary to conduct static network analyses. This includes deciding how to divide the brain into nodes such as by anatomically individuating regions and providing criteria that define which nodes are connected by edges and the strength of those connections such as by using functional connectivity analysis. In addition to this, to create a temporal series of networks researchers must also decide how to individuate snapshots in time which involves choosing or devising an analytic procedure that extracts a 'moment' or window of time from the data over which to calculate a static network.

Once a network representation of data has been created its properties and features can be derived. With a static network a researcher can identify properties of specific nodes, such as their participation coefficient, or examine how nodes within the network group together into communities. Communities are collections of nodes that have stronger connections with each other than they do with nodes outside the community. Identifying community structure requires assigning a community

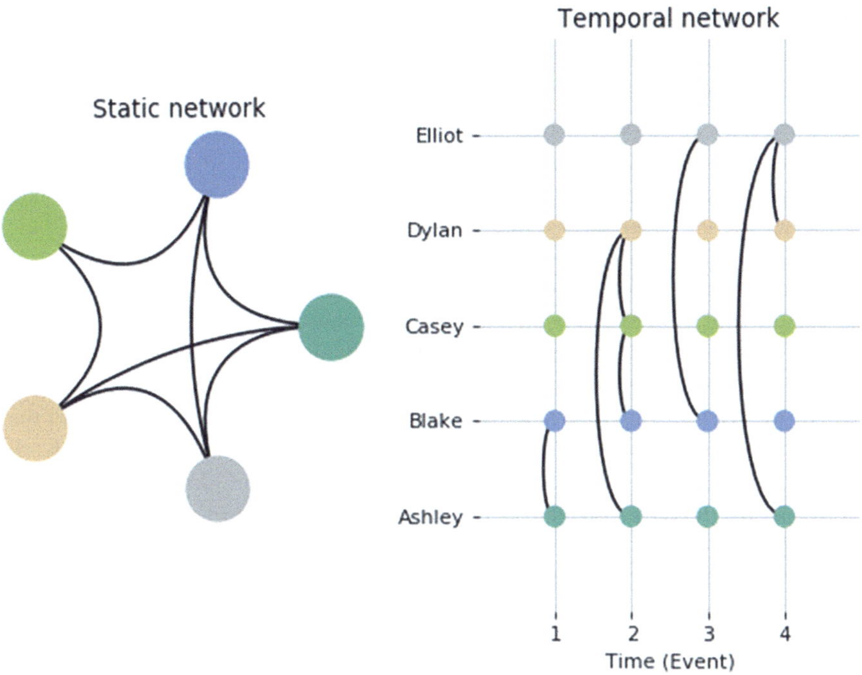

Fig. 8.1 Static and temporal networks
Caption: The left image shows a static social network. Each circle is a node that represents an individual, and each black line is an edge connecting two individuals. The right image shows a temporal network with the same nodes. The horizontal lines correspond to the nodes of the network and each vertical slice is a temporal snapshot. Along the vertical slices black lines indicate which nodes are connected during that snapshot. At time 1 only Blake and Ashley are connected, and at time 3 only Blake and Elliot are. The temporal network representations are also ordered in time, and so the network shown at time 1 occurred prior to the network shown at time 4. (Source: https://teneto.readthedocs.io)

identity to each node within a network. With a temporal network a researcher can further investigate how properties of nodes and the structure of the network change from snapshot to snapshot.

In the rest of this section I consider two forthcoming contributions to network neuroscience methods. The first is a critique of how the participation coefficient, a static network measure, is commonly used in temporal network analyses (Thompson et al. 2020). This case shows how the absence of epistemic friction can lead to the misuse of analysis methods. The second case examines the development of the temporal community by trajectory clustering (TCTC) method for inferring community structure directly from time series data (Thompson et al. *pre-print*). I use this case to highlight how anticipating and reducing epistemic friction facilitates the development and uptake of new methods. I will highlight other instances of epistemic friction along the way.

8.5.1 Participation in Time: Looking for Friction

Identifying parts of a network that may be particularly sensitive to damage, or otherwise play an important role in facilitating communication between and the coordination of disparate parts of the brain is a valuable use of network analyses. A node that has these properties is classified as a hub. One way to identify a hub is to calculate the 'degree' of a node, which, conceptually, involves counting how many edges it has. While its degree gives a sense of how strongly connected a node is, it doesn't account for the diversity of those connections. Power and colleagues noticed this and showed that node degree is more impacted by the overall size of the network than it is the communicative role a node plays within that network (Power et al. 2013). They proposed that the participation coefficient should be used instead of node degree to identify hubs, in part because it is sensitive to the diversity of the connections that a node has (see also van den Heuval and Sporns 2013). This work establishes the participation coefficient as a good measure for identifying hubs and estimating the degree to which a network is integrated.

The participation coefficient is one property of nodes within a network that has proven to be particularly insightful for cognitive neuroscientists. The participation coefficient of a node can be calculated if the network has been subdivided into communities. The participation of a node is measured by contrasting how many of its edges connect it to nodes that are part of communities outside of its own (Guimerà and Nunes Amaral 2005). A node with a zero-participation coefficient has no edges that connect to nodes outside of its community. That is to say, it is only interacting with nodes that are within its own community. A node with a participation coefficient of 1 has at least one edge connecting to every other community in the network.

If the participation coefficient provides some evidence that a given node or region is a hub, then evaluating how participation coefficients change over time could help zero in on more specific functional attributions for regions of the brain. Indeed, several influential articles have used the participation coefficient in temporal networks to address questions about the role of network-level integration and segregation in task performance (e.g. Betzel et al. 2015; Shine et al. 2016; Pedersen et al. 2017; Fukushima et al. 2018).

Consider Shine and colleagues work examining how the dynamics of network-level activity relates to the successful performance of cognitive tasks (2016). The broad theoretical aim of their investigation is to provide evidence that global network integration is important for effective cognitive performance (p. 544). A network is more integrated when there are more connections between its parts and segregated when parts of the network are relatively isolated from one another. To estimate how network integration changes over time they needed a network measure that reflects the strength of between region connectivity. Citing the work of Power and colleagues as justification for the decision, they choose to use the participation coefficient for this. This decision is an example of the community acceptance of a measure reducing friction during analysis.

This is not to say that the decision to use the participation coefficient in this way was made lightly by Shine and colleagues. I am merely noting that the existence of evidence than an analysis method tracks properties of interest, such as results showing that node participation corresponds with the hub-status of the node, helps researchers to choose a method or parameter value more rapidly than if there were no empirical results to appeal to or consider.

Shine and colleagues compared how the participation coefficients of nodes changed over time and found that they tended to increase during tasks. From this they inferred that "... that the brain transitions into a state of higher global integration in order to meet extrinsic task demands" (p. 546). A forthcoming critique of the participation coefficient as used in temporal network analysis raises a subtle problem for this interpretation (Thompson et al. 2020).

To conclude from a change in participation coefficient between network snapshots that the network is more integrated it must be assumed that differences in participation between network snapshots are comparable. However, different network snapshots are typically allowed to have different community structure. That is, from one snapshot to the next the overall number and distribution of communities can change. This is a problem because the participation coefficient of a node depends on the community identity of it and its neighbors. In other words, when the community structure of a network is allowed to vary over time, then the participation coefficient of a node becomes sensitive to its own connectivity and to the overall community structure of the network. In the paper critical of how participation is measured in temporal networks (Thompson et al. *preprint a*), the authors estimate, using fMRI data, that if community structure was held fixed across snapshots, then 66% of nodes change their participation in the opposite direction compared to when community structure is allowed to vary. The problem, put simply, is that the participation coefficient applied to a temporal network is sensitive to more properties of the data than its interpretation as evidence of network integration allows.

The participation coefficient is a well-established measure of network integration. In fact, it is one of the more well established and widely used analysis techniques amongst those that were used in the paper summarized above (Shine et al. 2016). It is not surprising that comparisons of participation coefficients through time have not yet been explicitly tested to verify that the measure was sensitive to only between community connectivity. There is no salient, *a priori* reason to doubt the efficacy of the participation coefficient in a network analysis context, especially given that there is consensus amongst network neuroscientists that the participation coefficient is a good indicator of network integration. What is more surprising is that it was closely examined at all.

While identifying a low friction decision can explain how the participation coefficient has been systematically misapplied in temporal analyses, a moment of high epistemic friction was the occasion for this critique being conceived.[3]

In a 2017 article outlining temporal extensions for measures from static network theory for fMRI researchers was published by the lead author of the participation coefficient critique (Thompson et al. 2017). Two of the measures presented in that paper were criticized during the author's dissertation defense for being classified as 'temporal' while failing to leverage temporal information in the data. There was nothing inherently wrong with the measures, only that they were mislabeled as 'temporal', and that this mislabeling may lead to misuse of the measures. The problem with calling them 'temporal' is that events are ordered in time, and neither of the two measures criticized are sensitive to that ordering.

When the network neuroscience community began to use the participation coefficient as part of temporal network analysis, the researcher decided to add the ability to calculate the participation coefficient into a software package they created and maintain. They had, due to that comment made during their defense, formed a habit of checking measures and algorithms more carefully before incorporating them into the package or using them in their research. In checking into the participation coefficient, they noticed that differences between temporal networks might make differences in participation difficult to interpret. The end result of this investigation was the critique partly summarized above, and the creation of a method for calculating node participation that is less sensitive to changes in community structure over time (Thompson et al. 2020).

Friction appears throughout this discussion. Low epistemic friction during analysis decisions partially explains why a method was misapplied, and higher friction in a different circumstance led to research revealing those interpretive errors. The critique itself will, if it has an impact once it is published, become a source of friction for researchers interested in node participation in temporal networks.

While reducing epistemic friction by appealing to literature exploring the utility of the participation coefficient contributed to the misuse of the method, pursuing the goal of reducing friction can, in different circumstances, be productive. This is the focus of the next case.

8.5.2 Dynamic Communities: Reducing Friction in Practice

Each discrete analysis step distorts data. The more steps there are in an analysis procedure, the more opportunities there are for noise to compound and interfere with the final results. Creating temporal network representations from BOLD signal data and assigning a community identity to each node has three steps. The first is

[3]The remainder of this section is partially autobiographical in content. The information reported here was obtained through conversations and collaborations with the scientist discussed.

to define the nodes of the network, such as by dividing the brain into anatomically differentiated regions. Then, edges between those nodes need to be determined. The BOLD signal is time series data and is what might be called 'node collected' because it is a continuous measure of changes in voxels, and groups of voxels correspond with nodes in brain networks. The alternative is 'edge collected' data, which describes a situation in which measurements directly pertain to the edges that exist between nodes. Counting interactions between friends in a social network is edge-collected data since the observations are about the connections between friends. When dealing with node collected data researchers have to use data manipulations like functional connectivity analysis to infer edges and their weights. Once edges are inferred, the third step is to assign the nodes to communities.

Temporal Communities through Trajectory Clustering (TCTC) is a new method for identifying how community structure changes in time that performs the last two steps of this process in a single transformation (Thompson et al. *pre-print*). TCTC groups nodes together into communities when their corresponding time series' fall within the same trajectory. For a group of nodes to be part of the considered part of the same trajectory the correlations between their BOLD signals must meet four criteria set which correspond with the algorithm's parameters. Two parameters, the tolerance rule and distance rule, control how much error is allowed. The size-rule determines the smallest community size, and the time rule determines how long a community-like arrangement of nodes has to be in synchrony to count as a community (Fig. 8.2).

As described, it seems like TCTC is poised to make the epistemic situation of network neuroscience worse.[4] Afterall, while the method may eliminate a step from the analysis process and so eliminate a source of error, there are four parameters that researchers have to specify in order to apply TCTC. This has the potential to increase the number of decisions researchers have to make during analysis. Furthermore,

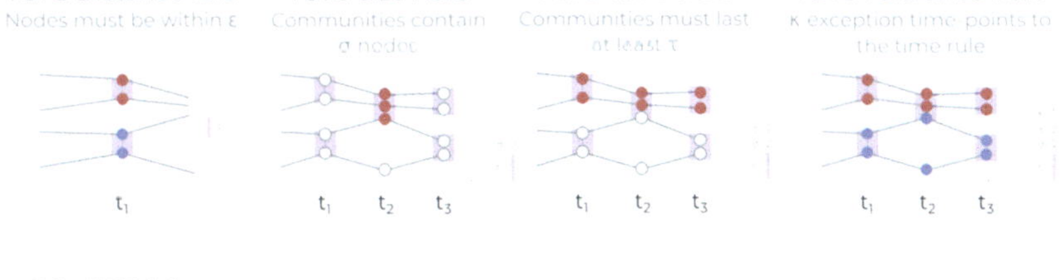

Fig. 8.2 TCTC four parameters
Caption: To be classified as a trajectory by TCTC a collection of nodes must satisfy each of the four depicted rules. In the above diagrams each rule is illustrated with four time series (the black lines) at one or three discrete time points. The boxes indicate where the distance rule is successfully satisfied. White nodes indicate when no trajectory is identified. (Source: Thompson et al. Pre-Print. Used with Permission under CC-BY-NC-ND 4.0 International license)

[4]I owe thanks to an anonymous reviewer for raising this challenge.

TCTC is not a wholly new method, but an alternative approach for performing analyses that network researchers already have protocols for. Creating it increases analytic flexibility as it is another method a researcher may choose to use.

If this method is to improve the epistemic situation of network neuroscience then it must be shown that the it has advantages over existing methods, and it needs to offer some epistemic benefits to compensate for worsening the problems that follow from analytic flexibility. Epistemic friction provides useful handles for evaluating the epistemic potential of new methods like TCTC. Method development can be framed as a process of anticipating and reducing epistemic resistance. Furthermore, one of the advantages TCTC has over a competitor method is that it can reduce a particular source of epistemic friction in analysis. I consider each of these in turn.

TCTC is designed to identify temporal dynamics in community structure. If the results of TCTC are averaged over time it should produce patterns similar to those generated by static community detection methods. To show that TCTC produces minimally reliable patterns it was applied to time-averaged open-access neuroimaging data. The analysis recovered time-averaged differences between sessions of resting state scans that were expected to be found in that data. Furthermore, the community structure differed when different tasks were compared. This shows that TCTC produces the expected when it is used to perform a time-averaged network analysis of an openly accessible fMRI dataset. A dataset that has been analyzed in hundreds, if not thousands, of studies. That is, TCTC produced results that are consistent with the currently accepted findings within the field.

This kind of demonstration provides a baseline level of confidence for a new method. Philosophers of science have identified similar bootstrapping practices in the development of new measurement technologies (Hacking 1981; Bechtel and Stufflebeam 1997), and so it is not surprising to find it playing a role in the development of new analysis methods. While it provides some confidence in the method's reliability, it is not itself sufficient to show that the method has epistemic advantages over alternatives. Afterall, the primary use for TCTC is to reveal temporal dynamics in community structure. Applying it to time-averaged data will not show that it can do this.

Recall that data patterns are informative about events underlying data to the degree that they are sensitive to causal dependencies that connect those events to the data they are derived from. Showing that a pattern in fMRI data is sensitive in this way can be done directly by conducting a multimodal study, such as using direct neural recordings in conjunction with fMRI. This is not common, especially for a new method, as it requires having access to appropriate materials and measurement technologies.

Another way to evaluate the sensitivity of an analysis method is through simulation. In a simulation data are fabricated with known internal structure or 'ground truth'. The method is applied to the fabricated data and ideally recovers the structure that was placed there. In the original draft of the TCTC paper simulations were not included in part because they do not accurately correspond with the epistemic circumstances of research. In a simulation the ground truth is known, while in most circumstances of research it is unknown. On one hand, simulating analyses can be useful for determining what kinds of patterns a method is sensitive to. On the other, it is difficult to generalize from successful simulation results to actual experimental conditions because of the lack of correspondence between the epistemic stances researchers have in each context. However, the first round of reviewers requested simulations and so they have since been included in the paper. In this case, the need to reduce friction between prospective users and the method outweighed the desire to avoid reducing friction in hypothetical scenarios of method misuse.

The primary case made for TCTC's results being informative and offering an advantage over currently used community detection methods is an argument that TCTC has less inherent friction than alternatives. The criteria TCTC uses for community detection, by design, refers to low level features of the network. To illustrate how this is an argument for pattern interpretability, consider the temporal extension of the Louvain algorithm for community detection, which is an alternative to TCTC (Mucha et al. 2010). This method has two parameters. The resolution parameter determines how communities are identified, and the coupling parameter determines how adjacent snapshots influence one another. When setting the optimization parameter for Louvain community clustering researchers are deciding how strong the overall network modularity should be (Meunier et al. 2009).

TCTCs parameters are grounded on facts about the relationships between the nodes themselves, not a meta-property of the communities or network such as modularity. The size and time rules, for example, are parameters that explicitly place limits on how small a community can be and how long a group of nodes need to be coordinated to count as a community. Additionally, because the parameters refer to low level properties of the network, if the investigators have a sufficiently large collection of data to do so robustly, a machine-learning inspired training protocol can be used to determine the optimal settings for those parameters empirically.

Revisiting the concerns raised above about analytic flexibility, TCTC, in addition to eliminating some sources of error by skipping the edge determination step and all of the parameter decisions that might go into that, offers researchers parameters that can be computationally optimized and concretely interpreted. While, from one point of view, these are additional degrees of freedom, they also, by being computationally optimized and directly interpretable, make it easier for researchers to reduce friction with the method when applying it. In a way this means that TCTC has lower epistemic friction for analysts than the Louvain algorithm because the parameters are easier to conceptually grasp and empirically determine values for. This is an epistemic advantage not because there is less resistance, but because the resistance analysts experience when being forced to decide parameter values has easier to traverse avenues for resolution. This suggests that, in addition to considering instances and sources of friction, it is important to evaluate if and how a source of friction can be and is overcome in practice.

A new method like TCTC is unlikely to be more than a curiosity if it doesn't offer something above and beyond interpretability. To receive uptake, it needs to create new opportunities for examining data. In terms of friction, it must reveal data patterns that can provoke frictional interactions with existing theories and judgements of data's evidential import. That is, it needs to transform data in a way that is meaningfully different from the available methods.

The most straightforward opportunity for creating this kind of friction that TCTC offers is that, unlike other methods for community assignment, it allows nodes to belong to multiple communities at once, or to belong to no community at all. This means that the algorithm isn't forced to 'make a decision' about the community identity assigned to nodes that, according to its criteria, have ambiguous community membership. Thus, TCTC could be applied to data that has been analyzed with less flexible community assignment criteria to identify nodes that may have indeterminate community identities. This may, depending on how such an analysis turns out, raise challenges for currently accepted theories and network-level explanations fMRI data.

Another potential source of friction that arises from TCTC is that the community dynamics it reveals correlate with trial-by-trial behavior. As a demonstration of this, TCTC was used to identify five community configurations that best explain the variance in BOLD signal data collected concurrently with the performance of a 2-back task. A 2-back task requires subjects to press a button if the stimuli they are presented with matches the one presented two trials earlier. It was found that many of the community configurations were associated with different behaviors. Some network configurations were associated with multiple behaviors at different times. For example, the same community configuration present earlier in the trial may increase reaction time accuracy and, later in the trial, increase accuracy. Further, multiple community configurations impact the same behavior. For example, multiple configurations, at different times during the trial, correlated with reaction time.

These results show that TCTC has the potential to create friction in the field for two reasons. Firstly, it shows that TCTC can access information at the scale of trial-by-trial behavior. This alone is remarkable for neuroimaging research where the standard practice is to average data across hundreds of trials to overcome the poor signal to noise ratio of the measurements. Secondly, these preliminary TCTC results introduce a new variable into the standard brain mapping formula.

Early fMRI research was characterized by spatial mappings in which the question to be answer was "where does this cognitive process occur?" More recently, techniques like temporal network analysis have allowed cognitive scientists to use fMRI to ask, "when does this cognitive process occur?", a question previously reserved for imaging methods with higher temporal resolution such as EEG. Through TCTC, it may become possible to examine the brain's role in cognition in terms of its parts, their internal temporal dynamics, and their overall network configuration. That is, to investigate when, where and what networks in the brain are doing with fMRI.

Just as data do not emerge from an experiment ready to use as evidence, data analysis methods rarely produce patterns that clearly indicate what causal factors played a role in shaping the data. As was the case with functional connectivity, these early demonstrations of TCTC will not be the last word on its epistemic utility. It will take a community of researchers trying to use the method and challenging it for the full scope of its assumptions and error characteristics to be determined. Whether or not that work is and can be done is contingent on the friction that the method induces as results using it are published, and the friction investigators encounter when trying to apply it.

8.6 Conclusion

The evidential import of data are assessed through their manipulation. The process of data analysis is epistemically challenging in part because data are causally separated from the events that they are intended to provide evidence for claims about. Experimental manipulations place researchers in epistemically advantageous positions by making contact with the objects and phenomena of interest. Data manipulations, on the other hand, are applied to material objects that are not in causal contact with the events they are used to learn about. I have argued that some of the inferential liabilities that go along with data manipulation are partly overcome through the occurrence of epistemic friction. Each of the instances of friction identified above included a reexamination of the epistemic circumstances of research. It is in this moment of reexamination that an analyst evaluates and reconsiders parameter choices, recognizes the importance of decisions already made and thought to be innocuous, and takes steps to eliminate the frictional interaction and continue with their work.

While the participation coefficient case suggested that low friction can lead to inferential errors, such as the misuse of an analysis method, the TCTC case showed how reducing friction is one way for a method to help move a field forward. By providing parameters that can be optimized to fit data and are more readily interpretable, TCTC makes it both easier to examine temporal dynamics in brain networks and easier to evaluate the significance of the patterns it isolates. Whether or not a data analysis procedure is epistemically advantageous is not a matter of abstract facts about the decision's researchers had to make, but a matter of how much friction was involved in each of those decisions, and how the investigators dealt with that friction.

I hope to have inspired interest in examining the circumstances of data analysis and discussing the positive epistemic roles played by data manipulations in neuroscience. Because, whether or not philosophers of neuroscience attend to them, data analysis methods will continue to have a substantial impact on the trajectory of research, especially as data become more accessible and analysis software becomes easier to use. How data are manipulated is a significant driver of progress in modern science. If philosophical analyses are to remain relevant and sensitive to current trends, we ought to attend as much to the epistemic characteristics of data analysis as we do to data production, measurement, and theory.

Acknowledgements I owe thanks to William Hedley Thompson for providing substantive comments on several drafts of this chapter, and the Poldrack lab at Stanford for the opportunity to be a member of their lab, as both an observer and collaborator. Adrian Currie, the editors of this book, and an anonymous reviewer provided incredibly helpful comments on an early draft. The National Science Foundation's STS program, and Social Sciences and Humanities Research Council of Canada provided financial support for this research.

References

Aktunc, M. E. (2014). Severe tests in neuroimaging: What we can learn and how we can learn it. *Philosophy of Science, 81*, 961–973.

Assaf, Y., & Pasternak, O. (2008). Diffusion tensor imaging (DTI)-based white matter mapping in brain research: A review. *Journal of Molecular Neuroscience, 34*, 51–61.

Bechtel, W. P., & Stufflebeam, R. S. (1997). PET: Exploring the myth and the method. *Philosophy of Science, 64*, S95–S106.

Betzel, R. F., He, Y., Rumschlag, J., & Sporns, O. (2015). Functional brain modules reconfigure at multiple scales across the human lifespan. *ArXiv*, (1510.08045v1). Accessed July 2019.

Bickle, J. (2016). Revolutions in neuroscience: Tool development. *Frontiers in Systems Neuroscience, 10*, 1–13.

Bissett, P. G., & Logan, G. D. (2011). Balancing cognitive demands: Control adjustments in the stop-signal paradigm. *Journal of Experimental Psychology: Learning, Memory, and Cognition, 37*, 392–404.

Boem, F., & Ratti, E. (2016). Towards a notion of intervention in big-data biology and molecular medicine. In *Philosophy of molecular medicine: Foundational issues in research and practice* (pp. 147–164). New York: Routledge, Taylor & Francis Group.

Borgerson, K. (2011). Ammending and defending critical contextual empiricism. *European Journal for Philosophy of Science, 1*, 435–449.

Bowring, A., Maumet, C., & Nichols, T. (2019). Exploring the impact of analysis software on task fMRI results. *Human Brain Mapping, 40*, 3362–3384. https://doi.org/10.1002/hbm.24603.

Buckner, R. (2003). The hemodynamic inverse problem: Making inferences about neural activity from measured MRI signals. *PNAS, 100*, 2177–2179. https://doi.org/10.1073/pnas.0630492100.

Carp, J. (2012). On the plurality of (methodological) worlds: Estimating the analytic flexibility of fMRI experiments. *Frontiers in Neuroscience, 6*, 149.

Chang, C., Liu, Z., Chen, M. C., Liu, X., & Duyn, J. H. (2013). EEG correlates of time-varying BOLD functional connectivity. *NeuroImage, 72*, 227–236.

Chirimuuta, M. (2013). Extending, changing, and explaining the brain. *Biology and Philosophy, 28*, 612–638.

Crandall, C. S., & Sherman, J. W. (2016). On the scientific superiority of conceptual replications for scientific progress. *Journal of Experimental Social Psychology, 66*, 93–99.

Currie, A. (2018). *Rock, bone, and ruin: An optimist's guide to the historical sciences*. Cambridge, MA: The MIT Press.

Currie, A., & Levy, A. (2019). Why experiments matter. *Inquiry, 62*(9–10), 1066–1090.

Datteri, E. (2009). Simulation experiments in bionics: A regulative methodological perspective. *Biology and Philosophy, 24*, 301–324.

Feest, U. (2017). Phenomena and objects of research in the cognitive and behavioral sciences. *Philosophy of Science, 84*, 1165–1176.

Feest, U. (forthcoming). Why replication is overrated. *Philosohphy of Science*. https://doi.org/10.1086/705451.

Friston, K. J. (1994). Functional and effective connectivity in neuroimaging: A synthesis. *Human Brain Mapping, 2*, 56–78.

Fukushima, M., Betzel, R. F., He, Y., & van den Heuvel, M. P. (2018). Structure – Function relationships during segregated and integrated network states of human brain functional connectivity. *Brain Structure and Function, 223*, 1091–1106.

Guala, F. (2002). Models, simulations, and experiments. In L. Magnani & N. J. Nersessian (Eds.), *Model-based reasoning: Science, technology, values* (pp. 59–74). New York: Kluwer.

Guimerà, R., & Nunes Amaral, L. A. (2005). Functional cartography of complex metabolic networks. *Nature, 433*, 895–900. https://doi.org/10.1038/nature03288.

Hacking, I. (1981). Do we see through a microscope? *Pacific Philosophical Quarterly, 62*, 305–322.

Haxby, J. V. (2012). Multivariate pattern analysis of fMRI: The early beginnings. *NeuroImage, 62*, 852–855.

Holme, P., & Saramäki, J. (2012). Temporal networks. *Physics Reports, 519*, 97–125.

Horikawa, T., & Kamitani, Y. (2017). Generic decoding of seen and imaged objects using hierarchical visual features. *Nature Communications, 8*, 15037.

Horowitz, B. (2003). The elusive concept of brain connectivity. *NeuroImage, 19*, 466–470.

Huettel, S., Song, A., & McCarthy, G. (2008). *Functional magnetic resonance imaging* (2nd ed.). Sunderland: Sinauer Associates.

Israel-Jost, V. (2016). Computer image processing: An epistemological aid in scientific investigation. *Perspectives on Science, 24*, 669–695.

Klein, C. (2010). Images are not the evidence in neuroimaging. *British Journal for the Philosophy of Science, 61*, 265–278.

Leonelli, S. (2016). *Data-centric biology: A philosophical study*. University of Chicago Press.

Lindquist, M. A., Meng Loh, J., Atlas, L. Y., & Wager, T. D. (2009). Modeling the hemodynamic response function in fMRI: Efficiency, bias and mis-modeling. *NeuroImage, 45*, S187–S198. https://doi.org/10.1016/j.neuroimage.2008.10.065.

Logothetis, N. K. (2008). What we can do and what we cannot do with fMRI. *Nature, 453*, 869–878.

Longino, H. (2012). *Studying human behavior: How scientists investigate aggression and sexuality*. Chicago: The University of Chicago Press.

Lurie, L., Kessler, D., Bassett, D., Betzel, R., Breakspear, M., Keilholz, S., Kucyi, A., Liegeois, R., Lindquist, M., McInstosh, A., Poldrack, R., Shine, J. M., Thompson, W., Beilczyk, N., Douw, L., Kraft, D., Miller, R., Muthuraman, M., Pasquini, L., Razi, A., Vidaurre, D., Xie, H., & V. Calhoun. Preprint. On the nature of resting fMRI and time-varying functional connectivity. *PsyArXiv* Preprints. https://wwww.doi.org/10.31234/osf.io/xtzre. Accessed July 2019.

Martin, C. B., Sullivan, J., Wright, J., & Köhler, S. (2018). How landmark suitability shapes recognition memory signals for objects in the medial temporal lobes. *NeuroImage, 166*, 425–436.

Mayo, D. (1996). *Error and the growth of experimental knowledge*. Chicago: Univeristy of Chicago Press.

McAllister, J. (1997). Phenomena and Patterns in data sets. *Erkenntnis, 47*, 217–228.

Medina, J. (2012). *The epistemology of resistance: Gender and racial oppression, epistemic injustice, and resistant imaginations* (Studies in Feminist Philosophy). New York: Oxford University Press.

Meunier, D., Lambiotte, R., Fornito, A., Ersche, K. D., & Bullmore, E. T. (2009). Hierarchical modularity in human brain functional networks. *Frontiers of Neuroinformatics*. https://doi.org/10.3389/neuro.11.037.2009.

Morgan, M. S. (2005). Experiments versus models: New phenomena, inference and surprise. *Journal of Economic Methodology, 12*, 317–329.

Mucha, P. J., Richardson, T., Macon, K., Porter, M. A., & Onnela, J. P. (2010). Community structure in time-dependent, multiscale, and multiplex networks. *Science, 328*, 876–878.

Murphy, K., & Fox, M. D. (2017). Towards a consensus regarding global signal regression for resting state functional connectivity MRI. *NeuroImage, 154*, 169–173.

Parke, E. C. (2014). Experiments, simulations, and epistemic privilege. *Philosophy of Science, 81*, 516–536. https://doi.org/10.1086/677956.

Pedersen, M., Omidvarnia, A., Jackson, G. D., Zalesky, A., & Walz, J. M. (2017). Spontaneous brain network activity: Analysis of its temporal complexity. *Network Neuroscience, 1*, 100–115.

Pessoa, L. (2014). Understanding brain networks and brain organization. *Physics of Life Reviews, 11*, 400–435.

Poldrack, R., & Gorgolewski, C. (2015). OpenfMRI: Open task sharing of fMRI data. *NeuroImage, 144*, 259–261. https://doi.org/10.1016/J.NEUROIMAGE.2015.05.073.

Poldrack, R. A., Mumford, J., & Nichols, T. (2011). *Handbook of functional MRI data analysis*. New York: Cambridge University Press.

Power, J. D., Schlaggar, B. L., Lessov-Schlaggar, C. N., & Petersen, S. E. (2013). Evidence for hubs in human functional brain networks. *Neuron, 79*, 798–813. https://doi.org/10.1016/j.neuron.2013.07.035.

Ritchie, J. B., Kaplan, D. M., & Klein, C. (2019). Decoding the brain: Neural representation and the limits of multivariate pattern analysis in cognitive neuroscience. *British Journal for the Philosophy of Science, 70*, 581–607.

Roskies, A. (2010a). Neuroimaging and inferential distance: The perils of pictures. In M. Bunzl & S. J. Hanson (Eds.), *Foundational issues in human brain mapping* (pp. 195–216). Cambridge, MA: The MIT Press.

Roskies, A. (2010b). Saving subtraction: A reply to Van Orden and Paap. *The British Journal for the Philosophy of Science, 61*, 635–665.

Roush, S. (2018). The epistemic superiority of experiment to simulation. *Synthese, 195*, 4883–4906.

Schölvinck, M. L., Maier, A., Ye, F. Q., Duyn, J. H., & Leopold, D. A. (2010). Neural basis of global resting-state fMRI activity. *PNAS, 107*, 10238–10243. https://doi.org/10.1073/pnas.0913110107.

Sher, G. (2010). Epistemic friction: Reflections on knowledge, truth, and logic. *Erkenntnis, 72*, 151–176.

Shine, M., Bissett, P., Bell, P. T., Koyejo, O., Gorgolewski, K. J., Moodie, C. A., & Poldrack, R. A. (2016). The dynamics of functional brain networks: Integrated networks states during cognitive task performance. *Neuron, 92*, 544–554.

Sullivan, J. (2018). Optogenetics, pluralism, and progress. *Philosophy of Science, 85*, 1090–1101.

Taylor P. A., Chen G. C., Glen D. R., Rajendra J. K., Reynolds R. C., & Cox, R. W. (2018). FMRI processing with AFNI: Some comments and corrections on "Exploring the Impact of Analysis Software on Task fMRI Results". *bioRxiv*. https://doi.org/10.1101/308643. Accessed 12 Apr 2019.

Thompson, W., Branefors, P., & Fransson, P. (2017). From static to temporal network theory: Applications to functional brain connectivity. *Network Neuroscience, 1*, 69–99.

Thompson, W., Wright, J., Shine, J. M., & Poldrack, R. A. Pre-Print. (2019). The identification of temporal communities through trajectory clustering correlates with single-trial behavioral fluctuations in neuroimaging data. *bioRxiv*. https://doi.org/10.1101/617027. Accessed 25 Apr 2019.

Thompson, W., Kastrati, G., Finc, K., Wright, J., Shine, J. M., & Polrack, R. A. (2020). Time-varying nodal measures with temporal community structure: A cautionary note to avoid misquantification. *Human Brain Mapping*. https://doi.org/10.1002/hbm.24950.

Uddin, L. Q., Mooshagian, E., Zaidel, E., Scheres, A., Margulies, D. S., Kelly, A. C., Shehzad, Z., Adelstein, J. S., Castellanos, F. X., Biswal, B. B., & Milham, M. P. (2008). Residual functional connectivity in the split-brain revealed with resting-state fMRI. *Neuroreport, 19*, 703–709.

Uttal, W. (2001). *The new phrenology*. Cambridge, MA: The MIT Press.

van den Heuval, M. P., & Hulshoff Pol, H. E. (2010). Exploring the brain network: A review of resting-state fMRI functional connectivity. *European Neuropsychopharmacology, 20*, 519–534.

van den Heuvel, M. P., & Sporns, O. (2013). Network hubs in the human brain. *Trends in Cognitive Sciences, 17*, 683–696.

Van Dijk, K. R., Sabuncu, M. R., & Buckner, R. L. (2012). The influence of head motion on intrinsic functional connectivity MRI. *NeuroImage, 59*, 431–438. https://doi.org/10.1016/j.neuroimage.2011.07.044.

Van Essen, D. C., Ugurbil, K., Auerbach, E., Barch, D., Behrens, T. E. J., Bucholz, R., Chang, A., Chen, L., Corbetta, M., Curtiss, S. W., Della Penna, S., Feinberg, D., Glasser, M. F., Harel, N., Heath, A. C., Larson-Prior, L., Marcus, D., Michalareas, G., Moeller, S., Oostenveld, R., Peterson, S. E., Schlagger, B. L., Smith, S. M., Snyder, A. Z., Xu, J., Yacoub, E., & WU-Minn HCP Consortium. (2012). The Human Connectome project: A data acquisition perspective. *NeuroImage, 62*, 2222–2231. https://doi.org/10.1016/j.neuroimage.2012.02.018.

Van Orden, G. C., & Paap, K. R. (1997). Functional neuroimages fail to discover pieces of mind in the parts of the brain. *Philosophy of Science, 64*, S85–S94.

Woodward, J. (2000). Data, phenomena and reliability. *Philosophy of Science, 67*, 163–S179.

Woodward, J. (2003). *Making things happen: A theory of causal explanation*. New York: Oxford University Press.

Wright, J. W. (2017). The analysis of data and the evidential scope of neuroimaging results. *British Journal for the Philosophy of Science, 69*, 1179–1203. https://doi.org/10.1093/bjps/axx012.

Wright, J. (2018). Seeing patterns in neuroimaging data. In C. Ambrosio & W. MacLehose (Eds.), *Imagining the brain: Episodes in the history of brain research* (pp. 299–323). Cambridge, MA: Academic.

Chapter 9
Neural Reuse and the Nature of Evolutionary Constraints

Charles Rathkopf

Abstract In humans, the reuse of neural structure is particularly pronounced at short, task-relevant timescales. Here, an argument is developed for the claim that facts about neural reuse at task-relevant timescales conflict with at least one characterization of neural reuse at an evolutionary timescale. It is then argued that, in order to resolve the conflict, we must conceptualize evolutionary-scale reuse more abstractly than has been generally recognized. The final section of the paper explores the relationship between neural reuse and human nature. It is argued that neural reuse is not well-described as a process that constrains our present cognitive capacities. Instead, it liberates those capacities from the ancestral tethers that might otherwise have constrained them.

9.1 A Latent Disagreement About Neural Reuse

One might think that each time an organism acquires a novel behavioral capacity, some correspondingly novel structure must have been wired together in its head. Neural reuse is the contrasting idea that novel capacities are often made possible by the redeployment of existing neural structures in new task domains. Here, I hope to identify a latent disagreement in the scientific discussion of neural reuse.

The disagreement has remained latent because it concerns the relationship between two background assumptions, which have themselves received little attention. The first assumption concerns the multiplicity of timescales at which neural reuse might occur. The second concerns the role of representation in theories of neural function. These two topics come together in a particularly interesting way in Stanislas Dehaene's work on reading acquisition. After introducing neural reuse more thoroughly, I will give a brief overview of Dehaene's theory, and draw from it a principle about how timescale and representational character are related. That

C. Rathkopf (✉)
Forschungszentrum Jülich GmbH, Institute of Neuroscience and Medicine, Ethics in the Neurosciences (INM-8), Jülich, Germany
e-mail: c.rathkopf@fz-juelich.de

© Springer Nature Switzerland AG 2021

F. Calzavarini, M. Viola (eds.), *Neural Mechanisms*, Studies in Brain and Mind 17,
https://doi.org/10.1007/978-3-030-54092-0_9

principle – which I call the content constraint view – is not the only way to conceive of the relationship between timescale and representational character. I sketch an alternative view of this relationship, and then work out three consequences of accepting that alternative view, each of which serves to refine our understanding of neural reuse.

In the final section of the paper, I explore a loftier and more speculative set of ideas about the relationship between neural reuse and human nature. It is argued that, if the view of neural reuse developed earlier in the paper is right, then neural reuse helps explain how human nature managed to acquire its uniquely open-ended character.

9.2 Reuse: A Central Theme, and Its Variations

Here, I use the term "reuse" in a maximally broad sense, intended to capture a common theme running through a complex and partially overlapping set of theories. Labels for these theories include "neural repurposing" (Parkinson and Wheatley 2015), "neuronal recycling" (Dehaene and Cohen 2007), "massive redeployment" (Anderson 2007), "cognitive recycling" (Barack 2017), and "neural exaptation" (Chapman et al. 2017). Neural reuse, in the maximally broad sense intended here, is entailed by each theory in this list. It can be defined as a commitment to two simple ideas. The first is that local neural structures contribute to multiple cognitive or behavioral tasks. The term "local neural structure" is meant to be quite inclusive. It covers everything from cytologically-defined microscale structures, such as cortical columns, all the way up to functionally defined cortical regions identified by means of brain imaging.

The second idea is that the cognitive or behavioral tasks to which a structure contributes must be conceptually distinct. If the latter function logically entails the former, the two functions are not conceptually distinct. A non-scientific example may be helpful here. Consider the following two claims. On Monday, my travel mug is used to transport coffee. On Tuesday, it is used to transport hot coffee. Because "transporting hot coffee" entails "transporting coffee," this is not a case of reuse in the relevant sense. To make this a case of reuse in the relevant sense, I would have to transport something conceptually unrelated, like soup.[1] Now let's consider a neuroscientific example. In task condition A, the supplementary motor area (SMA) subserves motor command preparation. In task condition B, the SMA subserves reaching movement preparation. Because preparation for a reaching movement is one kind of motor command preparation, these two functions are not conceptually distinct. The conceptual overlap between these two functions blurs the

[1] In this prosaic example, there is no deep truth about which functions are genuinely distinct, because the individuation conditions for the functions of a coffee mug are, presumably, a matter of convention rather than discovery.

distinction between the theory of neural reuse and the comparatively bland claim that neural function is subject to variation of some sort or another. In a review paper on the SMA that focuses on conceptual difficulties associated with theories of SMA function, Naschev et al. put the point thus: "Functional pleomorphism is conceptually problematic owing to the difficulty of explaining the process of switching between different neural functions" (Naschev et al. 2008). Another function sometimes ascribed to the SMA is the regulation of task-switching, which is arguably distinct from movement preparation, and would, therefore, support the case for neural reuse in that area.

The dual characterization provided thus far shows what the various theories of neural reuse have in common. They differ from one another in many dimensions, two of which are relevant here. The first has to do with timescale. What are the timescales at which neural reuse occurs? A view that is commonly assumed, if not explicitly defended, is that there are exactly two such scales: one phylogenetic and one ontogenetic (Gallese 2008; Anderson and Finlay 2014). Such an assumption appears to be held, for example, by Parkinson and Wheatley (2015), who divide their discussion of the topic into "neural repurposing across lifetimes" and "neural repurposing within lifetimes." It is also commonly assumed, if not explicitly defended, that the reuse process at the phylogenetic scale stands in a relatively harmonious relationship to reuse at the ontogenetic scale. At the very least, none of the existing literature explores the possibility that our description of neural reuse at one scale will carry implications for the viability of description at another. This assumption can be challenged. As I argue below, once we explore the possibility of additional timescales, the relations between these two default scales begin to look less harmonious.

Another dimension of difference between theories of neural reuse concerns the kinds of purposes, or functions, that a theory might describe at each scale. Even after we have restricted ourselves to a single scale in space and time, the varieties of neural function are many. Some functions are characterized in terms of proximate effects on other neural structures; others in terms of distal effects on behavior. Functions can also be distinguished with respect to the faculty to which they contribute: perception, memory, motor control, etc. The distinction I want to draw, which I take to be orthogonal both to the proximal/distal distinction, and to the choice of mental faculty, divides what I will call content functions from all others. A content function is any function in which the contribution a structure makes to the operation of the system of which it is a part involves the representation of an element in the task-environment of the organism.

Two components of this definition deserve some unpacking. The first is the concept of a neural representation. In most areas of neuroscience, the term "representation" is used liberally.[2] The concept I mean to invoke here has a more distinctive theoretical role. A pattern of activity only counts as a representation, in

[2] To see this, consider how difficult it is to design an experiment that might serve to falsify the claim that "x is a representation," where x is any pattern of neural activity you choose.

the sense I have in mind, if (i) it is correlated with some environmental parameter of relevance, and (ii) it plays a causal role in the cognitive process that enables the organism to achieve some behavioral goal, by acting as a signal that informs the activities of downstream neural mechanisms. This account of representation is incomplete, but useful. The first condition suffices to rule out neural activity that systematically influences behavior without targeting external properties. The second condition rules out what have been called idle correlations (Rathkopf 2017), which fail to figure in the representational activities of the organism because no mechanism exists that is capable of exploiting the correlation in order to direct behavior.

The second component in the definition of content function that deserves unpacking is the concept evoked by the phrase "element in the task-environment of the organism." To be an element in the task-environment of the organism is to be the kind of property to which the organism must at some point dedicate attention, in order to complete a particular task successfully. Consider, for example, the so-called fusiform face area (FFA) in humans. It has been described as cortical structure that is dedicated to the detection of faces (Kanwisher 2010). The representations of faces purportedly instantiated by that structure must be consulted before one can, for example, appropriately orient one's gaze toward a conversational partner. Faces, therefore, will commonly count as elements in the task environment of humans, and face-detection will commonly count as a content function.

The class of non-content functions will include both neural functions that do not demand representational characterization, along with neural functions that do, but which are only indirectly connected with what would ordinarily be countenanced as a task. As Phillip Haueis (2018) has recently argued, there are many kinds of representational activity in the brain that are only indirectly involved with the accomplishment of intuitively recognizable behavioral goals, and which, therefore, have only a tenuous connection to familiar, folk-psychological modes of description. Moreover, there are many neural activities that play roles that are both highly specific and vital to the life of the organism, but which do not admit of representational description at all. Pacemaker neurons, for example, dampen the dynamics of various neural networks by means of intrinsically modulated bursting activity (Ramirez et al. 2004). Purkinje cells in the cerebellum have been described as gain modulators, that multiply incoming signals from a wide variety of perceptual sources (Luque et al. 2019). Cases like these remind us that neural reuse need not, as a matter of definition, consist exclusively in transitions between content functions.

Thus far, I have introduced a very general notion of neural reuse, and introduced two ways to distinguish between the many kinds of neural function that might be involved in any given case of neural reuse. First, I distinguished between neural functions instantiated on task-relevant time scale and those instantiated on an evolutionary time scale. Second, I distinguished between content functions and non-content functions. The core insight in this essay is that these two distinctions are empirically linked. If we characterize the function of a local neural structure at the timescale of an individual task, we may find good evidence that it realizes a content function. If, however, we try to characterize its function on larger timescales, we are likely to find that the evidence for content functions disappears. Before I present the argument that shows how timescale and representational status are related, it will be helpful to examine a particular theory of neural reuse and its application to a particular cognitive phenomenon. For this purpose, I have chosen Stanislas Dehaene's theory of neuronal recycling and its application to literacy. Dehaene's theory is appropriate for the job, not only because of the strength of its influence, which is considerable, but also because it illustrates the logic behind a view of the relationship between biological evolution and mental content that is implicit in a lot of evolutionary psychology, but which, I'll argue, ought to be resisted.

9.3 The Paradox of Reading

In his book "Reading in the brain," Dehaene presents a theory of reading and reading acquisition. The book begins by introducing what Dehaene calls the reading paradox, which is most succinctly expressed in the following two sentences: "Nothing in our evolution could have prepared us to absorb language through vision. Yet brain imaging demonstrates that the brain contains fixed circuitry exquisitely attuned to reading (Dehaene 2009, p. 24)." Dehaene's version of neural reuse, which he calls the neuronal recycling hypothesis, is offered as a solution to this paradox. To understand his theory, then, we first need to understand this paradox in more detail, and some of the data that appear to generate it.

The reading paradox presents us with two claims that are, ostensibly, both true and mutually inconsistent. The first is about human evolution. We know from anthropological evidence that the earliest human writing systems appeared about 6000 years ago, in the form of Mesopotamian cuneiform (d'Errico and Colagè 2018). We also know from mutation frequency data that 6000 years is too short a period for substantial neurogenetic adaptations to have accumulated. We can be confident, therefore, that the capacity for literacy is not the direct product of a genetic mutation that has only recently swept through the human gene pool.

The second half of the paradox also deserves a closer look. What does it mean to say that "the brain contains fixed circuitry, exquisitely attuned to reading?" The circuitry to which Dehaene refers is a small, functionally defined cortical area located in the left ventral occipito-temporal junction. That area is now commonly labeled with a functional designation that Dehaene himself coined: the visual word form area, or VWFA. Dehaene ascribes two properties to this circuitry. He says that it is fixed, and that it is exquisitely attuned to reading. Let us first examine what he means by the latter. Dehaene's claim that the VWFA is exquisitely attuned to reading is what he takes to be the upshot of a family of interesting results from lesion and imaging data, which, when taken as a whole, suggest that, in literate adult subjects, the area is specialized for word recognition.

The following six pieces of evidence are commonly taken to provide support for this localizationist conclusion.

1. In normal literate subjects, the region is differentially responsive to written, but not spoken words (Dehaene and Cohen 2007).
2. Illiterate adults do not show responsivity to letters in VWFA, and ex-illiterate adults (people who first learned to read in adulthood) exhibit less responsivity than literates. (Thiebaut et al. 2012).
3. In blind subjects, the region is differentially responsive to words presented in Braille, but not to tactile control stimuli (Reich et al. 2011).[3]
4. Lesions to the area appear to result in pure alexia, a condition in which formerly literate subjects cannot understand written words, despite being able to understand and produce verbal speech at roughly normal levels of competency (Gaillard et al. 2006).
5. fMRI priming effects in this region are invariant to alternative representations of the same priming word. For example, the stimulus "RADIO" is an effective prime for "radio," whereas "oidar" is not (Dehaene and Cohen 2007).
6. The repetition suppression effect disappears in this region for mirror-images of words and individual letters. The visual system regards most objects as equivalent to their mirror-images. We learn to violate this rule when learning to read, in order to distinguish, for example, "b" from "d." That this region responds differently to mirror images suggests that the region is sensitive to words as meaningful units, rather than as linear strings of wiry objects (Dehaene and Dehaene-Lambertz 2016; Dehaene 2013).

These results provide strong evidence that the brains of literate adults contain an area with a response profile dominated by words and letters. If Dehaene's interpretation of the data is correct, then the overriding function of the VWFA is to represent words and letters. Since words and letters are elements of common human task environments, Dehaene's hypothesis describes a content function, in the sense defined above.

The apparently localized nature of word recognition is fascinating in its own right, but what exactly is its relevance to the paradox of reading? On Dehaene's view, it is a theoretical surprise that word recognition appears to be carried out in such a small and discrete cortical area. The sense of surprise is reinforced by the claim that this area is "fixed." This term refers to the fact that the spatial position of the area, despite being functionally rather than anatomically identified, is robust across individual subjects and language groups.[4] The combination of response-specificity and positional robustness characteristic of the VWFA is loosely analogous to the kinds of retinotopic maps found in early visual cortex. By analogy to areas like these, Dehaene expects that, in general, positionally robust, map-like circuits in human cortex will subserve capacities that emerged long ago and that are part of our biological, rather than cultural, heritage.

[3] Although this claim has recently been disputed, in light of new data. See Kim et al. (2017).

[4] Although see Coltheart (2014) for a somewhat deflationary interpretation of the degree of positional robustness that is actually licensed by the neuroimaging data.

Now that we have a firmer grasp on the meaning of the two claims involved in the paradox of reading, we can ask: is it reasonable to characterize them as a paradox? Perhaps not. If we streamline the wording a bit, the purported paradox juxtaposes the claim that (i) orthographic word identification is a localized brain function, with the claim that (ii) orthographic word identification could not have played a role in human evolution. From a logical point of view, these claims are not actually inconsistent. If their conjunction appears paradoxical, it is only because we have tacitly accepted a background assumption which says that localized content functions are necessarily driven by the genetic evolution of the species.

Like many assumptions lurking in the scientific background, this one arouses suspicion as soon as it is formulated explicitly and offered up for critical inspection. The assumption asks us to contrast evolved functions with learned ones. But, as developmental systems theorists have emphasized, this contrast is easily abused, because every neural function emerges from a process of biological development, and the distinction between development and learning is both highly theoretical and highly contested (Oyama 2000). Moreover, even on a thin conception of learning, there are no uncontroversial examples of content functions that develop in its absence. In light of the entangled nature of evolution and development, any theory that requires us to assign causal responsibility for a trait to one process or the other should at least be explicit about how the assignment should be carried out. Since the assumption in this case is merely implicit, no such instructions are provided. It is reasonable to suspect, therefore, that the conceptual foundations underlying the assumption are unstable. In Sect. 9.6, I'll argue that the assumption should be rejected. In the following section, however, we examine Dehaene's favored solution instead.

9.4 Neuronal Recycling as a Solution to the Paradox

Because Dehaene leaves untouched the assumption linking localization and evo lutionary provenance, the only way he can solve the paradox of reading is by showing that, contrary to first appearance, one of the two claims that comprise the paradox is not strictly true. Dehaene aims to undermine, or at least weaken, the claim about evolution. The theory of neuronal recycling says that, although natural selection cannot be directly responsible for having shaped a circuit dedicated to reading, natural selection is, nevertheless, responsible for having indirectly shaped the mechanism that enables us to read. Natural selection shaped a circuit for a particular function that is sufficiently close to reading, but which, unlike reading itself, reaches far back into human evolutionary history.

> Cultural acquisitions (e.g., reading) must find their "neuronal niche," a set of circuits that are sufficiently close to the required function and sufficiently plastic as to reorient a significant fraction of their neural resources to this novel use (Dehaene and Cohen 2007).

Here, and in other passages, Dehaene appeals to a principle of similarity between functions to explain what makes it the case that they share the same cortical fate. The similarity relation holds between an older function and a newer one. At this point, it will be useful to introduce a pair of terminological stipulations. In any case of neural reuse, whether it occurs on an evolutionary scale or not, I'll refer to the older function as the primary function, and the newer one as the secondary function. A core commitment of neuronal recycling can then be expressed as follows: primary functions are necessarily similar to secondary functions. When expressed this way, the obscurity of the claim looms large. Similarity with respect to what?

In Dehaene's 2009 book, as well as in many of the articles he has produced with various co-authors on the topic, including the 2007 article with Laurent Cohen, (from which the quote above is drawn) his answer to this question appears to be that the relevant kind of similarity is similarity with respect to content. Dehaene stresses that, according to neuronal recycling, cortical circuits are typically biased towards the representation of certain elements of the organism's task environment. These biases serve to constrain the range of cultural symbols humans can learn to use.

> According to this view, our evolutionary history, and therefore our genetic organization, specifies a cerebral architecture that is both constrained and partially plastic, and that delimits a space of learnable objects. New cultural acquisitions are possible only inasmuch as they are able to fit within the pre-existing constraints of our brain architecture (Dehaene 2008, p. 12).

What kinds of neural properties have the power to delimit the space of learnable objects, as Dehaene puts it? One might attempt to answer this question in terms of content-neutral limitations on the systems' capacity to process information. If the object is too complex for the perceptual system to discriminate, for example, it is not a learnable object. (This is, presumably, one reason that no written languages employ symbols with 1000 overlapping components.) However, this is not the kind of answer Dehaene has in mind. Dehaene's view seems to be that the limitation is neither merely perceptual, nor directly related to the complexity of the object. On Dehaene's view, we have an inherited "preference" for objects with particular semantic qualities. These content preferences are genetically entrenched, and it is in virtue of that entrenchment that the space of learnable objects is limited. On this view, unless some very sophisticated genetic engineering becomes a viable option, the space of learnable objects is destined to remain circumscribed.

This focus on evolutionarily entrenched content is one way of making sense of two bodies of evidence. The first body of evidence is the response specificity of the VWFA, which was described above. The second body of evidence is the fact that all known written languages employ characters with specific geometric similarities. For example, if you plot the distribution of the number of line crossings required to represent all of the written characters in all of the world's languages, you get a tight cluster around the number three (Changizi and Shimojo 2005). Dehaene also cites as evidence the (purported) fact that written characters in all human languages are necessarily composed of combinations of elementary shapes. Dehaene sees both bodies of evidence (response specificity and orthographic similarity across languages) as effects of a hidden common cause - the content bias in VWFA. The content bias is postulated, by means of an inference to the best explanation, precisely in order to account for both the neural and the anthropological data.[5]

To summarize the foregoing remarks, Dehaene's theory of neuronal recycling is offered as a solution to the paradox of reading. It counts as a solution because it purports to show that the evolutionary claim that constitutes the first half of the paradox is, despite its initial plausibility, wrong. Evolution did indeed "prepare us to absorb language through vision," but it did so indirectly. What I will *the content constraint view* is a theory about that process of indirect preparation. It can be split into two claims.

1. The primary evolutionary function of the VWFA is a content function.
2. Constraints on the range of secondary functions for which the VWFA can be "recycled" derive from the nature of the content targeted by its primary function.

In the following section, I provide reasons to think that the content constraint view is incorrect. In his most recent work on the topic, Dehaene et al. (2018) defend a view of the VWFA that is in tension with the content constraint view. One might worry, therefore, that I have been constructing a straw man. However, my motivation for articulating the view is not to weigh in on debates about the neural substrates of literacy. It is rather to articulate a conception of neural reuse in which content plays a central explanatory role, even on an evolutionary scale. The content constraint view is worth articulating not because it has arduous defenders who happen to be wrong, or because it has a severely detrimental effect on the design of new experiments, but because the consequences of rejecting it are theoretically interesting. Once we reject it, I'll argue, we see that theories of neural reuse, when pitched at an evolutionary scale, are more enigmatic than has been recognized thus far.

9.5 A Clash Between Timescales

The content constraint view describes a process that bridges two timescales. The primary function gets stabilized on an evolutionary timescale. It plays an important role in the selection history of the organism, and thereby leaves a trace on the genetic information transmitted across generations. That genetic information manifests itself in the form of a content bias, which is itself expressed by a particular local structure. The secondary function operates on a different timescale altogether. It gets stabilized on a developmental scale. The target of the secondary function is determined in part by developmental context and cultural input, but is also

[5]The anthropological data Dehaene offers as evidence of neural reuse may be not as straightforward as he sometimes makes it sound. Max Coltheart has argued that the uniformity to which Dehaene refers is simply not there (Coltheart 2014). I am sympathetic to Coltheart's concerns about the evidence, but would like to resist Dehaene's account on different grounds altogether. I will therefore just assume the evidence says exactly what Dehaene says it does.

constrained by the content bias in the circuit that subserves it. In what follows, the target of my attention is the nature of this purported constraint, and how it might have come about over evolutionary time.

The challenge I want to pose emerges from thinking about the evolutionary implications of another kind of neural reuse; one that unfolds more quickly than the kind Dehaene describes. This faster process, which I call task-scale neural reuse, is a phenomenon in which a local neural structure transitions from supporting one behavioral task to supporting another by means of a reconfiguration of its network of partnering structures. Such reconfiguration unfolds on a timescale relevant to individual cognitive and behavioral tasks, on the order of seconds or minutes. On this view, each structure supports different functions at different times, depending not only on the current perceptual input, but also on set of structures with which functional connectivity has been established.

The evidence for this architectural principle is multifaceted. One of the more significant sources of evidence comes from meta-analyses of brain imaging studies on humans. For example, Anderson et al. (2013) ask how many distinct tasks, drawn from distinct cognitive domains, are supported by each region of the brain. To estimate an answer to this question, they measure voxel-by-voxel diversity in data generated by a collection of over 2000 functional neuroimaging experiments. The analysis shows that even small regions of the brain contribute to multiple tasks both within and between cognitive domains (Anderson et al. 2013).

> The upshot: local neural structures are not highly selective and typically contribute to multiple tasks across domain boundaries. Because the domains are highly varied, the observations cannot be explained by the similarity of the task domains (Anderson 2014, p. 10).

This passage is particularly appropriate for our exposition of Anderson's view because it is explicit about the absence of an underlying similarity relation that could serve to unify or circumscribe the set of tasks that a given structure, could, in principle, be recruited to support. If the list of functions associated with each structure ranges across both tasks and cognitive domains, then no structure specializes in the representation of a particular element in a particular task-environment. In other words, no structure specializes in any particular content function. The anti-localizationist implications of task-scale neural reuse are well known, and detailed arguments to this effect can be found elsewhere Zerilli (2019).

There is also reason to believe that the distributed functional architecture implied by task-scale neural reuse has always been a feature of the human brain. Macaque cortex, for example, appears to implement a form of task-scale neural reuse (Iriki and Taoka 2012), and the last common ancestor of macaques and humans lived approximately 25 million years ago (Disotell and Tosi 2007). The idea that task-scale neural reuse is ancient in our lineage poses a direct threat to the content constraint view. To see this, we need only ask what justification we have for claiming that some neural structure has a primary function that can be characterized in terms of content. Typically, the biological justification for isolating one primary function from the myriad causal interactions in which a given structure may be engaged involves an appeal to natural selection. But if task-scale neural reuse is ancient, natural selection will have had little opportunity to tailor a structure for its capacity to contribute to any particular content function.

This argument shows that if we want to characterize the contribution of a neural structure to the capacities of an organism on an evolutionary scale, we cannot invoke any particular content-function. And this claim, in turn, conflicts with the content constraint view. If the evolution of local neural structures was not driven by the demands of dealing with particular kinds of content, then constraints on the range of secondary functions that those structures can come to realize are not accurately described as constraints on content. Of course, this argument does not show that the range of secondary functions a neural structure can come to support is *unconstrained*. Nor does it show that the operative constraints, whatever they are, are not bound up with the evolutionary history of the organism. It only shows that those constraints should not be described as a content-bias embedded in the physiology of local neural structures.

As mentioned above, recent work from Dehaene and colleagues on the constraints involved in letter recognition in the VWFA displaces the content constraint view, and is, therefore, no longer in tension with the apparent preponderance of task-scale neural reuse. The alternative view focuses on facts about connectivity, such as the relationship in the ventral stream between lateral position and degree of foveal input, or the question of whether a site projects to language areas. Similar facts about the connectivity profile of the VWFA had been discussed in earlier work (Dehaene 2009; Hannagan et al. 2015). However, in that earlier work, discussions of connectivity appear alongside claims about content bias in the VWFA. Facts about connectivity are framed as an explanation for why the VWFA appears where it does. This explanation of VWFA location appears to be offered as a *supplement* to the theory of content bias in the VWFA, rather than as a replacement for it. In the most recent work (Dehaene-Lambertz et al. 2018), the notion of content bias is simply left out. New longitudinal data allowed Deheane-Lambertz et al. to look back in time at the specific voxels in each subject that later came to be the site in which the VWFA emerged.[6] It turned out that, in pre-literate children, those voxels display far less stimulus preference than had previously been believed. In light of this new data, the 2018 paper suggests that the connectivity profile of the VWFA not only explains its location in cortex; it also generates the expected constraints on orthographic symbol use.

I'll now consider an objection that will likely have occurred to anyone familiar with research on object-selective cortex. Isn't the FFA a good example of a structure that has always been largely dedicated to one kind of content, and which, therefore, could have undergone selection for its capacity to represent faces? And if it did

[6]If you want to study the site at which the VWFA will appear in the brains of children who are currently pre-literate, you have to guess where it will appear in the future. Individual variability imposes a relatively low ceiling on the accuracy of such guesses. The Dehaene-Lambertz et al. (2018) study is the first to overcome this methodological difficulty.

undergo selection for its capacity to represent faces, shouldn't we say that the representation of face-like content is both the primary function of the area, and the source of at least some of the developmental constraints it confronts in modern humans? Two lines of response are available. One is that the FFA may simply be an exception. One could argue that task-scale neural reuse characterizes the functional architecture of *most* of the brain, but not the FFA. In fact, this suggestion is compatible with what I've said so far. The central claim in this section has a conditional form: *if* a structure has long been involved in the implementation of task-scale neural reuse, then it is unlikely that the structure was tailored by natural selection for the representation of some particular class of content. If the antecedent of the conditional goes unsatisfied in a particular case, the truth-value of the consequent is dialectically irrelevant. However, this response may not be the best one. The fact that the FFA might be an exception does nothing to show that an appeal to face-like content is the most appropriate way to articulate the nature of the developmental constraints on the capacities of the cortical site. In this connection, it is worth noting that, in order for past content to serve as causal constraint on the range of secondary functions a neural structure can acquire, the physiological properties underlying the content bias must be *canalized*. That is, the structure must end up acquiring those properties even in developmental environments that lack content-specific perceptual triggers. Without canalization in this sense, primary functions could not delimit the space of representational objects, as Dehaene puts it, because eventually, alternative cultural environments would emerge, and invite the development of alternative neural phenotypes. Is the FFA canalized in this sense? Until recently, this question had been impossible to answer. This changed in 2017, however, when Mike Arcaro and colleagues used welder's masks to raise three monkeys in a faceless environment. At 200 days after birth, which was the last time that imaging was done before exposing the monkeys to a normal social environment, the site corresponding to the FFA in those monkeys had not developed a preference for faces (Arcaro et al. 2017). This shows that, even in the case of the FFA, constraints on the development of cortical structures are not best articulated in terms of some pre-theoretically familiar class of representational content.

9.6 Three Consequences of the Clash

Here I will briefly draw out three conceptual consequences of the clash between timescales. The first consequence concerns the paradox of reading. Recall that the paradox of reading consisted of two explicit claims, and one implicit assumption. The first claim says that writing is too recent an invention for either writing or reading to have played a role in human genetic evolution. The second claim says that the word identification is localized to a particular cortical structure. The implicit assumption was that localized content functions are necessarily driven by the genetic evolution of the species, rather than by learning and development. In light of the clash between timescales, we can see that the assumption deserves to be rejected. Localization of content always depends on the task demands imposed by the developmental environment.

The second consequence of the clash concerns the character of ancient primary functions. The upshot of the previous section was that the kind of primary functions required by the content constraint view are not evolutionarily plausible. What then is the status of ancient primary functions more generally? This is a difficult question, but I think we can say this much: if the goal is to characterize just one function that captures the historical role played by a given structure, we will have to generalize over the wide variety of task-scale neural functions supported by that structure. According to this suggestion, ancient primary functions do exist, but are more abstract than the content-constraint view requires. Once we generalize over all possible task-scale functions, there is little reason to think that the resulting conception of neural function will be accessible by means of folk-psychological reasoning. If such abstract functions can be represented accurately, it will be by means of a more rarified and theoretical form of representation, perhaps one that draws on the language of computation. Only such an abstract conception of function could bring unity to the otherwise heterogeneous list of context-bound functions that a given structure will subserve over evolutionary history. Alternatively, one might say that the list of context-bound functions is not subject to *any* unifying principle, regardless of the degree of abstraction we are willing to adopt. The best one can do is to produce open-ended lists of context-bound neural functions. Context-bound functions (whether oriented toward a particular task or not) are useful for many scientific purposes (Burnston 2016), but they are too disparate to serve as a foundation for an ancient primary function. According to the context-bound list suggestion, nothing in nature satisfies the concept of ancient primary function.

Regardless of which view of ancient primary functions one prefers, the meaning of the claim that a neural structure has been subject to neural reuse on an evolutionary scale turns out to be far less transparent an idea than it had at first seemed. The need for a more abstract characterization of neural function threatens the coherence of evolutionary neural reuse, because, as discussed in Sect. 9.2, reuse demands a degree of conceptual distinctness between functions. If a cortical structure primarily performs an abstract function articulated in domain-neutral terms, such as, for example, gain modulation, then any apparently novel functional activity will count as an instantiation of the same function in a novel context, rather than as the realization of new function per se.

I suspect that the initially intuitive impression given by the idea of evolutionary neural reuse depends on the intuitive familiarity of the content functions that are mistakenly presumed to serve as the relata in the reuse relation. If reuse is imagined to be a transition between two content functions, both of which are accessible to folk-psychological reasoning, it will appear as though we already understand what is involved in a transition from primary to secondary functions (even if the observational consequences associated with the instantiation of either function are vague or indeterminate, and that, as a result, we cannot precisely specify the empirical content of transition events). However, once we take seriously the idea that ancient neural functions cannot be captured in terms of dedication to, or specialization in, any content-type that would be readily accessible from a folk-psychological stance, intuitions about the boundaries between neural functions wither away. As they wither, so does the intuitive status of evolutionary neural reuse itself.

How far should we take this skeptical reasoning? Should we go as far as to declare that any suggestion of evolutionary neural reuse is conceptually bankrupt? Certainly not. Reuse applies to the structures that compose the human brain just as it applies to every other biological trait. As Darwin put it: "Thus, throughout nature almost every part of each living being has probably served, in a slightly modified condition, for diverse purposes, and has acted in the living machinery of many ancient and distinct specific forms (Darwin 1877, p. 284)." An immediate implication of Darwin's assertion is that neural reuse, in particular, has been common. We can accept that implication without presuming that we already know what the relata of the neural reuse relation are. Moreover, as noted in the initial discussion of content functions, there are many kinds of non-content functions to which the argument developed here does not apply.

The third consequence of the clash is a rather subtle, but also rather useful disambiguation of a prediction Michael Anderson makes about the relationship between the evolutionary age of a neural function, and the amount of cortical real estate it recruits. The ambiguous form of the prediction is this: in both evolutionary and developmental time, newer functions will demand more cortical real estate than older functions. It is valuable to figure out exactly what this prediction says, because it is one of the central principles that lends falsifiable empirical content to the neural reuse framework. If we insist on agnosticism about the nature of the relata in the neural reuse relation, while remaining cognizant of the diversity of kinds of neural function, the ambiguity in Anderson's prediction becomes easy to see. The prediction can be interpreted in strong and weak forms. The weaker interpretation treats the two timescales independently, and can be expressed like this:

Weak prediction. It will typically be the case that, (i) for any given pair of functions characterized on a developmental timescale, F1 and F2, if F1 demands more cortical real estate than F2, then F1 will have developed later than F2, and (ii) for any given pair of functions characterized on an evolutionary timescale, F1 and F2, if F1 demands more cortical real estate than F2, F1 will have evolved later than F2.

The strong interpretation collapses the two timescales together. It can be expressed like this:

Strong prediction. It will typically be the case that if function F1 demands more cortical real estate than F2, it will have appeared after F2 both in developmental and evolutionary time.

The crucial feature of the strong interpretation is that it appeals to the same pair of functions on both scales. It is a neuroscientific application of the late nineteenth century biologist Ernst Haeckel's memorable pronouncement that "ontogeny recapitulates phylogeny."

In light of the clash between timescales, only the weaker of these two claims is justified. The primary functions that get stabilized on an evolutionary scale will be content-neutral. At the task-relevant scale, many of the functions temporarily instantiated by any given structure will indeed involve the representation of a particular kind of content. Typically, therefore, the functions recognizable at an evolutionary scale will not be recognizable at a task-relevant scale. If so, content-oriented neural functions comprise a domain in which, contra Haeckel, ontogeny does not recapitulate phylogeny. The content of cognition is less tethered by the capacities of our ancestors than a casual consideration of neural reuse would suggest.

9.7 Constraint and Liberation

Thus far, I have argued against one way to conceptualize evolutionary constraints on human brain function. Nevertheless, there is no denying that we have inherited specific neural structures from our ancestors, and that the capacities of those neural structures make our mental life possible. I would now like to ask whether there is some other, more general sense in which our mental life is constrained by the functional capacities of the brains of our ancestors, from whom the design of our brains is inherited.

To answer that question, it will help to articulate what a "constraint" amounts to, in the domain of brain evolution. To say that the ancient functional profile of a neural structure constrains its modern homologue is to say that the range of capacities associated with the modern structure is narrower than it would have been, had the ancient functional profile been different. But different in what way? Many alternative ancient functional profiles would surely have led to an alternative set of contemporary capacities, but not necessarily to a narrower one. What kind of alternative ancient functional profile must we imagine, in order to make plausible the idea that, had that alternative been profile been the actual one, we would today enjoy an even broader suite of cognitive capacities? Precisely because task-scale reuse has been part of our species for a long time, it is hard to know how to answer this question. Given the ancient provenance of task-scale neural reuse, neural structures have long been capable of realizing a diverse list of functions. Moreover, it is not at all clear that nature has imposed a theoretical upper limit on either the length or the diversity of that list. So neural reuse at the evolutionary scale has not clearly constrained us; or at least not in any way that we can confidently point to. The structures that compose our brains are constrained by their evolutionary history, but only in the non-committal sense in which every biological structure is "constrained" by its evolutionary history. Neural reuse does not entail some special, additional kind of constraint.

What about the opposite view? Is there any sense in which evolutionary neural reuse has helped to lift, or at least soften, some of the constraints on our mental life? Anderson (2014) predicts that the late-evolving capacities that are distinctive of human cognition require more extensive reuse of neural structures than older, less distinctively human capacities. Primary examples include the reuse of motor circuits for language (Pulvermüller 2005) and numerical cognition (Penner-Wilger and Anderson 2013). This suggests that, in comparison with other species, humans have an unusually amplified capacity to reuse neural structures for novel cognitive ends.

This idea is suggestive. In a poetic mood, one might even be tempted to say that that neural reuse has been a source of human freedom. This claim carries more philosophical baggage than the corresponding claim about constraint, but its intended meaning is not difficult to work out. Its meaning is approximately the inverse of the claim about constraint. To say that neural reuse has been a source of freedom is to say that our species, in virtue of having acquired an unusually amplified capacity for task-scale neural reuse, is capable of realizing a broader set of neural functions now than we would have been able to realize, had that amplified capacity for task-scale neural reuse never been acquired. The counterfactual invoked by this claim is easier to evaluate than the one invoked by the claim about constraint, since, in this case, the counterfactual refers to a comparatively close possible world in which only one property is absent. Moreover, in order to evaluate this counterfactual, one does not need to know exactly what our species would have looked like, had task-scale reuse not emerged. It would suffice to show that the cognitive repertoire of our species would have been radically smaller without it. Let us assume that, at the level of the whole organism, the number of cognitive tasks that a human can accomplish is a function of the number of tasks that local neural structures can support. Assume also that each task recruits a network of local neural structures. If these two assumptions are warranted, then the number of cognitive tasks that a human can possibly undertake will be a combinatoric function of the number of tasks each local structure can support. When viewed that way, task-scale neural reuse has exponentially increased the number of tasks we humans can undertake, and in that sense, has indeed been a source of human freedom.

Acknowledgements Thanks to Matteo Colombo, Philipp Haueis, and Lena Kästner for insightful feedback on my *Neural Mechanisms Online* talk, which was my first attempt to work out the issues discussed in this chapter.

References

Anderson, M. L. (2007). Massive redeployment, exaptation, and the functional integration of cognitive operations. *Synthese, 159*(3), 329–345.

Anderson, M. L., & Finlay, B. L. (2014). Allocating structure to function: the strong links between neuroplasticity and natural selection. *Frontiers in Human Neuroscience, 7*, 918.

Anderson, M. L., Kinnison, J., & Pessoa, L. (2013). Describing functional diversity of brain regions and brain networks. *Neuroimage, 73*, 50–58.

Arcaro, M. J., Schade, P. F., Vincent, J. L., Ponce, C. R., & Livingstone, M. (2017). Seeing faces is necessary for face-domain formation. *Nature Neuroscience, 20*, 1404–1412.

Barack, D. L. (2017). Cognitive recycling. *The British Journal for the Philosophy of Science, 70*(1), 239–268.

Bergeron, V. (2010). Neural reuse and cognitive homology. *Behavioral and Brain Sciences, 33*(4), 268–269.

Burnston, D. C. (2016). A contextualist approach to functional localization in the brain. *Biology and Philosophy, 31*(4), 527–550.

Changizi, M. A., & Shimojo, S. (2005). Character complexity and redundancy in writing systems over human history. *Proceedings of the Royal Society B: Biological Sciences, 272*(1560), 267–275.

Chapman, P. D., Bradley, S. P., Haught, E. J., Riggs, K. E., Haffar, M. M., Daly, K. C., & Dacks, A. M. (2017). Co-option of a motor-to-sensory histaminergic circuit correlates with insect flight biomechanics. *Proceedings of the Royal Society B: Biological Sciences, 284*(1859), 20170339.

Coltheart, M. (2014). The neuronal recycling hypothesis for reading and the question of reading universals. *Mind & Language, 29*(3), 255–269.

d'Errico, F., & Colagè, I. (2018). Cultural exaptation and cultural neural reuse: A mechanism for the emergence of modern culture and behavior. *Biological Theory, 13*, 1–15.

Darwin, C. (1877). *On the various contrivances by which British and foreign orchids are fertilised by insects*. London: John Murray.

Dehaene, S. (2008). Cerebral constraints in reading and arithmetic: Education as a "neuronal recycling" process. *The educated brain: Essays in neuroeducation*, pp. 232–247.

Dehaene, S. (2009). *Reading in the brain: The new science of how we read*. New York: Penguin.

Dehaene, S. (2013). Inside the letterbox: how literacy transforms the human brain. *In Cerebrum: the Dana forum on brain science, volume 2013. Dana Foundation, 2013*, 7.

Dehaene, S., & Cohen, L. (2007). Cultural recycling of cortical maps. *Neuron, 56*(2), 384–398.

Dehaene, S., & Dehaene-Lambertz, G. (2016). Is the brain prewired for letters? *Nature Neuroscience, 19*(9), 1192.

Dehaene-Lambertz, G., Monzalvo, K., & Dehaene, S. (2018). The emergence of the visual word form: Longitudinal evolution of category-specific ventral visual areas during reading acquisition. *PLoS Biology, 16*(3), e2004103.

Disotell, T. R., & Tosi, A. J. (2007). The monkey's perspective. *Genome Biology, 8*(9), 226.

Gaillard, R., Naccache, L., Pinel, P., Clémenceau, S., Volle, E., Hasboun, D., Dupont, S., Baulac, M., Dehaene, S., Adam, C., et al. (2006). Direct intracranial, fmri, and lesion evidence for the causal role of left inferotemporal cortex in reading. *Neuron, 50*(2), 191–204.

Gallese, V. (2008). Mirror neurons and the social nature of language: The neural exploitation hypothesis. *Social Neuroscience, 3*(3–4), 317–333.

Hannagan, T., Amedi, A., Cohen, L., Dehaene-Lambertz, G., & Dehaene, S. (2015). Origins of the specialization for letters and numbers in ventral occipitotemporal cortex. *Trends in Cognitive Sciences, 19*(7), 374–382.

Haueis, P. (2018). Beyond cognitive myopia: a patchwork approach to the concept of neural function. *Synthese, 195*(12), 5373–5402.

Iriki, A., & Taoka, M. (2012). Triadic (ecological, neural, cognitive) niche construction: a scenario of human brain evolution extrapolating tool use and language from the control of reaching actions. *Philosophical Transactions of the Royal Society, B: Biological Sciences, 367*(1585), 10–23.

Kanwisher, N. (2010). Functional specificity in the human brain: a window into the functional architecture of the mind. *Proceedings of the National Academy of Sciences, 107*(25), 11163–11170.

Kim, J. S., Kanjlia, S., Merabet, L. B., & Bedny, M. (2017). Development of the visual word form area requires visual experience: Evidence from blind braille readers. *Journal of Neuroscience, 37*(47), 11495–11504.

Luque, N. R., Naveros, F., Carrillo, R. R., Ros, E., & Arleo, A. (2019). Spike burst-pause dynamics of Purkinje cells regulate sensorimotor adaptation. *PLoS Computational Biology, 15*(3), e1006298.

McCaffrey, J. B. (2015). The brain's heterogeneous functional landscape. *Philosophy of Science, 82*(5), 1010–1022.

Naschev, P., Kennard, C., & Husain, M. (2008). Functional role of the supplementary and pre-supplementary motor areas. *Nature Reviews Neuroscience, 9*(11), 856–869.
Oyama, S. (2000). *The ontogeny of information: Developmental systems and evolution*. Durham/London: Duke University Press.
Parkinson, C., & Wheatley, T. (2015). The repurposed social brain. *Trends in Cognitive Sciences, 19*(3), 133–141.
Penner-Wilger, M., & Anderson, M. L. (2013). The relation between finger gnosis and mathematical ability: Why redeployment of neural circuits best explains the finding. *Frontiers in Psychology, 4*, 877.
Pulvermüller, F. (2005). Brain mechanisms linking language and action. *Nature Reviews Neuroscience, 6*, 576–582.
Ramirez, J.-M., Tryba, A. K., & Pena, F. (2004). Pacemaker neurons and neuronal networks: an integrative view. *Current Opinion in Neurobiology, 14*(6), 665–674.
Rathkopf, C. (2013). Localization and intrinsic function. *Philosophy of Science, 80*(1), 1–21.
Rathkopf, C. (2017). Neural information and the problem of objectivity. *Biology and Philosophy, 32*(3), 321–336.
Reich, L., Szwed, M., Cohen, L., & Amedi, A. (2011). A ventral stream reading center independent of reading experience. *Current Biology, 21*, 363–368.
Thiebaut de Schotten, M., Cohen, L., Amemiya, E., Braga, L. W., & Dehaene, S. (2012). Learning to read improves the structure of the arcuate fasciculus. *Cerebral Cortex, 24*(4), 989–995.
Zerilli, J. (2019). Neural reuse and the modularity of mind: Where to next for modularity? *Biological Theory, 14*(1), 1–20.

Chapter 10
Behavior Considered as an Enabling Constraint

Vicente Raja and Michael L. Anderson

Abstract Two fundamental challenges of contemporary neuroscience are to make sense of the scalar relations in the nervous system and to understand the way behavior emerges from these relations while at the same time affects them. In this paper, we analyze the notion of enabling constraint and the way it can frame the two kinds of relations involved in the challenges: of different neural scales (e.g., molecular scale, genetic scale, single-neurons, neural networks, etc.) and between neural systems and behavior. We think the notion of enabling constraint provides a promising alternative to other classic, mechanistic understandings of these relations and the different issues they raise for contemporary neuroscience.

10.1 Introduction

Humans cannot fly. This is a little bit disappointing, but we have largely made our peace with it (recurring dream motifs notwithstanding) and have invented planes. Planes allow us to fly, albeit with several mechanical, temporal, and legal restrictions. However, there are many other things we cannot do even with the help of science or technology. Just because the physical world is the way it is, we cannot become invisible by wearing Bilbo's ring. Other times we cannot perform an action just because we impose some restrictions in our own behavior, as when Socrates refused to escape from prison due to his moral principles. These situations speak to a common fact: what we can and cannot do is *constrained* in many different ways. Even more, such a fact is not reduced to what we can and cannot do, but it is a

V. Raja (✉)
Rotman Institute of Philosophy, University of Western Ontario, London, ON, Canada
e-mail: vgalian@uwo.ca

M. L. Anderson
Rotman Institute of Philosophy, University of Western Ontario, London, ON, Canada,

Department of Philosophy, University of Western Ontario, London, ON, Canada

Brain and Mind Institute, University of Western Ontario, London, ON, Canada

© Springer Nature Switzerland AG 2021

F. Calzavarini, M. Viola (eds.), *Neural Mechanisms*, Studies in Brain and Mind 17,
https://doi.org/10.1007/978-3-030-54092-0_10

general observation about living and non-living systems. All of them are constrained in one way or another and, thus, what they do and can do is also constrained. In other words, constraints are ubiquitous in the world, writ large, and affect almost every aspect of it.

Due to the ubiquity of constraints, it is fair to expect that cognitive systems are constrained in different ways—e.g., physically, biologically, socially, etc.—and that those constraints may play an important explanatory role in cognitive science. On the one hand, this expectation is trivial. It is well acknowledged within the cognitive sciences that several limitations in terms of cognitive, neural, and bodily abilities shape our cognitive states. On the other hand, the notion of constraint is not always acknowledged as part of the explanatory activity in cognitive science. For example, when Carl Craver (2008) distinguishes between the two traditions of understanding scientific explanation, the reductionistic one, identified with Hempel's deductive-nomological strategy, and the systemic one, identified with the mechanist strategy (Bechtel and Richardson 1993), the notion of constraint does not seem to be regarded as relevant. Reductionistic explanations appeal to notions such as derivability or reduction, while the systemic ones appeal to notions of constitution and componential activity. None of these notions seem to account for the role that, for instance, FAA rules play in our ability to fly: these rules are not reducible or derivable from the mechanics of planes, for example, nor apparently do the mechanics of planes have FAA rules as a constitutive component. However, FAA rules do constrain our ability to fly, and would be part of the explanation for any number of aviation-related phenomena. Of course, the fact that these two approaches to scientific explanation seem unable to explain the relationship between our ability to fly in planes and aviation rules might be irrelevant. At the end of the day, such a relationship might not be an explanandum of a scientific explanation.

The real problem arises when we find instances of scientific explanation that seem to require notions that are provided neither by the deductive-nomological strategy nor the mechanistic one. In the concrete case of cognitive science, Anderson (2015a) has proposed that some of the explanations in the field require the notion of *enabling constraint*, instead of, say, constitution or derivability, to fully account for the different relations between the components and events that are relevant to understanding the functioning and behavior of cognitive systems. In this paper, we further elaborate on the notion of enabling constraint and, specifically, on the way it can illuminate the relationship between neural activity and behavior.

Our main thesis is that behavior can be understood as an enabling constraint of neural activity. In this sense, we should not take behavior just as a product of the activity of the brain, but also as one of the events that allow for that very activity in the first place. In order to support our thesis, in Sect. 10.2, the notion of enabling constraint is characterized in detail. To do so, we analyze the notion of constraint in biology and offer an example of what an enabling constraint in neuroscience is. In Sect. 10.3, we directly address the claim that behavior is an enabling constraint of neural activity. We build upon the literature on self-organization and the "enslaved brain" (e.g., Van Orden et al. 2012; Dotov 2014) to understand the idea of constraint between different scales of a complex system. Then, we elaborate on the particularities that make behavior be an *enabling* constraint and not a constraint simpliciter. Finally, we explore some consequences that follow from our main thesis.

10.2 Enabling Constraints

Constraints are usually understood in terms of a limitation. The common understanding of the role of constraints in a system is that constraints reduce the functional diversity of the system. Namely, if we have two similar systems but only one of them is constrained, the unconstrained one will exhibit more degrees of freedom in its behavior precisely because of the lack of constraints. This notion of constraints notwithstanding, the notion we are defending here rests on the idea that some constraints are not just a limitation but the key to *enabling* some functions. In other words, some constraints act as conditions that allow systems to exhibit behaviors that could not be exhibited without the presence of these constraints. In this section we try to show how this can be.

10.2.1 What Is an Enabling Constraint?

Consider a biological system S.[1] Let us stipulate that input to or activity in S can result in one of some set of outcomes {O}.[2] With this as a background, we propose:

> **Constraint** $=_{\mathrm{Df}}$ A relationship between a biological system S and entities or processes {X} such that {X} changes the probability distribution over the members of {O}.

Note that these changes can be absolute in the sense that the constraints could reduce the probability of O_n to near[3] 0 or increase it to near 1, but in the general case we can speak of changes to the probability distribution over the elements of (the relative probabilities of outcomes in) {O}. Also note that we put no strictures on the possible organizational arrangements within {X}.

[1] Biological systems being the subset of functional systems of interest to us here.

[2] There are some potential complications here, since one could imagine that an input to or activity in S producing more than one outcome, such that the probability distribution over {O} need not sum to one. Here we will develop the theory in the context of the special case where there is a single such outcome, and thus the probability distribution across elements of {O} sums to 1.

[3] "Near" because, whether or not biological systems are deterministic, it is clear that they do not exert deterministic control over themselves, nor over one another, at any level of description.

Note further that the definition as stated speaks merely of outcomes and not, as might be expected, *functional* outcomes. This is because outcomes in the set {O} come in at least four categories: undetectable; useless; counterproductive; and productive. Put a different way, constraint(s) can render S, respectively: inert; non-functional; dysfunctional; and functional. Consider the case of general anesthesia, which imposes myriad constraints on patients, so that interventions such as making an incision that would normally result in dramatic responses, instead result in none. Anesthesia has rendered the patient *inert* with regard to a range of inputs and interventions. Similarly, consider a modification of an automobile such that the harder the accelerator is pressed, the tighter the brake calipers squeeze. Here this additional constraint added between sub-components of the car result in a system that *responds* to input—the engine revs, gasoline is burned—but these outcomes are to no avail from the standpoint of the automobile (or its operator); the car has been rendered *useless*.

Changing constraints in different ways can result in positive new capacities, but dysfunctional ones. Consider the famous rewired frogs from Ingle (1973). These frogs were subject to the unilateral removal of the optic tectum, resulting in the optic tract enervating the ipsilateral tectum instead. This reshuffling of functional constraints in the frogs' nervous system resulted in coherent behaviors, snapping away from prey objects and jumping towards predators, but behaviors which are *dysfunctional* in the frogs' actual circumstances. Rounding out the case, we can say that the functional constraints as they exist in the *normal* frog nervous system enable *functional* (in this case adaptive) .behaviors. We will call these outcomes *strictly functional*.

In the last paragraph we introduced the notion of a "positive" capacity, because we think there is a useful distinction to be made between negative and positive *constraints*. All constraints constrain, that is, limit possible outcomes or behaviors, and are in this sense negative. But some constraints in addition allow for, promote, or actualize functional outcomes that would not be possible in their absence. Constraints that result in inert or non-functional systems we call negative constraints (or "merely negative") while those that result in coherent behaviors we call positive.

One can think of the distinction between the dysfunctional and strictly functional outcomes that result from positive constraints by noting that for dysfunctional outcomes there is a nearby possible world in which the outcomes would be strictly functional—for the rewired frogs that would be a world in which its predator and prey animals were switched. Similarly, the distinction between dysfunctional and non-functional is that for the latter there is no such nearby possible world. Alternately, we could say that a negative constraint does not actualize any capacity (that could be useful in some imaginable circumstance) not already possessed by the system in the absence of said constraint.

Before offering a formal definition of enabling constraint, which will rely on the distinctions made above, it is worth considering whether those distinctions can in fact be maintained. Might it be that for *all* constraints there could be shown to be some positive outcome that emerges or is made possible, relative to some circumstance? Is the distinction between positive and negative constraints actually substantive, or merely pragmatic?[4]

Consider again the case of general anesthesia. In imposing myriad constraints on the patient does it not thereby enable fresh possibilities? You might at first think so. After all, it enables safe surgery, which is a vital positive outcome. More whimsically, it enables the patient to be carried about or tattooed without complaint. But note that none of these possibilities represent the emergence of a functional outcome *for the system itself*. What anesthesia allows is for things to be done *to the patient*; it does not enable, or make the patient more likely, to do things.

Similarly, the class of systems rendered *useless* by some constraint may respond to stimuli, but those responses have no function in the current or any neighboring context, and the constraint has thereby failed to actualize (although it has made more likely) any capacity (staying still, in the case of the modified automobile) not already possessed by the system in the absence of the constraint. We are hard-pressed to think of a case where it would be possible to redescribe this outcome in positive terms.

In contrast, it is quite clear that positive constraints allow the system to *do* things—make available or more probable functional outcomes that would be difficult or impossible in the absence of those constraints. We thus conclude that the distinction between positive and negative constraints is substantive and not merely pragmatic.

This being said, we suspect the distinction between dysfunctional and strictly functional outcomes (and, as we will soon see, between "merely positive" and "enabling" constraints), although substantive, will always also have a pragmatic, context-relative aspect. After all, a distinction between a counterproductive and a productive outcome will always depend on the circumstance in which the outcome arises.

Full discussion of the notion of function, and the differences between strictly functional and dysfunctional outcomes and behaviors, would take us very far afield from the purposes of the present paper. Here we simply note that we see no barrier to cashing out the distinction between these in terms of proper functions (Millikan 1989), that is, in evolutionary, developmental, or other historical terms. Nor, where appropriate, do we see an issue (for our purposes[5]) with cashing out the distinction in contextual terms, that is in terms of synchronic relations between elements of a larger system. Our frog example offers a well-known case of drawing the distinction in the first way. For the second we can imagine the utility (or disutility) of yelling "Fire!" in a crowded theatre to depend on the presence of fire. The outcome of the

[4]Thanks to Alessandra Buccella and Charles Rathkopf for pushing us on this issue, in their comments on an earlier draft of this paper.

[5]What we gesture at by saying this is that we are in the business of developing a conceptual framework that will support fruitful empirical investigation. The proposal does not have to cleanly adjudicate between all border cases to be epistemically and heuristically useful in scientific practice.

constraints imposed on the crowd by that exclamation will be dysfunctional in the absence and strictly functional in the presence of fire.

This, then, brings us at last to the formal definition of an enabling constraint:

Enabling constraint $=_{\mathrm{Df}}$ A positive constraint between S and {X} that results in strictly functional outcome(s) for S.[6]

Where S is the system under consideration, and {X} is the set of entities or processes impacting S. Three aspects of this definition are noteworthy. First, it is abstract enough to encompass *physical* constraints, but also more abstract (perhaps in some sense non-physical) constraints including power relationships, social structures and cultural conventions.[7] This being said, for the remainder of the paper we will restrict the discussion to physical relationships. We do so because what we are presenting here is the theoretical framework for an empirical project. This involves the development of methods for recognizing and measuring the existence of constraints in brain-body-environment systems. We think we know how to do this for physical systems—that is, we think we know how to establish and quantify the existence and effect of mutual constraints between perceptual information, behavioral dynamics, and brain dynamics (work that is currently in progress). However, we currently do *not* have any clear sense of how a social constraint could be measured (and we don't believe anyone else does either), although it is of course true that the effects of those constraints can be readily observed. This is a deeply interesting area for future research, as we (and others) extend Gibson's conceptual apparatus to support an understanding of social behavior, social affordances, and the like.

It is important to stress that specifically *physical* enabling constraints (i.e. those rooted in physical relationships between S and {X}) can nevertheless be described in terms of high-order variables.[8] For example, we can think of the transmission of Shannon information as a physical relationship between sender and receiver although the relevant variables are informational. Or we can understand the activation of a perceiver's mirror neurons when contemplating some behavior (Leonetti et al. 2015) as a physical relationship although the relevant variables used in the explanation appeal to behavior or perceptual information.

[6]It is important to flag that this is a significantly different, but we hope more precise and useful, definition than the one offered in Anderson (2015a: 12). Thanks to Alessandra Buccella, Charles Rathkopf, Michael Silberstein, and an anonymous reviewer for the various comments that motivated this revision.

[7]Thanks to Michael Silberstein for pressing us to clarify this in his comments on an earlier draft of the essay.

[8]We recognize that there are some authors, e.g. (Silberstein 2018, in press) who take such higher-order variables as order parameters or network topologies to be non-physical. We do not wish to enter into this debate here, and suspect neither resolution of it affects our conclusions here at all.

Second, the definition as it stands makes no mention of different levels of organization. This is intended as a direct contrast to the formalizations offered by the New Mechanists (Craver 2008; Craver and Bechtel 2007; Craver and Darden 2001) in which the elements of a system/mechanism that enable function must be at a *lower* spatial scale than the mechanism itself (see discussion in Sect. 10.2.3, below, and in Anderson 2015a, b; Kohler 2015). Enabling constraints can exist between any given {X} and S that are at the same or different levels or scales of description. The activity of a single neuron, for example, may be constrained by entities or processes that operate at lower scales, such as molecules or genes, at the same scale, such as neighboring neurons, and at higher scales, such as network dynamics or organism-environment interactions. In this sense, an adequate understanding of the notion of enabling constraint proposed here favors the idea of explaining cognitive systems as complex systems with more than one relevant scale of description, and more than one direction of possible interaction between them.

Third, enabling constraints play a *positive* role when constraining a system. Enabling constraints do not only limit the availability of functional outcomes in a system but can literally be the key to making some strictly functional outcomes available, and even possible or impossible. Such a positive role of enabling constraints parallels the positive role of developmental constraints in evolutionary processes proposed by some developmental biologists since the late 1970s. A brief review of that literature will help motivate our own project.

10.2.2 The Notion of Constraint in Biology

In 1979, Stephen J. Gould and Richard C. Lewontin published a highly influential paper in which they criticized the adaptationist program in evolutionary biology. The paper begins with a discussion on the spandrels of St. Mark's Cathedral in Venice to illustrate Gould and Lewontin's claims against the usual adaptationist procedure: breaking organisms into single traits and then proposing an adaptive story for each of them relying on the almost omnipotent power of natural selection (1979, p. 585). In opposition to this view, Gould and Lewontin favor what they take to be a more Darwinian and holistic view on evolution, in which not all organisms' traits are adaptive—as with the spandrels in St. Mark's Cathedral, which are not functional but the byproduct of the functional role of other elements of the cathedral—and in which natural selection is not the only explanatory tool to account for the appearance of those traits.

Among the different proposals made by Gould and Lewontin, we are interested in the notion of *developmental constraint*.[9] Put simply, developmental constraints are a way to refer to the influence of developmental mechanisms in evolution (Gould

[9]Although we focus on the notion of developmental constraint, Gould and Lewontin (1979) has remained influential within theoretical evolutionary biology in many ways—e.g., for the

1980; Amundson 1994; Schwenk 1994; Rausher et al. 2008). The underlying idea is that evolutionary processes are not just the product of natural selection, but also of developmental factors. For example, for a specific trait it is possible that the phenotypic variability developmental mechanisms *can* produce overrules the power of natural selection *to* produce a phenotype that would be a better environmental fit given the selection pressure. In such a case, the constraints imposed by development are at least as important as natural selection in understanding the evolutionary process.

Following Amundson (1994), developmental constraints have been understood in "negative" terms by adaptationists but in "positive" terms by developmental biologists. The adaptationist account of developmental constraints highlights their restricting influence on adaptation. In this view, constraints are just limitations of phenotypical variability imposed by the mechanisms of embryology. This notion is "negative" insofar as developmental constraints are characterized in terms of purely conservative forces that restrict the otherwise guiding force of adaptation (i.e., natural selection). For developmental biologists, however, developmental constraints do not have to do directly with adaptation but with the kinds of forms (i.e., structures, shapes) the mechanisms of embryology are able to produce. In other words, developmental biologists are less concerned with adaptation, and more with the way organismic forms are produced by developmental mechanisms. It is possible, nevertheless, that developmental constraints on forms affect adaptation: the forms generated by developmental mechanisms may be, subsequently, selected by an evolutionary process. In this view, developmental constraints are not regarded as mere limitations of adaptation, but as essential (positive) contributors to evolution. The forms generated by developmental mechanisms are the ones affected by natural selection but, at the same time, are the ones that make organisms sensitive or not to specific instances of selection pressure. For example, some selective forces may have no way to affect organisms just because of the lack of the right morphology. Holekamp et al. (2013) defend this position with regard to behavioral flexibility. For instance, they claim that because of the lack of manual dexterity in carnivores in comparison to primates:

> [M]utations, for example, in nervous system structure of function that might affect fitness in primates via modified use of hands, cannot affect fitness in carnivores, so the fitness landscape for carnivore behaviour is effectively limited by limb morphology (p. 5).

Although Holekamp et al. frame it in negative terms, it can also be framed in positive terms: primates' limb morphology allows them to be influenced by evolutionary forces to which carnivores are completely blind. In this sense, developmental constraints in the form of the limbs enable primates to open new evolutionary

developmental systems approach (Oyama 2000; Oyama et al. 2001) or the evo-devo discourse in biology (Brigandt 2015; Carroll 2008; Goodman and Coughlin 2000; Hall 2003; Held 2014).

directions. Similarly, the limb structure of carnivores opens them to adaptations affecting gait patterns, speed and agility not available to primates.[10]

In our opinion, such a positive understanding of developmental constraints places them among the enabling constraints of evolution.[11] Developmental constraints change the probability of the outcomes of evolutionary processes by actively providing them with a directionality—i.e., with a way in which natural selection itself may affect or not different organisms. In this sense, developmental constraints enable evolutionary processes to be the way they are. Now, the question is: Is it possible to find positive, enabling constraints in the cognitive sciences? In other words, is it possible to find cognitive mechanisms that are positively constrained by different entities or events? We think the answer to these questions is yes.

10.2.3 Enabling Constraints in Neuroscience

In the case of evolution, developmental constraints act as enabling constraints insofar as they are taken to be actively contributing to the evolutionary process itself. The main advocates of this idea claim that, therefore, developmental constraints are part of the *mechanism* of evolution (Gould 1980). However, it is worth noting that the notion of mechanism they are using is somewhat different from the most well-developed notion of mechanism in the cognitive sciences in the last decades (the one defended by new mechanists; see Bechtel 2009; Craver 2008; Craver and Darden 2001) and is surely more liberal. Minimally, it does not require a commitment to the notion of constitution, for example.[12] In the case of cognitive science, we understand enabling constraints as non-constitutive parts of cognitive mechanisms, at least in terms of the notion defended by new mechanists. Indeed, enabling constraints can be understood as capturing systemic interactions within organisms and between organisms and environments without the need for relationships of constitution.

According to Anderson (2015a), the stimulus-direction selectivity exhibited by the dendrites of Starburst Amacrine Cells (SACs) found in mammalian retina (Tauchi and Masland 1984) exemplifies the relations of enabling constraint in the

[10]Obviously, "not available" is a temporal notion, since over vast swaths of time, there may be no part of the morphological landscape truly inaccessible.

[11]Indeed, the name "enabling constraint" has been used to refer to the proposals of Stanley N. Salthe (1993) regarding the relationship between evolution and development in the form of self-organized processes (see, e.g., Juarrero 1999).

[12]Actually, as far as evolution is a process and not a system, the notion of mechanism proposed by new mechanists might even not apply to it. Otherwise, if the mechanism of evolution is identified with natural selection itself, it seems that developmental constraints are not a proper part of the mechanism and, therefore, are not a constitutive part of it although still actively contribute to evolution as a process. In this latter sense, the notion of enabling constraint applied to developmental constraints would be more similar to the notion as we will use it regarding cognitive science.

sense we propose. Put simply, SACs are starburst-shaped retinal cells with the neural body in the center and dendrites arrayed around it. SACs form dense, highly overlapping layers across the retina and are physically and functionally nested between bipolar cells and direction-selective ganglion cells (Masland 2005). In terms of function, SACs participate in motion detection/perception and optokinetic eye movements, among other things (Yoshida et al. 2001). Specifically, each individual SAC dendrite is sensitive to stimuli moving centrifugally across the cell away from the center, signaling detection with the release of neurotransmitter from the distal end of the dendrite (Euler et al. 2002). In this sense, SACs' dendrites are subparts of SACs that perform the function of stimuli-direction signaling.

The mechanism that allows SACs' dendrites to be directionally selective depends on properties of the dendrites themselves but also on the interaction between dendrites and the bipolar cells around them and between the individual dendrites of neighboring cells. Although we are not going into detail on the mechanism itself, two aspects of it are noteworthy.[13] First, the function of individual dendrites depends on interactions at its own scale by means of the inhibitory activity of other dendrites, that is, the mutual inhibition observed between overlapping dendrites is part of what enables direction selectivity. And second, the function of individual dendrites depends on the activity of other cells, such as the bipolar cells synaptically connected to SACs: bipolar cells successively synapse onto the dendritic process, resulting in passive reinforcement of excitatory input that preferentially promotes neurotransmitter release in response to motion in the centrifugal direction (Demb 2007; Lee and Zhou 2006). That is, part of the explanation of the dendrite's function is the *spatial arrangement* of the surrounding SACs and bipolar cells, something that is not a property of the dendrite nor of the surrounding cells. In this sense, the function of SACs' dendrites as stimuli-direction selectors must be understood as a product of the proper activity of individual dendrites plus their interactions with other elements of the nervous system. That is, some of the constraints are *external* to the system in question, whereas the new mechanists generally envision the relevant functional parts to be internal to the system.

As already noted, Anderson (2015a) presents this example as a way to illustrate a kind of relationship between parts of cognitive systems that mechanisms cannot accommodate if they are understood as precisely formalized by the new mechanists. First, the new mechanists' notion requires components of mechanisms to be of a lower scale than the system that instantiates the mechanism itself. Regarding the stimuli-direction sensitivity of SACs' dendrites, this means that the components of the mechanism that allow for such a function must be spatial sub-parts of the SACs' dendrites themselves. Therefore, other dendrites or bipolar cells cannot be components of the mechanism as they are of equal or higher scale than individual SACs'

[13] For a detailed description of the mechanism, see Anderson (2015a).

dendrites.[14] In addition, the new mechanists' notion of mechanism requires that functional explanation be fully grounded in the components of the mechanism and the interactions between them. For this reason, and given that neighboring dendrites are not proper components of any given individual SAC dendrite, new-mechanist explanations do not seem to naturally capture the emergence of direction-selectivity from interacting structures here.

In contrast, the role of other dendrites in the stimuli-direction selectivity of individual SACs' dendrites is quite naturally understood in terms of the notion of enabling constraint. The mutual constraint observed between SAC dendrites changes the functional outcomes in the dendrite from signaling motion in any direction (which the SAC dendrites do in isolation) to signaling motion only in a specific direction. This function could not be accomplished without the constraint exerted between cells. In this sense, neighboring SAC dendrites exert a *positive, enabling* constraint on one another. We would like to suggest that that the notion of an enabling constraint may be a fruitful partner to the new mechanist account, allowing one to characterize functional relationships in a broader array of systems than can be captured by mechanism alone.

Still, one might sense not partnership but rather competition. Consider that for the new mechanists, the bipolar cells would be understood as merely providing the *input* to the system of interest, relegating the spatial arrangement of those cells (and the surrounding SACs) to the context or background conditions necessary for the mechanism in the dendrite to operate. One willing to take that stance might not see the attraction, much less the necessity, of adopting the language of constraint. Laying out and adjudicating this debate in detail deserves a paper of its own, but a few words on the subject are perhaps in order.

First, it is worth recalling that, historically, the attempt to neatly divide cause from context or explanation from background conditions has been quite fraught (Mackie 1965; Van Fraassen 1977). But even for one optimistic about the prospect of being able to confidently identify background conditions as such, the current case does not offer the most favorable grounds. For it seems—to us at least—that things like the spatial arrangement of bipolar cells and the mutual inhibition between SAC dendrites, far from merely defining the context within which the mechanism explaining direction selectivity operates, are in fact the very things that *explain direction selectivity*. They are a vital part of what one would need to understand to understand the emergence of direction selectivity in the dendrite at all. In this sense, they are quite unlike, say, the fact that for the mechanism to operate there needs to be glycolysis, and the Krebs cycle, and the right sort of diffusion gradients, and protein folding, etc. The operation and arrangement of SACs and bipolar cells can't be screened off nearly so easily. Our view, roughly, is this: if there are elements that cannot be screened off from even the strictly local explanation/understanding

[14]On the assumption that the system that exhibits direction-selectivity here *is* the dendrite. There are some subtleties to be considered regarding how best to define the functional system in this case. For discussion see Anderson (2015a, b; Köhler 2015), and this section, below.

of a phenomenon, but nor can they be part of the mechanism (because, for instance, they are at the wrong spatial scale, or are the wrong sort of thing), then we need to offer an alternative explanatory relationship for these elements. Enabling constraint is our candidate explanatory relationship.[15] Indeed, as suggested by a reviewer of this essay, the notion of enabling constraints offers a way of characterizing the boundaries of mechanisms in a more principled way than limiting them to strict spatial sub-parts, but without opening it up to the vagaries of generalized background conditions.

A natural follow-up reply might be to accept the validity of this argument, but deny a premise: that the elements in question can't be part of the mechanism. Perhaps, one might argue, we simply initially identified the mechanism itself at the wrong spatial scale, and in fact SACs and bipolar cells are all part of a larger mechanism for direction selectivity.

A well-worked out example of such a response has been offered by Kohler (2015) and countered by Anderson (2015b); the reader is directed to those articles for detailed discussion. Here we simply offer two summary points. First, if one redefines the mechanism in this way, one can no longer say (according to the rules of neomechanism) that it is the dendrite that exhibits the target explanandum, direction selectivity, and it is far from clear, in that case, what exactly *does* exhibit direction selectivity. Second, redefining the boundaries of a mechanism so as to make new mechanistic explanations *always* apply surely, at some point, risks looking dogmatic rather than scientific. Better, we think, to be open to multiple explanatory frameworks and adopt the one that best fits the case at hand.

To conclude this section: although we think there may be many systems whose function-structure relationships are well-captured by new mechanism, we think the application of enabling constraints in the explanations in cognitive science may allow functional characterization of a broader range of systems. Sometimes entities that are not components of a system nevertheless help fix the function of that system. Capturing the role of such non-constitutive elements in helping fix the function of a given system is important to developing fuller explanations of function-structure relationships than can be captured by componential thinking alone. Enabling constraints entail the description of systems at different scales—e.g., the function of SACs' dendrites considered at the scales of individual dendrites, dendritic interactions, and cellular interactions—and offer a way to understand scalar relations in those systems, without supposing there needs to be a strict functional hierarchy with only bottom-up determinations of functional outcomes. These scalar relations are especially interesting in the cognitive sciences as different disciplines interact while approaching similar cognitive phenomena at different levels of description: molecular underpinnings of the nervous system, single-neuron activity, neural networks, motor behavior, social interactions, etc. What is

[15]An alternate response might be to accept that they are part of the context, but to insist that the context operates precisely via constraint (see, e.g. Silberstein (2018, in press) on contextual constraint).

the relationship between single-neuron activity and network dynamics? Do they constrain each other? And what about the relationship between neural activity and behavior? Behavior is commonly understood just as an outcome of neural activity. We think instead that behavior and neural activity constrain each other: neural activity enables behavior, and behavior is an enabling constraint for neural activity.

10.3 Brain and Behavior

Our definition of enabling constraint aims to capture those relations between different aspects of cognitive systems that cannot be well accommodated within frameworks based on notions like constitution or based on purely serial/linear notions of cognitive activity. The problems of the notion of constitution have been illustrated with the stimuli-direction sensitivity of individual SACs' dendrites: some entities and processes needed for that function to be accomplished—e.g., the activity of bipolar cells, the inhibition from neighboring dendrites—can hardly be characterized as constitutive components of the mechanism instantiated by individual SACs' dendrites. A strict mechanist approach to cognition may also hide other assumptions. One of these assumptions is that cognitive activity may be understood in serial/linear terms in which there is some input to a mechanism that consequently performs a function and provides an output. An example of this assumption is the characterization of behavior as an outcome of an internal—usually neural—mechanism that executes a given behavior given some perceptual input and some functional goal.[16] In this sense, the relationship between neural activity and behavior is not one of constitution but of realizer and outcome.

The realizer-outcome view of the relationship between neural activity and behavior has been thoroughly criticized throughout the history of psychology and the cognitive sciences. The criticism has ranged from general attacks on the stimulus-response framework and its inability to capture the organic character of cognitive activities (Dewey 1896; Holt 1915; Gibson 1966)[17] to specific attacks on

[16]Notice that this fact may be true even for those mechanisms that include some kind of feed-forward model to reflect the current behavioral and perceptual outcomes of the behavioral mechanism on its future input as the behavioral output serially precedes the future input. See Pickering and Clark (2014).

[17]Put simply, the criticism counters the idea that cognitive activity starts with stimulation (e.g., visual stimulation) and ends up with a response (e.g., some movement of the limbs). On the contrary, critics claim, we must acknowledge the role of the "response" in the "stimulation" itself: cognitive activities are organic cycles of interdependent perception and action. In this sense, behavior is not just an outcome of neural activity.

the idea of the brain as a central controller of behavior and on the failure to provide a successful explanation of the emergence of the latter (Bernstein 1967; Turvey 1977; Gibson 1979; Meijer and Roth 1988; Kelso 1995).[18] More recently, the relationship between neural activity and behavior has been further analyzed and problematized in the neurosciences (Kelso et al. 2013; Krakauer et al. 2017; Pillai and Jirsa 2017; Raja 2018).

The relative success of these criticisms of the realizer-outcome view of the relationship between neural activity and behavior has prompted the appearance of a different understanding that may be summarized in J. J. Gibson's famous motto: "behavior is regular without being regulated." (1979, p. 225). Since the 1980s, a growing group of cognitive scientists has aimed to describe behavior in terms of the regularities in the dynamics of organism-environment interactions and not in terms of the outcome of the central controlling/regulatory activity of the brain (e.g., Kugler et al. 1980; Beer 1995, 2003; Kelso 1995; van Gelder 1998; Warren 2006).[19] In this sense, behaviors are taken to be activities of multiscale complex systems that can be captured at the scale of regular dynamical patterns of organism-environment interactions. These regularities are partially enabled by the dynamics of neural activity and, at the same time, constrain those very neural dynamics. Thus, behavior is not the outcome of some set of neural realizers, but an ongoing event occurring at a specific scale of a cognitive system (i.e., the scale of organism-environment interactions) that maintains a complex, circular relationship with other scales (e.g., the scale of neural activity).

As we see it, the notion of enabling constraint may shed light on such a complex, circular relationship between behavior and neural activity, and especially on its more challenging aspect: the way in which behavior constrains neural activity. The fact that neural activity partially enables behavior is a safe claim for any philosopher or neuroscientist. However, the complementary claim that behavior constraint neural activity may be not straightforwardly accepted.[20] In the following, we describe the nature of such a constraint and provide reasons for thinking of it as an enabling one.

[18]An example of this criticism is the supposed in-principle inability of a theory entailing a central controller to account for the *coordination* of all the effectors of a system as complex as the body of a human being to generate the desired behavior. The issue has been labeled as "the Charles V problem" in the literature on motor control (Meijer 2001).

[19]Importantly, the reader can remain agnostic regarding which alternative for the explanation of the relationship between behavior and neural activity is the correct one. For our purposes in this paper, we only need to acknowledge that the alternative, dynamical view of that relationship is a reality in the cognitive sciences.

[20]Especially if the realizer-outcome view of the relationship between behavior and neural activity is accepted.

10.3.1 The Arguments from Self-Organization and from Dynamical Systems

Generally speaking, those cognitive scientists that oppose the realizer-outcome view of the relationship between neural activity and behavior take cognitive systems to be self-organized complex systems which can be described at many spatiotemporal scales. For this reason, an adequate explanation of cognitive phenomena involves descriptions of cognitive activities at the neural scale (e.g., Anderson 2014; Tognoli and Kelso 2014), at the scale of the body (e.g., Kelso et al. 1981; Haken et al. 1985), and at the scale of organism-environment interactions (e.g., Fajen and Warren 2003; Warren 2006; Chemero 2009). However, an adequate explanation of cognitive phenomena cannot stop there and requires a story about the relations between these scales (Juarrero 1999; Van Order et al. 2003; Riley and Van Orden 2005; Raja and Anderson, 2019). The characteristic scalar properties of self-organized systems provide a way to understand these relations.

The study of self-organized complex systems yields the consistent observation of scale-free spatiotemporal regularities in their behavior, which are usually understood as fractal relationships between scales (see Bak 1990; Juarrero 1999; Riley and Van Orden 2005; Kuznetsov et al. 2013). Put simply, what we observe in the behavior of complex systems is that the value (power) of some of their variables increases or decreases at the different spatiotemporal scales in which their behavior is occurring (frequency) following a power law (Bak et al. 1987). This is the case, for example, of the Koch snowflake, in which the star-like or snowflake-like structure is the same one across scales. To be so, some of the variables of the structure, such the length of the lines or the area of the formed triangles, must increase or decrease with the scale of measurement. This is precisely what allows for finding the same structures at different scales. The relationship between power and frequency is a *scale-free* regularity insofar as it does not depend on the scale of measurement.

Another example of this fact is a tree. Branches stem from the trunk of the tree (scale 1). Then, smaller branches stem from bigger branches (scale 2). Then, even smaller branches stem from these branches (scale 3). And so on (scale 4 and following). Branches at different scales have different lengths and radiuses, but the relationship between length/radius (power) and the number of branches (frequency) is scale-free: length/radius proportionally decreases with the increment in the number of branches at each scale despite their initial values and the scales of measurement. In this sense, trees exhibit scale-free (or fractal) structure, as the same kind of relationship may be found regardless of the scale of measurement.

This kind of scale-free organization is taken to be a typical signature of self-organized systems (Bak 1996; Jensen 1998) and, therefore, of cognitive systems (see Van Orden et al. 2003; Stephen and Dixon 2009a). The usual reason given for this fact is that self-organized systems undergo transitions in the dynamics and structure of slower temporal scales that re-organize the dynamics and structure of faster temporal scales; a fractal structure is a consequence of such a re-organization (Stephen and Dixon 2009b). For example, interactions between the neurons of a brain region drive the emergence of patterns of synchronous or asynchronous behavior at the scale of the network of neurons that, at the same time, influence the behavior of single neurons themselves. In this sense, the activity at a slower temporal scale (i.e., synchrony/asynchrony at the network scale) re-organizes the activity at a faster temporal scale (i.e., firing behavior of single neurons). The fundamental consequence of this way of characterizing self-organized complex systems is that faster temporal scales are dependent on slower temporal scales. In terms of behavior and neural activity, the entailment is that changes in the dynamic organization of behavior re-organize the structure of neural activity, as behavior is at a slower temporal scale.

A related way to understand the scalar relations in cognitive systems comes from the work on *synergetics* developed by H. Haken (1973, 1977). Although Haken and others who work in a paradigm inspired by synergetics (e.g., Kelso 1995; Kugler et al. 1980, 1982) do not explicitly refer to fractality or scale-free organization, the underlying idea is similar[21]: in complex systems (including cognitive systems), the organization at higher spatial scales (usually the slower temporal scales) constrains the activity of lower spatial scales (usually the faster temporal ones). The key concept in synergetics is the *order parameter*. The order parameter is Haken's notion for the low-dimensional patterns that emerge from the collective behavior of the different components of a system. For example, the relative phase in the firing of neurons is the order parameter that defines the different modes of organization of the dynamics of the neural network they constitute (Bressler and Kelso 2016). In this sense, different values of relative phase define the states of the whole network in terms of synchrony and asynchrony.

What is interesting about order parameters is that they cannot be derived from the activity of the individual components of the systems. In the previous example, the order parameter (relative phase) is not a property of any single neuron but of the neural network itself. For this reason, it is said that the whole system (the neural network) imposes *order* on the behavior of its individual components (single neurons) through the order parameter. In other words, the dynamics of neural networks captured in terms of order parameters constrain the dynamics of single neurons. As in the case of the scale-free organization of systems, the idea behind order parameters is that at lower spatial scales (usually the faster temporal scale, e.g., single neurons firing), an entity's behavior is dependent on higher scales of collective behavior (usually the slower temporal scale, e.g., neural network dynamics). More concretely, in terms of behavior and neural dynamics, the entailment is that the dynamics of behavior impose order on the dynamics of neural activity (e.g., Kelso et al. 2013; Pillai and Jirsa 2017).

Both in the case of fractality and in the case of synergetics, the overall consequence with regard to the relationship between behavior and neural activity in cognitive systems is that behavior is not just enabled by neural dynamics but also

[21] Indeed, it is usual to understand both approaches as part of the same tradition and, generally, as part of the toolbox of nonlinear methods for the cognitive sciences (Riley and Van Orden 2005).

plays a central role in constraining those dynamics. In this sense, both approaches highlight the fact that in addition to the generally recognized influence of neural activity on behavior, behavior also influences neural activity. This influence is often conceptualized as a limitation or constraint in which "blue-collar brains" work under the government of behavior (Van Orden et al. 2012) or in which behavior put reins on the brain (Dotov 2014). On these views, the slower temporal scales of behavior constrain the variability of faster temporal scales of neural activity. Moreover, order parameters that emerge at higher scales of collective behavior reduce the degrees of freedom of the behavior of lower scales of componential behavior. That is: behavior is said to restrict or enslave neural activity. In this literature, then, behavior is depicted as a purely *negative* constraint on neural activity.

10.3.2 From Enslaving to Enabling

We are sympathetic to the understanding of the relationship between behavior and neural activity depicted in the previous section. However, we want to put forward a more radical thesis: that behavior is not just a negative constraint of neural activity, but a positive one. We want to claim that behavior doesn't just enslave neural dynamics but enables them. Ultimately, we want to claim that behavior is an enabling constraint of neural activity and that this follows directly from the very characterization of cognitive systems as self-organized multiscale complex systems we have offered. There are two senses in which behavior may be understood in this way. On the one hand, in the sense of self-organization, Van Orden et al. (2012) point out that behavior unfolds at a slower temporal scale than neural activity and, therefore, behavioral changes occur on a slower path than neural changes and many of the latter can occur within a stable state of the former. In this sense, behavior provides the context and history in which the dynamics of neural activity make functional sense. Thus, behavior allows for some neural activities that would be impossible in its absence. On the other hand, in the ecological sense, behavior provides what is needed for neural activity to be in contact with the environment. In this sense, without the *specific* provisions of behavioral dynamics, neural activities could not develop in the way they do and sometimes would not even be possible. Thus, behavior is a general condition of possibility of neural activity as such. Let's explore these two senses in some more depth.

After presenting examples of scale-free properties of neural dynamics through the analysis of EEG recordings, Van Orden et al. (2012) elaborate on several consequences of it. The first consequence is that the scale-free properties of neural dynamics suggest these dynamics are constrained by slower temporal dynamics. For this reason, explanations of behavior based just on neural dynamics are not enough: the interaction between scales affects neural dynamics and this fact must be reflected in our explanations. The second consequence is that a strong distinction between "behavior" and "brain activity" should be questioned. A better, more useful distinction is between the slower and faster temporal scales of the activity of organisms. The third and more important consequence is that slower temporal scales may be understood as playing the role of "context" or "memory" for faster temporal scales:

> Very slowly changing constraints could appear to be static if seen from the perspective of a very rapidly changing process. But the slow and fast changes are of course concurrent. On the one hand, concurrence allows very slowly changing constraints to serve a kind of memory function for more rapidly changing constraints. Slowly changing constraints remind a rapidly changing process of the constraints coming from the slow timescale, which may change only slightly, or not at all, from the constraints on previous cycles. Slower changes are in this way a means for faster changes to "remember" what they need to know about the status of all the more slowly changing constraints in the system. (Van Orden et al. 2012, p. 6).

To understand the way slower temporal scales (e.g., behavior) may serve as memory for faster temporal scales (e.g., neural dynamics) we need to recall the general properties of dynamical systems. Along with initial conditions and parameters, changes in dynamical systems depend on their own history. Namely, the present state of a dynamical system depends on its previous states. This is the most basic sense in which neural systems depend on their own history. However, as neural systems are nested within organisms and within organism-environment systems, these higher-order systems also participate in the history of neural dynamics. And they do so through constraining them. The constraints behavior imposes on neural dynamics limit the degrees of freedom available to the latter. In this sense, behavior restricts the variability of neural dynamics. But, importantly, it does so by maintaining (relatively) fixed the temporal context of the changes of neural dynamics and, therefore, acting as a kind of memory (or context) for those dynamics: changes at the temporal scale of neural dynamics are framed within the more stable states at the temporal scale of behavior. In virtue of this relative temporal stability, when the changes of neural dynamics occur, the temporal scale of behavior maintains information about the history of the system (memory) and about the possibilities available in the present (context).

This consequence of the scale-free properties of cognitive systems opens a new way to think about how cognitive systems deal with environmental states and information not currently present in the ongoing organism-environment interaction (Sanches de Oliveira and Raja 2018). Some of the information not currently available for the cognitive system is conserved in the slower temporal dynamics of the system allowing the faster temporal scales to manage it and thereby to exhibit a whole new specific set of functional outcomes. In other words, the fact that the slower temporal dynamics of behavior constrain the faster temporal dynamics of neural activity provides the latter with input in terms of the history and the current state of the whole cognitive system that would be unavailable without such a constraint. Therefore, behavior acts as an *enabling constraint* of neural activity by changing its functional outcomes via the constraining process.

A different way in which behavior may be taken to be an enabling constraint on neural activity has to do with the general input availability for the neural system. Among those approaches that reject the realizer-outcome view of the relationship between behavior and neural activity, (at least) those based on ecological psychology (Gibson 1966, 1979; Chemero 2009) have supported the idea that the trade-off between perception and action may be described in terms of informational control (Warren 2006).[22] Put simply, informational control posits that the regularities in the transformations of energy arrays as organisms move around are used to control that very moving around. These regularities are taken to be the *informational variables* used for the control of action (Segundo-Ortin et al. Forthcoming). For example, organisms are surrounded by ambient light that is structured around them in specific ways depending on the position of the source(s) of light and the layout of the environment. The structured ambient light is the relevant energy array for visual perception and, therefore, the information for visually controlled actions may be found in its transformations. These transformations are known as *optic flow* (Gibson 1958; Warren 1998). Different movements lead to different patterns of optic flow and different patterns of optic flow contain regular changes and stabilities that inform about movements themselves and help control them. For instance, a centrifugal optic flow—i.e., when the points of the optical field move from the center to the edges of the field—informs about forward locomotion and, therefore, maintaining that kind of optic flow helps to control the forward steering of the locomotion itself.

There is an interesting property of optic flow, however: it is a feature neither of the environment, nor of the structured ambient light, nor of the cognitive system alone. Chairs and tables, for example, are out there in the environment. Even ambient light is out there. However, optic flow is not just out there. Cognitive systems must move to generate optic flow. Nevertheless, optic flow is not "in" the cognitive system. If we move around with our eyes closed, there is no optic flow for us. Even more, if we move around in an environment in which there is no structured ambient light (e.g., in a dark room or in a foggy room), there is no optic flow for us either. Thus, optic flow is neither a property of the environment or of the organism, but a property of the organism-environment system that is generated due to the behavior of the organism in the environment. The actions of the organism (e.g., walking, turning, or staying still) generate specific patterns of optic flow given the structure of ambient light in the environment.

A consequence of such an understanding of perceptual information is that optic flow is *required* for perception and, therefore, the behavior of organisms in their environment is *required* for perception. If we accept that neural systems respond to perceptual information in their activity, we have to also accept that optic flows are necessary for at least some of the activities neural systems perform.[23] But optic flow is constrained by (i.e., shaped by) behavior, so if optic flow is an important variable for neural systems, that variable is constrained by behavior. In this sense, behavior constrains neural activity in a way that is necessary for the activity itself:

---

[22]It's important to note that the new mechanists have also acknowledged the inadequacy of the realizer-outcome approach, as they characterise cyclic and oscillatory mechanisms, such as circadian rhythms (Bechtel and Abrahamson 2013).

[23]The best way to describe the sensitivity of neural systems to perceptual information is still an open question. We take the concept of ecological resonance to be a good candidate to explain that sensitivity (Raja 2018; Raja and Anderson 2019).

behavior constrains the optic flow needed for the neural activity that enables visual perception. Without optic flow, neural activity would not accomplish its function in visual perception, and without behavior there would not be optic flow. Behavior is an *enabling constraint* of neural activity. Without behavior there would be no proper input for neural systems and they could not function in relevant ways (e.g., as part of the visual system).

We think these two examples—slow temporal scales providing memory and context for faster temporal scales, and behavior providing proper variables for neural systems—are two examples of the way behavior may be an enabling constraint for neural dynamics. By the two processes just detailed, behavior plays a fundamental role in the probability of the functional outcomes of neural activity, allowing for new functional outcomes and even for their functionality simpliciter. Of course, this is not to say that the relationship between behavior and neural activity is unidirectional. Nobody can neglect the role of neural activity as one of the main contributors to behavioral activity. We are obviously not denying the influence of neural activity on behavior, but are rather highlighting the influence of behavior on neural activities. Behavior considered as an enabling constraint for neural activity helps us better understand the complex scalar relations typical of cognitive systems and allows for a more complete understanding of cognitive activities.

10.4 Conclusion

In this paper, we have proposed "enabling constraint" as a potentially fruitful notion for the cognitive sciences to account for those scalar relations both within cognitive systems and between cognitive systems and the environment that cannot be easily accommodated by other notions such as constitution or derivability. More concretely, we have put forward the thesis that behavior may be considered as an enabling constraint of neural activity. The notion offers cognitive science a way to capture an important aspect of the relationship between behavior and neural activity, and the positive restriction the former imposes in the functional outcomes of the latter.

Elaborating on the positive role of the concept of developmental constraint in biology, we have proposed that enabling constraints are constraints that change the probability of functional outcomes of a system. We showed the way in which behavior may be seen as constraining neural activity in self-organized complex systems due to their scale-free properties. Finally, we have offered two instances in which behavior may be considered a positive constraint on the functional outcomes of neural activity: both framing it and making available new functional possibilities. In at least these two ways, behavior can be considered an enabling constraint of neural activity.

References

Amundson, R. (1994). Two concepts of constraint: Adaptationism and the challenge from developmental biology. *Philosophy of Science, 61*, 556–578.

Anderson, M. L. (2014). *After phrenology: Neural reuse and the interactive brain*. Cambridge, MA: MIT Press.

Anderson, M. L. (2015a). Beyond componential constitution in the brain: Starburst Amacrine Cells and enabling constraints. In T. Metzinger & J. M. Windt (Eds.), *Open MIND: 1(T)*. Frankfurt am Main: MIND Group. https://doi.org/10.15502/9783958570429.

Anderson, M. L. (2015b). Functional attributions and functional architecture. In T. Metzinger & J. M. Windt (Eds.), *Open MIND: 1(T)*. Frankfurt am Main: MIND Group. https://doi.org/10.15502/9783958570757.

Bak, P. (1990). Self-organized criticality. *Physica A, 163*, 403–409.

Bak, P. (1996). *How nature works: The science of self-organized criticality*. New York: Copernicus.

Bak, P., Tang, C., & Weisenfeld, K. (1987). Self-organized criticality: An explanation of 1/*f* noise. *Physical Review Letters, 59*(4), 381–384.

Bechtel, W. (2009). Constructing a philosophy of science of cognitive science. *Topics in Cognitive Science, 1*, 548–569.

Bechtel, W., & Abrahamsen, A. A. (2013). Thinking dynamically about biological mechanisms: Networks of coupled oscillators. *Foundations of Science, 18*(4), 707–723.

Bechtel, W., & Richardson, R. C. (1993). *Discovering complexity: Decomposition and localization as strategies in scientific research*. Cambridge, MA: The MIT Press.

Beer, R. D. (1995). A dynamical systems perspective on agent-environment interaction. *Artificial Intelligence, 72*, 173–215.

Beer, R. D. (2003). The dynamics of active categorical perception in an evolved model agent. *Adaptive Behavior, 11*(4), 209–243.

Bernstein, N. A. (1967). *The co-ordination and regulation of movements*. Oxford: Pergamon Press. (Original work published in Russian 1957; it is a volume edited by Bernstein himself).

Bressler, S. L., & Kelso, J. A. S. (2016). Coordination dynamics in cognitive neuroscience. *Frontiers in Neuroscience*. https://doi.org/10.3389/fnins.2016.00397.

Brigandt, I. (2015). From developmental constraints to evolvability: How concepts figure in explanation and disciplinary identity. In A. C. Love (Ed.), *Conceptual change in biology* (pp. 305–325). Boston: Springer.

Carroll, S. B. (2008). Evo-devo and an expanding evolutionary synthesis: A genetic theory of morphological evolution. *Cell, 134*(1), 25–36.

Chemero, A. (2009). *Radical embodied cognitive science*. Cambridge, MA: MIT Press.

Craver, C. F. (2008). *Explaining the brain: Mechanisms and the mosaic unity of neuroscience*. Oxford: Oxford University Press.

Craver, C. F., & Bechtel, W. (2007). Top-down causation without top-down causes. *Biology and Philosophy, 22*(4), 547–563.

Craver, C. F., & Darden, L. (2001). Discovering mechanisms in neurobiology: The case of spatial memory. In P. K. Marchamer, R. Grush, & McLaughlin (Eds.), *Theory and method in the neurosciences*. Pittsburgh: University of Pittsburgh Press.

Demb, J. B. (2007). Cellular mechanisms for direction selectivity in the retina. *Neuron, 55*(2), 179–186. https://doi.org/10.1016/j.neuron.2007.07.001.

Dewey, J. (1896). The reflex arc concept in psychology. *Psychological Review, 3*, 357–370.

Dotov, D. G. (2014). Putting reins of the brain: How the body and the environment use it. *Frontiers in Human Neuroscience, 8*, art. 795.

Euler, T., Detwiler, P. B., & Denk, W. (2002). Directionally selective calcium signals in dendrites of starburst amacrine cells. *Nature, 418*(6900), 845–852. https://doi.org/10.1038/nature00931.

Fajen, B. R., & Warren, W. H. (2003). Behavioral dynamics of steering, obstacle avoidance, and route selection. *Journal of Experimental Psychology: Human Perception and Performance, 29*, 343–362.

Gibson, J. J. (1958). Visually controlled locomotion and visual orientation in animals. Reprinted in E. S. Reed & R. Jones (Eds.; 1982), *Reasons for realism* (pp. 148–163), Hillside: Lawrence Erlbaum.

Gibson, J. J. (1966). *The senses considered as perceptual systems*. Boston: Houghton Miffin.

Gibson, J. J. (1979). *The ecological approach to visual perception*. Boston: Houghton Miffin.

Goodman, C. S., & Coughlin, B. C. (2000). The evolution of evo-devo biology. *Proceedings of the National Academy of Sciences USA, 97*(9), 4424–4456.

Gould, S. J. (1980). The evolutionary biology of constraint. *Daedalus, 109*(2), 39–52.

Gould, S. J., & Lewontin, R. C. (1979). The spandrels of San Marco and the Panglossian paradigm: A critique of the adaptationist programme. *Proceedings of the Royal Society of London. Series B. Biological Sciences, 205*, 581–598.

Haken, H. (1973). *Synergetics: Cooperative phenomena in multi-component systems*. Berlin: Springer.

Haken, H. (1977). *Synergetics: A workshop*. Berlin: Springer.

Haken, H., Kelso, J. A. S., & Bunz, H. (1985). A theoretical model of phase transitions in human hand movements. *Biological Cybernetics, 51*, 347–356.

Hall, B. K. (2003). Evo-devo: evolutionary developmental mechanisms. *International Journal of Developmental Biology, 47*(7–8), 491–495.

Held, L. I. (2014). *How the snake lost its legs: Curious tales from the frontier of evo-devo*. Cambridge: Cambridge University Press.

Holekamp, K. E., Swanson, E. M., & Van Meter, P. E. (2013). Developmental constraints on behavioural flexibility. *Philosophical Transactions of the Royal Society B, 368*, 20120350.

Holt, E. B. (1915). *The Freudian wish and its place in ethics*. New York: Henry Holt and Company.

Ingle, D. (1973). Two visual systems in the frog. *Science, 181*(4104), 1053–1055.

Jensen, H. J. (1998). *Self-organized criticality: Emergent complex behavior in physical and biological systems*. Cambridge: Cambridge University Press.

Juarrero, A. (1999). *Dynamics in action: Intentional behavior as a complex system*. Cambridge, MA: The MIT Press.

Kelso, J. A. S. (1995). *Dynamic patterns*. Cambridge, MA: MIT Press.

Kelso, J. A. S., Holt, K. G., Rubin, P., & Kugler, P. N. (1981). Patterns of human interlimb coordination emerge from the properties of nonlinear, limit cycle oscillatory processes: Theory and data. *Journal of Motor Behavior, 13*, 226–261.

Kelso, J. A. S., Dumas, G., & Tognoli, E. (2013). Outline of a general theory of behavior and brain coordination. *Neural Networks, 37*, 120–131.

Kohler, A. (2015). Carving the brain at its joints. In *Open MIND*. In T. Metzinger & J. M. Windt (Eds.), *Open MIND: 1(T)*. Frankfurt am Main: MIND Group. https://doi.org/10.15502/9783958570627.

Krakauer, J. W., Ghazanfar, A. A., Gomez-Marin, A., MacIver, M. A., & Poeppel, D. (2017). Neuroscience needs behavior: Correcting a reductionist bias. *Neuron, 93*, 480–490.

Kugler, P. N., Kelso, J. A. S., & Turvey, M. T. (1980). On the concept of coordinative structures as dissipative structures I: Theoretical lines of convergence. In G. E. Stelmach & J. Requin (Eds.), *Tutorials in motor behavior* (pp. 3–37). Amsterdam: North Holland.

Kugler, P. N., Kelso, J. A. S., & Turvey, M. T. (1982). On coordination and control in naturally developing systems. In J. A. S. Kelso & J. E. Clark (Eds.), *The development of movement control and coordination* (pp. 5–78). New York: Wiley.

Kuznetsov, N., Bonnette, S., & Riley, M. A. (2013). Nonlinear time series methods for analyzing behavioral sequences. In K. Davis et al. (Eds.), *Complex systems in sport* (pp. 83–102). London: Routledge.

Lee, S., & Zhou, Z. J. (2006). The synaptic mechanism of direction selectivity in distal processes of starburst amacrine cells. *Neuron, 51*(6), 787–799. https://doi.org/10.1016/j.neuron.2006.08.007.

Leonetti, A., Puglisi, G., Siugzdaite, R., Ferrari, C., Cerri, G., & Borroni, P. (2015). What you see is what you get: Motor resonance in peripheral vision. *Experimental Brain Research, 233*, 3013–3022.

Mackie, J. L. (1965). Causes and conditions. *American Philosophical Quarterly, 2*(4), 245–264.
Masland, R. H. (2005). The many roles of starburst amacrine cells. *Trends in Neurosciences, 28*(8), 395–396. https://doi.org/10.1016/j.tins.2005.06.002.
Meijer, O. G. (2001). Making things happen: An introduction to the history of movement science. In M. L. Latash & V. M. Zatsiorsky (Eds.), *Classics in movement science* (pp. 1–57). Champaign: Human Kinetics.
Meijer, O. G., & Roth, K. (1988). *Complex movement behaviour: 'The' motor-action controversy*. Amsterdam: North-Holland.
Millikan, R. (1989). In defense of proper functions. *Philosophy of Science, 56*(2), 288–302.
Oyama, S. (2000). *The ontogeny of information: Developmental systems and evolution*. Cambridge: Cambridge University Press.
Oyama, S., Griffiths, P. E., & Gray, R. D. (2001). Introduction: What is developmental systems theory? In S. Oyama, P. E. Griffiths, & R. D. Gray (Eds.), *Cycles of contingency: Developmental systems and evolution* (pp. 1–11). Cambridge, MA: The MIT Press.
Pickering, M. J., & Clark, A. (2014). Getting ahead: Forward models and their place in cognitive architecture. *Trends in Cognitive Science, 18*(9), 451–456.
Pillai, A. S., & Jirsa, V. K. (2017). Symmetry breaking in space-time hierarchies shapes brain dynamics and behavior. *Neuron, 94*, 1010–1026.
Raja, V. (2018). A theory of resonance: Towards an ecological cognitive architecture. *Minds and Machines, 28*(1), 29–51.
Raja, V., & Anderson, M. L. (2019). Radical embodied cognitive neuroscience. *Ecological Psychology, 31*(3), 166–181. https://doi.org/10.1080/10407413.2019.1615213.
Rausher, M. D., Lu, Y., & Meyer, K. (2008). Variation in constraint versus positive selection as an explanation for evolutionary rate variation among anthocyanin genes. *Journal of Molecular Evolution, 67*, 137–144.
Riley, M. A., & Van Orden, G. C. (2005). *Tutorials in contemporary nonlinear methods for the behavioral sciences*. http://www.nsf.gov/sbe/bcs/pac/nmbs/nmbs.jsp
Salthe, S. N. (1993). *Development and evolution: Complexity and change in biology*. Cambridge, MA: The MIT Press.
Sanches de Oliveira, G., & Raja, V. (2018). The cognition-perception distinction across paradigms: An ecological view. In T. T. Rogers, M. Rau, X. Zhu, & C. W. Kalish (Eds.), *Proceedings of the 40th annual conference of the cognitive science society* (pp. 2403–2408). Austin: Cognitive Science Society.
Schwenk, K. (1994). A utilitarian approach to evolutionary constraint. *Zoology, 98*, 251–262.
Segundo-Ortin, M., Heras-Escribano, M., & Raja, V. (forthcoming). Ecological psychology is radical enough: A reply to radical enactivists. *Philosophical Psychology*.
Silberstein, M. (2018) Contextual emergence. In A. D. Carruth & J. T. M. Miller (Eds.), Special issue of *Philosophica* on emergence. (Vol. 91 pp. 145–92.
Silberstein, M. (in press). Constraints on localization and decomposition as explanatory strategies in the biological sciences 2.0. In: F. Calzavarini & M Viola (Eds.), *Neural mechanisms: New challenges in the philosophy of neuroscience*. Springer.
Stephen, D. G., & Dixon, J. A. (2009a). Dynamics of representational change: Entropy, action, and cognition. *Journal of Experimental Psychology: Human Perception and Performance, 35*(6), 1811–1832.
Stephen, D. G., & Dixon, J. A. (2009b). The self-organization of insight: Entropy and power laws in problem solving. *The Journal of Problem Solving, 2*(1), 72–101.
Tauchi, M., & Masland, R. H. (1984). The shape and arrangement of the cholinergic neurons in the rabbit retina. *Proceedings of the Royal Society of London. Series B. Biological Sciences, 223*(1230), 101–119. https://doi.org/10.1098/rspb.1984.0085.
Tognoli, E., & Kelso, J. A. S. (2014). The metastable brain. *Neuron, 81*, 35–48.
Turvey, M. T. (1977). Preliminaries to a theory of action with reference to vision. In R. Shaw & J. Bransford (Eds.), *Perceiving, acting, and knowing: Toward an ecological psychology* (pp. 211–265). Hillsdale: Erlbaum.

Van Fraassen, B. (1977). The pragmatics of explanation. *merican Philosophical Quarterly, 14*, 143–150.
van Gelder, T. (1998). The dynamical hypothesis in cognitive science. *Behavioral and Brain Sciences, 21*(5), 615–665.
Van Orden, G. C., Holden, J. G., & Turvey, M. T. (2003). Self-organization of cognitive performance. *Journal of Experimental Psychology: General, 132*(3), 331–350.
Van Orden, G. C., Hollis, G., & Wallot, S. (2012). The blue-collar brain. *Frontiers in Psychology, 3*, art. 207.
Warren, W. H. (1998). Visually controlled locomotion: 40 years later. *Ecological Psychology, 10*(3–4), 177–219.
Warren, W. H. (2006). The dynamics of perception and action. *Psychological Review, 113*(2), 358–389.
Yoshida, K., Watanabe, D., Ishikane, H., Tachibana, M., Pastan, I., & Nakanishi, S. (2001). A key role of starburst amacrine cells in originating retinal directional selectivity and optokinetic eye movement. *Neuron, 30*(3), 771–780. https://doi.org/10.1016/S0896-6273(01)00316-6.

Part III
Metaphysical Challenges

Chapter 11
Your Brain Is Like a Computer: Function, Analogy, Simplification

Mazviita Chirimuuta

Abstract The relationship between brain and computer is a perennial theme in theoretical neuroscience, but it has received relatively little attention in the philosophy of neuroscience. This paper argues that much of the popularity of the brain-computer comparison (e.g. circuit models of neurons and brain areas since McCulloch and Pitts, Bull Math Biophys 5: 115–33, 1943) can be explained by their utility as ways of simplifying the brain. More specifically, by justifying a sharp distinction between aspects of neural anatomy and physiology that serve information-processing, and those that are 'mere metabolic support,' the computational framework provides a means of abstracting away from the complexities of cellular neurobiology, as those details come to be classified as irrelevant to the (computational) functions of the system. I argue that the relation between brain and computer should be understood as one of analogy, and consider the implications of this interpretation for notions of multiple realisation. I suggest some limitations of our understanding of the brain and cognition that may stem from the radical abstraction imposed by the computational framework.

11.1 Preamble: Leibniz the Inventor

Many histories of computation begin with the unrealised ambition of Gottfried Leibniz to devise a "universal characteristic", a symbolic language in which factual propositions could be represented and further truths inferred by means of a mechanical calculating device (Davis 2000). Amongst the twentieth century pioneers of computer science and artificial intelligence who took Leibniz for an

M. Chirimuuta (✉)
History & Philosophy of Science, University of Pittsburgh, Pittsburgh, PA, USA
e-mail: mchirimu@exseed.ed.ac

© Springer Nature Switzerland AG 2021

F. Calzavarini, M. Viola (eds.), *Neural Mechanisms*, Studies in Brain and Mind 17,
https://doi.org/10.1007/978-3-030-54092-0_11

inspirational figure[1] were Warren McCulloch and Walter Pitts (Lettvin 2016: xix). Single cell neurophysiology and the engineering of digital computers both grew into maturity in the early 1940's, and significantly influenced one another (Arbib 2016). Cybernetics – the study of information flow and self-regulation in all systems, living and manufactured – was the natural product of these interconnected developments,[2] while McCulloch and Pitts (1943) opus – "A Logical Calculus of the Ideas Immanent in Nervous Activity" – could plausibly be received as the fruit of Leibniz's 270 year old insight that one and the same power of reasoning may inhabit the living man and the mechanical device (Morar 2015:126 fn11).

By showing that, under certain assumptions, small assemblies of connected neurons could be taken to operate as logic gates, McCulloch and Pitts were able to claim that the brain *is* – not metaphorically or analogously – a computer. However, the prospect that logic by itself would be all the theory needed to understand the brain turned out to be a mirage. According to the recollections of neurophysiologist Jerome Lettvin, the results of detailed observation of the responses of neurons in the frog's retina left Pitts severely disillusioned because the peculiarities of neuronal behaviour did not make sense from a purely logical point of view.[3]

Following the early literalism, and the subsequent apprehension that the nervous system is more tangled than the crystalline ideals of logicians would have it, the relation between brain and computer has been left under-specified. Computer models of neural systems are more than mere models in the sense of simulations, like weather models, that represent but do not re-enact the processes of nature. Instead, neural circuits, and the computational models of them, are thought by the scientists to *be doing the same thing* – processing information (Miłkowski 2018).[4] At the same time, many have voiced the concern that the electronic computer is a mere metaphor for the biological brain, one that places a conceptual box around neuroscientists' thinking and should be discarded along with the hydraulic model of the nervous system, and the image of the cortex as a telephone exchange (Daugman 2001). In this paper I account for the tenacity of the idea of brain as a computer by appealing to its usefulness as a means of simplifying the brain. I will take the brain-computer relationship to be one of analogy, whereby comparisons are drawn

[1]See Morar (2015) on Leibniz's invention of a mechanical calculator for the four arithmetical functions, and the history of reception of Leibniz's contributions in this area.

[2]See Kline (2015) and Pickering (2010) for overviews of the cybernetic movement in the USA and UK, respectively.

[3]"up to that time [of results of Lettvin et al. (1959)], Walter had the belief that if you could master logic, and really master it, the world in fact would become more and more transparent. In some sense or another logic was literally the key to understanding the world. It was apparent to him after we had done the frog's eye that even if logic played a part, it didn't play the important or central part that one would have expected." Lettvin, interviewed in Anderson and Rosenfeld (1998: 10)

[4]I do not mean to suggest that there is a uniform opinion amongst neuroscientists on what the nature of neural information processing is. Views on this have certainly differentiated since McCulloch and Pitts.

between electronic systems – engineered to be somewhat functionally similar to biological ones – and the vastly more complex organic brain.

My analogical interpretation will be presented as an alternative to the literal interpretations of neural-computational models which presume that the running of the model is a more or less accurate reproduction of a computation first instantiated in biological tissue. In order to pre-empt the worry that there is no substantial difference between the literal and analogical interpretations, I specify at the outset that I am not defining analogies as homomorphisms that obtain between the brain and its model.[5] For on that definition the analogical relationship would amount to the instantiation of the same structure (i.e. function computed) in the neural system and the model. It would follow, on the assumption of a "mapping" account of computational implementation, that there would be no daylight between the literal and analogical interpretations of neurocomputational models, because the literal interpretation just is the claim that neural system and its model compute (approximately) the same function. On my conception, to say that a model should be interpreted analogically is to say that the target is *like* the model is some way that may turn out to be dependent on the interests of the scientists, and the techniques they employ. The crucial point, to be defended in Sect. 11.3, is that the structure in the brain found to be to be relevantly similar to the model is not assumed to be an inherent, human-independent fact about the brain.

In Sect. 11.2 I describe how the brain-computer analogy permits scientists to draw a distinction between the aspects of neuro-anatomy and physiology that are "*for information processing*", as opposed to "*mere metabolic support*". The analogy offers answers to the question of what neural mechanisms are *for*, which are left hanging if one takes the brain only to be an intricate causal web, and one neglects the functional perspective afforded by thinking of the brain as an organic computer. This makes research in neurobiology more efficient by channelling the possibly endless delineation of biochemical interactions along the paths carved out by hypotheses arrived at by reverse engineering the information-processing functions of the neurons. Yet, the empirical successes of this research programme that are made possible because of this gain in efficiency do not warrant the conclusion that the neural systems themselves compute the functions specified in the model, or that the brain itself is a computer.

[5]In this I am following the definition of analogies in the philosophy of science literature on analogical reasoning. As Dardashti, Thébault, and Winsberg (2017) put it, instances where an isomorphism obtains are a subset of all the cases of analogies in science, and they support stronger inferences than the other cases. See Knuuttila and Loettgers (2014: 87) for further discussion of why analogical reasoning in science goes beyond the isolation of structures that map from model to target.

11.2 Simplification and the Computational Brain

As stated above, my view is that the relationship between brain and electronic computer, neural physiology and patterns of activation in a circuit board, should be interpreted as one of analogy. This is in contrast with the view that the brain *is literally* a kind of computer, and that neural circuits are one of many potential realisers for the coding schemes discovered by computational neuroscientists, and sometimes implemented by AI engineers when aiming at biological realism. In Sect. 11.3 I give a proper elaboration of this contrast, and state some advantages of my own interpretation. The claim of this section is that a major benefit of computational theory in neuroscience is the simplification of the brain that it affords. What I say here is neutral between the literal and analogical interpretations of computational models of the brain (regardless of whether the modellers whose work I discuss themselves understand their models more literally or analogically).

We have noted already that the earliest hopes for a computational theory of the brain – McCulloch and Pitts' plan for neural reverse engineering on the assumption that the brain is a computing machine and made up of neuronal logic gates (Piccinini 2004) – were defeated by the unruliness (with respect to McCulloch and Pitts' logically derived expectations) of the responses of actual neurons to visual stimulation. Given these initial disappointments, one might ask how it was that computationalism still went on to become the dominant theoretical framework for neuroscience.[6] This is a broad question which deserves a complex answer, referring to historical and sociological factors, and to differences between sub-specialities within the science. However, for the purposes of this paper, I offer a simple answer, that boils down only to one characteristic of computationalism – that it provides neuroscientists with a very useful, possibly indispensable, means to simplify their subject of investigation. More specifically, my claims are (1) that computationalism permits a distinction between the functional (information processing) aspects of neural anatomy and physiology and what is there merely as metabolic support, thereby justifying the neglect of countless layers of biological complexity; and (2) that computational theory, in giving the specification of neural functions, provides an ingredient lacking in purely mechanistic approaches to neurobiology, without which it would be far more difficult to separate relevant from irrelevant causal factors and hence to state when the characterisation of a mechanism is sufficiently complete.

[6]Note that this should not be confused with the issue of whether the dominant mode of explanation in neuroscience is mechanistic or computational. Those on the mechanist side of this debate, such as Kaplan (2011), acknowledge the importance of computationalism in theoretical neuroscience, and argue furthermore that computational models provide mechanistic explanations. Another point is that those promoting dynamical systems theory as a better theoretical framework than computationalism for some neural systems (e.g. Shenoy et al. 2013) do not dispute the dominance of computationalism in neuroscience as it stands.

11.2.1 The Isolation of the Functional

It should not be news to anyone who has observed the practice of science that part of the task (and art) of devising a new experiment or explanation is the drawing of a distinction between the target of investigation and the additional factors that can reasonably be classified as background conditions. For a system of any complexity (which is all of the systems studied in biological science), the outcome of the endeavour largely turns on the aptness of the distinction. As the neurologist Kurt Goldstein (1934/1939) argued, all of the supposed "background" factors within an organism are highly relevant to the behaviour of the whole creature, in ways that most of experimental biology ignores; yet even if one acknowledges the lack of an absolute distinction between target and background, it is still usually appropriate for the biologist to train her attention selectively on the target, as one does with a visual image affording figure-ground separation.

My contention here is that much of the value that the computational framework provides to neuroscience is in the distinction it supports between the function of a neural system (information processing), which provides the target of investigation, and the residual features that can be placed in the background as mere metabolic support.[7] The classic characterisation of the neuron as a device which gathers inputs at the dendrites, calculates a function and delivers an output (a number of spikes sent down the axon) is the most prevalent way that this distinction has been put to use in neuroscience. While this picture is much broader than McCulloch and Pitts' (1943) formalism, they can be credited with disseminating the idea that the single neuron is an input-output device, and giving neuro-modellers an excuse for abstracting away from most of the cell biology underling the reception and generation of action potentials:

> The liberating effect of the mode of thinking characteristic of the McCulloch and Pitts theory can be felt on two levels. ..... On the local level it eliminates all consideration of the detailed biology of the individual cells from the problem of understanding the integrative behaviour of the nervous system. This is done by postulating a hypothetical species of neuron defined entirely by the computation of an output as a logical function of a restricted set of input neurons. (Papert 2016: xxxiii)

The utility of this simple picture goes a long way to explaining the persistence of the "neuron doctrine"—the thesis that neurons are the functional unit of the nervous system, whose job it is to receive, process and send information—in the face of some countervailing empirical findings (Bullock et al. 2005).[8]

---

[7] Haueis (2018) also discusses the distinction between cognitive and non-cognitive functions of the nervous system.

[8] Cao (2014) recommends going beyond the neuron doctrine to consider synapses and glia also as functional units of the nervous system. This raises the question of the technical feasibility of gathering synapse-resolution data of neural responses, and attempting to model the brain in such a fine-grained way (noting that each cortical neuron receives, on average, tens of thousands of inputs). If the neuron doctrine provides a "good enough" framework for modelling the brain, especially useful for the activation patterns associated with observable behaviours (perception,

The strategy, just outlined, for isolating the functional begins with the concrete neural system and abstracts away from it all features classified as non-functional, metabolic support. Another modus operandi is to start with the specification of a cognitive task (such as detection of edges in a photograph), consider what computations would be needed to achieve the task, and then to build an artificial system (i.e. a computational model) that performs it. With the model in place, the final step is to use it as a template or map when looking for activation and connectivity patterns in the brain that are responsible for the performance of this task. This strategy is described by Lettvin, in response to the criticism that computational models used in neuroscience – such as connectionist networks – lack similarity to neural systems:

> But, even if ideally one could record from any element or part of an element in situ, it is not in the least obvious how the records could be interpreted.[9] To a greater degree than in any other current science, we must know what to look for in order to recognize it.....
>
> This is where a prior art is needed, some understanding of process[10] design. And that is where AI, PDP, and the whole investment in building [neurocomputational models of intelligence] enter in. Critics carp that the current golems do not resemble our friends Tom, Dick, or Harry. But the brute point is that a working golem is not only preferable to total ignorance, it also shows how processes can be designed analogous to those we are frustrated in explaining in terms of nervous action. It also suggests what to look for. Lettvin (2016:xvii–xviii)[11]

If anything, the problem of "knowing what to look for" is more acute now than when Lettvin wrote this. In the last ten years, the increase in the variety of tools and methods for observing neural activity (from single cells to whole brains) has surprised and delighted many. However, the downside of these advances is that they bring to light kinds of complexity that were not previously apparent, especially at sub-cellular scales. This is how neuroscientist Yves Frégnac describes the situation:

---

learning, decision making) which involve large populations of neurons, then there is little reason to attempt the impossible and replace neurons with synapses as the fundamental signalling systems, even if one acknowledges that in the brain much information processing does occur within synapses. Below I take up the issue of the importance of these details that are relegated to the background in the classic neuro-computational picture.

[9]A point made vivid by Jonas and Kording (2017).

[10]Lettvin often uses this word in his characterisation of the 'engineering-stance' in neuroscience. It should not be confused with the notion of 'process models' in psychology, or other kinds of mechanistic models.

[11]Pickering (2010: 6) takes this methodology to be the standard practice for cybernetics in neuroscience, though many of the artificial devices were not computer programmes:

> Just how did the cyberneticians attack the adaptive brain? The answer is, in the first instance, by building electromechanical devices that were themselves adaptive and which could thus be understood as perspicuous and suggestive models for understanding the brain itself. The simplest such model was the servomechanism—an engineering device that reacts to fluctuations in its environment in such a way as to cancel them out. A domestic thermostat is a servomechanism; so was the nineteenth-century steam-engine 'governor' which led Wiener to the word 'cybernetics.'

> Each overcoming of technological barriers opens a Pandora's box by revealing hidden variables, mechanisms, and nonlinearities, adding new levels of complexity. By reaching the microscopic-scale resolution, advanced technologies have unveiled a new world of diversity and randomness, which was not apparent in pioneer functional studies using spike rate readout or mesoscopic imaging of reduced sensitivity. (Frégnac 2017: 471)

He points to the need for a greater understanding of how mesoscopic and macroscopic regularities emerge from the processes observed microscopically. But a wider point is that if artificial systems, sharing none of the microscopic details of the neural ones, can be built to duplicate some specific functions,[12] then one has an acceptable excuse for keeping shut the Pandora's box of sub-cellular neurobiology.

11.2.2 Mechanism and Function

In response to a criticism of the mechanistic account of explanation, which takes issue with the favouring of more detailed descriptions of mechanisms as providing better explanations than less detailed, 'sketchy' ones, Craver and Kaplan (2018) emphasise that their account has never favoured more detailed descriptions, per se, but has only suggested that models describing more of the *relevant* details have the edge over more abstract ones. But this immediately raises the question of how the scientist comes to know how to distinguish the relevant from the irrelevant factors. In any biological system, the nervous system especially, one finds a densely interconnected causal web with many layers of structural intricacy, and patterns of effect across various spatial and temporal scales. Craver and Kaplan appeal to a "mutual manipulability" criterion that is clear and unobjectionable in principle.[13] However, if their norms for explanation are to be considered in practice it becomes hard to see how only the causal factors in a neural system relevant to a particular phenomenon – as opposed to background factors not constitutive of the mechanism itself – could be isolated if only the mechanistic perspective is employed. An individual neuron will have thousands of feasible targets or 'handles' for experimental manipulation – for example, the different kinds of ion channels, which could be blocked on select portions of the membrane; the various different receptors that could be agonised or antagonised; the countless proteins transcribed in the cell which could be targets

[12] I am alluding here to multiple realisation – a topic to be discussed directly in Sect. 11.3. But the point can still be made without supposing there are cases in which one would want to say that an artificial and a neural system are two different realisers of *the same* function. Consider just the comparison between a fairly abstract and a highly detailed model of a neural circuit (e.g. a model where neurons are just represented as a time series of spike rates, and a 'compartment model' which represents some of the anatomical structure of the neuron). If the former is an equally good working model of the function of interest, then it is a reasonable working assumption that the behaviour of the neural system can be understood without reference to sub-cellular structure.

[13] "A factor is constitutively relevant when (ideal) interventions on putative component parts can be used to change the explanandum phenomenon as a whole and, conversely, interventions on the explanandum phenomenon as a whole can produce changes in the component parts" (Craver and Kaplan 2018: 20).

of genetic manipulation. One needs to multiply this list of causal variables by 10 or by 100 if the system comprises a small population of neurons. One faces a combinatorial explosion of experiments that would be needed to determine the independent causal relevance of each of these factors in a putative mechanism. But of course neuroscientists do not plan sequences of experiments according to brute force search! When designing an experiment with the aim of determining which of the many causal variables present in a system are crucial to its behaviour (given a certain explanatory question), how does a neuroscientist know which ones to select from an inexhaustible list? One should think of hypotheses regarding the information processing functions of neuronal structures as heuristics that drastically reduce this search space.

For example, at a fairly high level of abstraction, only net excitation minus inhibition is the causal factor relevant to determining whether a neuron's firing rate will increase or decrease. This abstraction disregards the kinds of neurotransmitters found at the synapse, receptor types, and location of synapses.[14] And of course this is the kind of abstraction fostered by the neuron doctrine and fundamental to McCulloch and Pitts' vision of the brain as a computer in which the logic gates are built from neurons.[15] In essence, without any prior assumption in place about what the neuron's function is, and what aspects of physiology and anatomy are relevant to it, the search for relevant causal factors would have to proceed by brute force or be guided by pure prejudice. This indicates that the functional, informational processing perspective on neural systems is an indispensable complement to the mechanistic approach in neurobiology. Another way to make this point is just to say that the boundaries around neural mechanisms are not simply there in the brain, discoverable through a small enough number of causal experiments. There are many justifiable ways for the neuroscientist to carve up the subsystems of the brain into mechanisms, and separate them from background conditions. The computational perspective is one approach that has suggested to scientists a particularly fruitful set of delineations. I return to this point in Sect. 11.3.4.

The difference between the physicist's and the engineer's perspectives on nature is a useful analogue to the difference between mechanistic and computational perspectives in neuroscience (Fairhall 2014). When one considers the structures

[14]Craver and Kaplan (2018: p. 19 fn 16) appeal to the purely causal notion of "screening off" in order to address the question of why complete (ontic) explanations do not end in quarks. The idea is that "low-level differences" will be ignored if they "make no relevant difference once the higher-level behaviour is fixed." I would like to point out that for the kind of abstractions I mention here, screening off should not be expected to occur – i.e. these excluded details do causally affect neuronal behavior in ways that are not fully summarized by the "higher level" variables of net excitation and inhibition, because of non-linearities in the behaviour of the cell. This suggests that a search for "relevant details" that proceeded only by the method of searching for "higher level" causal variables to replace "lower level" ones would not result in the abstractions found to be most useful in computational neuroscience.

[15]There is latitude here in the abstracting assumptions. I have described a case where total inhibition is subtracted from total excitation, whereas McCulloch and Pitts (1943: 118) posit that inhibitory input at any one synapse will cancel out the effects of excitation.

of the brain as a physical system, it is a web of causal interactions in which considerations of function are alien; in contrast, the notions of design and function are inherent to the engineering perspective, from which it is natural to regard the brain as a target of reverse-engineering (Sterling and Laughlin 2015). The mechanistic approach is supposed only to decompose a system into its structures and causal interactions, showing how their interaction brings about or constitutes the phenomenon which identifies the mechanism. On the computational approach, one begins with the consideration of what the neural system is *for*, and the question of how that function is achieved is addressed only after this. When dealing with complex, biological systems, any attempt to employ only the function-less physics stance would quickly get one lost amongst tangled causal details. This is a point made by the neurologist Francis Walshe:

> The modern student finds it difficult to see the wood for the trees ... He does not always have a synoptic concept of the nervous system in his mind ... If we subject a clock to minute analysis by the methods of physics and chemistry, we shall learn a great deal about its constituents, but we shall not discover its operational principles, that is, what makes these constituents function as a clock. Physics and chemistry are not competent to answer questions of this order, which are an engineer's task ... Both modes have their place and limitations; and they complement one another. (Walshe 1961: 131)[16]

It is the task of theory in science to provide the "synoptic concept" of a subject matter, and in neuroscience the computational theory is best developed, though I do not claim that this is the only possible theory of the nervous system.[17]

A wrinkle in the comparison I have drawn between the physicist's approach and mechanistic perspective in biology is that a mechanistic investigation *does* incorporate a notion of function or purpose, that is completely alien to physics. This is because without such a notion one actually cannot delineate a mechanism – mechanisms are mechanisms *for* the phenomena they produce or constitute (Craver and Kaplan 2018: 23 fn19). Within the mechanistic outlook this notion of function

[16]See also Knuuttila and Loettgers (2014: 79) on the contrast between physics and engineering based approaches within synthetic biology research. One might also be reminded of the so-called "design stance" (Dennett 1987).

[17]And I certainly am not claiming that the computational perspective should float free from experimentally derived facts regarding neural mechanisms. Theorising unconstrained by experimental results risks producing elegant models that do not apply to the actual brain.

has an ambiguous status, resulting in a curious tension.[18] On the one hand, purpose or function cannot be thought of as an *inherent* feature of the mechanism in question (which is, officially, just a purposeless causal web of processes which take place according to the laws of physics and chemistry); on the other hand, mechanisms are thought of as defined by the things that they do, which is normally understood as the purpose served in the context of the tissue, organ, or organism. This difference is papered over with the thought that one can gesture at Darwinian adaptation and the notion of selected functions to bridge this gap – even if, in reality, no-one ever attempts to show that every system classified as a mechanism has actually been a target of natural selection, and so has a "proper function". And in fact Craver and Darden (2013; 53–54) deny that the "phenomena" by which mechanisms are identified need be proper functions.

In relation to this, Jerome Lettvin makes the very interesting point that the engineering approach is prominent in biology precisely where there is a vacuum left following biologists' attempt to adhere strictly to physical-chemical (and hence purpose-less) perspectives when conceptualising their subject matter:

> Ever since biology became a science at the hands of biochemists it has carefully avoided or renounced the concept of purpose as having any role in the systems observed.... Only the observer may have purpose, but nothing observed is to be explained by it. This materialist article of faith has forced any study of process out of science and into the hands of engineers to whom purpose and process are the fundamental concepts in designing and understanding and optimizing machines. (1998:13)

Lettvin goes on to say that, "we had better use the process [i.e. functional characterisation] to tell what to look for in the mechanism rather than the other way round." (1998:17).

With this in mind, we can appreciate that cybernetics, the scientific movement in which McCulloch and Pitts were players, and from which today's computational neuroscience descended, was self-consciously a science of finality in a mechanistic world. And it was possible for cybernetics to develop as a science of finality because engineering was very well represented in this interdisciplinary research field. Cyberneticians took the design stance in biology, both in the hope of gaining scientific insights, and in order to receive inspiration for the design of intelligent artificial devices. Thus Rosenblueth, Wiener, and Bigelow (1943: 23) simply

[18]See Canguilhem (1965/2008) for many remarkable thoughts on the relationship between the mechanistic and finalistic perspectives on nature. The problematic idea that there is an exclusive rather than complementary relationship between mechanism and teleology is evident in the description by Craver and Tabery (2017) of mechanism as a self-contained "scientific worldview":

> Some have held that natural phenomena should be understood teleologically. Others have been convinced that understanding the natural world is nothing more than being able to predict its behavior. Commitment to mechanism as a framework concept is commitment to something distinct from and, for many, exclusive of, these alternative conceptions. If this appears trivial, rather than a central achievement in the history of science, it is because the mechanistic perspective now so thoroughly dominates our scientific worldview.

redefine "teleology" as "purpose controlled by feed-back", and thereby avoid any reference to final causation.[19]

11.3 Two Interpretations of the Brain-Computer Relationship

The building of machines in order to elucidate processes underlying vital functions, including cognition, is a strategy that goes back at least to the automaton-makers of the eighteenth century.[20] But an open question here is whether, in order to understand the efficacy of this pattern of investigation, one must resort to a literal interpretation of the artificial models (computer programs or other devices) as duplicating and thereby bringing to light the *same* process or function as it occurs in the living system, or if one can still make sense of the research strategy by taking the machine-organism relationship as one of analogy.[21] That is, by saying that the organism is *like* the machine in some to be determined way, but making salient the

[19]It is worth quoting Rosenblueth, Wiener and Bigelow (1943: 23) at length:

> Teleology has been interpreted in the past to imply purpose and the vague concept of a "final cause" has been often added. This concept of final causes has led to the opposition of teleology to determinism. A discussion of causality, determinism and final causes is beyond the scope of this essay. It may be pointed out, however, that purposefulness, as defined here, is quite independent of causality, initial or final. Teleology has been discredited chiefly because it was defined to imply a cause subsequent in time to a given effect. When this aspect of teleology was dismissed, however, the associated recognition of the importance of purpose was also unfortunately discarded. Since we consider purposefulness a concept necessary for the understanding of certain modes of behavior we suggest that a teleological study is useful if it avoids problems of causality and concerns itself merely with an investigation of purpose.

Note also that Francis Walshe, quoted above on the complementary relationship between the physicist's and engineer's stances in neuroscience, was quite critical of Rosenblueth et al.'s paper, highlighting the mismatch between the operation of feedback in the cerebellum and in the artificial system, which, he argues, means the literal interpretation of the cybernetic model is not warranted (Walshe 1951). See also Mayr (1988: 46) for the argument that control via negative feedback is not sufficient to capture the range of behaviours described as teleological, pace Rosenblueth et al.

[20]As Canguilhem (1963: 510) describes, "texts, taken from Quesnay, Vaucanson and Le Cat, do not indeed leave any doubt that their common plan was to use the resources of automatism as a dodge, or as a trick with theoretical intent, in order to elucidate the mechanism of physiological functions by the reduction of the unknown to the known, and by complete reproduction of analogous effects in an experimentally intelligible manner."

[21]A potential misinterpretation of Sect. 11.2 may push one towards the literal interpretation. If one thinks that the brain – like a digital computer designed to be indifferent to e.g. variation in magnetic grains in a hard drive – is a device that "ignores its own complexity", then an abstract computational description of the system can be equally, literally true of the brain as of the machine. However, the point of Sect. 11.2 is to explain how and why neuroscientists ignore the complexity of the brain, leaving it a live possibility that those details *do* matter to cognition in animals (see Sect. 11.3.3).

numerous differences (disanalogies) that limit the appropriateness of the machine-organism comparison to the narrow domain of the phenomena explicitly modelled.

Theoretical neuroscience has benefitted from a strategic vagueness on this point – the difficult question of whether the differences between brains and computers are significant disanalogies which restrict the scope of the comparison of the two kinds of system has been deferred indefinitely. According to Lettvin, McCulloch was under no illusion that neural assemblies share all the properties and behaviours of digital logic gates. However, the comparison was appropriate because, Lettvin (2016: xviii–xix) asserts, "there are properties of such connected systems that are more or less independent of the intrinsic nature of the nonlinear elements used, whether gates or neurons". The latitude in the "more or less independent" here is useful for the scientist because the observation of relative independence provides clues to the scientist about which causal factors do not need to be made the target of an experiment, and which details may safely be left out without foreclosing on the possibility that the independence may turn out to fail in some circumstances, and that those neglected details might later be the subject of experiment and modelling.

Even while noting ambiguities like these within the writings of computational neuroscientists, I do think that the literal interpretation is the majority view within the discipline – given enough latitude in the notion of computation in play. Complaints from neuroscientists that the brain is *not* a computer usually just make the point that the brain is not a digital, serial machine, while still asserting that the brain is a kind of computer. Marcus (2015: 209) nicely expresses this position:

> it is obvious that brains (especially those of vertebrates) *are* computers, in the sense of being systems that operate over inputs and manipulate information systematically. Brains might not be (purely) *digital* computers, their memories may operate under different principles, and they may perform different sorts of operations on the information they encode, but they surely encode information .... Computers are, in a nutshell, systematic architectures that take inputs, encode and manipulate information, and transform their inputs into outputs. Brains are, so far as we can tell, exactly that.

Many go further in asserting that any disanalogies between information processing as it occurs in electronic and neural tissue do not present an obstacle to the deployment of computational simulations of the brain to provide explanations of cognitive capacities, and the eventual reproduction of those capacities in machines.[22] I will now provide some exposition of this literal way of interpreting computational models of the brain, before offering an alternative that centres on the notion of analogy.

[22]This strong view is best exemplified in the work of researchers at the interface between neuroscience and the deep learning style of AI, such as Hassabis et al. (2017) and Yamins and DiCarlo (2016). It subscribes to the *computational theory of mind* much discussed in the in philosophy of mind, psychology, and cognitive science. In this paper I do not say anything directly about the interpretation of computational models in branches of cognitive science other than neuroscience. However, there are certainly implications to the extent that my account causes trouble for the computational theory of mind.

11.3.1 The Literal Interpretation: Formal Realism

One point that can be derived from the above discussion of the relationship between the physical and engineering approaches, and the mechanistic and computational perspectives that go with them (Sect. 11.2.2), is that the engineering approach in contemporary biology is a distant echo of the Aristotelian tenet that living systems cannot be understood without a first regard to their purposes and their *forms* (patterns of organisation). These notions of form and finality were, according to popular history, banished from science in the seventeenth century and then, after a long wandering in exile, put mercifully to death by Darwin. Yet, as various philosophers and historians of biology have argued, these ideas are ever present in modern biology, even if going by different names (Allen et al. 1998). I argued above that cybernetics can be understood as a kind of neo-Aristotelian research programme, in that it restores a place for finality in the science of living systems. Some advocates of functionalism in the philosophy of mind have emphasised the Aristotelian aspects of the theory (Nussbaum and Putnam 1992). Although this connection can sometimes be overstretched (Burnyeat 1992), I give the name *formal realism* to the literal stance towards neuro-computational models, which itself can be thought of as a tenet of functionalism.[23]

In Aristotle's hylomorphism – as applied to living beings – the explanation of how the body is able to do what it does (achieve its ends) is put in terms of the presence of a form inherent in the matter, which together comprise the body. Forms can be thought of, generally, as patterns or principles of organisation, so that when one takes the literal interpretation of computational models of the brain as a modern version of hylomorphism, the relevant forms are computational functions,[24] not "souls" or "animae", and the neural realiser is the matter made intelligent by the presence of the form. Thus the modern formal realist takes computation to be the

[23]Another tenet of functionalism is the classic account of multiple-realisation which gives the abstract computational "level" of neuro-modelling a robust ontological interpretation. Elsewhere I call this approach MR 1.0 and argue that it be replaced with an ontologically modest view, MR 2.0, which treats the computational as a level of explanation rather than a level of being (Chirimuuta 2018b). MR 2.0 is consistent with the analogical interpretation of computational models offered below (Sect. 11.3.2); indeed, the analogical interpretation is intended to be an elaboration of some of the ideas presented in my earlier paper.

[24]We might also consider here the bivalence of the word "function", which has both a mathematical and a biological sense (Longuenesse 2005: 93). Interestingly, the two meanings coincide in formal realism, where the function is at once the mathematical operation computed by the neurons, and the biological purpose of this activity. Note that because the relevant forms in computational neuroscience are mathematical ones, formal realism here has a Platonic as well as an Aristotelian feel: the underlying order of the brain is a mathematical one. Elsewhere I say more about the Platonic dimension (Chirimuuta 2020).

essence or principle responsible for cognition and underlying intelligent behaviour. So even though the neuroscientists who work in the computational tradition and offer literal interpretations of their models, and any philosophers following in attendance,[25] would not embrace any characterisation of themselves as adherents to an Aristotelian metaphysics, to the extent that that their research treats computation as the essence of cognition and intelligence, the label of formal realism is apt.

Hylomorphism does not entail multiple realizability – the notion that the one and the same form can inhere in radically different kinds. However, when the relevant forms are mathematical functions, multiple realizability is inevitable because of the fact that the same computation (e.g. multiplication of 653×10) can in principle be performed by a variety of physical realisers, including an artificial computer (mechanical or electronic) or biological tissue. The picture of an abstract mathematical form, finding itself realised in an array of material substrates – breathing intelligence into them, one might say – has had long appeal. According to Morar (2015:126) this is what occurred to Leibniz after his encounter with the famous adding-subtracting machine invented by Pascal:

> As Leibniz came out through the door of Louis XIV's library after seeing the Pascaline, he left behind all of his previous ideas of what a new type of calculator could look like, but not his goals. He had begun thinking about building a machine since at least 1670, two years before he came to Paris, and the challenge was clear: if mortal man had the power to transpose in 'yellow brass' the faculty of mathematical reasoning, there could be no doubt that God had been able to house a 'more general spirit' into the body of animals, giving them life.

While I do not suppose that any defender of formal realism in computational neuroscience owes us an elaborate metaphysics of an Aristotelian or Leibnizian sort, I will say that the view does bring up some challenging metaphysical questions, as well as empirical ones. The view seems to presuppose a realism about mathematical form which is normally associated with a Platonism – where mathematical abstracta exist outside space and time. At the same time, mathematical operations are taken to be realized in the material brain, which is located in time and space. Are we to think these mathematical forms as inhering in material objects, in the way that Aristotle's notion of form brought Platonic ideas down to earth? The standard answer to this question is to point to the concept of implementation. The pressing challenge, then, is to give an account of the implementation of computational functions in concrete material that does not imply pancomputationalism (Putnam 1988), while showing how the computational level of explanation is autonomous from the implementational one (Ritchie and Piccinini 2018). I do not mean to suggest that attempts to solve these problems are all hopeless. But one of the selling

[25] Examples of formal realism in philosophy are Egan (2017), Shagrir (2010) and Shagrir (2018).

points of my alternative interpretation is that it does not have the burden of needing to solve such problems, as I will explain in Sect. 11.3.4.

Another issue, noted above, is that the view implies the multiple realisability of computations underlying intelligence, and hence multiple realisation as an empirical fact. Polger and Shapiro (2016) present a thorough case that the evidence for multiple realisation is lacking, contrary to the expectations of functionalist philosophers of mind. Of course others have a different opinion, and it is not obvious that the challenges are insurmountable (Aizawa 2018). I am not claiming that the formal realism is untenable just because of the empirical case that has been made against MR. However, the fact that this challenge exists does provide motivation for the development of an alternative which does not need to meet this demand.

11.3.2 The Analogical Interpretation: Formal Idealism[26]

According to Cassirer, the felt need for an explanation of the applicability of mathematics in empirical science that did not depend on any dogmatic metaphysical assertions was Kant's first step along the road to his critical philosophy (Seidengart 2012: 141). To advance towards an alternative to the literal interpretation of computational models in neuroscience, I suggest that we re-tread this path. While the formal realist takes for granted the brute existence of mathematical forms, which are realised equivalently in brains or computers, the *formal idealist*[27] takes the mathematical forms represented in computational models of the brain not to be straightforward discoveries regarding mathematical structure or information processing in the brain, but constructs developed through an arduous process of experimentation, model building, and analogical reasoning. This Kant-inspired proposal is that the mathematical structures which make the brain intelligible to us, as an organ whose function is to process information, are to some extent imposed by us onto the neural system and should not be taken as straightforward discoveries

[26]The analogical interpretation should be understood in the specific sense described here, not to be confused with the "analog-model" account of the brain (Shagrir 2010), which I classify as a formal realism. The reader here may be reminded of the philosophical discussion, responding to Putnam (1988), over whether the computational mappings of state transitions are arbitrarily up to human observers, or constrained by the causal structure of the implementing system. The important difference with my discussion is that it is centred on scientific practice. While no geologist has claimed that their lumps of rock implement finite-state automata, many neuroscientists claim to have discovered functions implemented in the brain. Thus I am starting with the claim of formal realism as it has been put forward from the science, and my alternative to it is shaped by considerations of modelling practice within the science.

[27]Kant (1929: B519, note a) gives "formal idealism" as a gloss for "transcendental idealism". The former term draws attention to the point that the idealism in Kant's philosophy is restricted to the way that our knowledge of nature is *formed* or *structured* by our cognitive capacities rather than a structure pre-given in things-in-themselves.

Earth (Source)		Mars (Target)
	Known Similarities	
Orbits the sun Has a moon Revolves on axis Subject to gravity		Orbits the sun Has moons Revolves on axis Subject to gravity
	Inferred Similarity	
Supports life	==>	*May* support life

Fig. 11.1 A schematic for analogical reasoning, after Bartha (2016)

of mathematical forms inherent in the system.[28] Since, by hypothesis, our neuro-computational models are not discoveries of the inherent computational capacities of the brain, but are as abstract and idealised as any other models in science, an analogical interpretation of these models is more appropriate than a literal one.

In the classic account, Hesse (1966) charts the structure of analogical reasoning in science using diagrams which compare two systems (the analogue source and target) along vertical and horizontal axes. For example, the analogical inference that Mars, because of its similarities with the Earth, *may* support life is depicted in Fig. 11.1.

Figure 11.2 offers an example, based on research published by Mante et al. (2013) on perceptual decision making in the prefrontal cortex.[29] The researchers gathered both neurophysiological and behavioural data from monkeys performing a task in which stimuli varied either in colour or in direction of motion, and depending on a contextual cue the monkey had to report on either one of these stimulus dimensions. They also trained a recurrent neural network (RNN) model to perform a virtual equivalent of the experimental task. Through reverse engineering of the trained RNN, the researchers formulated an explanation of how the network was able to accomplish this kind of decision making, turning on the fact that there is a *line attractor* in the low dimensional state space of the network which allows for integration of context dependent information. The researchers observed a number of similarities between the trained RNN and the prefrontal cortex (see Fig. 11.2). On the basis of this it is possible to make the analogical inference that the process underlying the context-dependent perceptual decision, discovered by reverse engineering the RNN, may also occur within the cortex.

This inference is put forward not as conclusive proof, but as a plausible explanation of the biological function that also serves as a hypothesis for future

[28] See also Chirimuuta (2020) for an argument against formal realism, based on the existence of empirically adequate but incompatible mathematical models of certain brain areas.

[29] For a more lengthy discussion of this research and the explanations it affords see Chirimuuta (2018a).

Computer (Source) *RNN model*		**Brain (Target)** *Prefrontal Cortex*
	Observed Similarities	
Makes context-dependent perceptual decision. Irrelevant sensory information is represented in the population. In the 3D state space, the angle of the 'choice' axis is fixed in relation to the 'colour' and 'motion' axes.		Makes context-dependent perceptual decision. Irrelevant sensory information is represented in the population. In the 3D state space, the angle of the 'choice' axis is fixed in relation to the 'colour' and 'motion' axes.
	Inferred Similarity	
There is a line attractor in the state space, which explains integration of information.	==>	*May* be that there is a line attractor in the state space, which explains integration of information.

Fig. 11.2 *Prospective* pattern of analogical reasoning

experimental testing. Because of the forward looking aspect of this kind of analogical reasoning, I call it *prospective*. It should be noted that the authors of this research present the RNN as a literal representation of the coding that occurs in the prefrontal cortex, such that the reverse engineering that leads to the discovery of how the task is performed in the model is thereby a discovery of the biological process. In contrast, the analogical interpretation is more tentative than this, being sensitive to the open possibility that future discoveries of dissimilarities between brain and model will call into question the validity of the analogical inference.

Figure 11.3 presents a more elaborate kind of analogical reasoning in neuroscience, that I call *abstractive*. This example is taken from David Marr and Shimon Ullman, whose approach to computational modelling in neuroscience has been highly influential.[30] Because of the "behavioural" similarity observed across the systems (the ability to detect edges), and the similarities in patterns of activation in response to edges, the analogical inference is made that neurons in the cat's early visual system – *retinal ganglion cells* (RGC) and neurons in *lateral geniculate nucleus* (LGN) – can be modelled as computing a *Laplacian of Gaussian* function.[31]

[30] Marr and Ullman (1981); Marr (1982: 54–65).

[31] Marr (1982:64) makes the stronger (but hedged) claim that these neurons *are* computing the function: "it is not too unreasonable to propose that the $\nabla^2 G$ function is what is carried by the X cells of the retina and lateral geniculate body, positive values being carried by the on-center X cells, and negative values by the off-center X cells." This amounts to a formal realism, so I do not

Computer (Source) *Laplacian of Gaussian Model*		**Brain (Target)** *LGN or RGC neurons in cat*
	Observed Similarities	
Detects edges in a photo. Characteristic peaks of model output for onset and offset of edges.		Responds to moving edges. Average increases in neural activity for onset and offset of edges.
	Observed Dissimilarities	
Peaks for onset and offset are *symmetrical*. Implemented in digital computer.		Peaks for onset and offset are *asymmetrical*. [Ignored] Is an electrically excitable cell.
	Inferred Similarity	
Model computes Laplacian of Gaussian function.	==>	RGC and LGN neurons can be modelled as computing Laplacian of Gaussian function.
	Abstractive Inference	
==>	*Differences in implementation are not relevant to the particular capacity here investigated.*	

Fig. 11.3 *Abstractive* pattern of analogical reasoning

In addition to the observation of similar overall behaviour, the dissimilarities in the material substrates of the systems may also be noted and the *abstractive inference* made that these dissimilarities are *not relevant* to the scientist's investigation of the capacity for edge detection.[32] The possibility of this kind of abstraction is a precondition for Marr's (1982: 25) distinction between the levels of computational theory and algorithm, and that of implementation. This kind of abstractive inference fits with my account of how it is that computational models aid neuroscientists in the simplification of the brain – the abstractions discussed above can be licensed by this sort of analogy. But by putting this account of abstraction and simplification in the context of a non-literal, analogical approach to interpretation of neuro-computational models, there is no commitment made here to "computational

propose my weaker interpretation of the case as one proposed by Marr himself – see Egan (2017) and Shagrir (2010) for discussions of this example which instead endorse the literal interpretation. That said, I do think Marr can be read as making the abstractive inference. A short biographical note: I first heard of this example during an undergraduate lecture by the late and much missed Tom Troscianko. Intrigued by the idea that the retina does calculus, I decided to do my final year research project with him, and then went on to do graduate research with one of his collaborators. I am still wondering....

[32]NB – the inference is not that the differences in implementation are irrelevant *tout court*, but that they can reasonably be ignored for this kind of investigation of this particular capacity.

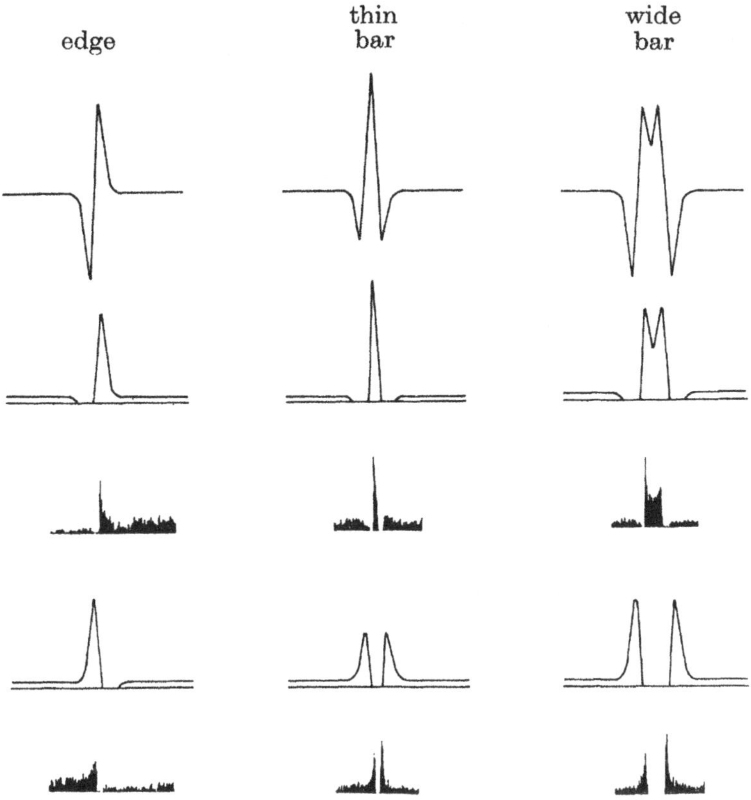

Fig. 11.4 Comparison between Laplacian of Gaussian model and neural data. The neural data indicate an unequal treatment of light vs. dark edges and bars that is not captured by the model. From Marr and Ullman (1981): 165; Marr (1982): 65

essentialism" about the brain, or to the idea that all the information processing that occurs in the brain *must* be multiply realisable.

The terminology of formal realism versus idealism helps to illuminate the distinction between literal and analogical interpretations. According to formal idealism, the relevant similarities between the model and target are not simply there, waiting to be discovered by the scientist but are in some respect constructed, or massaged out of equivocal data. Some details from our example will reinforce this proposal. Figure 11.4 is the figure provided in order to illustrate the correspondence between the Laplacian of Gaussian model and the neural data (Marr and Ullman 1981: 165; Marr 1982: 65). If one examines the average neural traces depicted here, and in addition the data presented in the original neurophysiology papers from which these examples were taken (Rodieck and Stone 1965: Figures 1 and 2; Dreher and Sanderson 1973), it is striking that there is a pattern of the neural response that goes un-noted by Marr and is not captured by the model – the asymmetry of peak response, depending on the polarity of the visual stimulus, and whether the bar stimulus is being swept onto the neuron's receptive field, or leaving the field. For example, the first column of Fig. 11.4 shows that a light edge on grey background generates more neuronal response than a dark edge, whereas the model response is exactly equal. The general point is that the positing of an analogy – here that the same pattern of activation occurs in the model as for the neurons – requires selective attention to certain similarities, and the ignoring of dissimilarities. This is a matter of judgment of the scientist, and the data do not usually, by themselves, force one interpretation over all others – Marr *could* have taken the asymmetry to be a relevant part of the neuronal behaviour, and come up with a mathematical model that captured this.[33] One should not think of the structure described in any particular model as simply duplicating a structure that is pre-existing in nature, as a formal realist would assert.

Formal idealism does not suppose that the finding of structure in a target of investigation is purely "made up" and then projected onto the data, but takes it to be the result of the researcher's experimental interaction with the target, such that the human-dependent element of the structure can never be fully removed. One might be reminded of the way that the visual system finds shapes in what might appear as very disordered stimuli, as demonstrated with certain images in Gestalt psychology. While visual Gestalts are in most cases formed involuntarily, I emphasise that the scientist has a certain amount of latitude and choice in the determination of the patterns which are the target of modelling, because these depend on methods of data collection, data processing (at minimum, averaging) and style of representation.

Another way of describing the difference between formal realism and idealism, is that in the first case the abstractions of computational neuroscience are presented as if the work of the researchers has been to pare away all the extraneous neurobiological details, in order to find the essence (form) of the brain qua information processor. This is something like picking all the leaves off a tree and asserting that the bare trunk and branches are the essential structure of the tree. In contrast, the formal idealist does not assert that the computation described in the model is an essential feature of the neural circuit. The abstractions introduced by the model are taken to be there for the convenience of the scientist (i.e. to provide an economical representation which does not overload the scientist with a million details), rather than a means by which the true structures of the brain are revealed. A botanist would not insist that the leafless form is the essential structure of a tree, given

[33]One might be remined of Kripke's plus/quus argument that any finite series of observations of a natural system can in principle be modelled by quite different mathematical functions. (I thank Brian McLaughlin for this observation.) However, my argument should really be taken as one grounded in the concerns of scientific practice, where Kripke's in principle alternative models would be ruled out for pragmatic reasons for they add mathematical complexity without improving the fit to the dataset. My point is, in essence, that as a consequence of the complexity of the neural events, and thus of the datasets gleaned from them, the determination of signal versus noise is not unambiguous and for that reason the datasets afford numerous plausible mathematical descriptions. Marr treated the asymmetry in the responses as noise and left it out of his model; another scientist would have been equally justified in treating it as signal, a feature to be included in the model.

the importance of the leaves in the life of the tree; nonetheless, a pared down representation would be useful, and good enough, for many purposes.

11.3.3 Why Formal Idealism?

Formal idealism is a doctrine of restraint: it declines to infer from the success of the computational approach in neuroscience that the brain *really is* a computer, an organic device performing calculations to which the scientist's models provide a closer or wider approximation. But one must acknowledge that the literal interpretations of computational models offered by formal realism are particularly tempting in neuroscience. In other disciplines, like physics and chemistry, non-literal interpretations of computational models are more the norm. Canguilhem (1963: 514–515) notes how in physics the analogical use of mathematical models does not invite one to project the ontology of the analogue-source on to the analogue-target, a caution that is often lacking when such models are used in biology.

His point is that the use of an inorganic system as the analogue source for an organic target carries with it a promise of a "reduction" of the organic to the inorganic – i.e. the making sense of the organic in perspicacious physical terms – which is why the literal interpretations are so alluring. Canguilhem goes on to say that cybernetic models are a good example of this tendency, especially when the model's actions (e.g. in a robot), tend to simulate or mimic natural behaviour.[34] In other words, formal realism offers the promise that it is possible to devise quantitative, formal, and perspicacious models for whatever it is that the nervous system does. When this interpretation holds sway, there is a tendency to downplay the disanalogies between brains and man-made computational systems (even if the official doctrine is that the brain is *not* like a PC), and to keep the details relegated to "mere metabolic support" on the side lines of neuroscientific investigation.

The neurophysiologist Lord Adrian (1954) once quipped that, "[w]hat we can learn from the machines is how our brains must differ from them."[35] One very significant point of difference is that the hardware of electronic computers is engineered *not* to undergo material changes with use, whereas there is an inherent tendency for biological cells, whose material constitution is changing as they metabolise, to undergo use-based plasticity (Chirimuuta 2017; Godfrey-Smith 2016). Thus it should not surprise us that the plasticity shown by the brain, with ordinary development and deliberate learning is very much unlike what is seen in computational machines, even in artificial neural networks designed to simulate

[34]"Despite their great degree of mathematical complexity, it does not appear that cybernetic models are always safe from this accident. The magical aspect of simulation is strongly resistant to the exorcism of science." Canguilhem (1963: 515); Cf. Dreyfus (1972: 79–80).

[35]Quoted approvingly by Canguilhem (1963: 516).

synaptic plasticity (Lake et al. 2017). The usefulness of engineering-analogues for understanding the "principles of neural design" (Sterling and Laughlin 2015) is tempered by the way that they impose an engineer's template in which structure-function relationships are fixed and transparent, and where use-dependent change is conceptualised as perturbation demanding mitigation, not a background fact of life. It could be that this very basic difference between organic and artefactual intelligence is one of the reasons why expert systems in AI, impressive as they are, have so far not made steps towards generalisation.[36]

11.3.4 Some Worries, and How to Avoid Them

Above I stated that one of the selling points of formal idealism is that it allows one to account for the usefulness and explanatory value of computational models in neuroscience, without burdening oneself with the need to subscribe to a theory of implementation. The formal realist claims that a brain area implements some computations specified by scientists. The *triviality objection* to the computational theory of mind asks what entitles one to say that the brain implements *those* ones, but not any of the countless other computations that also map onto a physical system like the brain (Sprevak 2018). The formal realist must appeal to a theory of implementation which would allow her to rule out the trivial computations, but retain the claim that the brain does implement certain computations. The formal idealist is not faced with this challenge, because she is not claiming that the brain implements any computations, but that it is useful to model the brain *as if* it is computing. Compare our case with the interpretation of the liquid drop model of the atomic nucleus (Morrison 2011). A literalist, like our formal realist, would say that the nucleus simply *is* a liquid drop. She may then be pressed to explicate what it is that makes liquids different from solids, and what the liquidity of the nucleus consists in. Someone following my manner of interpretation can merely say that the nucleus is *like* a liquid drop in some way, that making this comparison is useful to nuclear physics, and put questions about the metaphysics of liquidity to one side. All that needs to be assumed is that some things are uncontroversially and pretheoretically liquid drops, or computers, and since the actual focus of discussion is on atomic nuclei and brains, theoretical enquiries about the nature of liquidity and computation are tangential.

It is to be noted, of course, that some current theories of implementation have been tailored to address the question of how the brain can be said to compute

[36]Of course other disanalogies are most likely relevant here, such as the "noisiness" of neural components in comparison with electronic ones. Also, the embodiment of organic intelligence, whereas most expert systems are disembodied, not capable of acting in the physical world. But note that embodied AI systems (e.g. autonomous cars) have also proved to be limited in their operation outside of controlled conditions, suggesting that embodiment by itself doesn't overcome the obstacles to creating a general AI.

biologically relevant functions, and of course the formal realist may refer to them (see e.g. Ritchie and Piccinini 2018). I will point out that no theory of implementation is uncontroversial, and appealing to such a theory cannot by itself make the case for the formal realism over my preferred view. One argument for formal realism might be to say that if the computational description is a useful simplification – a good analogy – it must be that it does a good job of capturing the structure of the target system. That, then, is reason to think that the system is literally computational. Conversely, if the target system is not literally computational, then the computational approach must provide a "poor" simplification, and a misleading analogy.

But this argument is simply assuming that models work – provide useful simplifications – to the extent that they faithfully represent structures that are there in the target system, an assumption which is at odds with so much work in the philosophy of science on modelling, abstraction and idealisation. So many models that scientists employ, such as the liquid drop model of the nucleus, represent their target in ways known to be false. This does not detract from their utility, as means for prediction or simplification of the subject matter, but it does mean that we should be wary about making metaphysical claims about the nature of the target on the basis of them. There is no reason to think that models in neuroscience work any differently. As I have argued elsewhere, the computational approach is one modelling perspective, that must make certain idealising assumptions; it holds its own for certain applications, but there are other quantitative approaches in theoretical neuroscience that are complementary to it (Chirimuuta 2020). The existence of multiple, complementary perspectives is another good reason to avoid literal interpretations of any of the models proposed.

11.4 Coda: Leibniz the Biologist[37]

An important supplement to the observations offered above, of Leibniz as an inventor of the computational theory of mind, is to note his views on the difference between man-made machines and living beings. He held that organic bodies were machines, but ones of infinite complexity. For unlike inorganic artefacts, the component parts of animal machines are themselves machines, and the parts of those smaller machines are also machines, *ad infinitum*.[38] Leibniz was inspired here by the recent discoveries of microscopists (Cassirer 1950), and his picture of living systems

[37] Of course this label is anachronistic. The word "biology" was first used in 1766, fifty years after the death of Leibniz (Smith 2011: 1).

[38] As Smith (2011: 100) relates, "the animal body is not a 'mere' machine but a special kind of machine, a 'more exquisite' or 'more divine' machine, ... This is the machine of nature, or the organic body, whose exquisiteness resides in the fact that it remains a machine in its least parts, which is to say that there is no stage in its decomposition at which one arrives at nonmachinic components."

as comprising tiny machines telescoped one inside the other is not so different from that of a contemporary biologist.

I have argued in this paper that computational models, which take the workings of neural systems to be essentially like those of man-made devices – thus rejecting Leibniz's distinction between "divine machines" and human built ones – have been so useful to neuroscientists precisely because they remove from consideration the levels of complexity that Leibniz took to be crucial to the workings of nature. It is not too fanciful to consider the intricacies of synaptic behaviour – far more than the passive signal transmission of classical neural-computational theory (Grant 2018) – as a modern illustration of this idea of Leibniz. It remains to be seen whether the mysteries of biological cognition will open up to an approach which takes organic intelligence on its own terms. But the replacement of formal realism with an approach which pays attention to the various modes of analogy and disanalogy between brains and computers, will at least help philosophers avoid any false directions indicated by overreaching, literal interpretations.

Acknowledgments I am most grateful to audiences at the Ludwig Maximilian University (Workshop on Analogical Reasoning in Science), Rutgers University (Center for Cognitive Science Colloquium), University of Edinburgh, and the 2019 Workshop on the Philosophy of Mind and Cognitive Science (Valparaíso, Chile) for very thoughtful discussions of this paper. Furthermore, I owe much to comments from Cameron Buckner, Philipp Haueis, Brendan Ritchie, Bill Wimsatt and an anonymous referee.

References

Adrian, E. D. (1954). Address of the President Dr E. D. Adrian, O.M., at the anniversary meeting, 30 November 1953. *Proceedings of the Royal Society of London B, 142*, 1–9.

Aizawa, K. (2018). Multiple realization and multiple "ways" of realization: A progress report. *Studies in History and Philosophy of Science, 68*, 3–9.

Allen, C., Bekoff, M., & Lauder, G. (Eds.). (1998). *Natures purposes: Analyses of function and design in biology*. Cambrdige, MA: MIT Press.

Anderson, J. A., & Rosenfeld, E. (Eds.). (1998). *Talking nets: An oral history of neural networks*. Cambridge, MA: MIT Press.

Arbib, M. A. (2016). Afterword: Warren McCulloch's search for the logic of the nervous system. In W. S. McCulloch (Ed.), *Embodiments of mind*. Cambridge, MA: MIT Press.

Bartha, P. (2016). Analogy and analogical reasoning. In *The Stanford Encyclopedia of Philosophy*.

Bullock, T. H., Bennett, M. V. L., Johnston, D., Josephson, R., Marder, E., & Field, R. D. (2005). The neuron doctrine, redux. *Science, 310*, 791–793.

Burnyeat, M. F. (1992). Is an Aristotelian philosophy of mind still credible (A Draft). In M. C. Nussbaum & A. O. Rorty (Eds.), *Essays on Aristotle's de anima*. Oxford: Oxford University Press.

Canguilhem, G. (1963). The role of analogies and models in biological discovery. In A. C. Crombie (Ed.), *Scientific change*. New York: Basic Books.

Canguilhem, G. (1965/2008). Machine and Organism. In P. Marrati & T. Meyers (Eds.), *Knowledge of life*. New York: Fordham University Press.

Cao, R. (2014). Signaling in the brain: In search of functional units. *Philosophy of Science, 81*, 891–901.

Cassirer, E. (1950). *The problem of knowledge: Philosophy, science, and history since Hegel*. New Haven: Yale University Press.

Chirimuuta, M. (2017). Crash testing an engineering framework in neuroscience: Does the idea of robustness break down? *Philosophy of Science, 84*, 1140–1151.

Chirimuuta, M. (2018a). Explanation in computational neuroscience: Causal and non-causal. *British Journal for the Philosophy of Science, 69*, 849–880.

Chirimuuta, M. (2018b). Marr, Mayr, and MR: What functionalism should now be about. *Philosophical Psychology, 31*, 403–418.

Chirimuuta, M. (2020). Charting the heraclitean brain: Perspectivism and simplification in models of the motor cortex. In M. Massimi & C. McCoy (Eds.), *Understanding perspectivism: Scientific challenges and methodological prospects*. New York: Routledge.

Craver, C. F., & Darden, L. (2013). *In search of mechanisms*. Chicago, IL: Chicago University Press.

Craver, C. F., & Kaplan, D. M. (2018). Are more details better? On the norms of completeness for mechanistic explanations. *British Journal for the Philosophy of Science, 71*, 287–319.

Craver, C. F., & Tabery, J. (2017). Mechanisms in science. In *The Stanford Encyclopedia of Philosophy*. Stanford: Stanford University.

Dardashti, R., Thébault, K. P. Y., & Winsberg, E. (2017). Confirmation via analogue simulation: What dumb holes could tell us about gravity. *British Journal for the Philosophy of Science, 68*, 55–89.

Daugman, J. G. (2001). Brain metaphor and brain theory. In W. Bechtel, P. Mandik, J. Mundale, & R. S. Stufflebeam (Eds.), *Philosophy and the neurosciences: A reader*. Oxford: Blackwell.

Davis, M. (2000). *The universal computer: The road from Leibniz to Turing*. New York: W. W. Norton & Company.

Dennett, D. C. (1987). *The intentional stance*. Cambridge, MA: MIT Press.

Dreher, B., & Sanderson, K. J. (1973). Receptive Field Analysis: Responses to Moving Visual Contours by Single Lateral Geniculate Neurones in the Cat. *The Journal of Physiology, 234*, 95–118.

Dreyfus, H. L. (1972). *What computers can't do: A critique of artificial reason*. New York: Harper & Row.

Egan, F. (2017). Function-theoretic explanation and the search for neural mechanisms. In D. M. Kaplan (Ed.), *Explanation and integration in mind and brain science*. Oxford: Oxford University Press.

Fairhall, A. (2014). The receptive field is dead. Long live the receptive field? *Current Opinion in Neurobiology, 25*, ix–xii.

Frégnac, Y. (2017). Big data and the industrialization of neuroscience: A safe roadmap for understanding the brain? *Science, 358*, 470–477.

Godfrey-Smith, P. (2016). Mind, matter, and metabolism. *Journal of Philosophy, 113*, 481–506.

Goldstein, K. (1934/1939). *The organism: A holistic approach to biology derived from pathological data in man*. New York: American Book Company.

Grant. (2018). Synapse molecular complexity and the plasticity behaviour problem. *Brain and Neuroscience Advances, 2*, 1–7.

Hassabis, D., Kumaran, D., Summerfield, C., & Botvinick, M. (2017). Neuroscience-inspired artificial intelligence. *Neuron, 95*, 245–258.

Haueis, P. (2018). Beyond cognitive myopia: A patchwork approach to the concept of neural function. *Synthese, 195*, 5373–5402.

Hesse, M. B. (1966). *Models and analogies in science*. Notre Dame, Indiana: Indiana University Press.

Jonas, E., & Kording, K. (2017). Could a neuroscientist understand a microprocessor? *PLoS Computational Biology, 13*, e1005268.

Kant, I. (1929). *The critique of pure reason*. Basingstoke: Palgrave.

Kaplan, D. M. (2011). Explanation and description in computational neuroscience. *Synthese, 183*, 339–373.

Kline, R. R. (2015). *The cybernetics moment: or why we call our age the information age*. Baltimore, MA: John Hopkins University Press.
Knuuttila, T., & Loettgers, A. (2014). Varieties of noise: Analogical reasoning in synthetic biology. *Studies in History and Philosophy of Science, 48*, 76–88.
Lake, B. M., Ullman, T. D., Tenenbaum, J. B., & Gershman, S. J. (2017). Building machines that learn and think like people. *Behavioral and Brain Sciences, 40*, 1–72.
Lettvin, J. (2016). Foreword to the 1988 reissue. In W. S. McCulloch (Ed.), *Embodiments of mind*. Cambridge, MA: MIT Press.
Lettvin, J. Y., Maturana, H. R., McCulloch, W. S., & Pitts, W. H. (1959). What the frog's eye tells the frog's brain. *Proceedings of the IRE, 47*, 1940–1959.
Longuenesse, B. (2005). *Kant on the human standpoint*. Cambridge: Cambridge University Press.
Mante, V., Sussillo, D., Shenoy, K. V., & Newsome, W. T. (2013). Context-dependent computation by recurrent dynamics in prefrontal cortex. *Nature, 503*, 78–84.
Marcus, G. (2015). The computational brain. In G. Marcus & J. Freeman (Eds.), *The future of the brain*. Princeton: Princeton University Press.
Marr, D. (1982). *Vision: a computational investigation into the human representation and processing of visual information*. San Francisco: W. H. Freeman.
Marr, D., & Ullman, S. (1981). Directional selectivity and its use in early visual processing. *Proceedings of the Royal Society of London B, 211*, 151–180.
Mayr, E. (1988). The multiple meanings of teleological. In E. Mayr (Ed.), *Toward a new philosophy of biology*. Cambridge, MA: Belknap Press of Harvard University Press.
McCulloch, W. S., & Pitts, W. (1943). A logical calculus of the ideas immanent in nervous activity. *Bulletin of Mathematical Biophysics, 5*, 115–133.
Miłkowski, M. (2018). From computer metaphor to computational modeling: The evolution of computationalism. *Minds and Machines, 28*, 515–541.
Morar, F.-S. (2015). Reinventing machines: the transmission history of the Leibniz calculator. *British Society for the History of Science, 48*, 123–146.
Morrison, M. (2011). One phenomenon, many models: Inconsistency and complementarity. *Studies in History and Philosophy of Science, 42*, 342–351.
Nussbaum, M. C., & Putnam, H. (1992). Changing Aristotle's mind. In M. C. Nussbaum & A. O. Rorty (Eds.), *Essays on Aristotle's de Anima*. Oxford: Oxford University Press.
Papert, S. (2016). Introduction. In W. S. McCulloch (Ed.), *Embodiments of mind*. Cambridge, MA: MIT Press.
Piccinini, G. (2004). The first computational theory of mind and brain: A close look at mcculloch and pitts's "Logical Calculus of Ideas Immanent in Nervous Activity". *Synthese, 141*, 175–215.
Pickering, A. (2010). *The cybernetic brain: Sketches of another future*. Chicago: Chicago University Press.
Polger, T. W., & Shapiro, L. A. (2016). *The multiple realization book*. Oxford: Oxford University Press.
Putnam, H. (1988). *Representation and reality*. Cambridge, MA: MIT Press.
Ritchie, J. B., & Piccinini, G. (2018). Computational implementation. In M. Sprevak & M. Colombo (Eds.), *The Routledge handbook of the computational mind*. London: Routledge.
Rodieck, R. W., & Stone, J. (1965). Response of cat retinal ganglion cells to moving visual patterns. *Journal of Neurophysiology, 28*, 819–832.
Rosenblueth, A., Wiener, N., & Bigelow, J. (1943). Behavior, purpose and teleology. *Philosophy of Science, 10*, 18–24.
Seidengart, J. (2012). Cassirer, reader, publisher, and interpreter of Leibniz's philosophy. In R. Kroemer & Y. C. Drian (Eds.), *New essays in Leibniz reception: In science and philosophy of science* (pp. 1800–2000). Springer: Basel.
Shagrir, O. (2010). Brains as analog-model computers. *Studies in History and Philosophy of Science, 41*, 271–279.
Shagrir, O. (2018). The brain as an input–output model of the world. *Minds and Machines, 28*, 53–75.

Shenoy, K. V., Sahani, M., & Churchland, M. M. (2013). Cortical control of arm movements: A dynamical systems perspective. *Annual Review of Neuroscience, 36*, 337–359.

Smith, J. E. H. (2011). *Divine machines: Leibniz and the sciences of life*. Princeton: Princeton University Press.

Sprevak, M. (2018). Triviality arguments about computational implementation. In M. Sprevak & M. Colombo (Eds.), *Routledge handbook of the computational mind*. London: Routledge.

Sterling, P., & Laughlin, S. (2015). *Principles of neural design*. Cambridge, MA: MIT Press.

Walshe, F. M. R. (1951). The hypothesis of cybernetics. *British Journal for the Philosophy of Science, 2*, 161–163.

Walshe, F. M. R. (1961). Contributions of John Hughlings Jackson to neurology. *Archives of Neurology, 5*, 119–131.

Yamins, D. L. K., & DiCarlo, J. J. (2016). Using goal-driven deep learning models to understand sensory cortex. *Nature Neuroscience, 19*, 356–365.

Chapter 12
The Mind-Body Problem 3.0

Marco J. Nathan

Abstract This essay identifies two shifts in the conceptual evolution of the mind-body problem since it was molded into its modern form. The "mind-body problem 1.0" corresponds to Descartes' ontological question: what are minds and how are they related to bodies? The "mind-body problem 2.0" reflects the core issue underlying much discussion of brains and minds in the twentieth century: can mental states be reduced to neural states? While both issues are no longer central to scientific research, the philosophy of mind ain't quite done yet. In an attempt to recast a classic discussion in a more contemporary guise, I present a "mind-body problem 3.0." In a slogan, this can be expressed as the question: how should we pursue psychology in the age of neuroscience?

12.1 Introduction

The "mind-body problem"—the hallowed task of characterizing the relation between the mental and the physical—lies at the core of the philosophy of mind. Still, its nature remains baffling. What exactly makes it a problem? What would constitute a viable solution? When did the issue arise? How did it evolve over time? And why is it still troubling after all these years?

The mind-body problem is typically presented as a single, monolithic, perduring puzzle that has framed discussions of mental states, at least, since Descartes molded the question into its current form.[1] This essay examines, and, ultimately, rejects

[1]It is not trivial to find explicit statements of this assumption, partly because the mind-body problem is well-known and contemporary authors seldom bother to present it in full detail. Here are some representative quotes: "[T]he persuasive imagery of the Cartesian Theater [the idea of a centered locus of consciousness in the brain] keeps coming back to haunt us—laypeople and

M. J. Nathan (✉)
University of Denver, Denver, CO, USA
e-mail: marco.nathan@du.edu

© Springer Nature Switzerland AG 2021

F. Calzavarini, M. Viola (eds.), *Neural Mechanisms*, Studies in Brain and Mind 17,
https://doi.org/10.1007/978-3-030-54092-0_12

this presupposition. Over time, the content of the mind-body problem has shifted substantially. The inquiries driving contemporary philosophy of mind are not the original ones troubling Descartes. This point is not especially original, as prominent scholars such as Kim (1999, 2011) and Heil (2013), have advanced analogous points. It should also not come as a real shock, given that almost four centuries have passed since the publication of the *Meditations*, in 1641. More controversially, I suggest that twenty-first-century research has moved away from the theoretical discussions that framed the interface between psychology and neuroscience just a few decades ago. Thus, on the widespread assumption that philosophy and science do—and ought to—mutually inform one another, the mind-body problem requires a makeover. It is time to update our philosophical agenda.

This article is structured as follows. §2 kicks off the discussion by introducing what I call the "mind-body problem 1.0." This is Descartes' ontological question: what are minds and how are they related to bodies? After briefly surveying Descartes' well-known proposal, its shortcomings, and the main alternatives, I conclude that this issue was never solved. Rather, it was "dissolved," that is, recast in a related but different form, when people realized that neither substance monism nor substance dualism tell us much about the nature of mind. This reformulation, which I call the "mind-body problem 2.0," is presented in §3. The mind-body problem 2.0, simply put, is the core issue underlying much discussion of brains and minds in the century just passed: can mental states be reduced to neural states? Just like version 1.0, the mind-body problem 2.0 is no longer central to twenty-first-century scientific research. The main culprit, I maintain, is the lack of a clear and coherent framework for characterizing reduction. My argument consists of two main steps. First, §4 provides a succinct overview of how reduction has been conceived in the philosophy of science, since the "classical" model of the 1960s. Second, §5 maintains that it is time to move away from questions of reduction, which are less substantive, more terminological than it is often assumed. Similar observations have triggered provocative proclamations of the philosophy of mind being over. Such obituaries strike me as premature. Philosophy of mind ain't quite done yet. In an attempt to recast the traditional heart of the subfield, the mind-body problem, in a more contemporary guise, §6 poses a "mind-body problem version 3.0." In a slogan, this can be expressed by the question: how should we pursue psychology in the age of neuroscience? Finally, §7 wraps up the discussion with concluding remarks.

Before moving on, a few preliminary clarifications are in order. First, the discussion in the ensuing pages admittedly presupposes a modest form of methodological naturalism, according to which philosophical and scientific analyses are mutually relevant. Critics who view philosophy as a purely "armchair" intellectual endeavor,

---

scientists alike—even after its ghostly dualism has been denounced and exorcized" (Dennett 1991, p. 107). "The mind-body problem was posed in its modern form only in the seventeenth century, with the emergence of the conception of the physical world on which we are now all brought up" (Nagel 1995, p. 97). "What exactly are the relations between the mental and the physical, and in particular how can there be causal relations between them? (. . .) This is the most famous problem that Descartes left us, and it is usually called the 'mind-body problem'" (Searle 2004, p. 11).

insulated from empirical observations, will be likely left unmoved. Second, at the same time, my goal is not to eschew philosophical problems and replace them with scientific ones. My aim is rather to show how classic philosophical problems, appropriately revamped, are still quite pertinent to empirical inquiries. Third, and relatedly, some readers may wonder about the advantages of characterizing modern ventures into the philosophy of psychology and neuroscience as variants of the old "mind-body problem." Once we recognize that we have moved away from Cartesian concerns, why not dismiss the mind-body problem as a historical relic of a bygone time? My response, in brief, is that the overarching moniker provides a useful guideline to appreciate the historical continuity across the field. Even though version 3.0 is different from both 2.0 and 1.0, treating them a family of issues pertaining to the relation between the mental and the physical at large helps us see how each problem rises from the ashes of its predecessor. Fourth, and finally, although much of the ensuing discussion covers well-known terrain, the overarching aim of this essay is not merely, or even primarily, expository. My goal is to provide a critical diachronic overview and a fresh diagnosis of past issues. This rational reconstruction suggests an alternative trajectory for the future of the philosophy of mind.

12.2 The Mind-Body Problem 1.0

Our journey begins by revisiting an old story. This is the tale of how Descartes provided the original formulation of the modern mind-body problem, setting the stage for subsequent discussions over the centuries to come.

Descartes lived most of his life in the seventeenth century, a time of profound change across the sciences. Setting nuances aside, natural philosophy was in the process of moving away from the teleological worldview inherited from Aristotle and subsequently developed by medieval scholastics, heading towards the mechanistic *Weltanschauung* pioneered by Galileo. Descartes, who was a fine man of science, enthusiastically endorsed the in-principle possibility of subsuming the physical universe under deterministic laws, eschewing any reference to goals, purposes, or other forms of teleology. At the same time, as a deeply religious and moral man, Descartes was troubled by the thought that humans might be nothing more than complex machines.

Some readers might feel inclined to brush off Descartes's qualms with uncompromising materialism as a legacy of a bygone time, a pernicious combination of religious dogmatism and factual ignorance. Yet, such interpretation would be both uncharitable and inaccurate. First, from a historical perspective, Descartes was very much on top of the science of his time, as witnessed by his notable contributions to various fields, such as mathematics, physics, and physiology. Second, from a conceptual standpoint, Descartes' rationale for eschewing radical physicalism was hardly antiscientific. Simply put, he realized that the behavior of conscious and unconscious entities is not explained in the same way. Inanimate objects typically obey strict physical equations or mechanistic law-like generalizations.

Animate organisms, in contrast, are subsumed under intentional, goal-directed, or teleological descriptions, such as those commonly found in current psychology, sociology, economics, and related fields. The psychological explanation of an agent pouring herself a glass of water because she intends to quench her thirst looks nothing like the mechanistic account of why a glass shatters when it falls to the ground. This discrepancy, no less evident today than it was in the 1600s, raises obvious follow-ups. What underlies the difference? What exactly distinguishes animate organisms from inanimate objects?

Descartes' proposal is so famous that a few brief remarks should suffice. Human beings, he claimed, are not purely material. We have both extended bodies and minds. Given our *res cogitans plus res extensa* composition, our behavior will be the resultant of mental and physical causes. Then what characterizes these substances? Do they interact? If so, how? This was the birth of the mind-body problem or, more precisely, what I call the "mind-body problem 1.0." Descartes' concerns were primarily ontological. Mental states cannot be analyzed physically because they are not material things at all. A mind, for him, is a *sui generis* kind of substance: *res cogitans*.[2]

In the 1600s, the ontology of mind was truly an open issue. Interactionism was hardly an *ad hoc* stipulation. It was a fecund speculative hypothesis. Sure, Elizabeth of Bohemia was quick to pinpoint troubling aspects. Yet, after Descartes' *Meditations*, Hobbes, Spinoza, Malebranche, Leibniz, and other prominent philosophers and scientists debated whether there is a substance, a force, an *élan vital* distinguishing animate from inanimate entities.

Things changed. By the late 1800s, empirical evidence against substance dualism had rapidly mounted. With the eclipse of vitalism in biology, Descartes' research program regressed and was eventually replaced by forms of substance monism, which became the default ontology against which to address psycho-physical relations and related methodological issues. By the mid-1900s, most scholars viewed minds either as physical systems or as being realized by such systems. As we'll see, this includes authors with Cartesian inclinations, who replace substance dualism with alternative frameworks, such as property dualism or panpsychism. The vast majority of scientists and philosophers found the case against *res cogitans* overwhelming.

This suggests that Descartes' original question, "mind-body problem 1.0," has finally been answered. In a sense, it has. Minds are no longer characterized as ontologically distinct. Yet, rejecting *res cogitans* evidently tells us little about the nature of mind. Substance monism, alas, leaves ample room for disagreement regarding which properties constitute or instantiate mental states. In particular, it does not constrain how psychological systems must relate to their physical substrate. In this other sense, Descartes' problem was never solved. It was *dissolved*, recast in a related albeit novel guise.

[2]Descartes's conception of substance was strikingly nuanced (Rodriguez-Pereyra 2008).

This conceptual shift can be clearly seen in mid-twentieth century philosophy of mind. Ryle (1949, pp. 21–22) famously wrote that "[Descartes] had mistaken the logic of his problem. Instead of asking by what criteria intelligent behaviour is actually distinguished from non-intelligent behaviour, he asked 'Given that the principle of mechanical causation does not tell us the difference, what other causal principle will tell it us? He realized that the problem was not one of mechanics and assumed that it must therefore be one of some counterpart to mechanics." Ryle's reconstruction is accurate. Except that Descartes was hardly mistaken about the logic of his problem. Rather, he was raising an issue about ontology. Ryle and many of his Anglo-Saxon colleagues, in contrast, had an altogether different question in mind.

12.3 The Mind-Body Problem 2.0

The previous section surveyed Descartes' groundbreaking speculations regarding the relation between the mental and the physical. Descartes' concern was, first and foremost, an ontological one. Given that minds cannot be physical entities, he wondered, what kinds of substance could they be? His query was answered, once and for all, at the turn of the twentieth century, when vitalism was expunged from biology and most scholars, scientists and philosophers alike, embraced forms of substance monism. Still, and this is the crucial point, settling the issue revealed very little about the nature of mind. What kind of physical systems are minds or instantiate them? How should mental states be studied? What distinguishes conscious organisms from inanimate objects? To address these questions, which lie at the heart of the philosophy of mind, Descartes' problem—the mind-body problem 1.0— had to be reformulated. How, exactly? To guide our discussion, let's look at some influential theories of mind articulated in the century just passed.

One of the first notable materialist theories of mind developed in the twentieth century, popular among both philosophers and psychologists, is *behaviorism*, which analyzes mental states as complex clusters of dispositions to behave.[3] A few years later, the *type-identity theory*, pioneered by U.T. Place (1956) and J.J.C. Smart (1959), purported to identify mental states and brain states at the type level. In response to objections to type-identity theory, especially those pertaining to the multiple-realizability of psychological kinds (Putnam 1967; Fodor 1974),

[3]To be sure, psychologists and philosophers had different agendas. To reflect this divergence, it is common to distinguish two strands of behaviorism (Fodor 1981). First, "philosophical" (also known as "logical" or "analytic") behaviorism is associated with a thesis about the nature of mind and the meaning of mental states. Second, "psychological" or "methodological," behaviorism emerged from an influential scientific methodology applied to psychology. For the sake of simplicity, I shall not distinguish between the two variants.

functionalism refined behaviorist insights by characterizing mental states as causal roles determined by inputs, outputs, and various other kinds of internal connections with other mental states (Putnam 1965; Armstrong 1981).[4] An alternative path was explored by Donald Davidson (1970), who posited an identity of mental and physical states at the token level—whence the name *token-identity theory of mind*—and described their relation in terms of supervenience. Another view is *eliminative materialism*, a position principally advocated by Patricia and Paul Churchland (1981, 1986), which treats mental states as theoretical entities posited by folk psychology. Commonsensical as it may seem, they argue, folk psychology is a flawed theory of mind and, as such, it is not a candidate for integration. Rather, it should be eliminated and replaced by a mature neuroscience, which will turn out to be more predictive, explanatory, and connected to other fields of science. A final noteworthy approach is a revamped version of dualism. *Property dualism* agrees with Descartes that there is a real distinction between mental and physical attributes. Yet, it does not view *res cogitans* and *res extensa* as mutually exclusive. One and the same substance may have both physical and mental properties. While substance dualism today has few, if any, proponents, property dualism is still advocated in the philosophical literature (Jackson 1982; Chalmers 1996).

A comprehensive overview of these, and other, influential theories of mind lies beyond the scope of this essay. For present purposes, the crucial matter—pun intended—is pinpointing the main matter of contention underlying all these approaches. As soon as this is done, it immediately becomes evident how irrelevant Descartes' question, the mind-body problem 1.0, has become. The main reason is not that most parties involved, including property dualists, eschew substance dualism. This is true, but of marginal significance from our contemporary standpoint. After all, even some of Descartes' contemporaries questioned his ontology. Whether minds are mental substances or physical substances has become utterly tangential to the twentieth-century debate. The modern focus is on the issue of *reduction*. Can mental states be reduced to brain or, more generally, physical states? If so, how? If not, why not? Behaviorism, type-identity theory, and eliminative materialism all answer in the positive: psycho-neural reduction is feasible, at least as a matter of principle. Token-identity theory, functionalism, and property dualism answer in the negative, claiming that any such reduction is doomed to failure.[5] In short, the in-principle reducibility of mental states to physical states, or the impossibility thereof,

[4]Again, I am unabashedly clashing together several variants of functionalism, such as Putnam's "psycho-functionalism" and Armstrong's "a priori functionalism" (Block 1978).

[5]As Matteo Colombo has brought to my attention, the mind-body problem 1.0 could also be framed as a matter of reduction. On this reading, Descartes may be interpreted as providing a negative argument: minds cannot be reduced to bodies because they are altogether different substances. This is an effective strategy to bring Descartes into modern debates, finding some narrative continuity in the last four hundred years of philosophy of mind. Still, this operation should be understood, from our contemporary perspective. From historical standpoint, Descartes' target was not reduction. He was interested in ontological questions about the nature of minds and their interactions with bodies.

lies at the core of twentieth-century philosophy of mind. This is what I call the "mind-body problem 2.0."

If mind-body 2.0—that is, psycho-neural reduction—is so central to contemporary philosophy of mind, one could legitimately ask, what can be said about its success or failure after decades of extensive debate? Even a cursory look at the specialized literature reveals that clear-cut, conclusive answers are nowhere to be found. Of course, we have made significant progress in the discovery of psycho-neural mechanisms underlying higher and, especially, lower cognition. But have these findings advanced the *tout court* reduction of mental states to brain states? If so, the news has not been broken, as there seems to be no more consensus today than there was in the 1950s.

When confronted with this lack of resolution, many scientists and philosophers interested in the nature of mental states tend to justify the situation by appealing to the intricacy of the subject matter. The human brain is the most complex organic structure discovered so far in the universe, composed of billions of cells and an astronomical number of possible connections among them. No wonder that solving the dispute is so darn hard! I have no quibbles with any of the premises. Studying the human brain is, indeed, frustratingly difficult. Nevertheless, I am skeptical of the diagnosis. The complexity of the structures under investigation is not the principal cause of the lack of tangible progress when it comes to the mind-body problem 2.0. The main culprit, as I will go on to argue, is the notion of *reduction* itself.

Before moving on, I should clarify what distinguishes my view from similar positions in the literature. Over the last few decades, there has been no shortage of philosophical attempts to explain away the mind-body problem. Notably, Chomsky (2000, 2002) has written extensively on the topic, arguing convincingly that contrary to common wisdom, the mind-body problem "did not disappear because of inadequacies of the Cartesian concept of mind, but because the concept of body collapsed with Newton's demolition of the mechanical philosophy" (2002, p. 71). I am, indeed, quite sympathetic to Chomsky's remarks. Yet, I want to draw attention to a different issue, that applies not much to Descartes' original position—the "mind-body problem 1.0"—but to question of the reducibility of mental states. This corresponds to what I call the "mind-body problem 2.0." It should be evident how my attempt to undermine the very question of reduction puts me at odds with both traditional reductionist and antireductionist philosophical perspectives.

Spelling out the argument involves breaking it down into two main steps. First, §4 takes a detour into the history of the philosophy of science, focusing on how the concept of reduction has morphed since the collapse of the classic model of reduction, endorsed by logical positivism. Next, §5 explains why focusing on the issue of reduction—the crux of the mind-body problem 2.0—might have been a red herring driving philosophers down a wrong path.

Relatedly, eliminative materialists prefer to talk about "elimination" as opposed to "reduction." Yet, the former concept can be straightforwardly treated as a limiting case of the latter.

12.4 Pegs, Holes, Atoms: An Overview of Reduction

What does it mean to "reduce" a theory, a concept, or a law of nature? Positivist philosophy of science had a clear, unequivocal answer, whose locus classicus became Nagel's (1961) *The Structure of Science*. Simply put, from the standpoint of logical empiricism, to reduce a theory T_1 to a theory T_2 is to show that the laws of T_1 can be deduced from the laws of T_2 via "bridge principles" that translate the concepts of T_1 into the vocabulary of T_2.[6]

It is worth stressing that reductionism, as constructed by Nagel, is ontologically neutral, in the sense that it is independent of physicalism or dualism. Now, most philosophers of science since the mid-twentieth century, friends and foes of positivism alike, have been thoroughgoing materialists. Nevertheless, Cartesians too could conceive of reduction—or, presumably, lack thereof—along Nagel's lines. Again, this goes to show how much the debate had shifted from Descartes' original formulation of mind-body 1.0.

Influential as it was, Nagel's reductionism was eventually eroded by powerful objections. Setting details aside, the main problem involved the lack of bridge laws. With the possible exception of a few hackneyed examples, the multiple-realizability of kinds across the special sciences implies that there are not enough connecting principles for classical derivational reduction to take flight as a general model of science (Putnam 1967; Fodor 1974).

Does this mean that the mind-body problem 2.0—the question of psycho-neural reduction—has finally been answered in the negative? Not at all. The collapse of classical reductionism hardly signaled the end of reductionism *tout court*. The old positivist model has been reformulated and replaced by a novel, more promising framework, intended to avoid the shortcomings of its illustrious predecessor. The new wave of reductionism is packaged as a set of epistemic questions concerning explanation. Can we provide micro-depictions of all macro-events? And are the resulting lower-level explanations invariably deeper than their higher-level counterparts? Reductionists typically answer both questions in the positive, advocating a form of epistemic fundamentalism. Whereas it is often epistemically necessary or pragmatically convenient to stick to macro-descriptions, the neo-reductionist story goes, adding detail always increases the depth of coarser descriptions.

Readers will likely guess my follow-up. Can we describe every scientific event at more fundamental, finer-grained levels? And it is really the case that these micro-depictions invariably enhance explanatory power? Despite valiant attempts to resolve the conundrum, clear-cut answers are still wanting. Why is this so? The main reason, as we'll see in §5, involves the vagueness of reduction. Despite the appearance of substantive disagreement, both parties end up talking past each other.

[6] To be sure, Nagel's own conception of reduction was subtler, and its proper interpretation remains a matter of controversy (Fazekas 2009; Klein 2009). Nevertheless, for present purposes I am less interested in Nagel's actual views, and more in how his model of reduction was received and discussed within philosophy (Fodor 1974; Kitcher 2003).

In order to get there, however, we need to continue to follow the unravelling of this longstanding debate.

In an influential article published over four decades ago, Putnam (1975) argued that traditional discussions of the mind-body problem rest on a misleading assumption. The problematic presupposition in question is a conditional premise: if we accept that human beings are purely material entities, then there must be a *bona fide* physical explanation of our behavior. Physicalists, Putnam noted, use this premise in a *modus ponens* inference:

- (a) Humans are purely material beings.
- (b) If humans are purely material beings, then there must be a *bona fide* physical explanation of our behavior.
- (c) ∴ There must be a physical explanation of our behavior.

Dualists, in contrast, embed the conditional (b) in a *modus tollens* inference, rephrased here in the subjunctive mood, to enhance readability:

- (b) If humans were purely material beings, there would be a *bona fide* physical explanation of our behavior.
- (c*) There is no physical explanation of our behavior.
- (a*) ∴ Humans are not purely material beings.

These two arguments advance diverging conclusions. Yet, physicalists and dualists alike accept the conditional premise. This, Putnam maintains, is a mistake. Both parties miss the mark, as (b) should be rejected as unsound.

In support of his conclusion, Putnam presents a suggestive analogy (Fig. 12.1). He considers a rigid board with two holes: a circle exactly one inch in diameter and a square one inch high. Now, take a cubical peg just under one inch high. The peg will go through the square hole. However, it will not go through the round hole. How do we explain these elementary observations?

Putnam sketches two types of explanations. The first begins by observing that the board and the peg are rigid lattices of atoms. If we compute the astronomical number of all physically possible trajectories of the peg, we will eventually discover that no trajectory passes through the round hole, whereas at least one trajectory, likely more, passes through the square hole. An alternative explanation begins in exactly

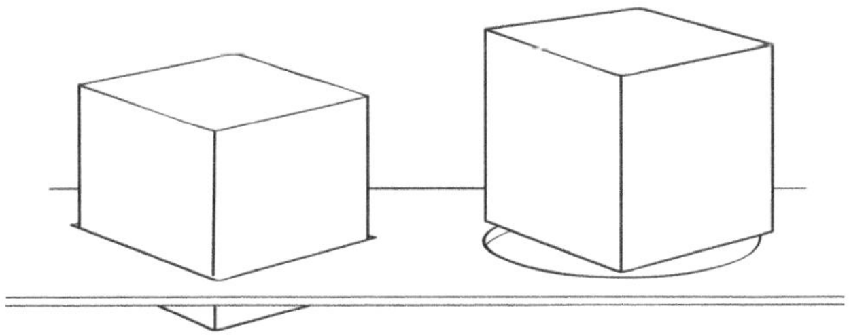

Fig. 12.1 Putnam's square-peg-round-hole example

the same way, by noting that the board and the peg are rigid systems. Yet, instead of comparing trajectories, it points out that the square hole is slightly larger than the cross section of the peg, whereas the round hole is smaller. Call the former kind of explanation "physical," "lower-level," or "micro" and label the latter one "geometrical," "higher-level," or "macro." The question is: are both explanations adequate? If not, why not? And, if so, which one is better and why?

Putnam contends that the geometrical explanation is objectively superior. (Actually, Putnam goes as far as claiming that the physical explanation is not explanatory at all, but I set this more controversial thesis to the side.) The reason is that, whereas the physical description only applies to the specific case at hand, the geometrical story generalizes to similar structures. To illustrate, an exhaustive listing of all trajectories will only account for why *this* peg will or will not go through these particular holes. In contrast, the geometrical account captures why *no* square peg will go through a hole smaller than its cross-section. As Putnam (1975, p. 297) puts it, "in terms of real life disciplines, real life ways of slicing up scientific problems, the higher-level explanation is far more general, which is why it is *explanatory*."

The significant philosophical moral drawn by Putnam from this intuitive toy example is the *explanatory autonomy of the mental from the physical*. Higher-level explanations, regardless of whether they involve pegs and holes, or psychological states cannot—and should not—be explained at lower levels, in terms of neuroscientific, biochemical, or physical properties.

Putnam's argument has left a mark by firing up a longstanding debate. Philosophers started asking: is it really the case that macro-explanations are objectively superior to their micro-level counterparts? Antireductionists answered in the positive. In the philosophy of mind, authors such as Fodor (1968), Davidson (1970), Jackson (1982), Yablo (1992), Chalmers (1996), Hornsby (1997), and Burge (2007, 2013) have buttressed various arguments supporting the autonomy of mental states from underlying neural ones. Reductionists beg to disagree. Scholars like Paul and Patricia Churchland (1981, 1986), Bickle (1998, 2003), and Kim (1999) have countered that micro-explanations are the key to deepen our understanding of the mind.[7]

Obviously, at the most general level, the question of reduction must be understood as a matter of principle, not practice. Current physics is not even close to replacing biology, psychology, economics, or any other special science. We lack the understanding of subatomic systems and, especially, the computing power required to approximate the perfect vision of a "Laplacian Demon." Still, reductionists claim, in theory, it would be possible to provide micro-explanations to replace

[7]An analogous, equally heated debate emerged in the philosophy of science. Putnam's square-peg example was developed and extended to real-life scientific scenarios in biology (Kitcher 2003), psychology (Fodor 1974), and the social sciences (Garfinkel 1981). Post-positivist neo-reductionists disagreed. Authors such as Waters (1990), Sober (1999, 2000), Rosenberg (2006), and Strevens (2008) stressed that, while micro-explanations are often unnecessarily complex or anti-economical, they do emphasize crucial details that are typically presupposed implicitly or taken for granted at the macro-level.

and improve current macro-depictions. Antireductionists reject this conclusion. Addressing biology, psychology, or economics in physical terms, they argue, would not deepen these inquiries.

In short, Putnam revamped the debate on epistemic reductionism across the sciences by questioning the possibility of explaining higher-level states at more fundamental levels. Fellow antireductionists follow suit and embrace the explanatory autonomy of the mental, motivated by structurally analogous arguments. Contemporary reductionists retort that these considerations miss the mark. Lower-level explanations, they suggest, always enhance the explanatory power of coarser depictions. But was this the right direction to point the discussion? Is explanatory reduction the core issue underlying the square-peg-round-hole scenario and its implications for the philosophy of mind? As we shall now see, there are reasons to be skeptical.

12.5 Some Bugs in the Mind-Body Problem 2.0

Let's take stock. §4 retraced the origins of reduction, the conceptual core of the mind-body problem 2.0 and of much discussion in twentieth-century philosophy of mind. The issue driving the debate is whether breaking down macro-explanations into micro-explanations invariably increases explanatory power. Epistemic reductionists argue in the positive. Antireductionists answer in the negative. How much progress have we made towards a solution?

To get started, consider the current state of psycho-neural reduction. While recent advancements in cognitive neuroscience have yielded a plethora of results, much remains unknown. Reductionists typically stress the remarkable successes with sensory systems and various domains of lower cognition such as early vision, pain, and taste, as evidence for the power and promise of decomposition strategies. Antireductionists rejoin that comparable achievements cannot be boasted for language processing, decision making, and other domains of higher cognition, especially consciousness. Despite this divergence, both parties agree that knowledge of the structure and location of psycho-neural mechanisms implementing and computing cognitive functions has increased exponentially. The philosophical debate hinges on whether or not it is possible to enhance the power of higher-level explanations via lower-level descriptions. This dichotomy, note, mirrors Putnam's square peg round hole scenario. But is this the proper analogy to draw?

To assess the prospects of psycho-neural reduction, understood along the lines just delineated, it is instructive to compare it with the corresponding debate over reductionism in the life sciences. This mirroring is enlightening because neural mechanisms are still relatively obscure, due to the complexity of the system under study. In contrast, biologists have a clearer picture of the implementation of functional structures at the molecular level. We already know quite a bit about, say, how important phylogenetic adaptations are transmitted across generations and develop at the ontogenetic level.

These considerations suggest that the case for or against reductionism is closed, or is close to being settled, in the life sciences. After all, if the crucial issue is whether all macro-biological explanations can be strengthened at the micro-biological level, having concrete case studies to assess should provide decisive evidence, one way or the other. To be sure, the fate of reductionism *tout court* depends on much more than a handful of successful or failed stories. Even the accomplished reduction of, say, evolution to molecular genetics would still fall short of an overarching reductionism. Nevertheless, it would provide strong evidence in favor of reductionism as a "working hypothesis."

Unfortunately, the jury is still out, and any verdict is far from reached. The status of epistemic reductionism in genetics, ontogeny, evolution, and other branches of biology remains as open and controversial as ever (Sarkar 1998; Sober 2000; Kitcher 2003; Rosenberg 2006; Dupré 2012; Griffiths and Stotz 2013). Sophisticated antireductionists acknowledge the success of molecular biology. Still, they stress how so-called "molecular" explanations consistently appeal to structural and functional concepts, and holistic states of systems. This, antireductionists claim, shows that the appearance of reduction is nothing but a smoke screen. Modest reductionists, in contrast, appreciate the importance of functional and dispositional properties in genetic and other lower-level explanations. Yet, they contend that all these seemingly higher-level concepts belong to the domain and vocabulary of molecular biology, broadly construed. Thus, the debate ultimately hinges not on the nature and depth of explanations, which are widely agreed upon, but on whether these explanations should be labelled as "molecular." As a result, discussants talk past each other, making the dispute more terminological and less substantial than is typically assumed (Nathan 2012, under contract).

With this in mind, let us return to psychology and neuroscience. Does the current psycho-neural interface vindicate or thwart reductionism? Do the brain sciences have the conceptual resources to describe all mental events in neural terms? And does this enhance their explanatory power? Well, much depends on how one characterizes the levels and vocabularies in question. Unsurprisingly, modest reductionists tend to presuppose a generous, ecumenical conception of "lower-level" descriptions. This includes functional, dispositional, and structural concepts, typically found at higher levels in the scientific hierarchy. In turn, sophisticated antireductionists, for the most part, agree on the importance of this explanatory apparatus. Yet, they are less liberal on what can be categorized as "lower-level," "micro," "neural," or "molecular." As in the biological case, discussants talk past each other and quibble over labels, making the debate semantic, as opposed to substantive.

What moral should we draw from all of this? The take-home message is that philosophers of mind, psychology, and neuroscience should learn the hard lesson from their colleagues in biology. Important as they are, empirical discoveries concerning where and how cognitive functions are implemented in the brain are unlikely to solve any longstanding philosophical dispute over the mind-body problem. The reason is not the complexity of the human mind and brain—which, I emphasize once again, should not be questioned. The real problem is the nature of reduction which, contrary to common wisdom, turns out to be a murky construct. This lack of clarity becomes especially evident when one asks the question: does cognitive neuropsychology fit in better with reductionism or antireductionism? The answer is along the lines of both or neither. But does it even matter? Current psychology and neuroscience seem perfectly compatible with both stances, which is precisely what one could expect in the case of a merely verbal disagreement.

In conclusion, following this stalemate, the philosophical debate over reductionism has lost traction, as witnessed by the lack of resolution, coupled with the shortage of novel insight. Conceptual progress requires recognizing that the crux of the mind-body problem is independent of the muddled status of reduction. As Putnam noted long ago, the main issue is the question of explanatory autonomy. Since the 1970s, autonomy and reduction have been viewed as contradictory. Many scholars, and Putnam himself was no exception, view autonomy as the rejection of reduction and reduction as the denial of autonomy. This, we shall now see, is a consequential mistake.

12.6 The Mind-Body Problem 3.0

How should the mind-body problem be repackaged in the twenty-first century, given that its conceptual core—the bridge between the mental and the physical—is independent of both ontology and reduction, that is, versions 1.0 and 2.0? This section proposes a new framework for recasting Descartes' dilemma, so as to reflect real, substantive disagreement across current scientific inquiry, while maintaining some continuity with its roots in early modern philosophy.

To get started, let's return to the relation between psychology and neuroscience. On the one hand, evidence concerning brain activity deepens, in several ways, higher-level psychological explanations. Discovering the inner workings of brain processes and networks sheds much light on why the mind works the way it does: its abilities, biases, and computational limitations. On the other hand, it seems equally undeniable that psychological descriptions and explanations enjoy an autonomy of sorts, in the sense that they can be established, corroborated, and explained without the aid of neuroscience or any other more fundamental discipline. Allow me to briefly elaborate.

Consider some well-known psychological generalizations, such as the widespread tendency of subjects confronted with cases of moral decision making—such as the famous "trolley problems"—to follow consequentialist rules, unless doing so involves using other people directly as a means. Or take the "endowment effect," which captures how the price that subjects are willing to accept to part with goods vastly exceeds the price they are willing to pay to acquire the same goods, violating core tenets of expected utility theory. In both cases, the generalizations themselves can be expressed, confirmed, refined, and explained, in purely psychological terms. To wit, the former generalization is typically accounted for by appealing to negative emotions clashing with consequentialist reasoning, whereas the latter effect is often taken to depend on loss aversion. From this standpoint, learning more about the mechanisms which compute these cognitive patterns is no more necessary than physical details in Putnam's square peg. Yet, the appropriate moral is not that fMRI cannot contribute to the study of higher cognition. *Au contraire*, so-called "reverse inferences" play a crucial role deepening these explanations by discriminating between competing psychological hypotheses (Del Pinal and Nathan 2013; Nathan and Del Pinal 2016).

How can both points be maintained simultaneously? How can psychology be autonomous, while depending on the underlying neural substrate? How can we use neuroscience to advance the study of the mind, without threatening the independence of higher explanatory levels? These are the pressing questions pertaining to the mind-brain relation. Putnam had the right insight when he pointed the discussion towards autonomy. The crucial mistake was turning the issue of autonomy into a debate about reduction.

The central philosophical question underlying current neuropsychological debates, I maintain, is how to pursue the scientific study of the mind in the age of neuroscience. This is what I call the *mind-body problem 3.0*.

What makes 3.0 different from the previous versions 1.0 and 2.0? My goal is to sketch a constructive framework for recasting old questions in a new guise, thereby avoiding the thorny issues of ontology and reduction.

The main hang-up can be posed in the form of a dilemma. On the one hand, neural details are crucial for understanding the structure, implementation, and behavior of psychological systems.

This was the main insight of twentieth-century physicalism. At a bare minimum, brains set boundary conditions and constraints on what minds can or cannot do and why this is the case. But this being so, in what sense is the higher-level truly autonomous? On the other hand, if one begins by stressing the autonomy of psychology, it becomes hard to see why neural details should matter at all.

There is a simple way out of this impasse. The problem with traditional formulations of materialism, including reductionist and antireductionist approaches alike, is presupposing, more or less explicitly, that higher-level explanations and their lower-level counterparts have the same explananda, the same objects of explanation. In essence, what discussants failed to recognize is that questions at different levels and with varying scope are, effectively, different questions. Some illustrations should help make the point clearer.

First, let's return to the square-peg-round-hole scenario. From a metaphysical standpoint, board and peg supervene in their atomic structure. This is true, albeit uncontroversial and inessential to the main point of contention. Descartes' ontological concerns, mind-body 1.0, have long been put to rest, for good reason. The relevant issue is whether these micro-details enhance the power of the explanation of the system's behavior. Putnam's deep insight was recognizing that much depends on what we are trying to explain. If the explanandum is *that* the square peg will not pass through the round hole, then the micro-details can be effectively black-boxed.

In contrast, if we are trying to capture *why* this is so, then looking at the physical structure of the system, down to its subatomic properties, becomes relevant.

Now, apply this perspective to the psycho-neural interface. Are brain-level details relevant to the study of the mind? The short answer is that it depends. Some cognitive inquiries are framed in a way that makes them perfectly autonomous, in the sense that they can be confirmed, refined, and explained without the aid of physical, molecular, or even neural details. Cognitive hypotheses regarding the engagement of negative emotions in trolley problems or loss aversion in economic decision making do not require the aid of neuroimaging or other neuroscientific techniques. Still, this is not to deny that there are other, equally important questions to ask about how these higher-level functions are implemented, processed, or realized at more fundamental levels. It is here that neural details may become indispensable.

The main point—stressed, in different ways, in both the scientific (Marr 1982) and the philosophical literature (Garfinkel 1981)—is that translating higher-level questions into lower-level ones, or vice versa, may yield different inquiries. Failure to recognize this has generated confusion. The misleading assumption, shared by reductionists and antireductionists alike, is that higher- and lower-level explanations are in competition. They are not. Borrowing a Kuhnian metaphor, explanations with different scope are typically *incommensurable*. Because of their different targets, any attempt to rank them in terms of explanatory power turns into an exercise in futility.

In short, Putnam's contribution was recognizing that higher-level explanations are epistemically "autonomous" from lower-level ones. His mistake, which wreaked much havoc in subsequent discussion in the philosophy of science and mind, was turning this into a vindication of antireductionism. Putnam, and many philosophers after him, identified autonomy with antireductionism. These concepts, I maintain, should not be conflated. Whereas reductionism and antireductionism disagree on whether more fundamental depictions should invariably be preferred over less-fundamental ones, both stances presuppose—indeed, require—convergence in the objects of explanations. Autonomy, in contrast, gains traction by rejecting this presupposition and embracing a form of epistemic incommensurability (Nathan, under contract).

Before moving on, let me address how the present proposal fits in with two current debates in the philosophy of mind. First, readers may note some analogies between my presentation of the mind-body problem 3.0 and the "mechanistic turn" in the philosophy of neuroscience, which was born out of a reaction to the traditional reductionism vs. antireductionism divide (Bechtel and Richardson 2010). While the new wave of mechanistic philosophy is too sizable a movement to present, let alone assess, in a few statements, I should stress that, despite its popularity, it has not yet escaped the grip of the mind-body problem 2.0. Over the last few years, mechanistic accounts of explanation have been criticized based on the allegation that they are committed to the unpalatable tenet that adding *any* kind of detail about a mechanism will improve an explanation (Batterman and Rice 2014; Chirimuuta 2014; Levy 2014). Neo-mechanists have responded by explicitly distancing themselves from this "more details are better" stance and replacing it with the thesis that that only relevant details improve an explanation (Baetu 2015; Boone and Piccinini 2016a; Craver and Kaplan 2018). My present perspective can be squared with the mechanistic joinder. First, not all details are relevant or helpful for every explanation. Second, *which* details matter will crucially depend on the explanandum at hand. Third, and finally, determining which details are relevant to an explanatory task is no simple task (Krickel and Kohar this volume). Yet, to avoid the grip of reduction, it is crucial to stress the incommensurability of explanations at different levels, a point seldom stressed explicitly, to the best of my knowledge.

Second, as mentioned at the outset, the current discrepancy between traditional philosophy of mind and ongoing debates in the cognitive neurosciences has not gone unnoticed. For instance, Chemero and Silberstein (2008, p. 1) maintain that "The philosophy of mind is over."[8] Boone and Piccinini (2016b) take the argument one step forward by suggesting that cognitive science itself, as traditionally conceived, is currently in the process of being replaced by cognitive neuroscience. As a result, the old debate between reductionism and autonomy has faded into the background, replaced by a focus on multilevel mechanistic explanations.[9] In contrast, I have tried to stress here the continuity between past and present debates. Nevertheless, the obvious differences between these proposals and the perspective defended here should not be overstated. Chemero and Silbertein's holistic cognitive science, Boone and Piccinini's multilevel mechanistic explanation and my attempted differentiation between autonomy and antireductionism all share a common assumption: in some form or another, philosophy still has an important role to play. The fundamental question of contemporary philosophy of mind is how to pursue the scientific study of the mind in the age of neuroscience. This, in essence, is the mind-body problem 3.0.

[8]Chemero and Silberstein motivate their provocative claim as follows: "The two main debates in the philosophy of mind over the last few decades about the essence of mental states (they are physical, functional, phenomenal, etc.) and over mental context have run their course. Positions have hardened; objections are repeated; theoretical filigrees are attached. These relatively armchair discussions are being replaced by empirically oriented debates in philosophy of cognitive and neural sciences" (2008, p. 1).

[9]"The scientific practices based on the two-level view (functional/cognitive /computational' vs. neural/mechanistic/implementation) are being replaced by scientific practices based on the view that there are many levels of mechanistic organization. No one level has a monopoly on cognition proper. Instead, different levels are more or less cognitive depending on their specific properties. The different levels and the disciplines that study them are not autonomous from one another. Instead, the different disciplines contribute to the common enterprise of constructing multilevel mechanistic explanations of cognitive phenomena. In other words, there is no longer any meaningful distinction between cognitive psychology and the relevant portions of neuroscience—they are merging to form cognitive neuroscience" (Boone and Piccinini 2016b, p. 1510).

12.7 Concluding Remarks

Time to pull some strings together. I distinguished three variants of the mind-body problem. Version 1.0 reflects Descartes' ontological quandary: what kind of substances are minds and how are they related to bodies? Version 2.0 tacitly underlies much twentieth century philosophy of mind: can mental states be reduced to brain states? Neither variant has been solved. Both have been dissolved, recast. Finally, I advanced a revamped "mind-body problem, version 3.0," In a slogan, this is the question of how to pursue psychology, the modern science of the mind, in the age of neuroscience, the science of the brain. How should these two disciplines inform each other?

Undoubtedly, many readers will remain unpersuaded by my contemporary reformulation of the mind-body problem. Property dualists maintain that the gap between physical and phenomenal properties is ontological, not merely epistemic (Chalmers 1996). Hylomorphists are concerned with the ontological irreducibility of macro-causes (Jaworski 2016; Koslicki 2018). These issues seem orthogonal to the problem of explaining the metaphysical relationship between higher-level properties and their micro-base. Version 3.0 implicitly strips the mind body problem of all its metaphysical underpinnings, including the insistence of some new-wave mechanists, in an ontic approach to explanation (Craver 2007). Does this unduly narrow its scope?

My response is that it might be time to reshuffle the deck. Perhaps, issues which, prima facie, appear to be ontological in character could be fruitfully repackaged as questions of explanation and methodology. This proposal is supported by the observation that current scientific research can be made consistent with virtually all combinations of materialism, dualism, reductionism, and antireductionism. To illustrate, most contemporary scholars presuppose some variety of materialism—and yours truly is no exception. Yet, it has been pointed out that, in principle, current psychology could be reconciled with various forms of dualism (Chalmers 1996). Similarly, nothing substantial hinges on whether or not psychology is "reducible" to neuroscience. As noted, the answer depends on how exactly one conceives of reduction and how broadly the domain of lower-level theories is defined. Even contemporary uses of neuroimaging are compatible with various forms of antireductionism and ontological dualism (Del Pinal and Nathan 2013; Nathan and Del Pinal 2016). Paraphrasing Wittgenstein, these are matters of expression, not facts of the world. After centuries of discussion, it might be time to abandon old ontological questions and try out something new.

My aim here transcended mere exposition and historical reconstruction. Both my *pars destruens* and *pars construens* advance critical analyses and suggestions for moving forward. Still, the succinct remarks contained in this article, by themselves, admittedly fall way short of a *solution* to the mind- body problem 3.0. Follow-ups await. Which inquiries should be prioritized? How does one determine whether a question is best explained at higher or lower levels? How much detail is relevant? Do explanatory standards cut across domains? Can we provide effective mappings of scientific ontologies at different steps of the hierarchy? Providing answers requires a painstaking combination of empirical and conceptual work. In addition, we saw that there are various alternative proposals for addressing the relation between psychology and neuroscience along the lines suggested here. Should we opt for a holistic approach? A focus on multi-level mechanisms? A revamped version of autonomy? Something altogether different? While this is not the appropriate venue for weighting these options, it seems to me that reframing the mind-body problem in a way that avoids getting entangled in verbal disputes concerning ontology and reduction is a step in the right direction.

I conclude by stressing two features of the present proposal. First, traditionally, the family of issues underlying the "mind-body problem" has been concerned with the exceptionality of human cognition. Cartesian dualists resist the identification of mind and matter by treating the former as ontologically distinct from anything else in the physical universe. Mid-twentieth-century reductionists, like Place and Smart, have advocated the treatment of psycho-neural reduction as a scientific hypothesis. Non-reductive physicalists, such as Nagel and Davidson, have responded by emphasizing features of the mental that make it unique. The same kind of tension can be found within the 3.0 version too. On the one hand, some philosophers might view the relation between psychology and neuroscience as a general issue in the philosophy of science. Just like there is a mind-body (*qua* psychology-neuroscience) problem, there is a biology-chemistry problem, an economics-sociology problem, etc. From a methodological perspective, all these interfaces are on a par. Others will disagree, for instance, by emphasizing features of the mental—a *sui generis* normativity, a "hard" problem of consciousness, or something along these lines—that make the mental special, or otherwise exceptional.

Second, contrary to versions 1.0 and 2.0, mind-body 3.0 raises a problem that is central to contemporary scientific agendas. The outcome of the debate on whether and how psychology and neuroscience can mutually inform each other will likely determine how the study of mental and neural structures will be approached—and funded—over decades to come. It is crucial for philosophy to keep asking the right questions and focus on substantive conceptual and empirical issues that are central to core scientific practice, like it has done for much of its history. The disconcerting alternative is for philosophical analysis to become irrelevant and fade into oblivion.

Acknowledgments The author is grateful to Bill Anderson, John Bickle, Fabrizio Calzavarini, Matteo Colombo, Guie Del Pinal, Carrie Figdor, Matteo Grasso, Philipp Haueis, Mika Smith, Marco Viola, and two reviewers for constructive comments on various versions of this essay, and to Stefano Mannone for designing the image. Earlier drafts were presented at the University of Milan, Mississippi State University, the University of Turin Neural Mechanisms Webinar Series, and the University of Denver. All audiences provided valuable feedback.

References

Armstrong, D. M. (1981). *The nature of mind*. Ithaca: Cornell University Press.

Baetu, T. M. (2015). The completeness of mechanistic explanation. *Philosophy of Science, 82*, 775–786.

Batterman, R. W., & Rice, C. C. (2014). Minimal model explanations. *Philosophy of Science, 81*, 349–376.

Bechtel, W., & Richardson, R. C. (2010). *Discovering complexity: Decomposition and localization as strategies in scientific research* (2nd ed.). Cambridge: MIT Press.

Bickle, J. (1998). *Psychoneural reduction: The new wave*. Cambridge, MA: MIT Press.

Bickle, J. (2003). *Philosophy and neuroscience: A ruthlessly reductive account*. Dordrecht: Kluwer.

Block, N. (1978). Troubles with functionalism. In C. Savage (Ed.), *Perception and Cognition* (pp. 261–325). Minneapolis: University of Minnesota Press.

Boone, W., & Piccinini, G. (2016a). Mechanistic abstraction. *Philosophy of Science, 83*, 686–697.

Boone, W., & Piccinini, G. (2016b). The cognitive neuroscience revolution. *Synthese, 193*, 1509–1534.

Burge, T. (2007). *Foundations of mind*. Oxford: Oxford University Press.

Burge, T. (2013). Modest dualism. In *Cognition through understanding. Philosophical essays* (Vol. 3, pp. 471–488). Oxford: Oxford University Press.

Chalmers, D. J. (1996). *The conscious mind: In search for a fundamental theory*. New York: Oxford University Press.

Chemero, A., & Silberstein, M. (2008). After the philosophy of mind: Replacing scholasticism with science. *Philosophy of Science, 75*, 1–27.

Chirimuuta, M. (2014). Minimal models and canonical neural computations: The distinctness of computational explanation in neuroscience. *Synthese, 191*, 127–153.

Chomsky, N. (2000). *New horizons in the study of language and mind*. Cambridge: Cambridge University Press.

Chomsky, N. (2002). *On nature and language*. Cambridge: Cambridge University Press.

Churchland, P. M. (1981). Eliminative materialism and the propositional attitudes. *The Journal of Philosophy, 78*(2), 67–90.

Churchland, P. (1986). *Neurophilosophy*. Cambridge: MIT Press.

Craver, C. F. (2007). *Explaining the brain: Mechanisms and the mosaic unity of neuroscience*. New York: Oxford University Press.

Craver, C. F., & Kaplan, D. M. (2018). Are more details better? On the norms of completeness for mechanistic explanation. *British Journal for the Philosophy of Science, 71*(1), 287–319.

Davidson, D. (1970). Mental events. In L. Foster & J. Swanson (Eds.), *Experience and theory* (pp. 79–101). London: Duckworth.

Del Pinal, G., & Nathan, M. J. (2013). There and up again: On the uses and misuses of neuroimaging in psychology. *Cognitive Neuropsychology, 30*(4), 233–252.

Dennett, D. C. (1991). *Consciousness explained*. Boston: Little Brown and Co..

Dupré, J. (2012). *Processes of life: Essays in the philosophy of biology*. New York: Oxford University Press.

Fazekas, P. (2009). Reconsidering the role of bridge laws in inter-theoretic relations. *Erkenntnis, 71*, 303–322.

Fodor, J. A. (1968). *Psychological explanation*. Cambridge: MIT Press.

Fodor, J. A. (1974). Special sciences (or: The disunity of science as a working hypothesis). *Synthese, 28*, 97–115.

Fodor, J. A. (1981). The mind-body problem. *Scientific American, 244*, 114–123.

Garfinkel, A. (1981). *Forms of explanation*. New Haven: Yale University Press.

Griffiths, P., & Stotz, K. (2013). *Genetics and philosophy: An introduction*. Cambridge: Cambridge University Press.

Heil, J. (2013). *Philosophy of mind: A contemporary introduction*. New York: Routledge.

Hornsby, J. (1997). *Simple mindedness: In defense of naive naturalism in the philosophy of mind*. Cambridge: Harvard University Press.
Jackson, F. (1982). Epiphenomenal qualia. *The Philosophical Quarterly, 32*, 127–136.
Jaworski, W. (2016). *Structure and the metaphysics of mind: How hylomorphism solves the mind-body problem*. Oxford: Oxford University Press.
Kim, J. (1999). *Mind in a physical world*. Cambridge: MIT Press.
Kim, J. (2011). *Philosophy of Mind*. Boulder: Westview.
Kitcher, P. (2003). *In Mendel's mirror. Philosophical reflections on biology*. New York: Oxford University Press.
Klein, C. (2009). Reduction without reductionism: A defense of Nagel on connectability. *The Philosophical Quarterly, 59*(234), 39–53.
Koslicki, K. (2018). *Form, matter, substance*. Oxford: Oxford University Press.
Krickel, B., & Kohar, M. (this volume). *Compare and contrast: How to assess the completeness of mechanistic explanation*.
Levy, A. (2014). What was Hodgkin and Huxley's achievement? *British Journal for the Philosophy of Science, 65*, 469–492.
Marr, D. (1982). *Vision: A computational investigation into the human representation and processing of visual information*. New York: Freeman.
Nagel, E. (1961). *The structure of science*. New York: Harcourt Brace.
Nagel, T. (1995). Searle: Why we are not computers. In *Other minds* (pp. 96–110). New York: Oxford University Press.
Nathan, M. J. (2012). The varieties of molecular explanation. *Philosophy of Science, 79*(2), 233–254.
Nathan, M. J. (under contract). *Black boxes: How science turns ignorance into knowledge*. New York: Oxford University Press.
Nathan, M. J., & Del Pinal, G. (2016). Mapping the mind: Bridge laws and the psycho-neural interface. *Synthese, 193*(2), 637–657.
Place, U. T. (1956). Is consciousness a brain process? *British Journal of Psychology, 47*, 44–50.
Putnam, H. (1965). Brains and behaviour. In R. Butler (Ed.), *Analytical philosophy* (Vol. 2, pp. 24–36). Oxford: Blackwell.
Putnam, H. (1967). Psychological predicates. In W. Capitan & D. Merrill (Eds.), *Art, mind, and religion* (pp. 37–48). Pittsburgh: University of Pittsburgh Press.
Putnam, H. (1975). Philosophy and our mental life. In *Mind, language, and reality* (pp. 291–303). New York: Cambridge University Press.
Rodriguez-Pereyra, G. (2008). Descartes' substance dualism and his independence notion of substance. *Journal of the History of Philosophy, 46*(1), 69–90.
Rosenberg, A. (2006). *Darwinian reductionism: Or how to stop worrying and love molecular biology*. Chicago: University of Chicago Press.
Ryle, G. (1949). *The concept of mind*. London: Hutchinson & Co..
Sarkar, S. (1998). *Genetics and reductionism*. Cambridge: Cambridge University Press.
Searle, J. R. (2004). *Mind: A brief introduction*. New York: Oxford University Press.
Smart, J. (1959). Sensations and brain processes. *Philosophical Review, 68*, 141–156.
Sober, E. (1999). The multiple realizability argument against reductionism. *Philosophy of Science, 66*, 542–564.
Sober, E. (2000). *Philosophy of biology* (2nd ed.). Boulder: Westview.
Strevens, M. (2008). *Depth. An account of scientific explanation*. Cambridge: Harvard University Press.
Waters, C. K. (1990). Why the anti-reductionist consensus won't survive: The case of classical Mendelian genetics. *Proceedings to the Biennial Meeting of the Philosophy of Science Association*, 125–39.
Yablo, S. (1992). Mental causation. *Philosophical Review, 101*, 254–280.

Chapter 13
Psychoneural Isomorphism: From Metaphysics to Robustness

Alfredo Vernazzani

Abstract At the beginning of the twentieth century, Gestalt psychologists put forward the concept of psychoneural isomorphism, which was meant to replace Fechner's obscure notion of psychophysical parallelism and facilitate the search for the neural correlates of the mind. However, the concept has generated much confusion in the debate, and today its role is still unclear. In this contribution, I will attempt a little conceptual spadework in clarifying the concept of psychoneural isomorphism, focusing exclusively on conscious visual perceptual experience and its neural correlates. Firstly, I will outline the history of our concept, and its alleged metaphysical and epistemic roles. Then, I will clarify the nature of isomorphism and rule out its metaphysical role. Finally, I will review some epistemic roles of our concept, zooming in on the work of Jean Petitot, and argue that it does not play a relevant heuristic role. I conclude suggesting that psychoneural isomorphism might play a role in robustness analysis.

13.1 Introduction

At the beginning of the twentieth century, the Gestalt psychologists put forward the concept of *psychoneural isomorphism* (Köhler 1929), the claim that the "mind" and the "neural" are isomorphic. The Gestaltists' aim, as we will see, was that of replacing the vague concept of psychophysical parallelism that constituted the philosophical foundation of much of early nineteenth century psychophysics. Yet, the concept has never been fully clarified and in contemporary contributions it still represents a source of puzzlement[1]. It is far from clear in what sense the mental and

[1]Some contemporary contributions include: Bridgeman (1983), Lehar (1999, 2003), Noë and Thompson (2004), O'Regan (1992), Palmer (1999).

A. Vernazzani (✉)
Institut für Philosophie II, Ruhr-Universität Bochum, Bochum, Germany
e-mail: alfredo-vernazzani@daad-alumni.de

© Springer Nature Switzerland AG 2021
F. Calzavarini, M. Viola (eds.), *Neural Mechanisms*, Studies in Brain and Mind 17,
https://doi.org/10.1007/978-3-030-54092-0_13

the neural would be isomorphic, or what theoretical purpose the concept is supposed to play. I set out to provide a conceptual roadmap of psychoneural isomorphism, one that can be used to dispel some misunderstandings and help the reader to frame the concept in the correct way, identifying the alleged roles of isomorphism with particular reference to contemporary debates.

In §2, I briefly reconstruct the history of our concept, locate it in the contemporary debate, and highlight its alleged roles. In §3, I make some conceptual clarifications, specifying what an isomorphism is and under what conditions we can properly speak of isomorphism. After distinguishing between an ontic and an epistemic reading, I turn to the ontic reading in §4, and on the epistemic reading in §5, taking as case-study Petitot's morphodynamical models. My contention is that while isomorphism arguably does not play the roles it has been traditionally associated with, there is an overlooked option that isomorphism might play a role in robustness analysis.

13.2 Why Psychoneural Isomorphism?

13.2.1 Historical Overview

The concept of psychoneural isomorphism was introduced in response to the nineteenth century debate about the philosophical foundations of psychology and psychophysics. In order to clarify our concept, it will be helpful to briefly sketch out the ideas of some key figures (§2.1.1) in response to which Köhler (§2.1.2) introduced the concept of isomorphism.

13.2.1.1 From Fechner to Müller

Over the course of his career, Gustav Fechner, the father of modern psychophysics, held different views. In his dissertation *Praemissae ad theoriam organismi generalem*, he stated that «parallelismus strictus existit inter animam et corpus, ita ut ex uno, rite cognito, alterum construi possit» (quoted from Heidelberger 2000, p. 53) [A tight parallelism holds between soul and body, such that from one, properly understood, the other one may be constructed][2]. This proposition anticipates his core metaphysical view, the "identity perspective" (*Identitätsansicht*) according to which the soul and the body are but aspects of the same substance[3]. This view underwent significant changes over the years. If in 1851 Fechner could still say

[2]All translations in this chapter are mine.

[3]Fechner's identity view owns a debt to Schelling, whose work exerted an influence on Fechner via Lorenz Oken's lectures, a *Naturphilosoph* whose ideas had been significantly inspired by the philosopher from Leonberg.

that the soul's and the body's processes are «im Grunde nur dieselben Processe» [basically the same processes], later Fechner drew closer to an objective idealism (*objektiver Idealismus*) according to which the fundamental layer of reality is spiritual. It was, however, ultimately his philosophical commitment to the deep unity of soul (or mind) and body that led him to articulate a mathematical approach to psychophysics.

Fechner exerted a considerable influence over his contemporaries and the younger generation of psychologists in Germany. With the exception of Helmholtz and his followers, who espoused a form of dualism, most psychologists adopted the *Identitätsansicht* as a heuristic method, i.e. as a conceptual bridge that might help investigate the brain, or sought to replace it with better conceptualizations. Mach (1865) belonged to this second group of researchers. Inspired by Fechner, he formulated a "principle of equivalence" [*Princip der Entsprechung*], according to which for every psychological event there must be a corresponding physical event and that identical psychological events must correspond to identical physical events. In the revised 1900 edition of his *Analyse der Empfindungen* (1886) he argued that the «guiding principle for the study of sensations» [*leitender Grundsatz für die Untersuchung der Empfindungen*] would be the «principle of complete parallelism of the psychical and the physical» [*Princip des vollständigen Parallelismus des Psychischen und Physischen*]. Mach meant this to be a «heuristic principle» [*heuristisches Princip*], which constitutes the «necessary presupposition of exact science» [*notwendige Voraussetzung der exakten Forschung*; 1886/1922, p. 50; in Scheerer 1994, p. 320]. Later, in the 1906 edition of his magnum opus, Mach stated to be looking for «similarity of form» [*Formähnlichkeit*] between the physical and the psychical and vice-versa (Heidelberger 2000).

In the pages of his work dedicated to the forerunners of psychoneural isomorphism, Köhler (1929) did not discuss Mach's ideas—oddly enough, since the development of Gestalt psychology was heavily influenced by Mach and Ehrenfels (Greenwood 2015, pp. 326–327). Köhler, however, discussed Hering's principle of parallelism and, most importantly, George Müller's psychophysical axioms (1896). Müller was Friedrich Schumann's teacher and mentor, who later became Carl Stumpf's collaborator and one of Wertheimer's teachers in Berlin. Schumann later moved to Frankfurt, where he hired as assistants both Köhler and Koffka. A fervent admirer of Fechner, Müller put forward five axioms with the explicit purpose of replacing the notion of psychophysical parallelism and provide a better heuristic principle (1896, p. 4). It is not possible to discuss the axioms in detail here, suffice to say that the first three axioms established a correspondence between mental, conscious states and their variations with underlying psychophysical processes.

13.2.1.2 Köhler

In his 1929 book, Köhler launched an attack against Behaviorism and pointed up the importance of first-person approaches to the study of the mind. According to Köhler, psychologists should investigate the «terra incognita» that lies between sensory stimulation and overt behavior:

> To the degree to which the interior of the living system is not yet accessible to observation, it will be our task to invent hypotheses about the events which here take place. For much is bound to happen between stimulation and response. (Köhler 1929, p. 51).

Köhler was aware of the limitations of early twentieth century neuroscience, and introduced a principle that, exploiting the dependence relation of the mind upon brain processes (ivi, p. 57), could be used to infer something about the latter given that the «[e]xperienced order in space is always structurally identical with a functional order in the distribution of the underlying brain processes», he called this principle «*psychophysical isomorphism*» (ivi, pp. 61–62). Although Köhler did not clearly define psychophysical (or "psychoneural", as it came to be called) isomorphism, he clearly understood this as a contribution to the debate sparkled by Fechner's *Identitätsansicht*.

Indeed, Köhler's terminology is often unclear. For example, in his 1938 book, he defined psychoneural isomorphism a "postulate" for the formulation of empirical hypotheses (Luccio 2010, p. 228). The terminology is unfortunate, for a postulate is a proposition that is assumed as true, but in several passages, Köhler seemed less committed towards isomorphism. In a late work, Köhler described the principle not as a postulate, but as «an hypothesis which has to undergo one empirical test after the other» (quoted in Scheerer 1994, p. 188). But Köhler also persistently confused metaphysics with heuristic assumptions, and in later works, he seemed to embrace psychoneural isomorphism for the sake of a monistic metaphysics:

> For instance, if the comparison were to show that, say, in perception, brain processes with a certain functional structure give rise to psychological facts with a different structure, such a discrepancy would prove that the mental world reacts to those brain processes as a realm with properties of its own—and this would mean dualism (Köhler 1960, quoted in Luccio 2010, p. 241; my emphasis).
>
> ... [monism] would become sensible precisely to the extent that isomorphism can be shown to constitute scientific truth. (Köhler 1960, quoted in Scheerer 1994, p. 189).

We will later (§3) examine whether psychoneural isomorphism supports a monistic account of the mind-brain relation.

13.2.2 Psychoneural Isomorphism Today

Glossing over some further developments both within and beyond Gestalt psychology[4], I want to draw the reader's attention to two recent debates in which

[4]Noteworthy is the development of second-order isomorphism that would hold between «(a) the relation among alternative external objects, and (b) the relations among their corresponding internal representations» (Shepard and Chipman 1970, p. 2). More recently, second-order isomorphisms have been exploited in research on dissimilarity matrices among internal representations (e.g.

our concept plays an important role: the problem of Naturalized Phenomenology (§2.2.1), and the search for neural correlates of consciousness (§2.2.2.).

13.2.2.1 Naturalized Phenomenology

Proponents of naturalized Phenomenology or Neurophenomenology argue that rigorous descriptions of our lived experience should complement the third-personal standpoints of the sciences of the mind since, it is argued, purely third-personal approaches cannot capture the nature of consciousness or conscious content. Such rigorous descriptions should be produced by subjects trained in Husserlian Phenomenology—with capitalized "P" to distinguish it from our conscious experience, or "phenomenology"— (Roy et al. 1999; Varela 1997).

The need for conceptually rigorous descriptions is motivated by the unreliability of naïve introspective reports (Schwitzgebel 2011; Vernazzani 2016). Rigorous phenomenological descriptions, so understood, would offer a precious help in theory construction as well as theory confirmation (Roy et al., p. 12), identifying the neural structures that might explain our conscious experience. In the current debate, these structures are known as the neural correlates of consciousness or NCCs (Chalmers 2000; Wu 2018). In order to shed light on the nature of the biological correlates of conscious experience, however, phenomenological descriptions must be naturalized. By "naturalized" Roy et al. mean «integrated into an explanatory framework where every acceptable property is made continuous with the properties admitted by the natural sciences» (1999, pp. 1–2). According to Roy et al., the naturalization can be achieved by means of mutual constraints in theory construction (cf. also Flanagan 1992, p. 11; Varela 1997). In turn, mutual constraints are made possible either by means of psychoneural isomorphism, or "generative passages" (Roy et al. 1999, p. 66–68). The former concept is quickly and obscurely dismissed, as Roy et al. suggest that «this isomorphic option makes the implicit assumption of keeping disciplinary boundaries» and urge that psychoneural isomorphism would entail some form of «psycho-neural identity» (ivi, p. 68). The latter concept makes reference to identification of «passages» that one could in principle mathematically specify to allow a transition from the "phenomenological" to the "neural." As Bayne (2004) rightly points out, however, it is far from clear whether such generative passages are substantially different, and in what sense, from a psychoneural isomorphism.

Kriegeskorte et al. 2008; Kriegeskorte and Kievit 2013). As Kriegeskorte et al. (2008, p. 4) remark, this approach is «complementary» to that of «first-order isomorphism», i.e. psychoneural isomorphism. In this contribution I will exclusively focus on psychoneural isomorphism and leave an exploration of the relations between first- and second order isomorphisms as an open avenue for future research.

13.2.2.2 The Search for Content-NCCs

The identification of neural structures or correlates related to conscious content (content-NCCs) represents a crucial goal in explaining our conscious perception. Focusing on conscious visual perception, an implicit assumption in much research is that there should be some sort of "matching" between experienced content and underlying neural representations in the content-NCCs (Crick and Koch 1995, 1998). Some researchers contend that the «proper form» (Pessoa et al. 1998, p. 726) of explanations in vision science necessarily involves talk about isomorphism between perceptual content and underlying neural processes (e.g. Fry 1948). Todorović argued that there must be an «identity of shapes of spatial distributions of percepts and the underlying neural activities» (1987, p. 548; cf. also Teller 1984).

Consider the case of the neon-color-spreading illusion, where a bright color seems to spread over a white background (cf. Bressan et al. 1997; Pessoa et al. 1998, pp. 730–731). An interesting question is whether it is a real phenomenon, or a theoretical artifact, produced by unreliable introspective reports (Dennett 1991; cf. §5.3). Assuming that it is a real phenomenon, the next question is whether a proper explanation of the neon-color spreading illusion (as well as other kinds of perceptual completion or filling-in; cf. Komatsu 2006; Pessoa and De Weerd 2003; Weil and Rees 2011) requires psychoneural isomorphism, i.e. a structural correspondence between percept and underlying neural activity. While some researchers opt for an isomorphic explanation (e.g. Von der Heydt et al. 2003, p. 107; Weil and Rees 2011, p. 41), not everyone agrees with this approach (e.g. Ratliff and Sirovich 1978), and Von der Heydt et al. contemplate an alternative explanation of filling-in of color by means of a tentative association of features by low-level mechanisms (symbolic filling-in theory).

The detailed explanation of filling-in phenomena is matter of empirical investigation, our concern here is whether, and in what sense, isomorphism plays an explanatory role, and whether we are justified in talking about isomorphism in the first place. Noë and Thompson (2004) and Thompson (2007) have argued that there is no isomorphism between perceptual content and content-NCCs, the latter being understood as receptive fields of single neurons, since perceptual content exhibits properties—such as being intentional or "world-presenting," and holistic—that the latter lack. Petitot (2008, p. 396) rejects this conclusion on the ground that Noë and Thompson have wrongly identified single cells' receptive fields as the neural correlates of perceptual content, and argues that an emergent morphology of larger populations of neurons exhibits an isomorphism with (conscious) perceptual content (§5.1).

13.2.3 The Roles of Psychoneural Isomorphism

From the foregoing cursory overview, we can identify several roles that psychoneural isomorphism is supposed to play:

- *Metaphysical role*. Several researchers think that there must be a close connection between the metaphysics of the mind-brain, and psychoneural isomorphism. Köhler and Petitot for example maintain that isomorphism supports a monistic mind-brain metaphysics, i.e. a version of the identity theory (§2.1.2)[5]. I call the claim that isomorphism supports monism "From Isomorphism to Monism" (I-M) (§4.1). Others, like Revonsuo (2000), claim that if the mental and the neural are identical, then they must be isomorphic. I call this "From Monism to Isomorphism" (§4.2).
- *Heuristic Role*. Another important role ascribed to isomorphism is as heuristic principle. Much of the debate stirred by Fechner hinges on the search for better ways to articulate a heuristic principle that could help bridging the mind-brain gap at a time when neuroscience was still in its infancy (§2.1.1). The guiding idea seems to be the following: If we do not have any kind of access to *b*, but we have access to *a*, and we know that, under some level of description, *b* is isomorphic to *a*, then studying the structure of *a* is enough to know the structure of *b*. It remains to be understood, however, whether, and to what extent, such a role might hold in the case of the mind-brain.
- *Explanatory Role*. One of neuroscientists' goals in searching for content-NCCs is explanatory (§2.2.2). As we have seen, it has been claimed that a content-NCC's neural representation must be isomorphic with the corresponding perceptual content. It is in virtue of this isomorphism that neurobiological models can be said to explain perceptual content (e.g. Noë and Thompson 2004; Pessoa et al. 1998; Roy et al. 1999). Here, the term "model" is ambiguous. On an epistemic account of scientific explanations, models are epistemic representations that might be used to explain a given phenomenon (Wright 2012; §3.3). On the ontic account, it is the very thing itself, e.g. a worldly mechanism, which is said to explain the phenomenon (Craver 2014).
- *Intertheoretic Role*. Rigorous phenomenological descriptions of perceptual content have been said to be isomorphic with models or descriptions of the underlying neurobiological activity. This is the core idea behind proponents of naturalized phenomenology's mutual constraints, which is supposed to serve in «theory confirmation» and «theory construction» (Roy et al. 1999, p. 12). Framed in these terms, psychoneural isomorphism serves an important role in intertheoretic integration, i.e. the problem of integrating different fields or theories (e.g. Darden and Maull 1977).

These different roles should be sharply distinguished. It is noteworthy that the isomorphic relata in these roles are different kinds of entities that require substantive additional qualifications. On the metaphysical role we are talking about the mind itself, whereas in its intertheoretic role we are arguably talking about models of

[5]There are of course further complications. One complication is represented by the externalist challenge, i.e. whether the brain or the "neural" is the sole substrate of the mind. Another complication is the kind of identity theory assumed, tokens or types. As we shall see (§3.3), we can put these complications aside.

Table 13.1 The different roles of psychoneural isomorphism assume ontologically distinct relata

Ontic isomorphism	Epistemic isomorphism
Metaphysical role	Explanatory (epistemic) role
Explanatory (ontic) role	Intertheoretic role
Heuristic role	Heuristic role

the mind. Accordingly, we can suggest the following partition between *ontic* and *epistemic* interpretations of psychoneural isomorphism (Table 13.1).

The ontic reading assumes that the isomorphic relata are worldly things, whereas the epistemic reading assumes the relata to be epistemic representations. The heuristic role can be placed on both sides.

In order to examine whether psychoneural isomorphism actually fulfills (some of) these roles, we must provide a robust definition of what an isomorphism is (§3.1), and then specify the nature of the relevant entities.

13.3 The Nature of Psychoneural Isomorphism

13.3.1 What Is an Isomorphism?

Unless the concept is used in a merely figurative way, the term "isomorphism" comes from mathematics[6]. More precisely, an isomorphism is a bijective homomorphism. A homomorphism is a function or map between two objects or domains that partially preserve their structures. Let us take two arbitrary domains *A* and *B* that are relational structures. A relational structure is a set *A* together with a family «R*i*» of relations on *A*. Two relational structures **A** and **B** are said to be similar if they have the same type. (I follow the convention of using a bold face **A** to refer to the relational structure, and the italics *A* to refer to the carrier set or domain). A homomorphism can be defined as follows:

> Let **A** and **B** be similar relational structures, with relations «R*i*» and «S*i*» respectively. A homomorphism from **A** to **B** is any function *m* from **A** into **B** satisfying the following condition, for each *i*: If $<a_1, a_2, \ldots a_n> \in Ri$, then $<m(a_1), m(a_2), \ldots m(a_n)> \in Si$. (Dunn and Hardegree 2001, p. 15)

[6]Brendan Ritchie has rightly pointed out to me that it is not clear whether Köhler understood isomorphism in the *mathematical* sense or rather in a more figurative and metaphorical sense. I have two replies to this. Firstly, although, as we have seen, Köhler did not give any clear definition, he meant this concept to provide a more rigorous and precise foundation than Fechner's notion of parallelism. This indicates that, arguably, he did not understood the concept as figurative or metaphorical. It should also be stressed that Köhler himself was certainly aware of the mathematical meaning of our concept, as he was trained in physics and mathematics under Max Planck. Secondly, and more importantly, the historical reception of our concept has clearly been interpreted it in the mathematical sense (e.g. Madden 1957; Lehar 2003).

B is the homomorphic image of **A** if there exists a homomorphism from **A** to **B**. Structural similarity admits different degrees. More informally, we can say that a homomorphic image **B** can be more or less structurally similar to **A**.

An isomorphism is a special kind of homomorphism. Thus, every isomorphism is also a homomorphism. For a homomorphism to be an isomorphism what is required is that the function m from **A** onto **B** must *completely* map the relational structure. In this sense, an isomorphism is a *bijective homomorphism*. A formal definition of isomorphism can now be given:

> A homomorphism h from **A** to **B** is said to be an isomorphism from **A** to **B** (between **A** and **B**) iff it satisfies the following conditions: (1) h is one-one; (2) h is onto. (Dunn and Hardegree 2001, p. 17)[7]

When a h satisfies these conditions, we may say for conciseness that **A** and **B** are *isomorphic*, or $\mathbf{A} \cong \mathbf{B}$. We can clarify our concept with the aid of an example. Consider the sequence of natural numbers $\mathbb{N}_0 = \{0, 1, 2, 3 \ldots +\infty\}$. This sequence is isomorphic to the sequence of annual time segments from 0 to positive infinity, i.e. we can specify a function from the set of annual time segments to $\mathbb{N}_0$ that is homomorphic, one-one, and onto. Formally, as we have seen, every isomorphism is a special case of homomorphism, but for clarity's sake, I will use the term "homomorphism" for functions that by definition are less then isomorphic.

What are A and B? So far, I have construed the carrier sets as domains and isomorphism as a function between domains. But an isomorphism might hold also between topological spaces (a "homeomorphism"), rings, vector spaces, categories, etc. Furthermore, notice that one can also have a homomorphic function from **A** to **A**, i.e. when the domain and its image are identical. This is an interesting case. A h from **A** to **A** is called an *endomorphism*, i.e. a homomorphism from **A** to **A**. If h is one-one and onto we get an *automorphism*, i.e. an isomorphism from **A** onto **A** (Cohn 1981, p. 49). This is important because it shows that even if domain and image are identical, the function does not *have* to be an automorphism[8]. We can illustrate this by means of an example. Consider a vector space V, an endomorphism from **V** to **V** is a linear map:

$$L : V \rightarrow V$$

An automorphism is an invertible endomorphism (an invertible homomorphism). However, if we assume a vector dimension dim V > 0, the endomorphism $L : V \rightarrow V$ with $v \mapsto 0$ will *not* be invertible, hence, it is not an automorphism, although domain and image are identical.

[7]Dunn and Hardegree define isomorphism by means of material implication. But usually, the concept is defined with a biconditional. I have rectified the quotation accordingly. Thanks to Christian Strasser for pointing this out to me.

[8]A limit case is the identity function, i.e. it is always possible to construct a function which returns the value used in the argument. Otherwise, however, the function must be specified (§4.2).

To sum up, these are the jointly necessary and sufficient requirements for an isomorphism:

(1) We must identify a domain A and its image B (the carrier sets).
(2) We must show that A and B contain elements which stand in some relation with each other, i.e. that **A** and **B** are relational structures, and what kind of structures they are.
(3) We must identify a homomorphism h from **A** to **B** which is one-one and onto.

Talk of isomorphism that fails to meet these requirements can only be understood metaphorically and will not be discussed here.

13.3.2 What Does Psychoneural Mean?

Having clarified what an isomorphism is, we must now turn to its qualification as "psychoneural." The choice of domains of isomorphism depends on our epistemic goals. The adjective "psychoneural" clearly suggests that our domain is the "psyche" or "mind" and the "neural" its image. Still, this leaves a great deal worth questioning.

Pribram (1983) stated that an isomorphism might hold between the brain and experience; or between the brain and the environment; or between all three. Arnheim argued that psychoneural isomorphism plays a fundamental role in conceptualizing the way we grasp other people's expressions (1949, pp. 58ff), with multiple domains being isomorphic. Madden (1957) distinguished between an isomorphism between stimuli and sensory responses; between receptor events and afferent neural processes; and between neural events and phenomenal (conscious) events. The latter pertains to what Fechner (1860) had called *internal psychophysics* (*innere Psychophysik*), i.e. the study of the relation between the brain and experience (*Erleben*). This is the isomorphism I will focus on, as it best captures Köhler's ideas, as well as the concept discussed in the contemporary debates, i.e. the neural correlates of consciousness and naturalized Phenomenology. Accordingly, I will have nothing to say about other putative isomorphisms, for example holding between retinal projection and the primary visual cortex V1[9].

[9] Similarly, I do not further discuss what we might call *implementational isomorphism*, i.e. the issue as of whether computational states are isomorphic to the underlying physical states or changes, for instance in the biological substrate (cf. Chalmers 2012; Miłkowski 2013; Piccinini 2015; Scheutz 2001).

There are some further clarifications in order. With reference to the discussions in §2.2.2 about naturalized Phenomenology as well as content-NCCs, it is clear that our concept is mostly discussed in relation to *conscious* visual *perceptual content*. With "consciousness" we shall understand the intrinsic or felt character of our mental lives, often characterized in terms of Nagel's construct "what-is-it-like-to-be" (1974). I shall occasionally refer to this aspect of consciousness as "phenomenology." This minimal characterization is neutral about whether there is an unbridgeable explanatory gap (Levine 1983) or whether consciousness might be reductively explained or ontologically reduced (e.g. Chalmers 1996). With "perceptual content" we shall understand, following the mainstream account of perceptual experience (e.g. Byrne 2001; Siegel 2010), the percipient's representational content at a given time[10]. I shall focus on "visual" perceptual contents, as the studies I refer to (§2 and §4) zoom in on this particular modality, but my considerations might be easily extended to all other perceptual modalities as well. Not all perceptual contents are conscious, so our first domain is the domain of a subject S's conscious perceptual content at a given time *t*. I call this the *phenomenological domain*, Ψ[11]. It is widely assumed that our conscious mental lives depend (at least partly) on some subset of neural activity. I call this subset of neural activity from which the phenomenological domain depends, the *neural image* or *domain*, ϕ. (In §4 I will return on the problem of the neural domain). I assume that the domains capture types, rather than token contents or neural structures.

Concerning the second requirement, it is assumed that the domains contain elements, and that such elements must stand in some relations with each others, i.e. these are not mere sets, but n-tuples of ordered elements. We can thus say that our domains carry, respectively, a *phenomenological relational structure* $\boldsymbol{\Psi}$ and a *neural relational structure* $\boldsymbol{\phi}$. Of course, one would have to specify such structures, but I will sidestep this issue for now. In general, determining, for instance, the nature of the phenomenological structure under examination will depend also by the specific nature of the domain and what sort of relations the elements in that domain might stand in. In order to satisfy the third requirement of isomorphism, we must specify a function *h* which is one-one and onto.

[10]Different accounts of the nature of perception may impose different constraints on the domains. Within a naïve realist account (e.g. Brewer 2011), for instance, the locution "perceptual content" does not refer to representational contents, but to the very things we are directly perceptually acquainted with. Accordingly, an isomorphism might hold between the observable aspects of things from a given standpoint, and the percipient's underlying neural activity.

[11]I assume a synchronic perspective, i.e. that of a subject ideally frozen at a time *t*. Alternatively, one could examine psychoneural isomorphism from a diachronic perspective, i.e. considering S's mental and neural states from t_1 to t_2.

13.4 Psychoneural Isomorphism and the Metaphysics of the Mind

The foregoing considerations still leave open the question of the epistemic or ontic interpretation (§2.3). A moment reflection suggests that this distinction is not completely straightforward and requires further clarifications. After analyzing the alleged metaphysical roles of isomorphism (§§4.1–2), I will argue that the ontic reading is untenable (§4.3).

13.4.1 From Isomorphism to Monism

Earlier (§2.3), I identified two distinct metaphysical roles, the first was an inference "From Isomorphism to Monism" (I-M). The idea, roughly, is that if we can specify a psychoneural isomorphism, we thereby have some evidence for the identity of these domains. An instance of this strategy can be traced back to Köhler:

> ... [monism] would become sensible precisely to the extent that isomorphism can be show to constitute scientific truth. (Köhler 1960, quoted in Scheerer 1994, p. 189; §2.1.2).

Another instance of this inference can be found in Petitot (2008) who, after showing that there is an isomorphism between morphological models *M* of the neurophysiology of the relevant functional architectures and morphological models *E* of Husserlian descriptions of the phenomenal relation between experienced space and quality (§5.1), comments that this would warrant a double-aspect theory. A double-aspect theory, in the words of Metzinger, amounts to the following claim: «[s]cientifically describing and phenomenally experiencing are just two different ways of accessing one and the same underlying reality» (2000, p. 4). In other words, Petitot thinks that if there is an isomorphism between *M* and *E*, then brain activity and the phenomenal experience must be identical (monism).

Let us first make a preliminary clarification about the nature of identity. A distinction can be drawn between two kinds of identity (Noonan and Curtis 2014): qualitative and numerical. For two things to be qualitatively identical under some respect is for them to possess the same property. Max Ernst's *L'Ange du Foyer* and Paul Nash's *Totes Meer* both share the properties of "being a painting," "being surrealist artworks", etc. Qualitative identity may be spelled out in different ways, depending on our assumptions about the metaphysics of properties (cf. Allen 2016). It is clear that two entities may be qualitatively identical with respect to some, or most, properties, without they being one and the same thing. Numerical identity is much stronger. If *a* and *b* are numerically identical it means that *a* just is *b*. Numerical identity implies total qualitative identity. The mind-brain monism presently discussed is a debate about numerical identity, whether, ultimately, the mind just *is* the brain (or, better, some subset of its neural activity). With these clarifications, we can now throw light on the inference I-M. At first, one might think that we are dealing with something like this:

> (I-M)-1: If $\boldsymbol{\Psi} \cong \boldsymbol{\phi}$, i.e. there is an h, such that h is one-one and onto between the given relational structures, then $\boldsymbol{\Psi}$ is qualitatively identical with $\boldsymbol{\phi}$.

(Recall that the boldface refers to a relational structure). Put in this way, the inference is just fine. If the two domains are isomorphic, they instantiate exactly the same mathematical, relational structure, hence, they are qualitatively identical in this respect. However, (I-M)-1 does *not* faithfully capture Köhler's and Petitot's thought, for what they refer to, when they talk about monism, is not a relation of qualitative identity with respect to relational structures, but of *numerical identity* between mind and brain, i.e. between what instantiate such structures. Hence, Köhler's and Petitot's idea might be better captured by:

> (I-M)-2: If $\boldsymbol{\Psi} \cong \boldsymbol{\phi}$, then Ψ is numerically identical with ϕ.

(Recall that the italics refer to carrier sets). Obviously, the consequent of (I-M)-2 naturally entails the following:

> If Ψ is numerically identical with ϕ, then $\boldsymbol{\Psi}$ is numerically identical with $\boldsymbol{\phi}$.

(That is, since the carrier sets are numerically identical, their relational structures must be numerically identical as well. This naturally follows from the application of Leibniz's law). (I-M)-2 is very different from (I-M)-1. The key difference is that in (I-M)-2 there is a jump from an antecedent, which expresses a mathematical function, to a consequent, which expresses a relation of numerical identity between carrier sets. There are two problems with (I-M)-2.

Firstly, the fact that there is an isomorphic function between the relevant domains does *not* justify the inference to numerical identity of the sets. Indeed, there are many examples of different domains or objects, mathematically described, which are numerically different. Put roughly, we can say that from the fact that two things instantiate the same property (e.g. "being blue") it obviously does not follow that they are numerically identical (e.g. your shirt and the sky). One can reply that my interpretation is uncharitable; perhaps, neither Köhler nor Petitot think that (I-M)-2 brings *conclusive* evidence for monism. Rather, their claims should be interpreted as saying that, *if* we could show that $\boldsymbol{\Psi}$ is not isomorphic with $\boldsymbol{\phi}$, they could not be numerically identical, again, in compliance with Leibniz's law. However, as we are about to see (§4.3), things are further compounded by multiple possible ways to mathematically describe the relevant structures.

Secondly, further reflection suggests that even (I-M)-2 does not faithfully capture Köhler's and Petitot's ideas. Let us zoom in on the consequent. The consequent expresses a relation of numerical identity between carrier sets. But carrier sets just cannot be the "neural" and the "mental," for sets are abstract mathematical entities. The carrier sets of the respective relational structures are just mathematical constructs, or, if we want, sets of symbols used to refer to worldly things in the world[12]. To make this point clear, consider the following example. Suppose you want to draw a list of all the people who sit in your living room right now. (Such a list might, of course, be empty). The list contains all and only the names of people in your living room, but the list contains obviously just names. The list might also be ordered, for example we may put the names in alphabetical order. However, what you would sort in this case are names, not real people in your living room. The most obvious implication of this problem is that talk about isomorphism is confined within mathematical entities, whereas talk about the alleged mind-brain identity refers to things in the world. I will further elaborate the consequence of this insight in §4.3.

13.4.2 From Monism to Isomorphism

A clear expression of this strategy can be found in Revonsuo:

> ... there must be isomorphism between one specific level of organization in the brain and phenomenal consciousness, simply because these boil down to one and the same thing. (2000, p. 67).

Once more, there is no further specification about the kind of identity assumed. Furthermore, Revonsuo does not discuss in what sense phenomenal consciousness would be structured, and thus fails to meet the second requirement of isomorphism (§3.1). We can abstract away from these issues, and zoom in, again, on the structure of this claim. Adopting our familiar terminology, the claim can be regimented as follows:

> (M-I): If the mind is the brain (*Monism*), i.e. Ψ is numerically identical with ϕ, then $\mathbf{\Psi}$ *must* be isomorphic with $\mathbf{\phi}$ (Automorphism).

Clearly, the implication allows for the consequent to be true even in the falsity of the antecedent: two numerically distinct things can be isomorphic. What is interesting is whether the consequent must follow from the antecedent. Now, as we have seen (§3.1), the fact that domain and image are numerically identical does not per se warrant that just any function h will be invertible, and thus an automorphism, for it is thoroughly possible that an h from $\mathbf{A}$ to $\mathbf{A}$ (or from $\mathbf{\Psi}$ to $\mathbf{\phi}$) will be a mere *endomorphism*. This was precisely the point illustrated by means of our example of endomorphism from a vector space $\mathbf{V}$ to $\mathbf{V}$. A cheap response may be that if domain and image are numerically identical, then it will always be possible to specify an identity function, i.e. a function whose output just corresponds to the input value. In such a case, however, psychoneural isomorphism will not be an interesting thesis, all it would give us is simply the value we already know. Beside the identity

[12]A further complication here is to determine which symbols stand in for worldly entities and which ones are merely internal to the representational system, but we can skip this complication here.

function, however, the exact function at stake must be further specified in order to see whether it is an automorphism or not. In other words, it is not obvious that given the numerical identity of the domains an automorphism follows. This may seem odd at first, but careful reflection suggests that the source of our puzzlement comes from mistaking the third requirement of isomorphism for a metaphysical intrinsic relation. An isomorphism, like any other morphism, is a mathematical function, it is a process we use to get an output once we fix a value chosen from the domain and as such it operates between abstract mathematical models, not things in the world.

13.4.3 Mathematical Models and Their Roles

The foregoing considerations put pressure against the ontic reading of isomorphism. Carrier sets, relational structures, and functions are abstract mathematical concepts. So, how can we make sense of psychoneural isomorphism in the first place? The short answer is, via mathematical models. Let our worldly things be the model's target. A mathematical model is an interpreted, idealized mathematical structure that stands in some representational relation to its targets and that can be studied to gain indirect insights about the targets they are about (Frigg and Hartmann 2009; Giere 1988; Weisberg 2013). It is only between such mathematical structures that we may find an isomorphism.

How should we model the targets? There are no strict rules for doing so. In general, mathematical models, just as other scientific models, are not meant to be mirror images of their targets. Models contain idealizations, abstractions (Weisberg 2007). Wisely contrived, such distortions enhance the epistemic power of our models (Elgin 2017, pp. 23–32). The way we build a mathematical model, just like any other model, and therefore what to leave out and what parameters should be idealized, depends on our *epistemic goals*. Usually, a model devised to maximize an epistemic goal does so at the expenses of other goals. Some models may have purely explorative value (e.g. Gelfert 2016 and below), others may maximize descriptive accuracy while having little predictive power, whereas other models provide scientific explanations. Scientific models play many other roles as well, but we will just focus on a basic distinction that is later going to play an important role (§§5.2–3). Some models play *explanatory* roles, others do not. Models of the latter kind are often called "phenomenological" (Frigg and Hartmann 2009; Hochstein 2013; Wimsatt 2007), but in order to avoid confusions with other uses of the term "phenomenology," I shall simply call them *non-explanatory* models.

Non-explanatory models have different uses. Batterman for instance examined the role of minimal models—i.e. highly idealized models—in statistical mechanics, and concludes that the best way to think of their role is «that they are means for extracting stable phenomenologies [i.e. regularities] from unknown and, perhaps, unknowable detailed theories» (2002, p. 35). Such regularities may then be used for computational or explanatory purposes (ivi, p. 37). Another fitting example comes from Bogen (2005). Following Mitchell's contention that the role of scientific generalizations «is to provide reliable expectations of the occurrence of events and patterns of properties» (2003, p. 124), Bogen argues that such models may be used to:

- Describe facts to be explained;
- Suggest constraints on acceptable explanations;
- Suggest and sharpen questions about causal mechanisms;
- Measure or calculate quantities;
- Support inductive inferences (2005, p. 401).

As an example, he considers the famous Hodgkin-Huxley equations of action potential, the pulse of electricity that traveling down the axon towards the synapse triggers the release of neurotransmitters. Studying the squid giant axon, Hodgkin and Huxley argued that the magnitude of the potassium current I_K which help repolarize the membrane varies with $\overline{g}_K$, the membrane's maximum potassium current conductance, a weighting factor (n^4), and a driving electrical force equal to the difference between E_m, the membrane potential, and the resting potential for potassium, E_k. The equation for the potassium is:

$$I_K = n^4 \, \overline{g}_K \; (E_m - E_K)$$

As Bogen argues, this equation incorporates the «qualitatively correct idea that I_K varies with $(E_m - E_K)$», yet he specifies that this model is also «quantitatively inaccurate to a significant degree». But in spite of its inaccuracies and poor *predictive* power, the model has played an important role for studying action potential. The mechanism governing action potential was, at that time, still unknown, and Hodgkin and Huxley meant their equations to be «empirical descriptions» of the target phenomenon (Hodgkin and Huxley 1952, p. 541; quoted from Bogen 2005, p. 404). The model served a useful *exploratory* role, describing the behavior of the phenomenon and, as Bogen says, indicated the «features of the phenomena of interest which mechanistic explanations should account for» (ivi, p. 403; cf. also Gelfert 2016, pp. 79–97).

The fact that different models embody different epistemic purposes directly bears on the case of psychoneural isomorphism, for when we construct a mathematical model our epistemic goals will determine which mathematical structure will be relevant, and accordingly, whether two models will be isomorphic or not. We can illustrate this point with an example. Suppose we take your coffee mug on the desk and that donut you bought for breakfast. How should the mathematical models capture the targets' structures? This depends on our epistemic goals. Within a classical geometrical model, clearly, the mug and the donut (a torus) do not have the same structure, hence our models will carry relational structures which clearly are not isomorphic. However, if we are interested in topological spaces (Munkres 2000, p. 76), things will be very different. It is one of the best-known examples in topology that a coffee mug is homeomorphic (i.e. topologically isomorphic, §3.1) to a torus (the donut). Of course, it depends on our epistemic goals what mathematical model we will have to construct: exploratory, descriptive, explanatory, etc. (We will further explore these considerations in §5), and in turn this will determine whether our models will be isomorphic or not.

Time to take stock, the correct analysis of psychoneural isomorphism must be an epistemic reading where mathematical models should be sharply distinguished from their targets. Let our targets be, as we have seen, S's conscious visual content $\mathfrak{C}$ at *t*, and the underlying neural structure $\mathfrak{N}$ that sustains it at the same time. A mathematical model E of $\mathfrak{C}$ will specify the carrier set Ψ together with a relational structure $\boldsymbol{\Psi}$ for epistemic purpose P; a mathematical model M of $\mathfrak{N}$ will specify a carrier set ϕ together with a relational structure $\boldsymbol{\phi}$ for an epistemic purpose P′. In the next section, I will focus on the alleged roles of isomorphism within the epistemic reading, focusing on Petitot's morphodynamical approach.

13.5 From Morphodynamics to Robustness

In this section, I will mainly focus on the work of the French mathematician and philosophers Jean Petitot. The are two reasons for this. Firstly, because he has provided the single most developed *mathematical* account of psychoneural isomorphism. Second, because in such an account Petitot has embraced all putative roles of psychoneural isomorphism, identifying his contribution as both in the project of naturalized Phenomenology (§2.2.1) and the search for the neural correlates of conscious content (§2.2.2). Thus, his work provides an ideal case-study for my purposes. Setting the ontic reading aside, we will now look closer (§5.1) at Petitot's approach and examine the alleged epistemic roles of isomorphism (§§5.2–3).

13.5.1 Petitot's Neurogeometry

Petitot is mainly interested in specifying the neurogeometry of the functional architecture of visual areas; more precisely the problem is the:

> …implémentation neuronale des algorithms de cette géométrie, le problem étant de comprendre comment les structures perceptives "macro" et leur morphodynamique peuvent émerger du niveau "micro" neuronal sous-jacent (2008, p. 22).
>
> [the neural implementation of this geometry's algorithms, the problem being that of understanding how the "macro" perceptive structures and their morphodynamics can emerge from the underlying "micro" neural level.]

This proposition is inscribed within the larger project of naturalizing Phenomenology (§2.2.1) in which Petitot occupies a unique position, having developed the best-articulated mathematical account of the program. According to the programmatic statements in Roy et al. (1999), the ontological divide between the "mental" and

Table 13.2 Conceptual and geometrical eidetics

Conceptual eidetics	Geometrical eidetics
Phenomenological descriptions	Morphodynamical models

the "physical" can be bypassed by means of a mathematatization of the two[13]. The mathematization, so understood, represents a way of naturalizing Phenomenology, i.e. a way of making Phenomenology continuous with the natural sciences. The mathematization follows these steps:

1. Phenomenological descriptions.
2. Mathematical models.
3. Naturalistic model.

The passage from (1) to (2) is achieved via a "theory-theory" or "model-model" «exact correspondence» (Petitot 1999, p. 343; §5.2). Phenomenological descriptions are conceptual, whereas the corresponding geometrical eidetics is expressed by means of a morphodynamical model (2008, p. 395). We thus obtain a schema captured in Table 13.2.

Let us consider an example. Petitot focuses on Husserl's insightful discussion of phenomenal saliency (*phänomenale Abhebung*), which enables the individuation of phenomena, i.e. appearances (cf. Husserl 1993, pp. 242–245). Phenomenal saliency is made possible by means of a distinction between distinct (*gesondert*) contents and fused (*verschmolzen*) contents. The process of fusion (*Verschmelzung*) creates a phenomenal whole; whereas the opposite process of distinction (*Sonderung*) demarcates the different parts. The *Sonderung* is based on the qualitative discontinuity of the "moments"—roughly, particularized properties—that compose the contents[14]. In short, the structure of visual appearances is based on the qualitative discontinuities between moments. These concepts, however, are not continuous with the concepts employed by neuroscientists in their researches, this is precisely why we need to bridge this conceptual gap.

Petitot argues that a mathematical translation of Husserl's analysis is possible (step 1–2). The relation between quality (e.g. color) and space corresponds to a (mathematical) category (Petitot 1999, p. 339; cf. also 1993, 2011, pp. 64–65), and in particular to a fibration or fibred space. A fibration is a differentiable manifold E endowed with a canonical projection $\pi : E \to M$ (a differentiable map) over another manifold M. $E_x = \pi^{-1}(x)$ of the points $x \in W$ by π are the "fibres" of the fibration, subspaces of E that are projected to points in M. A fibration must satisfy two axioms:

1. All the fibres E_x are diffeomorphic with a typical fiber F.

[13]It is a separate and interesting question to assess whether Phenomenology may be mathematized (e.g. Zahavi 2004).

[14]Mulligan (1999) has argued that Husserl's moments are trope-like entities.

2. The projection π is locally trivial, i.e. for every $x \in M$, there exists a neighborhood U of x such that the inverse image $E_U = \pi^{-1}(U)$ of U is diffeomorphic with the direct product $U \times F$ endowed with the canonical projection $U \times F \to U$, $(x, q) \to x$.

(A diffeomorphism is an isomorphism between manifolds, a topological space). This mathematical model would capture the relation between quality and extension in Husserl's Phenomenology (Husserl 1991, pp. 68–71; Petitot 2004). The base of the fibration is the extension and the total space is a sensible quality, say, color. With this mathematical model of $\mathfrak{C}$ (perceptual content, or better, an aspect of it) inspired by Phenomenological concepts, we have specified a carrier set and relational structure. We now need to move from step 2 to step 3. More precisely, we need to pin down some physical-mathematical model of the neural dynamics that implements the geometric description (Petitot 1999, pp. 338–343; 2008, pp. 380–381).

Petitot argues that one of the main problems of natural and computer vision is to understand «how signals can be transformed into geometrically well-behaved observables» (1999, p. 346), i.e. the process whereby an unstructured image I(x, y) becomes segmented. Perhaps the most widespread mathematical model for segmenting an image into distinct parts has been developed by Mumford (1994), and it is known as the Mumford-Shah model. There are alternative models as well, more local and based on anisotropic non-linear partial differential equations (Petitot 1999, p. 348; 2011, pp. 78ff). (I skip the mathematical details, the reader interested can find them in Petitot 1999, 2008, 2011, 2013). The relevant point is that the same fibration used to model the Phenomenological descriptions can be used to model the neurogeometry of the functional architecture of V1, the primary visual cortex. More specifically, Petitot develops his account basing on Hubert & Wiesel's (Bechtel 2001, pp. 232–234) discovery of the micromodules called hypercolumns (Petitot 2008, 2013, p. 75)[15]. We have thus achieved a genuine *psychoneural isomorphism* that respects all three requirements (§3.1):

> [. . .] l'accord entre le macro-niveau géométrique (morphologique) émergent M [...] et l'expérience phénoménale E [...] est extrêmement fort, beaucoup plus fort qu'une simple corrélation. C'est même la forme la plus forte possible de matching de contenus puisque, à la limite, c'est un isomorphisme» (2008, p. 367; emphasis added).
>
> [The matching between the emergent geometrical macro-level M (morphology) [. . .] and the phenomenal experience E [. . .] is very strong, much stronger than a simple correlation. It is the stronger possible kind of content matching since, at its limit, it is an isomorphism.]

What is, however, the epistemic achievement of such an isomorphism?

Firstly, Petitot contends that «[w]ith such a morphodynamical model we can easily *explain* the topological description physically» (2011, p. 69; emphasis added).

[15]Petitot points out that Noë and Thompson (2004)'s negative assessment of psychoneural isomorphism is largely based on their mistaken assumption that single cells would be the neural correlates of perceptual content. Petitot's neurogeometry is based instead on a morphodynamical analysis of larger population of neurons.

The model, apparently, extends its explanatory virtue also to the problem of subjective contours (the Kanizsa triangle, for example) or phenomena like the neon color spreading (§2.2.2.), the subjective impression of a color spreading across the four circles represented in the Neon Color Spreading (cf. Petitot 2003, 2013, pp. 81ff). In short, such models would explain «the structure of percepts» (Petitot 2013, p. 75). A mathematical (topological) or, as I shall say, following Haugeland (1998), *morphological* explanation ensues in virtue of the isomorphism. Morphological explanations are explanations «where the distinguishing marks of the style are that an ability is explained through appeal to a specified structure and to specified abilities of whatever is so structured» (ivi, p. 12). This corresponds to the *Explanatory Role* (§2.3. Table 13.1).

Secondly, Petitot contends that the mathematized Phenomenological descriptions enable us to bridge the conceptual gap between disciplines, using the first-person descriptions as constraints on the admissible naturalistic explanations and models (1999, p. 330). This contention exemplifies two further roles of isomorphism. The first is its *intertheoretic role*, i.e. the problem of showing how different disciplines interact (say, psychology and neuroscience). The second is the *heuristic* role since with the aid of mathematical models of first-person contents, it is claimed that we can guide the search for the underlying neural structures.

13.5.2 The Epistemic Roles of Psychoneural Isomorphism

I lump together the Intertheoretic and the heuristic role in §5.2.1; I turn then to the explanatory role in §5.2.2.

13.5.2.1 Intertheoretic and Heuristic Role

Coherently with the project of naturalizing Phenomenology, Petitot conceives phenomenological descriptions as playing a heuristic role in the search for content-NCCs. This would be possible in virtue of the isomorphism between models of neural activity and models of phenomenological descriptions. The relation between different models or theories—sometimes between different levels of description[16]—is known as the problem of *intertheoretic relations* (e.g. Danks 2014)[17]. The classical account of intertheoretic relation is *intertheoretic reduction* (e.g. Schaffner 1993).

[16]Some reductions are intra-level, as in the case of theories or models within the same level of description; our focus here is on models that belong to two different levels of description, i.e. the experienced or phenomenological, and the neural (Nickels 1973).

[17]Much of the philosophical literature has focused on relations between theories, but the same considerations apply to models as well.

Intertheoretic reduction follows the schema of deductive-nomological explanations (DN). On this account, scientific explanations are deductive arguments in which the explanandum features as the conclusion and among the premises there must be at least one law of nature (Hempel and Oppenheim 1948; Salmon 1989). An application of this model to intertheoretic reduction in neuroscience and its relation to DN explanations can be found in Churchland (1986). Such a reduction is a relation that obtains between models (or theories). Model reduction is sometimes thought to lead to ontological reduction of the models' targets. In order to clarify Churchland's thesis, and its relation with our main point, let us examine a classical example. Nagel (1961, pp. 338–345) distinguished between two types of reduction: homogeneous and heterogeneous. In the former case, the "primary" theory (reducing) and the "secondary" theory (reduced) share the same vocabulary. This allows a simple reduction where the reduced theory features as conclusion of a deductive argument. Things are different in the case of heterogeneous reduction. The classical example is the reduction of thermodynamics to statistical mechanics (ivi, pp. 339–345), where the former contains terms and concepts such as "temperature" that are absent in the primary theory. Without terminological consistency, it is not possible to establish a logical-deductive relation. In order to overcome this obstacle, Nagel (ivi, 353–354) introduced a «condition of connectability» that bridges the gaps between primary and secondary theory.

Back to Churchland. The reduction of the mental vocabulary to a neural vocabulary poses an obvious challenge of heterogeneity. In order to bridge the terminological and conceptual gap of the mental vocabulary within theory T_F, however, we can create a an isomorphic model of T_F, call it T_F*, which in turn can be deduced from a theory of neural processes T_N following the schema:

$$T_N \rightarrow T_F^* \cong T_F$$

(The arrow here does not represent the logical connective of material implication, but a deducibility relation). My contention is that Petitot's approach is strikingly similar to Churchland's. In Petitot's case a phenomenological, conceptual description D of $\mathfrak{C}$ (perceptual content), serves as base for creating a mathematical model E, which specifies a carrier set Ψ and a relational structure $\boldsymbol{\Psi}$. This is roughly an equivalent to the right hand side of Churchland's schema. In addition, from $\mathfrak{N}$ we get a mathematical model M that specifies a carrier set ϕ together with a relational structure $\boldsymbol{\phi}$. We thus obtain an isomorphism $\boldsymbol{\phi} \cong \boldsymbol{\Psi}$ of the respective neural and phenomenological models.

Can this account for the *heuristic* role of phenomenological descriptions? The short answer is "no", and there are two main reasons for this. First, because this isomorphism has a *reconstructive* character. The relevant neural structures underlying conscious perception must have been previously singled out in order for us to build a mathematical model thereof. The discovery of such structures does not rely on isomorphism or mathematical models, but is mostly achieved via careful selective interventions (Craver 2007; Woodward 2003) that uncover the constitutive or causally relevant components of the target system that bring about a change in the phenomenon, i.e. in our case conscious perceptual content. Second, because the isomorphism holds only between very specific mathematical models, and not between any mathematical model of the neural structures' activities or of perceptual content. And the choice of models, as we have seen (§4.3), largely depends on our epistemic goals. In general, and most of the time, cognitive scientists rely on a plurality of different models that serve different epistemic purposes and that target different facets of the phenomenon.

It may be argued that the approach vindicates the intertheoretic role of isomorphism. After all, as Petitot, Varela, Roy et al., and Köhler insisted (§2.1.2; §2.2.1), phenomenological descriptions are meant to deepen our understanding of how the brain works from a rigorous first-person perspective by putting constraints on models of the neural. Yet, there is a tension between Petitot's approach and the claim that phenomenological descriptions should put constraints on neural models. It lies in the fact that intertheoretic constraints are usually conceived as a better alternative to intertheoric reduction (e.g. Craver 2005, 2007, pp. 256ff; Danks 2014). While a lengthy discussion of this issue must be postponed to another contribution for reasons of space (cf. Vernazzani 2016), the following observation by Craver nicely summarizes the core issue: neuroscientists do not «create a homomorphic image of a phenomenon studied by those in another field» (2007, p. 266). The price of intertheoretic reduction is abstracting away from current neuroscience practice to achieve some sort of normative ideal, one that, perhaps, better suits more abstract epistemic purposes, like the quest for the unity of science (Oppenheim & Putnam 1958). As a regulative ideal, however, such intertheoretic reduction flies in the face of more local approaches.

13.5.2.2 A Morphological Explanation?

Petitot maintains that his account provides an explanation of perceptual content's structure. In his 2008, he even claims that the isomorphism between **M** and **E** bridges the explanatory gap (Levine 1983), showing why perceptual content has its consciously "felt" character.

The structure of the specific model of scientific explanation, however, is far from clear. Petitot seems to suggest that explanation has to do with deduction or derivation (Petitot 2008, p. 31). This draw his account close to what Haugeland called the «derivational-nomological» style of explanation, i.e. a «special case form of deductive-nomological explanation—where the distinction of the special case is that the presupposed regularities are expressed as equational relationships among quantitative variables, and the deduction is mathematical derivation of other such equations» (1998, p. 11; cf. §5.2.1). However, he also characterizes his approach as a kind of mathematical explanation or, to use Haugeland's terminology again, as a «morphological explanation» whose hallmark is that «an ability is explained through appeal to a specific structure and to specified abilities of whatever is so structured» (1998, p. 12). This is particularly clear in his characterization of the fibration structure of perceptual content as deriving mathematically from the structure of the underlying neurogeometry of V1 (Petitot 2013, 2008).

A first problem with Petitot's explanatory ambitions consists in the ambiguous nature of the explanandum. Put more informally, is the morphodynamical account supposed to provide an answer to the question "Why does the brain produce that particular perceptual content's structure?" or is it "Why do hypercolumns's neural activity mathematically necessitates that particular perceptual structure?" or, again, "Why is that particular perceptual content's structure conscious?" These are only a few possible questions; the point is that different ways of singling out the explanandum will require different explanations. Furthermore, whether isomorphism provides an explanation or not will also depend on the norms of explanation accepted within the given scientific community relative to a given account of explanation. For instance, in the case of neuroscience and cognitive science, several researchers (e.g. Bechtel and Richardson 2010; Craver 2007; Craver and Darden 2013; Miłkowski 2013) have convincingly argued that explanation is often (if not always) construed as a search for mechanisms, i.e. structured entities whose activity constitutes or causes the explanandum phenomenon. On this perspective, a mechanistic explanation of x will consist in determining all (and only) the constitutively or causally relevant components parts and operations of the responsible mechanism. Such explanations admit degrees of completeness, ranging from merely how-possibly sketches to how-actually, complete descriptions for the specific explanatory purpose (Craver and Darden 2013).

Morphological explanations' relation with mechanistic accounts of explanation has recently drawn some attention (e.g. Huneman 2018; Lange 2013; Levy and Bechtel 2013; Rathkopf 2015). My contention is that the explanatory role of isomorphism is unclear. Consider again our example of the mug and the donut: they are homeomorphic, but so what? No one would conclude that one thing explains the other or its structure. In our case, Petitot's model of the hypercolumns' activity may be a mathematically impressive achievement, but it does not help *explain* how the relevant neural structures' computations achieve this, not even if one assumes that they are numerically identical entities. In order to do so, one would have to clarify the nature of the relevant neural parts and operations and their organization. This brings us to a second problem.

As we have seen (§3.3), models are devised to serve some epistemic goal. In several passages, Petitot (2008) contends that his mathematical translation of phenomenological descriptions (steps 1–2) conforms to a rigorous specification of Marr's *computational* level. Marr described the scientists' task at the computational level as «an abstract formulation of what is being computed and why» (1977, p. 37; cf. also Marr 2010). The exact interpretation of Marr's computational level has been object of intense debate among philosophers and cognitive scientists (e.g. Shagrir and Bechtel 2017). According to some interpreters, the computational level consists solely in the specification of the task to be solved by the information-processing system (ivi, pp. 193–194). Egan (1991, 1995) has put forward an interpretation of Marr's computational level that consists in the mathematical specification of the function(s) computed (1995, p. 185). Petitot's own approach can be understood along the lines of Egan's interpretation, as his model provides a rigorous mathematical characterization of the problem to be solved, i.e. understanding how the neural system carries out the targeted function (Petitot 2008, p. 22). Put in these terms, however, and without disputing Marr's exegesis, the morphodynamical model E of perceptual content (its target) is a "non-explanatory" model that provides a mostly accurate *mathematical description* of the target.

Let us now turn to the mathematical model of the underlying neural activity, M. Such a model, once more, does not specify how the target neural structure actually performs the computations, but provides a mostly accurate mathematical model of the neural structure's activity as a whole. The isomorphism between M's and E's relational structures, in short, does not seem to embody any explanatory epistemic goal. This is not to say that M and E are theoretically idle, they might play a variety of different non-explanatory roles. I now turn to such roles.

13.5.3 Psychoneural Isomorphism and Robustness

In every explanation, including of course mechanistic explanation, the characterization of the explanandum phenomenon plays a central role (e.g. Bechtel 2013; Shagrir & Bechtel 2017). How to correctly characterize or describe conscious phenomena is anything but simple. One option is to construe $\boldsymbol{\phi} \cong \boldsymbol{\Psi}$ as a step in the phenomenal stabilization or robustness of the explanandum (e.g. Feest 2011; Wimsatt 2007). The notion of stabilization refers to:

- The processes and methods whereby scientists empirically identify a given phenomenon, and
- Gradually come to agree that the phenomenon is a stable, and robust feature of the world, rather than an artifact produced by an instrument, methodological assumptions and procedures, etc.

One (or perhaps, the *only*) way to determine the robustness of a given phenomenon is by means of multiple determinations in its identification, i.e. the convergent results between different methodologies or levels of analysis (cf. Hacking 1983; Wimsatt 2007, pp. 37–74). The need for robustness regarding mental perceptual phenomena, such as filling-in (§2.2.2) is motivated by the unreliability of first-person, naïve descriptions of one's own perceptual experiences (Dennett 1991; Schwitzgebel 2011). Notice that this was the very same motivation that grounds the introduction of rigorous first-person methodologies at the base of the project of naturalized Phenomenology (§2.2.1). Within this context the isomorphism $\boldsymbol{\phi} \cong \boldsymbol{\Psi}$—if achieved through independent modeling of each of the two targets—may provide evidence for the robustness of the phenomenon under scrutiny. It may do so by dint of the independent achievement of the same descriptions of the target phenomenon relying on different sources of data. Surely, the isomorphism may not be (and we should not expect it to be) the only method for achieving phenomenal stabilization, but it may represent a new and helpful conceptual tool in stabilizing the target phenomenon. This further role deserves to be further examined in subsequent studies.

13.6 Conclusion

In this contribution, I have provided a taxonomy of the putative roles of isomorphism and dispelled some misunderstandings regarding the relation between isomorphism and the metaphysics of the mind. I have urged that psychoneural isomorphism is better understood as a function between mathematical models, and stressed the importance of multiple, not always compatible, epistemic roles that such models embody. I have later reviewed the epistemic roles of psychoneural isomorphism, using the work of Jean Petitot as a case-study. While my conclusion has been largely negative, I hinted at a possible role of psychoneural isomorphism in phenomenal stabilization of perceptual content that should be object of further studies.

Acknowledgments For financial support, I would like to thank the Barbara-Wengeler-Stiftung and the Volkswagenstiftung's project "Situated Cognition. Perceiving the World and Understanding other Minds" led by Tobias Schlicht. For comments on earlier versions, many thanks also to Marcin Miłkowski, Albert Newen, Marco Viola, Fabrizio Calzavarini, Krys Dołega, Judith Martens, Elmarie Venter, Luke Roelofs, Antonio Piccolomini d'Aragona, Francesco Marchi, Brendan Ritchie, and Franceso Altiero.

References

Allen, S. (2016). *A critical introduction to properties*. London/New York: Bloomsbury.
Batterman, R. (2002). Asymptotics and the role of minimal models. *British Journal for the Philosophy of Science, 53*, 21–38.
Bayne, T. (2004). Closing the gap? Some questions for neurophenomenology. *Phenomenology and the Cognitive Sciences, 3*(4), 349–364.
Bechtel, W. (2001). Decomposing and localizing vision: An exemplar for cognitive neuroscience. In W. Bechtel, P. Mandik, J. Mundale, & R. Stufflebeam (Eds.), *Philosophy and the neurosciences* (pp. 225–249). Malden: Blackwell.
Bechtel, W. (2013). Understanding biological mechanisms: Using illustrations from circadian rhythm research. In K. Kampourakis (Ed.), *The philosophy of biology* (pp. 487–510). Dordrecht: Springer.
Bechtel, W., & Richardson, R. (2010). *Discovering complexity*. Cambridge, MA: MIT Press.
Bogen, J. (2005). Regularity and causality: Generalizations and causal explanation. *Studies in the History and Philosophy of Biology and Biomedical Science, 36*, 397–420.
Bressan, P., Mingolla, E., Spillmann, L., & Watanabe, T. (1997). Neon color spreading: A review. *Perception, 26*, 1353–1366.
Brewer, B. (2011). *Perception and its objects*. Oxford: Oxford University Press.
Bridgeman, B. (1983). Isomorphism is where you find it. *Behavioral and Brain Sciences, 6*, 658–659.
Byrne, A. (2001). Intentionalism Defended. *The Philosophical Review, 110*(2), 199–240.
Chalmers, D. (1996). *The conscious mind*. New York: Oxford University Press.

Chalmers, D. (2000). What is a neural correlate of consciousness? In T. Metzinger (Ed.), *Neural correlates of consciousness* (pp. 17–39). Cambridge, MA: MIT Press.
Chalmers, D. (2012). A computational Foundation for the Study of cognition. *Journal of Cognitive Science, 12*, 323–357.
Churchland, P. S. (1986). *Neurophilosophy*. Cambridge, MA: MIT Press.
Cohn, P. M. (1981). *Universal Algebra*. Dordrecht: Reidel.
Craver, C. (2005). Beyond reduction: Mechanisms, multifield integration, and the Unity of neuroscience. *Studies in History and Philosophy of Biology and Biomedical Sciences, 36*, 373–395.
Craver, C. (2007). *Explaining the brain*. New York: Oxford University Press.
Craver, C. (2014). The ontic account of scientific explanation. In M. I. Kaiser, O. R. Scholz, D. Plenge, & A. Hüttermann (Eds.), *Explanation in the special sciences* (pp. 27–52). Springer Verlag.
Craver, C., & Darden, L. (2013). *Search of mechanisms*. Chicago: University of Chicago Press.
Crick, F., & Koch, C. (1995). Are we aware of neural activity in primary visual cortex? *Nature, 375*, 121–123.
Crick, F., & Koch, C. (1998). Consciousness and neuroscience. *Cerebral Cortex, 8*, 97–107.
Danks, D. (2014). *Unifying the mind*. Cambridge, MA: MIT Press.
Darden, L., & Maull, N. (1977). Interfield theories. *Philosophy of Science, 44*(1), 43–64.
Dennett, D. (1991). *Consciousness explained*. Boston: Little, Brown, & co.
Dunn, M. J., & Hardegree, G. (2001). *Algebraic methods in philosophical logic*. Oxford: Clarendon Press.
Egan, F. (1991). Must psychology be individualistic. *Philosophical Review, 100*, 179–203.
Egan, F. (1995). Computation and content. *Philosophical Review, 104*, 181–203.
Elgin, C. (2017). *True enough*. Cambridge, MA: MIT Press.
Feest, U. (2011). What exactly is stabilized when phenomena are stabilized? *Synthese, 182*, 57–71.
Flanagan, O. (1992). *Consciousness reconsidered*. Cambridge, MA: MIT Press.
Frigg, R., & Hartmann, S. (2009). Models in Science. In E. Zalta (Ed.), *Stanford Encyclopedia of philosophy*. https://plato.stanford.edu/archives/spr2017/entries/modelsscience/
Fry, G. A. (1948). Mechanisms subserving simultaneous brightness contrast. *American Journal of Optometry and Archives of American Academy of Optometry, 25*(4), 162–178.
Gelfert, A. (2016). *How to do science with models*. Dordrecht: Springer.
Giere, R. (1988). *Explaining science*. London: University of Chicago Press.
Greenwood, J. (2015). *A Conceptual History of Psychology. Second edition*. Cambridge: Cambridge University Press.
Hacking, I. (1983). *Representing and intervening*. Cambridge, MA: Cambridge University Press.
Haugeland, J. (1998). *Having thought*. Cambridge, MA: Harvard University Press.
Heidelberger, M. (2000). Fechner und Mach zum Leib-Seele-Problem. In A. Arndt & W. Jaeschke (Eds.), *Materialismus und Spiritualismus: Philosophie und Wissenschaft nach 1845* (pp. 53–67). Hamburg: Meiner.
Hempel, C. G., & Oppenheim, P. (1948). Studies in the logic of explanation. *Philosophy of Science, 15*(2), 135–175.
Hochstein, E. (2013). Intentional models as essential scientific tools. *International Studies in the Philosophy of Science, 27*(2), 199–217.
Huneman, P. (2018). Diversifying the picture of explanations in biological sciences: Ways of combining topology with mechanisms. *Synthese*. https://doi.org/10.1007/s11229-015-0808-z.
Husserl, E. (1991). *Ding und Raum*. Hamburg: Felix Meiner Verlag.
Husserl, E. (1993). *Logische Untersuchungen, Vol. II/1*. Tübingen: Max Niemeyer Verlag.
Köhler, W. (1929). *Gestalt psychology*. Oxford: Liverlight.
Komatsu, H. (2006). The neural mechanisms of perceptual filling-in. *Nature Reviews Neuroscience, 7*, 220–231.
Kriegeskorte, N., & Kievit, R. A. (2013). Representational geometry: Integrating cognition, computation, and the brain. *Trends in Cognitive Sciences, 17*(8), 401–412.

Kriegeskorte, N., Marieke, M., & Bandettini, P. (2008). Representational similarity analysis – connecting the branches of systems neurosciences. *Frontiers in Systems Neurosciences*. https://doi.org/10.3389/neuro.06.004.2008.
Lange, M. (2013). What makes a scientific explanation distinctively mathematical? *British Journal for the Philosophy of Science, 64*, 485–511.
Lehar, S. (1999). Gestalt isomorphism and the quantification of spatial perception. *Gestalt Theory, 21*(2), 122–139.
Lehar, S. (2003). Gestalt isomorphism and the primacy of the subjective conscious experience: A gestalt bubble model. *Behavioral and Brain Sciences, 26*, 375–444.
Levine, J. (1983). Materialism and qualia: The explanatory gap. *Pacific Philosophical Quarterly, 64*, 354–361.
Levy, A., & Bechtel, W. (2013). Abstraction and the Organization of Mechanisms. *Philosophy of Science, 80*, 241–261.
Luccio, R. (2010). Anent isomorphism and its ambiguities: From Wertheimer to Köhler and Back to Spinoza. *Gestalt Theory, 32*(3), 208–234.
Mach, E. (1865). Über die Wirkung der räumlichen Vertheilung des Lichtreizes auf der Netzhaut. *Sitzungsberichte der kaiserlichen Akademie der Wissenschaften, Mathematisch-naturwissenschaftliche Classe, 52*(2), 303–322.
Madden, E. H. (1957). A logical analysis of 'psychological isomorphism'. *The British Journal for the Philosophy of Science, 8*, 177–191.
Marr, D. (1977). Artificial Intelligence: A Personal View. *Artificial Intelligence, 9*, 37–48.
Marr, D. (2010). *Vision*. Cambridge, MA: MIT Press.
Miłkowski, M. (2013). *Explaining the computational mind*. Cambridge, MA: MIT Press.
Mitchell, S. (2003). *Biological complexity and integrative pluralism*. Cambridge: Cambridge University Press.
Müller, G. (1896). Zur Psychophysik der Gesichtsempfindungen. Kap. 1. *Zeitschrift für Psychologie und Physiologie der Sinnesorgane, 10*, 1–82.
Mulligan, K. (1999). Perception, particulars and predicates. In D. Fisette (Ed.), *Consciousness and intentionality* (pp. 163–194). Dordrecht: Kluwer.
Mumford, D. (1994). Bayesian rationale for the Variational formulation. In B. M. ter Haar Romney (Ed.), *Geometry-driven diffusion in computer vision* (pp. 135–146). Dordrecht: Kluwer Verlag.
Munkres, J. (2000). *Topology*. Uppder Saddle River: Prentice Hall.
Nagel, E. (1961). *The structure of science*. New York: Harcourt, Brace & World.
Nickles, T. (1973). Two concepts of Intertheoretic reduction. *The Journal of Philosophy, 70*(7), 181–201.
Noë, A., & Thompson, E. (2004). Are there neural correlates of consciousness? *Journal of Consciousness Studies, 11*(1), 3–28.
Noonan, H., & Curtis, B. (2014). Identity. In E. Zalta (Ed.), *The Stanford Encyclopedia of philosophy*. https://plato.stanford.edu/archives/sum2014/entries/identity/
O'Regan, K. (1992). Solving the 'real' mysteries of visual representations: The world as an outside memory. *Canadian Journal of Philosophy, 46*(3), 461–488.
Oppenheim, P., & Putnam, H. (1958). Unity of science as a working hypothesis. *Minnesota Studies in the Philosophy of Science, 2*, 3–36.
Palmer, S. (1999). Color, consciousness, and the isomorphism constraint. *Behavioral and Brain Sciences, 22*, 923–989.
Pessoa, L., & De Weerd, P. (Eds.). (2003). *Filling-in: From perceptual completion to cortical reorganization*. Oxford: Oxford University Press.
Pessoa, L., Thompson, E., & Noë, A. (1998). Finding out about filling-in: A guide to perceptual completion for visual science and the philosophy of perception. *Behavioral and Brain Sciences, 21*, 723–802.
Petitot, J. (1992–1993). Phénoménologie naturalisée et morphodynamique: La fonction cognitive du synthétique 'a priori'. *Intellectica, 17*, 79–126.
Petitot, J. (1999). Morphological eidetics for a phenomenology of perception. In Petitot et al. (pp. 330–371).
Petitot, J. (2003). Neurogeometry of V1 and Kanizsa contours. *Axiomathes, 13*, 347–363.

Petitot, J. (2004). Géométrie et vision dans 'Ding und Raum' de Husserl. *Intellectica, 2*, 139–167.
Petitot, J. (2008). *Neurogéométrie de la vision*. Paris: Les Editions de l'École Polytechnique.
Petitot, J. (2011). *Cognitive Morphodynamics*. Bern: Peter Lang.
Petitot, J. (2013). Neurogeometry of neural functional architectures. *Chaos, Solitons & Fractals, 50*, 75–92.
Piccinini, G. (2015). *Physical computation: A mechanistic account*. New York: Oxford University Press.
Pribram, K. H. (1983). What is Iso and what is Morphic in isomorphism? *Psychological Research, 46*, 329–332.
Rathkopf, C. (2015). Network representation and complex systems. *Synthese*. https://doi.org/10.1007/s11229-015-0726-0.
Ratliff, F., & Sirovich, L. (1978). Equivalence classes of visual stimuli. *Vision Research, 18*(7), 845–851.
Revonsuo, A. (2000). Prospects for a scientific research program on consciousness. In Metzinger (pp. 57–75).
Roy, J.-M., Petitot, J., Pachoud, B., & Varela, F. (1999). Beyond the gap: An introduction to naturalizing phenomenology. In J. Petitot, F. Varela, B. Pachoud, & J.-M. Roy (Eds.), *Naturalizing phenomenology* (pp. 1–80). Stanford: Stanford University Press.
Salmon, W. (1989). *Four decades of scientific explanation*. Pittsburgh, PA: University of Pittsburgh Press.
Schaffner, K. F. (1993). *Discovery and explanation in biology and medicine*. Chicago: University of Chicago Press.
Scheerer, E. (1994). Psychoneural isomorphism: Historical background and current relevance. *Philosophical Psychology, 7*(2), 183–210.
Scheutz, M. (2001). Computational versus causal complexity. *Minds and Machines, 11*, 543–566.
Schwitzgebel, E. (2011). *Perplexities of consciousness*. Cambridge, MA: MIT Press.
Shagrir, O., & Bechtel, W. (2017). Marr's computational level and delineating phenomena. In D. Kaplan (Ed.), *Explanation and integration in mind and brain sciences* (pp. 190–214). New York: Oxford University Press.
Shepard, R., & Chipman, S. (1970). Second-order isomorphism of internal representations: Shapes of states. *Cognitive Psychology, 1*, 1–17.
Siegel, S. (2010). *The contents of visual experience*. New York: Oxford University Press.
Teller, D. (1984). Linking propositions. *Vision Research, 24*(10), 1233–1246.
Thompson, E. (2007). *Mind in Life*. Cambridge, MA: MIT Press.
Todorovic, D. (1987). The Craik-O'Brien-Cornsweet effect: New varities and their theoretical implications. *Perception & Psychophysics, 42*(6), 545–650.
Varela, F. (1997). Neurophenomenology: A methodological remedy for the hard problem. In J. Shear (Ed.), *Explaining consciousness*. Cambridge, MA: MIT Press.
Vernazzani, A. (2016). Fenomenologia naturalizzata nello studio dell'esperienza cosciente. *Rivista di filosofia, 107*(1), 27–48.
Von der Heydt, R., Friedman, H., & Zhou, H. (2003). Searching for the neural mechanism of colour filling-in. In Pessoa & De Weerd (pp. 106–127).
Weil, R., & Rees, G. (2011). A new taxonomy for perceptual filling-in. *Brain Research Reviews, 67*, 40–55.
Weisberg, M. (2007). Three kinds of idealization. *The Journal of Philosophy, 104*(12), 639–659.
Weisberg, M. (2013). *Simulation and similarity*. Cambridge, MA: MIT Press.
Wimsatt, W. (2007). *Re-engineering philosophy for limited beings*. Cambridge, MA: Harvard University Press.
Woodward, J. (2003). *Making Things Happen*. New York: Oxford University Press.
Wright, C. (2012). Mechanistic explanation without the ontic conception. *European Journal for Philosophy of Science*.https://doi.org/10.1007/s13194-012-0048-8.
Wu, W. (2018). The neuroscience of consciousness. In E. Zalta (Ed.), *The Stanford Encyclopedia of philosophy*. https://plato.stanford.edu/entries/consciousness-neuroscience/
Zahavi, D. (2004). Phenomenology and the project of naturalization. *Phenomenology and the Cognitive Sciences, 3*(4), 331–347.

Chapter 14
Folk Psychological and Neurocognitive Ontologies

Joe Dewhurst

Abstract It is becoming increasingly clear that our folk psychological ontology of the mental is unlikely to map neatly on to the functional organisation of the brain, leading to the development of novel 'cognitive ontologies' that aim to better describe this organisation. While the debate over which of these ontologies to adopt is still ongoing, we ought to think carefully about what the consequences for folk psychology might be. One option would be to endorse a new form of eliminative materialism, replacing the old folk psychological ontology with a novel neurocognitive ontology. This approach assumes a literalist attitude towards folk psychology, where the folk psychological and neurocognitive ontologies represent competing and incompatible ways of categorising the mental. According to an alternative approach, folk psychology aims to describe coarse-grained behaviour rather than fine-grained mechanisms, and the two kinds of ontology are better thought of as having different aims and purposes. In this chapter I will argue that the latter (coarse-grained) approach is a better way to make sense of everyday folk psychological practice, and also offers a more constructive way to understand the relationship between folk psychological and neurocognitive ontologies. The folk psychological ontology of the mental might not be appropriate for describing the functional organisation of the brain, but rather than eliminating or revising it, we should instead recognise that it has a very different aim and purpose than neurocognitive ontologies do.

14.1 Introduction

This chapter will introduce the threat posed to folk psychology by novel neurocognitive ontologies and respond to this threat by arguing that we should adopt a coarse-grained understanding of folk psychology. Rather than conceiving

J. Dewhurst (✉)
Munich Center for Mathematical Philosophy, Ludwig Maximilian University of Munich, Munich, Germany

© Springer Nature Switzerland AG 2021

F. Calzavarini, M. Viola (eds.), *Neural Mechanisms*, Studies in Brain and Mind 17,
https://doi.org/10.1007/978-3-030-54092-0_14

of folk psychology as aiming to literally describe the fine-grained structure of (neuro)cognition, we ought to understand it as aiming to interpret, predict, and explain the behaviour of whole persons by attributing dispositional states to them. Adopting this conception of folk psychology will insulate it from any threat posed by cognitive ontology revision, and offer a more appealing picture of the relationship between folk psychology and cognitive neuroscience.

In Sect. 14.2 I introduce what I mean by folk psychology, and in Sect. 14.3 I describe the ongoing 'cognitive ontology' debate, which has arisen due to the failure of our current mental categories to map neatly onto the structure of the brain. A common kind of response to this problem is to argue that we should adopt a novel cognitive ontology, i.e. a taxonomy of cognitive states that is better suited to the apparent functional architecture of the brain. In Sect. 14.4 I will describe how the adoption of a novel cognitive ontology might threaten folk psychology with a new form of eliminative materialism, based on the assumption that folk psychology aims to literally describe the same kinds of functions that cognitive neuroscience is interested in investigating. In Sect. 14.5 I will argue that we can avoid this threat by adopting an alternative coarse-grained approach, which conceives of folk psychology as aiming to describe the behaviour of whole persons, rather than the underlying neural mechanisms that generate this behaviour. Finally, in Sect. 14.6 I will consider the implications that adopting this approach would have for our understanding of the relationship between folk psychology and cognitive neuroscience. The real lesson we should take from cognitive ontology revision is not that folk psychological and neurocognitive ontologies are incompatible, but rather that they are best applied in different domains, and need not compete with one another.

14.2 Folk Psychology and Neuroscience

In the philosophical literature, 'folk psychology' is often conflated with 'propositional attitude psychology', and it is under this guise that the traditional debate about eliminative materialism has taken place. However, more recent work on folk psychology has drawn attention to the many other ways in which we might understand one another, including not only propositional attitudes but also other kinds of mental states (such as emotions or non-propositional attitudes), character traits, narratives, and normative constraints (see Spaulding 2018 for a general overview of these developments). In this section I will introduce both conceptions of folk psychology and the cognitive ontologies that they might support, and then consider how these could have influenced neuroscientific ontologies. This will set the scene for the next section, where I will discuss a recent debate about whether, and to what extent, our existing 'cognitive ontology' requires revision.

The term 'folk psychology' only became popular in the philosophical literature from around the 1980s onwards, following Paul and Patricia Churchland's arguments that it constitutes a primitive and largely unsuccessful theory of how the mind works, one that ought to be replaced with a new theory drawn from "the conceptual framework of a completed neuroscience" (Churchland 1981: 67; see also his 1979 and Churchland 1986). This "eliminative materialism" stood in contrast with Fodor's defense of folk psychology as a necessary framework for understanding the mind, which he thought that we have no conceivable alternative to (Fodor 1987: 132). Both Fodor and the Churchlands followed Lewis' earlier characterisation of common-sense psychology (as Lewis called it) as a proto-scientific theory. Lewis argued that our everyday language for talking about the mind could be treated "as a term-introducing scientific theory" (1972: 256), such that we could simply read off an ontology of mental states from the way that we talk about the mind.[1] Here Lewis focused primarily on the 'propositional attitudes', i.e. attitudes such as belief and desire that one can hold towards a proposition, and Fodor (1975) followed this approach in constructing his 'language of thought hypothesis', according to which cognition consists in the manipulation of folk-psychologically characterised propositional attitudes. A similar emphasis is found in the 'theory-theory' in social cognition, which argues that our understanding of other minds is guided by an implicit 'theory of mind', although not necessarily one identical to the language of thought hypothesis (see Gopnik and Wellman 1992; cf. Premack and Woodruff 1978, who first introduced the term 'theory of mind' to scientific psychology). For example, according to this theory I might come to attribute to you a belief about the location of some object based on theoretical inferences informed by your behaviour (looking in a certain place and seeming surprised, etc.). The theory-theory is based on a literal understanding of the common-sense propositional attitude theory invoked by Lewis, i.e. a very particular (and somewhat peculiar) philosophical interpretation of a much broader cultural practice of self- and other understanding.

It was this approach that framed the original debate over eliminative materialism in the 1980s, which took it for granted that folk psychology was in the business of attributing propositional attitudes to people in a proto-theoretical manner, such that it could be understood as a literally true or false theory, amenable to scientific investigation. The eliminativists, such as Paul and Patricia Churchland,[2] argued that our best neuroscience would demonstrate that this theory was false, whereas the realists denied that this could be possible, in Fodor's case going so far as to argue that folk psychology (and psychology more generally) was autonomous from neuroscience in a way that shielded it from empirical refutation at this level of

[1]Lewis explicitly denied that folk psychology originated as a theory of this kind, but rather followed Sellars (1956) in treating it as a "good myth" that might help us better understand the mind (Lewis 1972: 257).

[2]The other most notable eliminativist is Stich (1983), who later repudiated his version of the view due to concerns raised by Lycan (1988) about the reference of folk psychological terms. Eliminativsm also has historical antecedents in Feyerabend (1963) and Rorty (1965). Stich did acknowledge that folk psychology might be broader than just propositional attitude psychology, but glossed over this by suggesting all non-propositional mental terms could simply be restated in a propositional format (1983: 217).

analysis (Fodor 1974). Nonetheless, both the eliminativists and the realists took it for granted that folk psychology was trying to literally explain how cognition functions in a relatively fine-grained manner. In Sect. 14.4 I will argue that current debates over cognitive ontology revision pose a novel eliminativist threat, in a similar manner to the original eliminative materialism of the 1980s.

Subsequent work on folk psychology and social cognition has recognised that our common-sense understanding of other minds might not just consist in the attribution of propositional attitudes (see Lavelle 2019 for a general introduction to the topics discussed here). In social cognition, there has been an increased emphasis on non-theoretical means of understanding one another, such as simulation (Gordon 1986; Heal 1986), direct perception (Gallagher 2008a), and interaction (De Jaegher and Di Paolo 2007; Gallagher 2008b). Whether these are truly distinct from the theory-theory is a complicated question (see e.g. Lavelle 2012), and more recently there has been a shift towards endorsing some version of a hybrid theory that acknowledges the role of both theory and simulation in social cognition (see e.g. Mitchell 2005; Apperly 2008). There has also been a related shift away from focusing on just propositional attitude attribution and towards seeing folk psychology as a multifaceted phenomenon, consisting not just of propositional attitude attribution but also other means of understanding one another, such as character traits (Westra 2018), narrative structure (Bruner 1990; Hutto 2008), and normative constraints (Mameli 2001; McGeer 2007; Zawidzki 2013; Andrews 2015). This broader understanding of folk psychology, I will argue, might give us the resources to reconceive of the relationship between folk psychology and neuroscience in a way that avoids the threat of eliminative materialism posed by cognitive ontology revision.

Whether we take a broad or narrow view of folk psychology, we can ask what kind of ontology of mental states it provides us with, and thus what kind of theory of mind and cognition it entails. The ontological commitments of pre-theoretic folk psychology are at the very least unclear, and perhaps even simply indeterminate, but philosophers (and cognitive scientists) have nonetheless tried to interpret and 'clarify' them (and what we take folk psychology to be 'literally' committed to will depend on how this interpretation is carried out). Lewis proposed reading off a set of theoretical commitments from the "everyday platitudes" of common-sense psychology (1972: 252), which in contemporary philosophy of mind is often assumed to be reducible to some version of belief-desire psychology. A similar approach is reflected in Fodor's language of thought hypothesis, which expects the structure of cognition to match up (in some sense) with our everyday language for talking about the mind (understood by Fodor in narrow terms, i.e. propositional attitude attribution). It is important to note here that Fodor does *not* expect the reverse to be true, i.e. that folk psychology should conform to whatever our best scientific theory of mind and cognition is, but rather just that the folk theory is likely to be roughly correct in the first place.[3]

Classical eliminativism shares the realist assumption that we can simply read off an ontology of mental states from folk psychology, and then check whether this matches up with the empirical discoveries of our best cognitive neuroscience. A crucial difference is whether or not one expects this ontology to match up with the findings of neuroscience (in the case of the eliminativist) or with some more abstract psychological theory (in the case of the Fodorian realist), but the commitment to a 'literal' interpretation of folk psychology is apparent in both cases. Understood more broadly, folk psychology could be interpreted as supporting an ontology consisting of not just propositional attitudes, but also emotions, character traits, and perhaps even roles in a social narrative. There is then a further question of what kind of relationship this ontology has with the various cognitive scientific disciplines, including neuroscience.

Fodor saw folk psychology as being a precursor to our scientific *psychological* ontology, which he thought was wholly autonomous from neuroscience, whereas the eliminativists thought that our folk psychological ontology ought to be judged against our best neuroscience and revised or eliminated if it failed to match up.[4] The reality is probably somewhat more complex. On the one hand, it seems increasingly implausible that psychology could be wholly autonomous from neuroscience (see e.g. Boone and Piccinini 2016; cf. Piccinini and Craver 2011; Knoll 2018 for a dissenting opinion), meaning that revisions to our neuroscientific ontology might also entail revisions to our (folk) psychological ontology (see Sect. 14.4). On the other hand, the move away from a narrow understanding of folk psychology as *just* propositional attitude psychology means that we can now conceive of a more sophisticated relationship between folk psychology and neuroscience than mere one-to-one mapping.

In Sects. 14.5 and 14.6 I will argue that the dichotomy between folk psychological realism and eliminativism rests on the mistaken assumption that we should take folk psychology *literally*, i.e. understand it as being involved in the same kind of project as our scientific investigation of the mind and brain. We should adjust our perspective and reconceive of folk psychology as being in the business of interpreting the coarse-grained behaviour of whole persons, rather than the fine-

[3]This means that Fodor's position, *qua* revisions to our folk ontology, is actually quite similar to that which I will present in Sects. 14.5 and 14.6 of this chapter. I thank J. Brendan Ritchie for pressing me on this point.

[4]As noted above, Stich (1983) also endorsed a form of eliminativism, but he later realised that if one adopts a causal theory of reference then changes to the scientific ontology might instead give us reason to *revise* (rather than *eliminate*) the folk psychological ontology (see e.g. Stich 1996; cf. Lycan 1988). As I will argue in Sects. 14.5 and 14.6, I think this move misunderstands the relationship between scientific and folk ontologies in just the same way that (folk psychological) eliminativism does. This debate about "arguments from reference" (Mallon et al. 2009) dominated much philosophical discussion of folk psychology in the 1990s and 2000s, and I hope to bypass it entirely here by focusing more on the practical differences between scientific and folk ontologies.

grained mechanisms that generate that behaviour. From this alternative perspective it turns out that folk psychological and neuroscientific ontologies have such different aims, methods, and standards that it would be a mistake to directly compare them. This is not a new proposal, having antecedents in the idea that the application of folk psychological concepts within the context of neuroscience might constitute a kind of category mistake (see e.g. Bennett and Hacker 2003; cf. Ryle 1949). The novelty of my argument here is firstly in applying this idea to the specific case of cognitive ontology revision, and secondly in providing a distinctive kind of rationale for taking this approach, based not so much on linguistic or grammatical reasons, but rather on reasons to do with the nature of folk psychology itself, which seems more concerned with interpreting the behaviour of whole persons than with identifying the neural mechanisms responsible for that behaviour.

14.3 Ontology Revision in Cognitive Neuroscience

Aspects of our folk psychological ontology have historically influenced cognitive neuroscientific ontologies, even if sometimes only in a subtle and indirect manner. Many basic psychological concepts such as memory, attention, and belief originate in folk psychology, and although most of these concepts have undergone technical revision over the years, they still bear some traces of their folk psychological origins. When designing neuroimaging studies it is necessary to define a task that is intended to operationalise the cognitive function that you are attempting to investigate, and most of these tasks are broadly folk psychological in flavour, even if the functions they are intended to track are often more precisely defined. For example, a study investigating the neural correlates of written language processing might deploy a reading task, where 'reading' is understood as a single kind of cognitive function that is expected to map neatly onto a region of the brain. This kind of influence would be innocuous if it turned out that such categories were in fact appropriate for neuroscience, but as we will now see it is becoming increasingly clear that this might not be the case.

The term 'cognitive ontology' was coined by Price and Friston (2005), who use it to refer to the set of cognitive functions that we appeal to when conducting neuroimaging studies.[5] Ideally, they claim, this ontology should support a one-to-one mapping between functions and structures, such that "structures predict functions and functions predict structures" (Price and Friston 2005: 263). Each cognitive function (identified by getting a subject to perform a related task) should be correlated with activation in just one neural structure, and each structure should be implicated in just one kind of task or function. This would allow us to make clear statements about where in the brain each function is performed, vindicating

[5]They also refer to it as a 'functional ontology', but 'cognitive ontology' seems to be the terminology that is now used most commonly in the literature.

the accuracy of our cognitive ontology (and, if the two were identical, also our folk psychological ontology). There is perhaps an (implicit) assumption of mind/brain identity underlying this approach, and more generally the approaches to cognitive ontology discussed in this section, although interpreted cautiously their aim is only to establish *correlations* between structures and functions, not identity relations.[6] However, even under this more cautious interpretation, it is still assumed that there ought to be a correlation between cognitive functions and structures of the *brain*, rather than of the brain-and-body, or brain-body-and-world, or some other set of physical structures.

Unfortunately, it turns out that the cognitive ontologies applied in most neuroimaging studies do not support correlations of this kind. Typically we find cases of one-to-many mappings (where a single function appears to activate many structures), many-to-one mappings (where a single structure is implicated in many functions), and many-to-many mappings (where many different functions simultaneously cross-correlate with many different structures). Price & Friston see this as a problem, and one of the aims of their paper was to develop a way to revise our cognitive ontology in order to make it better match up with the structure of the brain. Their proposal is that we should develop a novel ontology by grouping together seemingly distinct functions that have similar activation profiles, coming up with more general labels for these new functions that captures their performance across different kinds of task. This would allow us to preserve one-to-one mapping at the expense of our original ontology, which would become subsumed under the new, more general functional categories.

To illustrate this approach, they focus on one example: the different kinds of function currently attributed to the left posterior lateral fusiform (LPLF). These include processing visual information about written words in reading tasks (Cohen et al. 2000); processing the visual attributes of animals in semantic categorisation tasks (Martin and Chao 2001); and processing visual/tactile information more generally (Amedi et al. 2002). The result is a case of many-to-one mapping, where a single structure (the LPLF) supports at least three different kinds of functional attribution. In order to avoid this, Price & Friston suggest reclassifying the function of the LPLF as 'sensorimotor integration', which they claim is able to accommodate each of the subsidiary functions attributed to it in different kinds of task. Their approach has since been criticised somewhat in the philosophical literature, with a common response being that 'sensorimotor' integration is just too broad a functional category to explain anything, and that we should instead attribute functions in a task or context sensitive manner (see e.g. Klein 2012; McCaffrey 2015; Burnston 2016). While Price & Friston acknowledge that there is a practical benefit to attributing more specific functions in the context of particular tasks, they still think it is beneficial to have a general functional category that preserves one-to-one mapping, such as 'sensorimotor' integration, because "it is more useful to label a

[6]See Towl (2011) and Nathan (this volume) for further discussion, and Vernazzani (this volume) for a historical perspective.

region with a function that explains all patterns of activation" (Price and Friston 2005: 268).[7] Insofar as their motivation here is primarily pragmatic, it could be seen as an example of McCauley and Bechtel's (2001) 'heuristic identity theory', which conceives of proposed "psycho-neural identities" as tools for generating new hypotheses and guiding experimentation in a manner that is fully compatible with a pluralistic cognitive ontology. However, for my purposes it is the general strategy and framing of the problem that is important, not the specific details of this case, and even the context sensitive mapping strategies will end up having counterintuitive consequences for our folk psychological ontology (which I will discuss in more detail in the next section).

Since Price & Friston first identified this problem, there have been many different proposals for how to resolve it, which can be broadly classified as 'top-down' (holding fixed our cognitive ontology and revising our understanding of neural structure) and 'bottom-up' (holding fixed our understanding of neural structure and revising our cognitive ontology). My focus here will be on the latter kind of approach, which if adopted would have the most significant impact on our folk psychological ontology (see McCaffrey and Machery 2016 for some general criticism of this kind of approach). In the rest of this section I will introduce two further bottom-up strategies for cognitive ontology revisions, each of which would threaten our folk psychological ontology in quite different ways. The first of these, advocated for by Russ Poldrack and colleagues, follows Price & Friston in aiming to preserve one-to-one mapping, while the second, developed by Michael Anderson, takes a more flexible approach based on the phenomenon of neural reuse, but nonetheless ends up with something very different to our current ontology.

Poldrack has proposed (and initiated) the development of a 'Cognitive Atlas', which aims to develop "a comprehensive, formally specified ontology of mental processes" (2010: 756), better suited for mapping to the structural organisation of the brain. This takes the form of an online database where different labs can upload their experimental protocols and results (www.cognitiveatlas.org), which can then be compared and analysed using data mining techniques. Poldrack and Yarkoni (2016) describe this approach in more detail, arguing that large-scale analyses of neuroimaging data can be used to overcome several challenges facing cognitive neuroscience, and emphasizing the role that "formal cognitive ontologies" can play in this process. They note that "all else being equal, we believe that a model of psychological processes that also maps systematically onto known biological structures is strongly preferable over one that does not" (*ibid*: 599); i.e., they give priority to biological or structural factors over functional or task-specific factors when determining their ontology.

[7] An alternative kind of response that I do not have space to consider here is to develop an ontology based on the evolutionary origins of these neural structures. Barrett (2012) proposes that the LPLF should be understood as performing "category specific object recognition", a functional attribution that he argues can accommodate the different kinds of task that this region is correlated with (see Rathkopf, this volume, for further discussion of this kind of evolutionary approach).

For example, Lenartowicz et al. (2010) analysed neuroimaging results associated with the construct 'cognitive control', finding five relevant key terms: working memory, response selection, response inhibition, task switching, and cognitive control itself (*ibid*: 682). In order to check which, if any, of these terms correspond uniquely to a neural structure, they performed a meta-analysis of neuroimaging studies in which they occur, and discovered that while there was a clear distinction between the patterns of activation corresponding to response selection on the one hand, and between working memory, response inhibition and cognitive control on the other, there was no clear distinction between the tasks associated with the latter group (the data corresponding to task switching was unclear). Based on this analysis they conclude that response selection is a distinct function associated with the precentral gyrus and middle frontal gyrus, and that cognitive control, response inhibition and working memory may together constitute a second distinct function associated with "a right-lateralized network involving frontal and subcortical regions" (Lenartowicz et al. 2010: 688). They acknowledge that their data was somewhat noisy, and so do not present these results as conclusive, but nonetheless take them to be indicative of the kinds of revision that we should make to our ontology of cognitive control. Based on an initial ontology of five functions, they end up with a reduced ontology of only two or three, by grouping three items together into a new functional construct. This is comparable to the way in which Price & Friston proposed grouping the various functions associated with the LPLF under a single unified function, and if applied more widely could have similarly major repercussions for our folk psychological ontology.

An alternative possibility, suggested by Figdor (2011), is that the kind of novel ontology proposed by Lenartowicz *et al* shouldn't be understood as a replacement for either our neurocognitive or folk psychological ontology, but rather as a kind of novel 'task ontology' that can be used to mediate between psychology and neuroscience. Once equipped with this improved task ontology, we could then begin the process of refining our cognitive ontology proper, by eliminating (or revising) "cognitive labels that cannot be stably operationalized" (*ibid*: 225), a process which is likely to involve modifications to some parts of the folk ontology, but could leave other parts intact. Figdor (2018) also argues for a literalist attitude towards the resulting ontology, which might qualify as a partial vindication of folk psychology, depending on how heavily it ends up being modified. Nonetheless, as I will argue in Sect. 14.5, if we reconceive of folk psychology in more coarse-grained terms then there is no need to worry about the eliminativist implications of cognitive ontology revision in the first place. Figdor's literalist project could happily coexist alongside this re-conception, as an alternative but non-competing way of making sense of the human cognitive system.

The third approach to cognitive ontology revision I want to introduce here is somewhat different, as it doesn't aim to map a single functional category onto each neural structure (thus preserving one-to-one mapping), but rather acknowledges that each neural structure might be implicated in multiple different tasks, and tries to construct an ontology that reflects this. Focusing on the phenomenon of neural reuse, Michael Anderson has proposed that we should characterise neural structures in terms of their 'personalities' rather than their functions, where personalities are understood as "the functional dispositions of individual regions, their underlying causal powers, and their propensities to cooperate with sets of other regions" (Anderson 2014: 114). A region that was previously identified as performing a single, discrete function might instead be characterised in terms of the general kind of contribution it makes to a wide range of tasks, where this contribution does not neatly correspond to anything that we might recognize as a cognitive function. More technically this proposal involves the generation of multidimensional "fingerprint plots" that represent the full range of functional properties associated with the brain (*ibid*: 118). These fingerprint plots closely resemble the diagrams used to represent human personality traits, and are intended to predict activation in a region across a wide range of tasks. For example, the plot for the left inferior parietal sulcus shows the most activation on inhibition tasks, somewhat less activation on vision, motor learning, observation, and preparation tasks, and so on. Rather than coming up with a novel functional description that predicts this behaviour, Anderson wants to give a multidimensional characterisation that accounts for the contributions of this region to a diverse range of tasks. Like Poldrack, he also suggests using statistical techniques to uncover the underlying dimensions that are principally responsible for a region's functional contributions, but these are also going to be unpredictable and opaque from a folk psychological perspective – i.e., Anderson does not envision dimension reduction as a route to the recovery of the folk psychological ontology, but rather as a tool for constructing an alternative. The envisioned outcome is an ontology of 'personalities' rather than functions, preserving one-to-one mapping at the expense of our pre-existing functional categories. Instead of saying that a structure performs a single function like 'word identification', each region of the brain will be given a complex, dispositional analysis that tells us the extent to which it is likely to be implicated in various kinds of task (for some examples see Anderson 2014: 118). The resulting ontology will look very different to that which we find in folk psychology, consisting of complex, multidimensional descriptions of dispositional properties, rather than simple functional attributions.

Regardless of what kind of solution one endorses to the problem of cognitive ontology revision, it seems likely that we will have to abandon, or at least revise, our existing cognitive ontology in response to it. In the next section I will consider what impact this might have on folk psychology itself, which is the source of the existing ontology, and thus might seem to be threatened by any potential revisions to it.

14.4 The Threat to Folk Psychology

Having presented three different approach to cognitive ontology revision, I will now consider the *prima facie* threat that such revision poses to folk psychology. As I suggested in the previous section, this threat arises because our existing cognitive ontology is at least somewhat inspired by folk psychology. If this ontology were successful, enabling one-to-one mappings between functions and structures, it could be seen to vindicate or naturalise our folk psychological categorisation of mental states and processes (at least under the literalist interpretation of folk psychology). However, if it is unsuccessful in the ways described in the previous section, requiring replacement or revision, then folk psychology might require a similar treatment. This would essentially constitute a novel form of eliminative materialism, with developments in cognitive neuroscience threatening to replace or revise our folk psychological ontology.[8]

The one-to-one mapping aspired to by Price & Friston is the ideal target aimed at by much contemporary cognitive neuroscience, at least implicitly. A typical approach to investigating the neural correlates of some cognitive function involves operationalizing that function with a particular task, and then measuring a subject's neural activity while they perform that task. The functions chosen for these studies typically still bear at least a passing resemblance to folk psychological categories, and the tasks that are intended to probe them are clearly inspired by a common-sense interpretation of the function. For example, in the LPLF studies described by Price & Friston, we see functions such as 'processing visual information about written words' and 'processing the visual attributes of animals' which, while expressed in a somewhat more technical manner than we might be used to, at least make some sense from a folk perspective. The hope of the folk psychological realist is that if these kinds of functions could be localised to discrete neural structures, then the folk psychological taxonomy of mental states and processes would be to some extent vindicated or naturalised.

In contrast, each of the proposals for cognitive ontology revision that I considered in the previous section would replace these (relatively) common-sense categories with something that it is much harder to make sense of from a folk psychological perspective. Price & Friston's proposal was to unite the various functions of the LPLF under the umbrella category 'sensorimotor integration', but even if this were more explanatory, it is not at all obvious that such a category has any place in our folk psychological ontology. Similarly, Poldrack's proposed Cognitive Atlas project could find patterns in large quantities of neuroimaging data that would not necessarily bear any direct resemblance to the kinds of behavioural patterns that folk psychology is sensitive to.[9] Based on such an analysis, Lenartowicz et al. (2010) proposed grouping together cognitive control, response inhibition, and working memory, each of which might individually make sense to folk psychology, but when combined do not appear to form a folk-psychologically meaningful cluster. Finally, Anderson's 'neural personalities' describe in quantitative terms the contributions made by each region to a diverse range of tasks, and certainly don't bear any close

[8]I have previously considered similar concerns arising from the predictive processing framework (Dewhurst 2017), and Clark (2019) considers whether this framework would entail the elimination of the folk psychological construct 'desire', responding in part to concerns raised by Klein (2018). Adopting the coarse-grained approach that I advocate here would dissolve concerns of this kind.

[9]Poldrack discusses some of these issues himself in a blogpost: http://www.russpoldrack.org/2016/04/how-folksy-is-psychology-linguistic.html

resemblance to folk psychological categories. So if we are going to have to replace our cognitive ontology with something resembling one of these proposals, then it seems like we will have to abandon the initial hope that we could vindicate or naturalise folk psychology by mapping the states and processes that it identifies on to the activity of neural structures.

Taken one step further, the failure of cognitive neuroscience to vindicate our folk psychological ontology could be used as the basis for a novel argument for eliminative materialism. If our most successful neuroscience requires a revised cognitive ontology composed of neural personalities, or novel categories such as a sensorimotor integration, then it suggests that our original, folk psychologically inspired ontology was also inaccurate, and perhaps deserving of elimination. At the very least this folk psychological ontology will require fairly radical revision if it is going to match up to the novel ontologies introduced in the previous section. Such a revision might subsequently influence how we conceive of ourselves and others, and how we go about predicting and explaining our everyday behaviour. Churchland suggests that, having rejected the folk theory of mind as inadequate, "one might learn to comprehend and report one's internal states and activities within a different and more adequate framework" (1979: 99), i.e. the framework provided by our best cognitive neuroscience. A proponent of a revised cognitive ontology might similarly suggest that we ought to start talking about one another's mental lives in terms of these new categories rather than those of folk psychology.

For the folk psychological realist this possibility will simply constitute a *reductio* of some aspect of the cognitive ontology project, and indeed there have been several alternative responses, such as suggesting that we might instead want to revise our understanding of the mapping relation towards something more context sensitive (Klein 2012; McCaffrey 2015; Burnston 2016; cf. Dewhurst 2019), or adopt a more flexible understanding of the functional structure of the brain (see e.g. Glymour and Hanson 2016). Nonetheless, for the realist who wants to try and identity one-to-one mappings between folk psychologically inspired cognitive functions and discrete neural structures, the kind of evidence appealed to by Price and Friston (2005) does seem to present a serious problem. Either they must accept that our folk psychological ontology is somewhat inaccurate compared with the functional structure of the brain, or they must give up on this particular kind of naturalisation project.

Both the realist and eliminativist interpretations of the relationship between folk psychology and cognitive neuroscience reflect a literalist approach towards our folk psychological ontology. That is, the hope that our folk psychological ontology might be naturalised or otherwise vindicated by neuroscience assumes that folk psychology was always aiming to *literally* describe what is going on in the head, and this assumption is similarly reflected in our feeling of disappointment when it fails to do so. Realist and eliminativist attitudes towards folk psychology can be seen as two sides of the same coin, sharing the basic assumption that the success or failure of folk psychology will depend on its eventual scientific vindication (or lack thereof). We must bear this assumption in mind when we are considering the impact of cognitive ontology revision on folk psychology. In the next section I will consider what an alternative might look like, and how it could make a difference to the implications of the cognitive ontology debate for folk psychology.

14.5 Towards a Coarse-Grained Folk Psychology

I am not the first to draw a connection between proposals for cognitive ontology revision and the threat of a novel eliminative materialism. In this section I will consider two previous engagements with this issue and argue that both point towards a similar solution: rather than embracing eliminativism as a consequence of cognitive ontology revision, we ought to adopt a more coarse-grained approach, where folk psychology is understood as aiming at predicting and explaining the behaviour of whole persons rather than saying anything about the functional organisation of their brains. I will now present this alternative picture of folk psychology and explain how it avoids the threat from cognitive ontology revision, before exploring its broader implications for the relationship between folk psychology and neuroscience.

The idea that our folk theories might not be in direct conflict with our empirical ones is of course not entirely novel. Similar proposals have been made previously with regard to e.g. emotion categories (Griffiths 1997), biological taxonomies (Dupre 1981), and concepts understood as psychological kinds (Machery 2009). More generally, anti-essentialist theories of natural kinds such as Boyd's (1999) homeostatic property cluster theory and Slater's (2015) stable property cluster theory would seem to support the idea that different 'kinds' of kinds might be appropriate in different social or epistemic contexts (cf. Ludwig 2017 on indigenous and scientific kinds). The attitude towards folk psychology and neurocognitive ontologies that I present here and in the next section is fully compatible with this general trend in the literature on natural kinds towards partial or local eliminativisms/revisionisms, where we can accept changes to our scientific ontology in some domain without thereby threatening the associated folk ontology.

Francken and Slors (2014, see also their 2018) describe how what they call "commonsense cognitive concepts" (i.e. folk psychological concepts) get incorporated into neuroscientific explanations, and how this might give rise to various kinds of problem. They identify an "implicit realism" about commonsense cognitive concepts as being the basis for this incorporation (*ibid*: 253–4), giving rise to the apparent dichotomy between folk psychological realism and eliminativism that I identified in the previous section. Their proposed solution is to instead adopt an 'interpretivist' approach, inspired by Davidson (1980) and Dennett (1987), whereby folk psychology is understood as tracking behavioural patterns rather than aiming to identify discrete states and processes in the brain. This would allow us to acknowledge the failure of folk psychological (or 'commonsense cognitive') concepts at accomplishing the latter task, while also preserving a positive role for folk psychology in interpreting the behaviour of whole persons, and thereby avoiding the eliminativism/realism dichotomy.

Murphy (2017a, see also his 2017b) paints a similar picture, distinguishing between three options that are available to us with regard to folk psychology and the cognitive ontology debate: integration, elimination, or autonomy. Integration is essentially what I have been calling literal realism, where folk psychology is assumed to make empirical claims about the structure of cognition and is therefore vulnerable to the mapping concerns raised by the likes of Price and Friston (2005). Elimination would be the consequence if integration fails, or requires such extensive revisions that our cognitive concepts no longer resemble their folk psychological origins in any meaningful way. Finally, autonomy offers a way out of the integration/elimination dichotomy, by conceiving of the role of folk psychology in a way that does not make it hostage to empirical success. This third option could be accomplished by adopting the interpretivist approach favoured by Francken and Slors (2014), which can help make sense of how folk psychology could be 'autonomous' from neuroscientific details but nonetheless predictive and explanatory of human behaviour. In the rest of this section. I will develop this approach in more detail, connecting it with contemporary dispositional approaches and arguing that it is compatible with a certain kind of (non-literal) realism about folk psychology.

As Francken and Slors (2014) note, their interpretivist proposal is probably best developed in Dennett's (1987) intentional stance approach, which conceives of folk psychology as being a particular kind of interpretive 'stance' that one can take towards a complex system, alongside the 'design' and 'physical' stances. The predictive and explanatory success of these stances, according to Dennett, depends on the existence of 'real patterns' in the behaviour of these complex systems, which can only be identified and acted on by interpreting them at a certain level of abstraction. So, the intentional stance (and thus folk psychology) succeeds by considering the coarse-grained behaviour of a whole person understood as a rational agent, rather than focusing on fine-grained neurophysiological details (which, indeed, we did not even have access to for most of our evolutionary and cultural history). Cognitive neuroscience, in contrast, might have more success by focusing on more fine-grained details, but this does not invalidate the intentional stance, or require that it should be revised in light of its failure to map onto the functional structure of the brain. Indeed, Dennett's approach can explain why our folk psychological ontology might be so different to the revised neurocognitive ontology, as there is no *prima facie* reason to think that the same kinds of concepts are going to be suited for picking up on real patterns at different levels of grain.

Interpretivism, including Dennett's intentional stance approach, also has a lot in common with dispositional approaches, which conceive of mental states (as attributed by folk psychology) as dispositions (behavioural or otherwise) rather than discrete entities. Schwitzgebel (2002) presents a modern defense of dispositionalism about belief, inspired by Ryle (1949), which allows for not only behavioural dispositions but also phenomenal and cognitive dispositions. This kind of account could be extended to other propositional attitudes and folk psychology more generally, and can explain the explanatory and predictive success of folk psychology without committing it to making empirical claims that might be at odds with our neurocognitive ontology. Both interpretivism and dispositionalism also enjoy some empirical support of their own, insofar as it seems like the folk might not actually be committed to making any claims about internal mental states (Curry 2018), and can make better sense of the language we use to express propositional attitudes (Matthews 2011, 2017, see also his 2007/2010).[10] Given that it can also avoid the threat posed by cognitive ontology revision, I think we have good reason to adopt this kind of approach towards folk psychology. The implication of the literal approach is that if folk psychology fails to match up with our best cognitive neuroscience, then it ought to be eliminated or revised. Yet even if this were the case, it seems obvious that 'the folk' (including cognitive scientists themselves) could carry on interpreting one another's behaviour in just the same way that they have been doing for millennia, putting pressure on the realist/eliminativist dichotomy posed by the literatist.

According to the coarse-grained account of folk psychology that I think we ought to endorse, when we attribute mental states to someone we are doing no more than saying that they are disposed to behave in certain ways. Crucially, we are making no commitment as to the structure of the mechanisms responsible for that behaviour, and as such folk psychology should not be taken to aim at literally describing these mechanisms. When I say that someone believes something, I just mean that they are likely to act as though it were true, and when I say that someone is brave or intelligent I am just making a general statement about the kinds of behaviour and competencies they are likely to exhibit. In neither case am I committing to any details about what is going on 'in the head', nor am I saying anything that could possibly be falsified by cognitive scientific discoveries. Even if it turned out that there was nothing resembling the structure of belief-desire psychology going on in the brain, or that character traits like bravery and intelligence were not stable scientific constructs, we could nonetheless go on attributing these concepts to one another in a meaningful manner. It might turn out that their meaningfulness has more to do with sociocultural constraints on behaviour than with anything mechanistic (see e.g. Zawidzki 2013), but it would nonetheless be meaningful. Understood in this way, folk psychology is in principle not vulnerable to scientific refutation, for so long as we continue to exhibit the correct behavioural dispositions, it will remain a coherent kind of social practice.

It is also important to note that both the interpretive and dispositional account of folk psychology can be construed as 'realist', albeit of a distinct form from the literal realism that I identified previously (i.e. the kind of realism that forces a dichotomy with eliminativism). Interpretivism about folk psychology is realist insofar as it claims that when we attribute mental states to one another, we are doing so on the basis of Dennettian real patterns, which is to say we are identifying and interpreting real patterns in the behaviour of those we attribute mental states to (see Dennett

[10] Although see Quilty-Dunn and Mandelbaum (2018) for some recent criticism of dispositionalism.

1981, 1991; cf. Ross 2000).[11] Similarly, dispositionalism about folk psychology is realist insofar as the dispositions we attribute to one another are just as real as any other dispositions, such as that of a soluble object (like a sugar cube) to dissolve when placed in water. Even if folk psychology does not correctly identify the fine-grained functional structure of the brain, it can nonetheless correctly identify behavioural patterns and dispositions which are just as real as those described by neuroscience.

14.6 The Relationship Between Folk Psychology and Neuroscience

Adopting a more coarse-grained approach to folk psychology can help us to re-evaluate the relationship between folk psychology and cognitive neuroscience. Rather than conceiving of neuroscience as aiming to identify the neural correlates of more-or-less folk psychological categories, we ought to instead conceive of it as aiming to uncover the complex mechanisms that give rise to the kind of *behaviour* that folk psychology describes, predicts, and explains.[12] There is no reason to think that these mechanisms will conform to the categories of folk psychology, but equally no reason for concern when they fail to do so. The two kinds of ontology (folk psychological and neuroscientific) are simply so different that we cannot (and should not) even try to directly compare them. McDowell (1994) makes a similar point about the relationship between subpersonal mechanisms and personal level perceptual experience: the former somehow enable the latter, but there is no conflict between the two, and no reason to think that one ought to be reducible to the other. Furthermore, there is also no reason to think that we should have any kind of privileged access to subpersonal mechanisms (either our own or those of other people), such that their structure might be reflected in our folk psychological categories.[13]

The literal interpretation of folk psychology, on the other hand, makes a commitment to certain kinds of empirical discoveries (i.e., neural structures with a functional architecture that can be mapped onto folk psychology). If our best cognitive ontology turns out to be radically different to the folk psychological ontology, then this might mean that the latter must be eliminated or revised. There are of course other options available to the literal realist. They could accept a limited amount of revision to the folk psychological ontology, stopping short of

[11]Whether or not Dennett himself should be interpreted as a realist is a complicated question which I do not intend to get into here. It is sufficient for my purposes that there is a sense in which his approach to folk psychology can be understood as realist.

[12]See Raja & Anderson (this volume) for further discussion of the relationship between neuroscience and behaviour.

[13]For more on the personal/subpersonal distinction, see Dennett (1969) and Drayson (2012, 2014).

full-blown eliminativism. They could also insist that it must be the neuroscience itself that is wrong, adopting a 'top-down' strategy and revising our interpretation of the neuroimaging data in order to match up with the folk ontology. There is a lot of interpretive work that must be done when conducting neuroimaging studies, all of which gives us some room for manoeuvre. For example, by switching to a network analysis of the functional relevance of neural activity (see e.g. Glymour and Hanson 2016; see also Wright, this volume), we could avoid the need to map cognitive functions directly onto neural structures, and thus perhaps preserve the neuroscientific relevance of the folk psychological ontology.

However, regardless of whether a strategy like this is successful, by moving to a more coarse-grained understanding of folk psychology we can avoid the threat of eliminativism entirely. One way to think of this approach is simply as a restatement of the idea that applying folk psychological concepts to neuroscience constitutes a category mistake (cf. Bennett and Hacker 2003), or that it somehow mixes up the kinds of language used to describe our manifest and scientific images of the world (cf. Sellars 1963). Our folk psychological ontology reflects the manifest image, our cognitive ontology reflects the scientific image, and there is no in-principle reason to think that they should be reconcilable. Of course, this approach would also rule out any straightforward reduction of the mental to the physical, although that is not to say that the mental states picked out by folk psychology are entirely independent of the physical states studied by cognitive neuroscience. There is more work to be done on how to make sense of this relationship in a naturalistic manner, but my own preferred approach is to see folk psychology as picking out (real) patterns in person-level behaviour that are generated by neuroscientific mechanisms (cf. Dennett 1991). Looked at in this way there is no need to eliminate, or even revise, folk psychology in response to developments in cognitive neuroscience, as it will remain just as good as it ever has been at picking out person-level patterns.[14] In some cases the folk *are* interested in something more fine-grained, such as when they pursue a clinical intervention from a neurosurgeon, but in these cases I think we should understand them as deferring to the expertise (and ontology) of the scientific community, rather than as adopting a more fine-grained ontology

The coarse-grained approach does still allow for a kind of *partial* eliminativism, which acknowledges the failure of folk psychological concepts at tracking fine-grained neuroscientific states and processes (i.e., the mapping problem), and allows that they might need to be revised, replaced, or eliminated from this explanatory context. Hence adopting this approach is compatible with calling for the revision of our cognitive ontology for neuroscientific purposes, and this might mean we will end up with a neuroscientific ontology that is very different to our folk psychological

[14] Which is not to say that it is *very* good at this. It is plausible that the success of folk psychology is at least somewhat overrated, especially when it comes to edge cases like mental illness and socially disruptive behaviour (see e.g. Matthews 2013 for some discussion of these issues, and the benefits of taking a dispositional approach to them). However, it is clearly successful at least some of the time, and the approach taken here can help make sense of how this could be true even if it fails to track the fine-grained structure of neural processing.

ontology. An ontology such as that envisioned by Price & Friston, Poldrack, or Anderson would likely only be translatable into folk psychological terms with considerable effort, and might not even be translatable in any meaningful sense at all. Rather than seeing this a problem to be avoided, I think we should instead try to come to terms with it, by being honest about the limitations of the scientific image for making sense of everyday experience, without thereby taking this to mean that everyday experience is somehow inexplicable. This is something that we have already had to deal with in other domains, such as physics, where our best ontology has no clear correspondence to everyday experience. The issue is perhaps more pressing when it comes to folk psychology, which is both more immediate and more personal than folk physics, but this doesn't mean that such an approach cannot be made to work.[15]

It is also possible that empirical and theoretical developments in cognitive neuroscience *will* eventually have an impact on the folk psychological ontology, in the same way that psychoanalytic concepts like 'the unconscious' entered the folk ontology during the twentieth century (cf. Richards 2000). Something similar may have happened in recent decades with the adoption (by the folk) of neurochemical terminology when describing and attributing certain kinds of mental states. e.g. statements like 'I'm not feeling great today, my serotonin levels are a bit low' (cf. Rodriguez 2006; Rose and Abi-Rached 2013; Francken and Slors 2018). It seems plausible (as Murphy 2017b suggests) that the most extensive changes to the folk ontology might come in response to psychiatric research, which can sometimes offer satisfying explanations for otherwise disturbing or inexplicable behaviour. In the non-pathological cases, where folk psychology is relatively successful at predicting and explaining behaviour, there is no need for it to conform to novel scientific categories, but in the pathological cases, where this behaviour is perhaps harder for it to explain, it might be more susceptible to influences from the cognitive scientific ontology (cf. Matthews 2013).

Here the difference from classical eliminativism is that these adjustments are not mandated or required by philosophers, but rather occur naturally as a process of linguistic or conceptual development, and may often not accurately reflect the neuroscience that they are inspired by. Murphy (2017a: 141) suggests that the real impact of cognitive neuroscience on folk discourse might have more to do with the ethical and political implications of our changing self-conception, but I think that such concerns are best addressed independently of empirical questions about what kind of ontology is best suited for scientific practice.[16] Adopting a more

[15]One strategy, which I will not pursue here, would be to use our neurocognitive ontology to *explain* why our folk psychological ontology is the way that it is, without treating such an explanation as a route to elimination or reduction. This would be a non-eliminativist version of the so-called 'illusionist' approach to conscious experience (see e.g. Frankish 2017), although as noted by Graziano (2016: 112–3), the label 'illusionist' might be somewhat misleading in this context.

[16]Knobe (2007) explores some ways in which moral judgements might both influence and be influenced by folk psychology, and suggests that neuroscientific concepts could not play the same

coarse-grained approach would allow us to keep questions about our neuroscientific ontology separate from questions about our folk ontology, even if the latter is sometimes informally influenced by the former (and *vice versa*, for better or worse). To be clear, my view is not that the folk psychological ontology is necessarily static and unchanging, but rather that changes to it are likely to be at best indirectly related to changes to our neurocognitive ontology, via the unpredictable medium of social and cultural interpretation of psychology, psychiatry, and neuroscience. Any deliberate attempts to change the folk psychological ontology are likely to be ineffective at best and/or to have unintentional (and potentially harmful) outcomes at worst.

How does all this relate to the broader conception of folk psychology introduced in Sect. 14.2? Character traits are plausibly just another kind of interpretation of whole persons, and already fitted well into the dispositional picture, as being 'brave', for example, can be understood in terms of being disposed to behave bravely when circumstances require it. Folk psychological narratives typically concern the actions of persons, not parts of their brain, and can again be understood as ways of interpreting and explaining those actions. Taking on a certain role in a narrative will also mean being disposed to behave in certain ways, and the social understanding that we can gain from these shared narratives does not depend on the structure of the (neural) mechanisms that generates that behaviour. Finally, the normative constraints imposed by folk psychology can help to ensure that our behaviour conforms to folk psychological expectations, regardless of the structure of subpersonal mechanisms. In this sense folk psychology can be thought of as a kind of self-fulfilling prophecy, sometimes acting to generate the very same behaviour that it predicted.[17] This also means that when we attribute folk psychological states to one another, we may have more than merely epistemic aims in mind (we might also be aiming to influence each other's behaviour, for example). Folk psychology has a broader social function that goes beyond mere prediction and explanation, and this function is not necessarily threatened by revisions to our neurocognitive ontology. Even if it turned out that the language of beliefs and desires, hopes and fears, norms and narratives, and so on, was completely unsuited to our analysis of neuroimaging studies, it would not stop being useful for our understanding of whole persons, and there would be no reason to think that we ought to revise or eliminate it.

By adopting a coarse-grained approach to the folk psychological ontology, we can effectively inoculate it against any eliminativist threat, including not only the

kind of role. For some recent considerations of the broader moral and social implications of contemporary neuroscience, see Caruso and Flanagan (2018).

[17]See Andrews 2015 for further discussion of what she calls "the folk psychological spiral", where our explanation of some unusual behaviour might commit us to acting more predictably in the future. Zawidzki (2013) presents a more general account of how what he calls "mindshaping" might help to regulate our behaviour in a way that makes predicting and explaining it computationally tractable. Understood in this way, folk psychological concepts would constitute socially constructed "human kinds", in Hacking's (1995) sense.

current threat from cognitive ontology revision, but also potential future threats from novel neuroscientific discoveries. At the same time, we ought to be sensitive to the misuse of folk psychological concepts within cognitive neuroscience, especially when such concepts do not pick out cognitive functions that map adequately onto the functional architecture of the brain. In such cases we should develop novel cognitive ontologies that better reflect this architecture, but doing so need not entail making any changes to analogous components of the folk psychological ontology. We can simply accept that the two ontologies have different targets (whole persons versus neural structures), and correspondingly different explanatory standards and predictive goals.

14.7 Conclusion

In Sect. 14.2 I introduced some different ways of understanding folk psychology and argued that the dichotomy between folk psychological realism and eliminativism depends on a fine-grained interpretation, where folk psychological concepts are understood as literally aiming to describe the mechanistic structure of cognition. In Sect. 14.3 I introduced the recent debate over cognitive ontology revision in neuroscience, and in Sect. 14.4 I demonstrated how some existing responses to this debate could threaten our existing folk psychological ontology. In Sects. 14.5 and 14.6 I presented an alternative approach to folk psychology and considered how this might change our understanding of the relationship between folk psychological and neuroscientific ontologies. I argued that, in order to avoid the threat of eliminativism posed by cognitive ontology revision, we ought to reject the fine-grained, literal understanding of folk psychology and instead adopt a coarse-grained approach, where folk psychology aims to predict and explain the behaviour of whole persons rather than tracking the mechanistic structure of cognition. Doing so would insulate folk psychology from the threat posed by cognitive ontology revision, and it can also help us to better understand the relationship between folk psychology and cognitive neuroscience, which should be seen as different levels of description rather than competing ontologies.

Acknowledgments Many thanks to Jonny Lee, Adrian Downey, E. Brown Dewhurst, and Carrie Figdor for providing helpful comments on earlier drafts, to J. Brendan Ritchie for his very helpful reviewer comments, and to Marco Viola and Fabrizio Calzavarini for hosting the *Neural Mechanisms* lecture series and editing this volume. Earlier versions of the material in this chapter have been presented at many workshops and conferences, including the *BSPS 2016 Annual Conference* in Cardiff, the *Early Career Mind Network Research Forum* in Durham in 2016, the "Symposium on Structure-Function Mappings in Cognitive Neuroscience" at the *14th Annual Conference of the Italian Society for Cognitive Science in Bologna in 2017*, and the *Colloquium on Consciousness and Cognition* at the Ruhr-Universität Bochum in June 2018.

References

Amedi, A., Jacobson, G., Hendler, T., Malach, R., & Zohary, E. (2002). Convergence of visual and tactile shape processing in the human lateral occipital complex. *Cerebral Cortex, 12*, 1202–1212.

Anderson, M. (2010). Neural reuse: A fundamental organizational principle of the brain. *Behavioural and Brain Sciences, 33*(4), 254–261.

Anderson, M. (2014). *After phrenology*. Cambridge, MA: MIT Press.

Andrews, K. (2015). The folk psychological spiral: Explanation, regulation, and language. *The Southern Journal of Philosophy, 53*, 50–67.

Apperly, I. A. (2008). Beyond simulation-theory and theory-theory. *Cognition, 107*(1), 266–283.

Barrett, H. C. (2012). A hierarchical model of the evolution of human brain specializations. *Proceedings of the National Academy of Sciences, 109*, 10733–10740.

Bennett, M. R., & Hacker, P. M. S. (2003). *Philosophical foundations of neuroscience*. Malden, MA: Blackwell Publishing.

Boone, W., & Piccinini, G. (2016). The cognitive neuroscience revolution. *Synthese, 193*(5), 1509–1534.

Boyd, R. (1999). Homeostasis, species, and higher taxa. In Wilson (Ed.), *Species: New interdisciplinary essays*. Cambridge, MA: MIT Press.

Bruner, B. (1990). *Acts of meaning*. Cambridge, MA: HUP.

Burnston, D. (2016). A contextualist approach to functional localization in the brain. *Biology and Philosophy, 31*(4), 527–550.

Caruso, G., & Flanagan, O. (Eds.). (2018). *Neuroexistentialism*. Oxford: OUP.

Churchland, P. M. (1979). *Scientific realism and the plasticity of mind*. Cambridge, UK: CUP.

Churchland, P. M. (1981). Eliminative materialism and the propositional attitudes. *Journal of Philosophy, 78*, 67–90.

Churchland, P. S. (1986). *Neurophilosophy: Toward a unified science of the mind/brain*. Cambridge, MA: MIT Press.

Clark, A. (2019). Beyond desire? Agency, choice, and the predictive mind. *Australasian Journal of Philosophy*. https://doi.org/10.1080/00048402.2019.1602661.

Cohen, L., Dehaene, S., Naccache, L., Lehericy, S., Dehaene-Lambertz, G., Henaff, M., & Michel, F. (2000). The visual word form area: Spatial and temporal characterization of an initial stage of reading in normal subjects and posterior split-brain patients. *Brain, 123*, 291–307.

Curry, D. S. (2018). Beliefs as inner causes: The (lack of) evidence. *Philosophical Psychology, 31*(6), 850–877.

Davidson, D. (1980). *Essays on actions and events*. Oxford: OUP.

De Jaegher, H., & Di Paolo, E. (2007). Participatory sense-making: An enactive approach to social cognition. *Phenomenology and the Cognitive Sciences, 6*(4), 485–507.

Dennett, D. (1969). *Content and consciousness*. Routledge and Kegan Paul.

Dennett, D. (1981). True Believers. In Haugeland (Ed.), *Mind Design*. Cambridge, MA: MIT Press.

Dennett, D. (1987). *The intentional stance*. Cambridge, MA: MIT Press.

Dennett, D. (1991). Real patterns. *The Journal of Philosophy, 88*(1), 27–51.

Dewhurst, J. (2017). Folk psychology and the Bayesian brain. In Metzinger & Wiese (Eds.), *Philosophy and predictive processing*. Frankfurt am Main: MIND Group.

Dewhurst, J. (2019). Context sensitive ontologies for a non-reductionist cognitive neuroscience. *Australasian Philosophical Review, 2*(2), 224–228.

Drayson, Z. (2012). The uses and abuses of the personal/subpersonal distinction. *Philosophical Perspectives, 26*(1), 1–18.

Drayson, Z. (2014). The personal/subpersonal distinction. *Philosophy Compass, 9*(5), 338–346.

Dupre, J. (1981). Natural kinds and biological taxa. *The Philosophical Review, 90*(1), 66–90.

Feyerabend, P. (1963). Mental events and the brain. *Journal of Philosophy, 60*, 295–296.

Figdor, C. (2011). Semantics and metaphysics in informatics: Toward an ontology of tasks. *Topics in Cognitive Science, 3*, 222–226.

Figdor, C. (2018). *Pieces of mind*. Oxford: OUP.
Fodor, J. (1974). Special sciences (or: The disunity of science as a working hypothesis). *Synthese, 28*(2), 97–115.
Fodor, J. (1975). *The language of thought*. Cambridge, MA: MIT Press.
Fodor, J. (1987). *Psychosemantics*. Cambridge, MA: MIT Press.
Francken, J., & Slors, M. (2014). From commonsense to science, and back. *Consciousness and Cognition, 29*, 248–258.
Francken, J., & Slors, M. (2018). Neuroscience and everyday life. *Brain and Cognition, 120*, 67–74.
Frankish, K. (Ed.). (2017). *Illusionism as a theory of consciousness*. Exeter: Imprint Academic.
Gallagher, S. (2008a). Direct perception in the intersubjective context. *Consciousness and Cognition, 17*, 535–543.
Gallagher, S. (2008b). Inference or interaction. *Philosophical Explorations, 11*(3), 163–174.
Glymour, C., & Hanson, C. (2016). Reverse inference in neuropsychology. *The British Journal for the Philosophy of Science, 67*(4), 1139–1153.
Gopnik, A., & Wellman, H. (1992). Why the Child's theory of mind really is a theory. *Mind and Language, 7*, 145–172.
Gordon, R. (1986). Folk psychology as simulation. *Mind and Language, 1*(2), 158–171.
Graziano, M. (2016). Consciousness engineered. *Journal of Consciousness Studies, 23*(11–12), 98–115.
Griffiths, P. (1997). *What emotions really are: The problem of psychological categories*. Chicago, IL: UCP.
Hacking, I. (1995). The looping effects of human kinds. In Sperber, Premack, & Premack (Eds.), *Causal Cognition*. Oxford: Clarendon Press.
Heal, J. (1986). Replication and functionalism. In Butterfield (Ed.), *Language, mind, and logic*. Cambridge: CUP.
Hutto, D. (2008). *Folk psychological narratives*. Cambridge, MA: MIT Press.
Klein, C. (2012). Cognitive ontology and region- versus network-oriented analyses. *Philosophy of Science, 79*(5), 952–960.
Klein, C. (2018). What do predictive coders want? *Synthese, 195*(6), 2541–2557.
Knobe, J. (2007). Folk psychology: Science and morals. In Hutto & Ratcliffe (Eds.), *Folk Pscyhology re-assessed*. Springer.
Knoll, A. (2018). Still autonomous after all. *Minds and Machines, 28*(1), 7–27.
Lavelle, J. S. (2012). Theory-theory and the direct perception of mental states. *Review of Philosophy and Psychology, 3*, 213–230.
Lavelle, J. S. (2019). *The social mind*. Routledge.
Lenartowicz, A., Kalar, D. J., Congdon, E., & Poldrack, R. A. (2010). Towards an ontology of cognitive control. *Topics in Cognitive Science, 2*(4), 678–692.
Lewis, D. (1972). Psychophysical and theoretical identifications. *Australasian Journal of Philosophy, 50*(3), 249–258.
Ludwig, D. (2017). Indigenous and scientific kinds. *The British Journal for the Philosophy of Science, 68*(1), 187–212.
Lycan, W. (1988). *Judgement and justification*. Cambridge: CUP.
Machery, E. (2009). *Doing without concepts*. Oxford: OUP.
Mallon, R., Machery, E., Nichols, S., & Stich, S. (2009). Against arguments from reference. *Philosophy and Phenomenological Research, 79*(2), 332–356.
Mameli, M. (2001). Mindreading, mindshaping, and evolution. *Biology and Philosophy, 16*(5), 595–626.
Martin, A., & Chao, L. L. (2001). Semantic memory and the brain: Structure and processes. *Current Opinion in Neurobiology, 11*, 194–201.
Matthews, R. (2007/2010). *The measure of mind: Propositional attitudes and their attribution*. Oxford: OUP.
Matthews, R. (2011). Measurement-theoretic accounts of propositional attitudes. *Philosophy Compass, 6*(11), 828–841.

Matthews, R. (2013). Belief and Belief's penumbra. In Nottlemann (Ed.), *New essays on belief.* Palgrave Macmillan.

Matthews, R. 2017. "The Elusive Case for Relationalism about the Attitudes." *Philosophy and Phenomenological Research*, online first. https://doi.org/10.1111/phpr.12380.

McCaffrey, J. (2015). The Brain's heterogeneous functional landscape. *Philosophy of Science, 82*(5), 1010–1022.

McCaffrey, J., & Machery, E. (2016). The reification objection to bottom-up cognitive ontology revision. *Behavioral and Brain Sciences, 39*, e125.

McCauley, & Bechtel, W. (2001). Explanatory pluralism and heuristic identity theory. *Theory & Psychology, 11*(6), 736–760.

McGeer, V. (2007). The regulative dimension of folk psychology. In Hutto & Ratcliffe (Eds.), *Folk psychology re-assessed.* Springer.

McDowell, J. (1994). The content of perceptual experience. *The Philosophical Quarterly, 44*(175), 190–205.

Mitchell, J. P. (2005). The false dichotomy between simulation and theory-theory. *Trends in Cognitive Science, 9*(8), P363–P364.

Murphy, D. (2017a). Brains and beliefs. In Kaplan (Ed.), *Explanation and integration in mind and brain science*. Oxford: OUP.

Murphy, D. (2017b). Can psychiatry refurnish the mind? *Philosophical Explorations, 20*(2), 160–174.

Piccinini, G., & Craver, C. (2011). Integrating psychology and neuroscience: Functional analyses as mechanism sketches. *Synthese, 183*(3), 283–311.

Poldrack, R. (2010). Mapping mental function to brain structure. *Perspectives on Psychological Science, 5*(6), 753–761.

Poldrack, R., & Yarkoni, T. (2016). From brain maps to cognitive ontologies. *Annual Review of Psychology, 67*, 587–612.

Premack, D., & Woodruff, G. (1978). Does the chimpanzee have a theory of mind? *Behavioral and Brain Sciences, 4*, 515–526.

Price, C. J., & Friston, K. J. (2005). Functional ontologies for cognition: The systematic definition of structure and function. *Cognitive Neuropsychology, 22*(3), 262–275.

Quilty-Dunn, J., & Mandelbaum, E. (2018). Against dispositionalism: Belief in cognitive science. *Philosophical Studies, 175*, 2353–2372.

Richards, G. (2000). Britain on the couch: The popularisation of psychoanalysis in Britain 1918-1940. *Science in Context, 13*(2), 183–230.

Rodriguez, P. (2006). Talking brains: A cognitive semantic analysis of an emerging folk neuropsychology. *Public Understanding of Science, 15*, 301–330.

Rorty, R. (1965). Mind-body identity, privacy, and categories. *The Review of Metaphysics, 19*(1), 24–54.

Rose, N., & Abi-Rached, J. M. (2013). *Neuro: The new brain sciences and the management of the mind*. Princeton University Press.

Ross, D. (2000). Rainforest realism: A Dennettian theory of existence. In D. Ross, A. Brook, & D. Thompson (Eds.), *Dennett's philosophy: A comprehensive assessment*. Cambridge, MA: MIT Press.

Ryle, G. (1949). *The concept of mind*. Hutchinson.

Schwitzgebel, E. (2002). A phenomenal, dispositional account of belief. *Nous, 36*, 249–275.

Sellars, W. (1956). Empiricism and the philosophy of mind. In Feigl & Scrivens (Eds.), *Minnesota studies in the philosophy of science*. University of Minnesota Press.

Sellars, W. (1963). *Science, perception, and reality*. New York: Humanities Press.

Slater, M. (2015). Natural kindness. *The British Journal for the Philosophy of Science, 66*(2), 375–411.

Spaulding, S. (2018). Mindreading beyond belief: A more comprehensive conception of how we understand others. *Philosophy Compass, 13*(11).

Stich, S. (1983). *From folk psychology to cognitive science*. Cambridge, MA: MIT Press.

Stich, S. (1996). *Deconstructing the mind*. Oxford: OUP.

Towl, B. N. (2011). Mind-brain correlations, identity, and neuroscience. *Philosophical Psychology, 25*(2), 187–202.
Westra, E. (2018). Character and theory of mind: An integrative approach. *Philosophical Studies, 175*(5), 1217–1241.
Zawidzki, T. (2013). *Mindshaping*. Cambridge, MA: MIT Press.

Part IV
Mechanistic Explanations

Chapter 15
Integration and the Mechanistic Triad: Producing, Underlying and Maintaining Mechanistic Explanations

Lena Kästner

Abstract Integration is a grand challenge for many contemporary research endeavors. Mechanistic explanations provide a multi-level approach especially suited to bring out different aspects of the causal-mechanical structure of the world. Yet, we encounter a triad of differently structured producing, underlying and maintaining mechanistic explanations. Understanding how the elements of this triad can be fruitfully and systematically linked, I suggest, may help drive scientific progress and integration. This paper discusses important conceptual ties between an explanandum and the metaphysical relations highlighted in the corresponding explanans: to explain how an *end product* or result is generated, scientists will usually search for the mechanism that produced it; to explain a *process*, they will typically search for the mechanism underlying it; and to explain how a system's *stable state* or *continuous behavior* is maintained, they search for the mechanism maintaining it. Appreciating these different projects, and understanding the connections between them, provides an important backdrop for explanatory integration. Besides, it allows us to reconcile apparently different conceptions of mechanistic explanations without heavy metaphysical baggage.

Keywords Integration · Mechanistic Explanation · Mechanism · Phenomena · Discovery · Integration · Maintaining Mechanism · Causation · Constitution

15.1 Introduction

As there is ever more specialization in the sciences, bringing together insights from multiple different perspectives—or *integration*—is becoming a crucial contemporary challenge (e.g. Green et al. 2015; O'Rourke et al. 2016). This is especially true for interdisciplinary research endeavors, such as, for example, evolutionary systems

L. Kästner (✉)
Department of Philosophy, Saarland University, Saarbrücken, Germany
e-mail: mail@lenakaestner.de

© Springer Nature Switzerland AG 2021
F. Calzavarini, M. Viola (eds.), *Neural Mechanisms*, Studies in Brain and Mind 17,
https://doi.org/10.1007/978-3-030-54092-0_15

biology, where insights about specific local mechanisms are being brought together with more general or global explanatory principles (cf. Wayne 2018). While there seems to be agreement that "integration is a generic combination process the details of which are determined by the specific contexts in which particular instances of integration occur" (O'Rourke et al. 2016, p. 67), there currently is no unequivocal philosophical account of what precisely integration is and how it works.

Recent debates about explanations in the philosophy of science capitalize on the potential of *mechanisms* to provide integrated multi-level or *mosaic* explanations (e.g. Craver 2007a, b). According to the mechanistic approach, scientists (at least in the life sciences) *explain phenomena by discovering the mechanisms responsible for them.* A range of different characterizations of mechanisms has been offered but a general consensus may be expressed as follows (see also Craver and Tabery 2015; Glennan 2017, p. 17; Illari and Williamson 2012, p. 120):

> A mechanism for a phenomenon consists of entities (or parts) whose activities and interactions are organized in such a way that they are responsible for the phenomenon.

While this unifying characterization usefully communicates the central tenets of the mechanistic approach it glosses over some important issues. For one thing, there is quite some discussion about what exactly "phenomena" are (e.g. Bogen and Woodward 1988; Feest 2016; Colaço 2018). I shall here side with the mainstream assumption that phenomena are the *explananda* of mechanistic explanations. For another, recent discussions about different kinds, types, or readings of mechanisms highlight that the mechanistic view provides shelter for a number of different specifications. Craver and Darden (2013, ch. 5), for instance, describe three *kinds* of mechanisms: mechanisms that produce, underlie, and maintain their phenomena, respectively. Kaiser and Krickel (2016) distinguish between causal and constitutive *readings* of mechanisms and Glennan (2017, ch. 5) enumerates a whole range of mechanism *types*. While all of these distinctions clearly have their merits, it is important to separate a number of different, though related, questions arising in their contexts. For instance, there are metaphysical questions about the nature of "being responsible for"; much of the contemporary mechanistic literature has focused on this metaphysical question of how mechanisms relate to their phenomena and how different metaphysical relationships—especially causation and constitution—can be identified within the mechanistic framework (e.g. Baumgartner and Gebharter 2015; Couch 2011; Fagan 2012; Harbecke 2010, 2015; Kästner 2017, Kästner and Andersen 2018; Kaiser and Krickel 2016; Krickel 2017; Leuridan 2012; Romero 2015).[1] Besides, there are also questions about the connection between ontology, discovery and explanation that are much less discussed (but see Kästner and Haueis 2019) but highly relevant for questions of integration. The focus of this paper shall therefore lie with them.

[1]The distinction between causal (etiological or productive) (e.g. Darden 2006, 2016; Darden et al. 2018) vs. constitutive or componential (e.g. Craver 2007a, b) mechanistic views parallels the distinction Salmon (1984) draws between constitutive and etiological explanations.

The central project of this paper is to illuminate how different mechanistic explanations can be linked up, or *integrated into a mechanism mosaic* (cf. Craver 2007a). To achieve this, I argue, we must understand what kinds of mechanistic explanations there are and how they relate to one another. My starting point will be Craver and Darden's (2013) recent treatment of mechanism discovery. In Sect. 15.2, I will briefly recap the discovery strategies associated with mechanisms producing, underlying, and maintaining their phenomena, respectively. I then argue, in Sect. 15.3, that whether scientists construct a mechanistic explanation that emphasizes causal, componential, and continuous aspects of the world will be determined by what kind of phenomenon they consider as their explanandum. To explain how an *end product* or final outcome is generated, scientists will usually search for the mechanism that produced it. To explain a *process*, they will typically search for the mechanism underlying it. And to explain a continuous rather than finite phenomenon, i.e. how a *property is kept stable* or a *continuous behavior is upheld*, they will look for the mechanism maintaining it. Based on these insights we begin to see how different kinds of mechanistic explanations hang together. Section 15.4 illustrates the relationship more clearly by looking at the example phenomenon of lactose metabolism in E. coli.

Before I begin, let me emphasize two things. First, I am not heading for a project in metaphysics. While some of the discussion I may also be read metaphysically, my concerns are about *explanation* and explanatory integration. I wish to remain agnostic about questions of ontology, as well as with respect to the question of whether (mechanistic) explanations are constrained primarily by ontic or epistemic norms (but see Kästner and Haueis 2019). Second, my contribution is meant to be constructive rather than controversial. By calling attention to some important ideas from recent mechanist philosophy and explicating them in some detail, I draw a picture of mechanistic integration that might well be instructive for discussions about integration more generally.

15.2 Mechanisms and Discovery: Glassboxing and the Mechanistic Triad

According to Machamer, Darden and Craver (Machamer et al. 2000), mechanisms constitute *nested hierarchies*, the levels of which "should be thought of as part-whole hierarchies" (p. 13, see also Craver 2007a, b). Scientists need not unfold this hierarchy completely, though. They may keep abstract descriptions or *mechanism schemas*.[2] A familiar graphical illustration of mechanisms due to Craver (2007a) is shown in Fig. 15.1. Much of the current discussion about mechanistic explanation

[2]Mechanism schemas are different from *mechanism sketches*, which contain missing pieces and black boxes we cannot (yet) fill in to yield a complete mechanistic explanation (e.g. Machamer et al. 2000; Craver 2007a; Craver and Darden 2013).

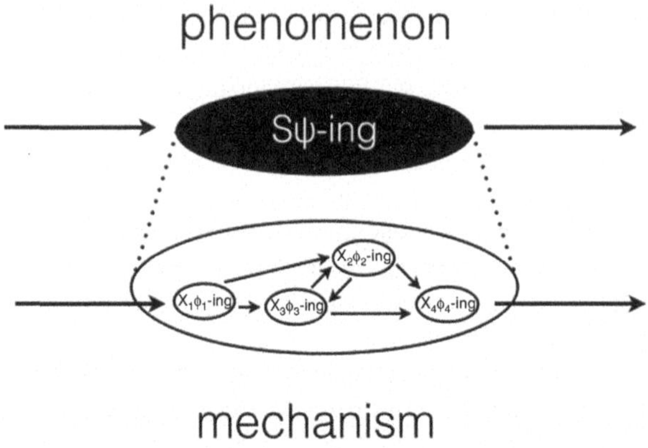

Fig. 15.1 A very well-known illustration of mechanisms. (See Craver 2007a, p. 121)

and mechanism discovery focuses on the idea that each of the components in a mechanism can itself be analyzed as a mechanism which has components that can be further mechanistically analyzed, and so on. Eventually, the whole thing bottoms out and the mechanism is transformed from an initial blackbox into a complete glassbox (Craver and Darden 2013) revealing nested mechanisms all the way down. On this glassboxing story, integration is primarily a matter of filling in the details of the mechanism using insights from different perspectives.

While glassboxing certainly is a vital part of mechanism discovery, it is neither unproblematic nor the full story. First, since mechanistic levels are strictly local there is no way to relate the subcomponents of two different components in a single larger mechanism, even if that larger mechanism has been successfully turned into a glass box (see Fazekas and Kertész 2011; Kästner 2018). Second, mechanism discovery is neither merely a downward-looking affair (e.g. Bechtel and Abrahamsen 2009), nor do more details always make for better explanations (Craver and Kaplan 2018). Indeed, it seems quite obvious that scientists often "look up and around, not just down" (Darden et al. 2018, p. 101). They study related phenomena, focus on different research questions and employ different methodologies and discovery strategies. Therefore, it seems only plausible to assume that successful mechanism discovery will need to combine insights gained through different discovery strategies. For descriptions of mechanisms can be provided at multiple levels and multiple degrees of abstraction (cf. Craver 2007a, ch. 7, Craver 2015; Glennan 2017, ch. 5) that will naturally require different tools and methodologies while differentially emphasizing various aspects of the causal-mechanical structure of the world.

Though they emphasize the role of glassboxing, Craver and Darden (2013) acknowledge that mechanism discovery is a complex and stepwise process throughout which mechanism sketches (and later schemas) as well as phenomenon characterizations are repeatedly revised in light of new insights about the inner workings of the mechanism (ibid., ch. 5). The overall structure of a mechanism schema for a given phenomenon, Carver and Darden suggest, is guided by "the decision about whether one is seeking a mechanism that produces, maintains, or underlies a phenomenon." (ibid., p. 65) The intended target of discovery, that is, shapes the discovery process (ibid., p. 15, see also Darden et al. 2018, p. 115). This lines up well with the idea shared among contemporary mechanistic philosophers that the same set of norms applies in mechanistic explanation and discovery. Indeed, discovery (a least in the life sciences) simply *is* the—often stepwise—development of mechanistic explanations (Bechtel and Richardson 2010, ch. 2); and mechanistic explanations simply *are* the product of—successive episodes of—discovery (Craver and Darden 2013, pp. 7, 65). Against this background then, it should not be surprising that the questions which guide mechanism discovery will have a significant impact on how the resulting mechanistic explanations are structured and what kinds of metaphysical relations they emphasize (see also Glennan 2017, pp. 93, 109). The question is how to piece different mechanistic explanations together to arrive at an integrated mosaic.

Before we can piece together the mosaic, though, we need to examine the pieces. Craver and Darden (2013) distinguish three discovery strategies, each of which they associate with a specific kind of mechanism being uncovered: If scientists search for a mechanism *producing* a phenomenon, they often start from the final product and search for the activities by which the mechanism's entities are transformed into the product. If they search for a mechanism *underlying* the phenomenon, scientists typically break down a system into its working parts to show how these parts are organized to give rise to the phenomenon to be explained. If scientists search for mechanisms *maintaining* a phenomenon, they search for factors that disturb the phenomenon as well as those correcting for the disturbances. The resulting mechanisms will vary accordingly (see Fig. 15.2). Note that despite their structural differences, all three kinds of mechanisms are captured by our consensus definition as "being responsible for" is deliberately ambiguous: it can refer to production, underlying or maintenance (see Sect. 15.1). However, little has been said so far about the *relations* between such different mechanisms. Yet, understanding how different mechanisms may be linked is the key to constructing an integrated mechanism mosaic.

Note that Craver and Darden talk about different kinds of *mechanisms,* not mechanistic *explanations*. I do not deny that their triad may be read metaphysically,[3] nor that a lot could be said about the metaphysics of different mechanisms. Nor do I doubt that the metaphysical structure of the world constrains mechanism discovery and explanation. In fact, I agree with Craver's (2013, p. 140) suggestion that one has to "carve mechanisms out of the busy and buzzing confusion that constitutes the causal structure of the world". Yet, my current project is not a metaphysical

[3]Indeed, there seem to be ontic commitments in the background (see Craver 2014) when Craver and Darden claim that "the intended target of the search—mechanisms—shapes the process" (2013, p. 15).

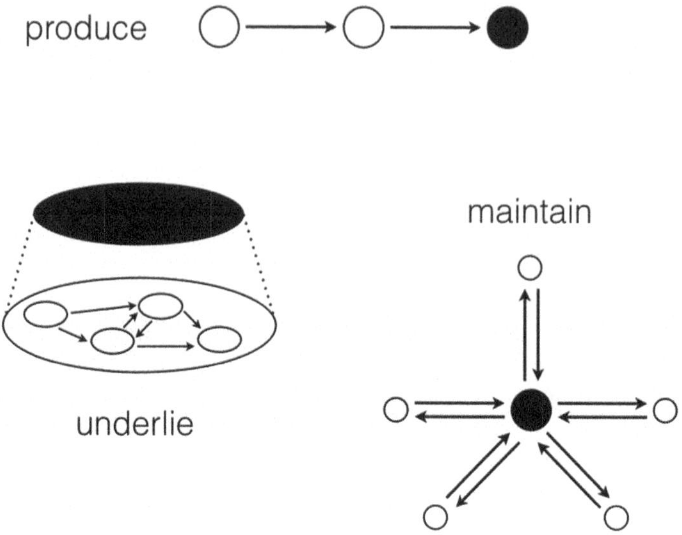

Fig. 15.2 Three kinds of mechanisms; black circles depict the phenomenon to be explained. (Adapted from Craver and Darden 2013, p. 66)

one: I am interested in how *explanations* describing producing, underlying and maintaining mechanisms are constructed and how they can be linked.

Many philosophers of science assume that different kinds or types of explanations serve to answer different research questions (e.g. Van Fraassen 1977; Giere 2006; Lange 2000; Salmon 1984).[4] This assumption is also inherent in the mechanistic view: mechanistic explanations are always mechanistic explanations *for a phenomenon* (e.g. Glennan 1996, 2017; Machamer 2004; Bechtel and Abrahamsen 2005; Craver 2007a, b; Craver 2013; Illari and Williamson 2012). Against this background, I suggest that whether a mechanistic explanation emphasizes causal (producing), componential (underlying), and continuous (maintaining) aspects of the world, respectively, crucially depends on how we specify the *explanandum*, viz. what kind of phenomenon we seek to explain.

In what follows, I shall examine which kinds of phenomena different mechanistic explanations serve to explain and how they are related. With this understanding in place, I take it, at least some mechanistic integration can be grounded in relations between the explananda of mechanistic explanations.

[4] I am using "kinds" as a non-technical notion throughout the paper to refer to different sorts, types, or classes of explanations and phenomena, respectively.

15.3 Explaining Different Phenomena

I suggest that the kind of mechanistic explanation researchers seek (i.e. the mechanism they carve out) depends on the nature of the phenomenon to be explained: whether it is an end product, a process, or a stable state or continuous operation being upheld. Depending on the exact research questions they ask, scientists may emphasize different aspects of the world, hence providing differently structured mechanistic explanations. For linguistic convenience, I shall at times simply talk of "mechanisms being discovered" rather than "mechanistic explanations representing mechanisms based on the outcomes of the discovery process". I begin by discussing underlying and producing mechanisms before I turn to maintaining ones.

15.3.1 Underlying and Producing

To discover underlying mechanisms scientists may primarily use decomposition strategies. Craver and Darden describe this as follows:

> [...] one typically breaks the system as a whole into component parts that one takes to be working components in a mechanism, and one shows how they are organized together, spatially, temporally, and actively such that they give rise to the phenomenon as a whole. (Craver and Darden 2013, pp. 65–66)

This is reminiscent of Craver's (2007a) view—as is the visual representation (see Sect. 15.2): Scientists employ intra- and interlevel manipulations to investigate what the relevant (lower-level) parts of a mechanism are and how they work together. Discovering underlying mechanisms is to identify the entities, their activities and their (spatio-temporal) organization. Taken together, these form the explanans. The explanandum is the complex (higher-level), more or less finite, *process* that occurs *while* the mechanism operates.[5] The overall phenomenon is implemented by the spatio-temporally organized acting entities or, put slightly differently, the causal interactions among mechanistic components (cf. Tabery 2004). For illustration consider the action potential: this phenomenon can be explained by the flowing of ions across membranes through voltage-gated channels. It consists in different phases we can describe in terms of the orchestrated activities of the participating entities (e.g. sodium influx during the rising phase, potassium efflux during the falling phase, etc.). Other prototypical examples include cognitive functions such as memory or spatial orientation that are explained in terms of neural processes like hippocampal long-term potentiation.

How about producing mechanisms? To discover a producing mechanism, Craver and Darden suggest,

[5]To pick up on Kaiser and Krickel's (2016) distinction: underlying mechanisms are of the constitutive kind while producing ones are of the causal kind.

> [...] one typically starts with some understanding of the end product and seeks the components that are assembled and the processes by which they are assembled and the activities that transform them on the way to the final stage. (Craver and Darden 2013, p. 65)

Notice the change in explanandum here. While underlying mechanisms are supposed to explain an overall process, the relevant explananda of producing mechanisms are final stages, *end products*, or *outcomes* of (supposedly causal) processes. Scientists seeking a producing mechanism may simply be looking for a causal sequence leading from one event to the next, eventually leading up to the phenomenon (the end product) of interest. For illustration consider a protein being synthesized. When we ask for an explanation of how the protein has been synthesized, we essentially ask how it has been produced. The explanation we expect in response will make reference to the relevant steps of protein synthesis. These include, very crudely speaking, transcription of DNA, followed by mRNA transferal from the nucleus to ribosomes, and translation of the mRNA into proteins. These different stages form a causal sequence that eventually results in the phenomenon we are trying to explain—it produces the end product. This reading of productive mechanisms squares well with Darden's (2006, 2008, 2018) account of mechanisms in which she emphasizes production. Explanations describing producing mechanisms are essentially causal in character; and they do not usually invoke multiple levels.[6]

The distinction between producing and underlying mechanisms mirrors the familiar distinctions between etiological and constitutive explanation (Salmon 1984; Craver 2007a) or causal and constitutive mechanistic explanations (Kaiser and Krickel 2016). Glennan's recent discussion of non-constituted and constituted phenomena (2017, ch. 5) also assumes such a distinction. However, it is important to notice that there is more to this difference than how the phenomenon-mechanism relation is construed (as causal or constitutive): explanations describing producing and underlying mechanisms, respectively, have *different explananda* (end products or overall processes). They offer responses to different kinds of questions, and it takes different discovery strategies to find them.

With the above in mind there are at least three plausible stories as to how underlying and producing mechanisms relate—and although most of my analysis falls directly out of the current literature on mechanisms, the relations in question are rarely made explicit. First, we may read Craver and Darden's description of how scientists discover producing mechanisms to be an account of how scientists study the *organization* of and causal relations among mechanistic components *within* an underlying mechanism. In this case, the causal explanation provided in terms of a productive mechanism could just be "plugged into" a (multi-level) mechanistic explanation. If this is correct, we may more adequately depict producing mechanisms as shown in Fig. 15.3. They essentially operate within underlying mechanisms. The underlying mechanism for the overall phenomenon (grey circle

[6]They may of course postulate causal connections between entities at different levels. However, this does not give them a systematic interlevel character.

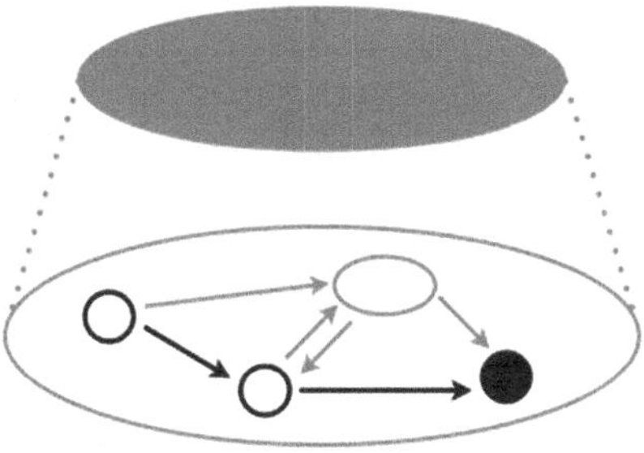

Fig. 15.3 Production within an underlying mechanism

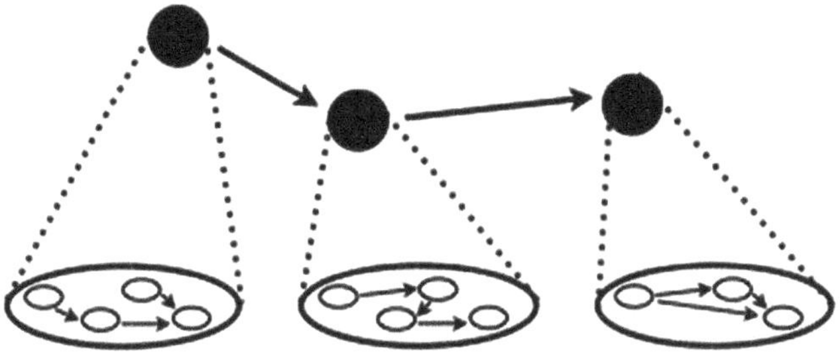

Fig. 15.4 Mechanisms underlying elements of productive mechanism

at the top) is spelled out by studying (some of) the productive stages within the mechanism (black). In studying the productive aspects within an underlying mechanism, scientists temporarily switch the explanandum: they focus on how a certain state or activity of a given component within the mechanism (black) is produced.

Second, each step in the causal sequence of a producing mechanism may be spelled out further by identifying the underlying mechanisms at each stage. If we want to explain how a protein was synthesized, for instance, we can seek the mechanisms underlying (i) transcription of DNA, (ii) mRNA transferal, and (iii) translation of the mRNA into proteins. This is illustrated in Fig. 15.4. Like in the first case, scientists here change the explanandum: to discover the underlying mechanisms at each stage they must ask how each of the processes occurring in (i)–(iii) are implemented rather than what is produced at the end of the sequence (thus all of the top-circles are black).

Finally, we may think that scientists searching for a producing mechanism are actually looking at different stages throughout the operation of an underlying mechanism over time; they are investigating how these different stages causally link up with one another. In this case, the explanandum is the behavior of the *same mechanism* at *different times*. We may depict this scenario as shown in Fig. 15.5. Still, the overall explanatory goal is similar to that in the second case: to analyze the mechanisms underlying a sequence of causally linked events. In both cases, underlying mechanisms essentially "fill in the details" of producing mechanisms.

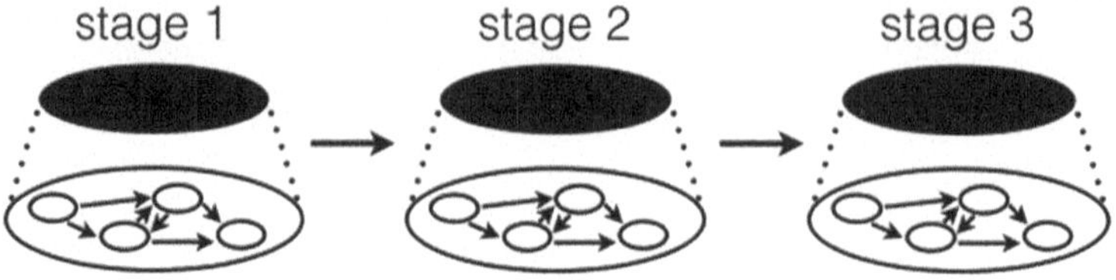

Fig. 15.5 Productive stages of underlying mechanism's operation

The major difference between these scenarios is how the information is integrated into a coherent picture. When spelling out a productive mechanism by discovering the underlying mechanisms at each step in the causal sequence, scientists offer a spatio-temporal decomposition of each step within a causal sequence. Each of these steps may be considered a black box that gets opened up. By contrast, when investigating the operation of a single mechanism over time, scientists look at the causal interactions among the same set of spatial parts (i.e. potentially relevant components) within a mechanism over time: they study the organization of entities and activities within the same mechanism while it produces the phenomenon at different stages (e.g. depolarization, rising and falling phases of the action potential). This serves to study temporal as well as spatial organization. Rather than offering a merely "downward-looking" decomposition, it highlights the dynamics of the internal workings of a mechanism producing the phenomenon to be explained.[7]

In the resulting representations of mechanisms of the overall phenomenon (e.g. the action potential) we will typically find insights gained from the different stages superimposed on a single underlying mechanism picture (I will return to this point when discussing maintaining mechanisms). When we aim to spell out the different steps of a causal sequence, by contrast, we typically find multi-level representations like the one shown in Fig. 15.6. Notice, though, that Fig. 15.6 is not the result of sole black box opening. It combines aspects of all three scenarios discussed here: The first scenario provides an analysis of some of the causal productive processes within a mechanism known to underlie a phenomenon. The second scenario helps to further spell out how the contributing processes themselves are mechanistically implemented; i.e. what mechanisms underlie them. The third scenario helps us to study how the system is organized over time. This latter information is often implicit in the structural representation of the mechanism.

To arrive at a multi-level mechanism like the one in Fig. 15.6, and eventually construct a larger mechanism mosaic, scientists must figure out how exactly to link the different mechanisms they discover. This incurs practical challenges such as switching between different descriptions and vocabularies and applying different

[7]Note that the relation I am after here is one between producing and underlying mechanisms for a phenomenon, *not* between the genesis of the mechanism responsible for the phenomenon and the operation of that mechanisms.

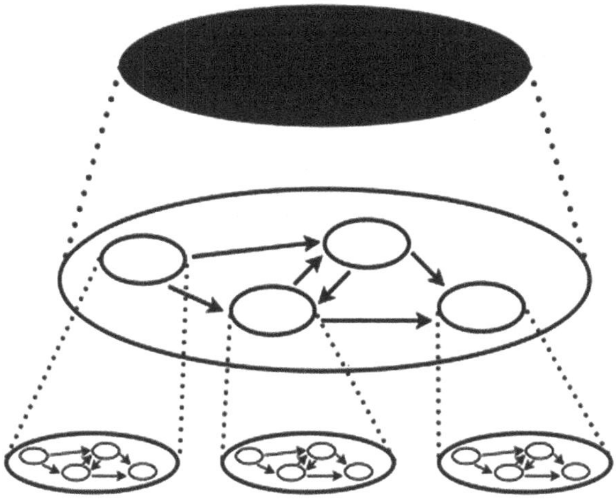

Fig. 15.6 Mechanisms produce and underlie phenomena, multi-level version

tools. But above all, it requires scientists to *explicate the different explananda* and phenomenon-mechanism relations clearly and recognize them across different studies and explanations.

When aiming to discover producing or underlying mechanisms, respectively, we focus on quite different research questions. Yet, their relation is highly systematic. For illustration consider the difference between explaining death (an end product) and dying (a process). If we look for how a phenomenon is produced, we essentially ask for the causes of an end product, or the stages through which its production proceeds. The operation of producing mechanisms temporally *precedes* the presence of the explanandum. If, on the other hand, we look for what underlies a phenomenon, we are asking for its implementational basis, for the operations that are carried out *while* the phenomenon occurs. So whether we will discover a producing or underlying mechanism will essentially be a matter of how we devise our research question. Consider the case of protein synthesis again. The mechanism producing a given protein is the (causal) sequence of events that eventually results in the protein. The mechanism that underlies protein synthesis (the process as a whole, not just the end product) encompasses all the different stages involved in synthesizing a protein. Similarly, we may consider the action potential. Craver and Darden explicitly say that "[t]he mechanism of the action potential [...] underlies or implements the phenomenon of the action potential; it does not produce it." (p. 19) This is obviously true in so far as we consider the action potential as whole as the explanandum, i.e. the whole process from when the membrane potential first deviates from resting state to when it has returned to resting state. However, we may also shift the explanandum and ask instead just how the brief sudden charge we recorded with an electrode in the neuron's axon was generated. In that case, we are no longer asking for an underlying mechanism but for a producing one—for the causes that lead to the electrical signal.

In summary, specifying the explanandum as a product or a process determines what kind of mechanistic explanation is sought. Depending on how scientists investigate, say, a man's dying of cancer, they may seek the mechanism underlying his dying or the mechanism that produced his death. Still, there is only one man and he dies only once! But one can carve up his dying differently, emphasize different aspects of the world (causal or constitutive) and draw the boundaries of the mechanism differently (hence including different sets of events in our explanations) depending on the exact explanatory target (product or process). Conceived this way, we can think of underlying and producing mechanisms as two sides of the same coin.

15.3.2 Maintaining

Producing and underlying mechanisms are familiar; maintaining mechanisms, by contrast, are currently understudied and only rarely feature in the mechanistic literature.[8] They are different from producing and underlying mechanisms in so far as their behavior is cyclic, dynamic, and continuing. Their representations seem to collapse multiple iterations of mechanism operation into a single diagram (see Fig. 15.2). As such, maintaining mechanisms are especially suited to explain regulatory processes. While a comprehensive treatment of maintaining mechanisms is beyond the scope of this paper, my aim here is to shed light on how we may link explanations describing maintaining mechanisms with those describing producing and underlying mechanisms. Painting with broad strokes, I suggest to view maintaining mechanisms as *special cases* of either producing or underlying mechanisms. Their key feature is that they are not linear and finite but cyclic and continuous. Explanations describing maintaining mechanisms can thus answer different kinds of questions. For instance, they are tuned to tell us what happens if a system has to deal with disturbances (in this sense we may consider them dispositional).[9]

Discovery of mechanisms maintaining their phenomena may be described as follows:

[8]But see discussions of modeling mechanisms using recursive Bayes nets, e.g. Casini et al. (2011), Clarke et al. (2014), and Gebharter and Kaiser (2014). Outside Bayesian models a notable exception is Bechtel's (2011) suggestion that "mechanistic explanation [...] must be extended to deal with biological mechanisms whose operations are not sequential but involve cyclic organization" (p. 554). Notice, however, that Bechtel is focusing on mechanisms *within* which there is a cyclic interaction of component yielding complex dynamic behavior. These could still qualify as underlying mechanisms in Craver and Darden's scheme, depending on how we read them.

[9]One might argue that maintaining mechanisms have a normative character distinguishing them from producing and underlying mechanisms; for they serve to keep something as it is supposed to be. For current purposes I will gloss over this issue.

> [...] one typically needs to characterize some process or property (the homeostatic point, shown in the center of the diagram) that is maintained at a given speed or level, one needs to recognize the forces that tend to move the system away from its homeostatic point, and one needs to characterize the process by which those divergences are detected and/or corrected. (Craver and Darden 2013, p. 66)

What is the explanandum in this case? I suggest that what is being maintained can be a *stable state* or a *continuous behavior*. The critic may object that continuous behaviors or processes are not really a homeostatic *point*. However, I take this to be a merely terminological concern. Once I have spelled out my reading of mechanisms maintaining a stable state it will, given what we have already learned about producing and underlying mechanisms, only be natural to include mechanisms maintaining continuously operating processes in the discussion. Whether it is a stable state or a continuous process being maintained, maintenance is achieved through feedback loops. Diverging forces are detected and counterbalancing forces are employed to correct for them. Together the different forces involved, including their detection and correction, make up the mechanism maintaining its phenomenon.[10]

For illustration of a stable state being maintained consider the resting membrane potential. Neurons at resting state are charged at about −70 mV. This negative charge of the intracellular fluid is due to different ion concentrations inside and outside the cell. Ions can permeate the cell's membrane only through specific channels. A few of these channels are open, though, allowing for some ions to leak through. For simplicity, let us just consider sodium (Na+) and potassium (K+)—two key ions in neural processing. There is lots of K+ but only little Na+ inside the cell, while there is lots of Na+ but only little K+ outside it. Hence, there is a diffusion force that pushes K+ out of and Na+ into the cell. Since ions can only pass through open channels, the leakage is very limited during rest where most channels are closed. The additional electrical potential (remember, the intracellular fluid is negatively charged; this is due to the presence of other molecules) leads both K+ and Na+ to leak into the cell. Again, this happens in a very limited fashion during rest. Still, K+ is leaking both in and out of the cell due the presence of both electrical and diffusion forces while Na+ leaks in one direction only. Although the overall leakage of Na+ is much less than that of K+, Na+ leakage is more severe and if there were no correction, the resting membrane potential would eventually disappear. In order to sustain it, the cell engages a so-called sodium-potassium pump. The pump basically is an ATP-fueled ion channel that actively exchanges two K+ ions from outside the cell with three Na+ ions from inside the cell. As the sodium-potassium pump counterbalances ion leakage, the resting membrane potential is maintained. The forces involved in this maintenance are electrical and diffusion forces as well as the sodium-potassium pump counteracting them.

[10]Obviously, before something can be maintained it has to be initially established. Therefore, the operation of maintaining mechanisms may require the previous operation of producing or underlying mechanisms.

Together they make up the (highly simplified) mechanism maintaining the resting membrane potential.

It is not only fixed states that are being maintained. Consider, for instance, circadian rhythms. Circadian rhythms are complex dynamic processes following roughly a 24-h cycle. They endogenously occur in almost all living things; they are probably best known as inner clocks regulating, among other things, sleep-wake cycles. Recent research in chronobiology aims to uncover the mechanisms maintaining sleep-wake cycles as we deal with disturbances such as artificial light or jet lag (e.g. Ohta et al. 2005; Reddy et al. 2002). Without going into the precise details of which genes are expressed and which proteins bind to which receptors it is clear that circadian rhythms are continuous processes. Organisms repeatedly progress through specified phases in an open-ended fashion. Notice the similarity of this to the case of the action potential considered in Sect. 15.3.1. In order to explain the action potential, scientists make reference to different phases (rising phase, peak, falling phase, etc.), each of which can be described by the orchestrated activities of participating entities. When explaining circadian rhythms, like when explaining action potentials, scientists look for an explanation of a process. However, unlike action potentials, circadian rhythms occur continuously; they are maintained over time.

Contrast this to the case of the membrane potential where the phenomenon to be explained is a (relatively) stable state. Similar to a protein that has been synthesized, the explanandum is the final stage or the *outcome* of the operation of a mechanism. It is simply that a maintaining mechanism will have to operate continuously, rather than once from beginning to end, to maintain the phenomenon (e.g. the membrane potential). Thus, the difference between mechanisms maintaining stable states and mechanisms producing phenomena can be construed as analogous to the difference between mechanisms maintaining continuous processes and mechanisms underlying phenomena. In both cases, maintaining mechanisms operate *open-endedly*; they simply keep going. Note, though, that this does not mean that everything a continuously operating mechanism does happens all the time. For illustration consider homeostasis: an infection may trigger a fever which results in a lot of sweating for the patient. A human being may experience infections several times in her lifetime, i.e. there is a sense in which fever and sweating are repetitive. Yet, they are only present if there also is a certain trigger or deviating force (the infection). If no such deviating force is present, no corrections need to be made. But just because one does not sweat (e.g. in winter), the continuous operation of the homeostasis mechanisms does stop.

Against this background, I suggest to view maintaining mechanisms as continuously operating versions of underlying and producing mechanisms, respectively. What distinguishes maintaining mechanisms from producing and underlying ones essentially is their cyclic, repeated, open-ended operation. They emphasize a third aspect of structure of the world, viz. continuity. Whether scientists will look for a maintaining mechanism, rather than a producing or underlying one, will thus—again—be a matter of how they specify the explanandum. One may even think of there being two dimensions along which to classify phenomena: a causal vs. constitutive dimension and a finite vs. continuous dimension (see Table 15.1).

Table 15.1 Phenomenon classification along two dimensions

Phenomena	Caused	Constituted
Linear, finite (specific)	Outcome/end product → producing	Finite overall process → underlying
Cyclic, continuous (general)	Stable state/property → (producing) maintaining	Continuous process → (underlying) maintaining

But is this really in line with Craver and Darden's mechanistic triad? After all, explanations describing maintaining mechanisms will often have quite a different structure from those describing producing or underlying mechanisms. However, I suggest that this is merely an artifact of collapsing the representations of maintaining mechanisms over time: multiple successive stages of mechanism operation are superimposed on one another all in the same spot. Once we transform the graphical representation and "spread" the maintaining mechanism over time, we can recognize the close resemblance with producing and underlying mechanisms, respectively.[11]

First, consider the "forces" moving the system away from and back to the equilibrium point. It seems plausible to assume that throughout discovery, scientists will decompose these feedback loops into causal chains consisting in multiple elements (see Fig. 15.7). This is already reminiscent of producing mechanisms. Production, however, is a linear, acyclic process while maintenance is cyclic. But we can depict the sequence of events occurring while a maintaining mechanism operates along a temporal axis (see Fig. 15.8). The explanandum (represented as big black dot) is produced repeatedly over time as it is maintained, it occurs over and over again. Once we see this picture, the resemblance between mechanisms producing their phenomena and mechanisms maintaining a stable state is immediately obvious. All that is needed to recognize this resemblance is attention to temporal order. This is not to say, of course, that we cannot or should not think of maintaining mechanisms as regulatory feedback networks or represent them in cyclic diagrams. In fact, I think, explanations describing maintaining mechanisms are particularly suited to explain stable states *because* they emphasize the continuous, open-ended, and cyclic aspects of the world.

It is worth adding another consideration: the forces at work in maintaining mechanisms may also interact with one another (dotted arrows in Figs. 15.9 and 15.10). This alteration does not affect the conception of maintaining mechanisms as repeatedly producing the homeostatic point over time; it simply adds shortcuts into the causal sequences considered before.

With this understanding of mechanisms maintaining stable states in place, let us turn to mechanisms maintaining continuous processes. Here the explanandum is the continuous behavior of a mechanism as a whole. It is, as in the case of underlying mechanisms, a temporally extended overall process—just that it is now repeated over and over again. The explanans are the forces underlying this behavior; they push the system away from and back to its stable behavior. We may thus consider

[11] There are actually different ways to achieve this transformation. But sketching one of them here shall suffice for illustration.

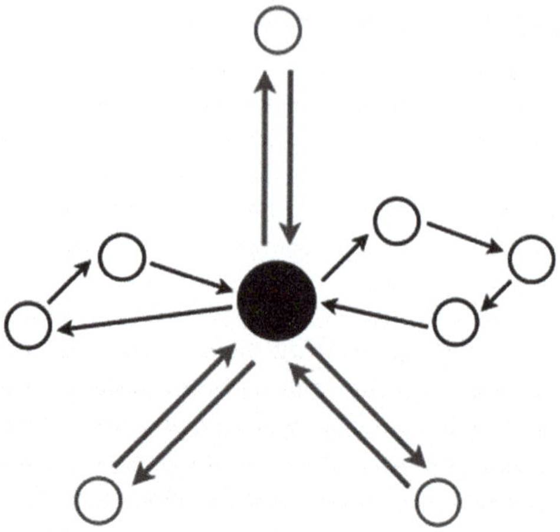

Fig. 15.7 Mechanisms maintaining phenomena where some detail is known about some of the relevant forces; diagram collapsed over time

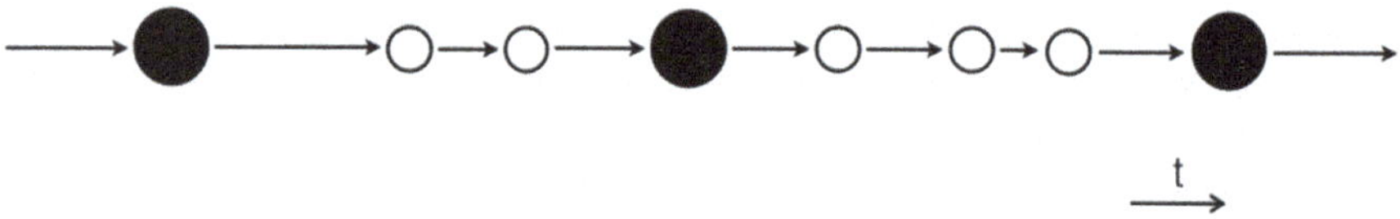

Fig. 15.8 Mechanisms maintaining phenomena (producing stable states) as they unfold over time. (Note that this picture is simplified. The different forces do not necessarily have to act sequentially but can also operate in parallel, not even necessarily at the same rate. (Thanks to an anonymous reviewer for pointing this out.) But this holds true for other causal mechanisms as well: there does not have to be a single straight causal chain leading up to the final product, there can be interferences at various stages, etc. For current purposes, however, we shall work with this simplified picture)

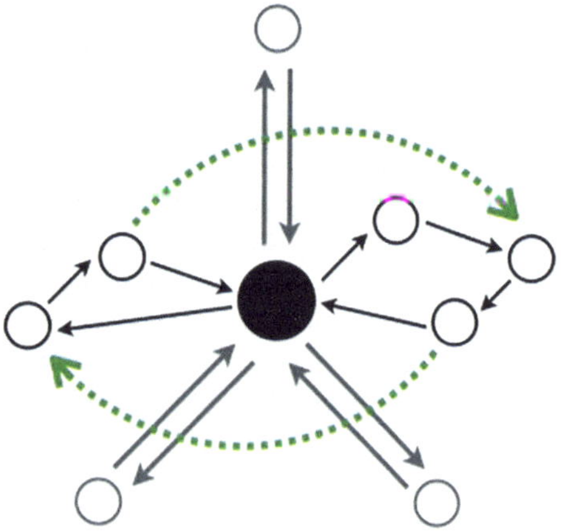

Fig. 15.9 Mechanisms maintaining phenomena with forces interacting (dotted arrows display interactions); diagram collapsed over time

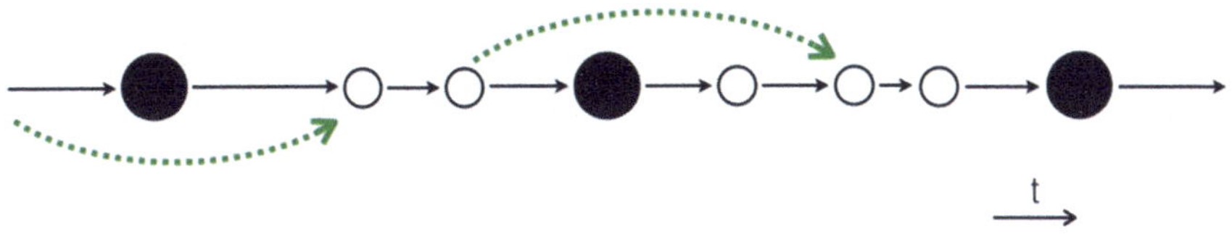

Fig. 15.10 Mechanisms maintaining phenomena (producing stable states) with forces interacting (green arrows display interactions) as they unfold over time

them the acting entities relevant to a mechanism's overall operation.[12] From here, it is only a very small step from talking about the "interacting forces" depicted in Fig. 15.9 to talking about "interacting components" in an underlying mechanism. All we need to acknowledge when we shift from explaining maintenance of a stable state to explaining maintenance of a continuous process is that we start focusing on components, i.e. the acting entities which are present simultaneously with the continuous process to be explained, rather than the (re-occurring) causes temporally preceding (the repeated instantiation of) the stable state to be explained. This shift in perspective is exactly analogous to the shift we observed when switching from producing to underlying mechanisms. As a result, we can picture maintaining mechanisms as shown in Fig. 15.11. The top-level represents the phenomenon, viz. the continuous overall behavior of the mechanism (corresponding to Craver and Darden's homeostatic point). At the level below we find the interacting components in the underlying mechanism (corresponding to Craver and Darden's forces). Each of these components can, of course, be further mechanistically analyzed such that the components in the resulting submechanisms correspond to the elements in the causal chains looping back and forth between the phenomenon in Fig. 15.9. Again, as with the difference between producing and underlying mechanisms, the difference between the two readings of maintaining mechanisms as producing maintaining mechanisms and underlying maintaining mechanisms is primarily one of how we specify the explanatory target—whether we consider a continuous behavior (i.e. a process) or stable state (i.e. a product) to be the explanandum (see Table 15.1).

The graphical transformations I presented visualize my central claims in this section. As scientists shift from one way of looking at the world to another, they may shift from explaining a product to explaining a process or a homeostatic point; and they may do so for any part of a mechanism up and down causal chains and componential hierarchies. Still, there is just one set of "goings-on"

[12]The notion of force seems much more abstract than that of an entity. But given that entities in mechanistic explanations can be fairly abstract (remember that all of this is about mechanism schema construction), acting entities here should not be taken to be in any way more concrete or material than forces. After all, all of this can be black boxes and filler terms that are merely functionally described.

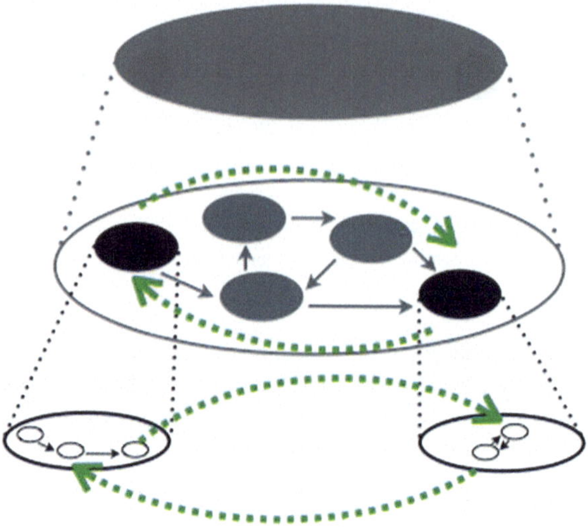

Fig. 15.11 Mechanisms maintaining phenomena (underlying continuous processes)

in the world, different aspects of which are emphasized in scientific explanations depending on how exactly the explanandum is specified. To integrate the insights that individual explanations record, we must recognize how their explananda relate, "translate" some of them, and identify intersections (such as shared components). But before illustrating how the mechanistic triad works together in explanation and discovery, let me address two possible worries regarding my construal of maintaining mechanisms.

15.3.3 Two Worries About Maintaining ... and Why Not to Worry

The first worry is that in maintaining mechanisms there are *forces shifting the phenomenon* away from and back towards its equilibrium (even if it is a continuous process) while in Fig. 15.11 there is nothing pointing directly at or away from the phenomenon. My response is that this impression is misguided, albeit perhaps an artifact of the graphical representation. It is not the case that the forces no longer act on the phenomenon. Their influence is now *implicit* in the underlying relation. To be sure, Fig. 15.11 does no longer depict this influence using solid arrows. Instead, we see interlevel phenomenon-mechanism relations; they are depicted by the usual ellipses connected with dotted lines.[13] So if the objection is that the relation between

[13]An alternative way to think about disturbing forces is to include them in the setup conditions of the mechanism or the phenomenon description. Analogously, correcting forces may be considered the entities and activities in the mechanism underlying the phenomenon. In this case, too, Craver and Darden's forces are implicit in the new figure.

phenomenon and forces (now pictured as components in the mechanism) has gone missing, it is simply wrong. But, the opponent might continue, the kind of relation was changed from causal to componential. This, however, is not an objection. It is precisely the point of acknowledging that mechanisms can be viewed differently, emphasizing different kinds of relations. I acknowledge that Craver and Darden's original diagram of maintaining mechanisms expresses a rough intuition, viz. that a phenomenon is upheld over time as various forces act on it. This intuition can be captured, as demonstrated above, both in terms of continuous producing and continuous underlying mechanisms. Besides, I have argued that as scientists shift from searching for a producing mechanism to searching for an underlying mechanism, they essentially change the explanandum. This, in turn, is accompanied by a shift from searching for causes to searching for components. Thus, it comes as no surprise that underlying maintaining mechanisms postulate componential rather than causal relations between forces and phenomenon. This is a feature of my proposal, not a bug. I do concede, however, that this reading of maintaining mechanisms inherits a problem from its bigger brother: underlying maintaining mechanisms and underlying mechanisms alike face yet unresolved challenges when it comes to characterizing the precise nature of the constitutive relation between a phenomenon and its mechanisms (Sect. 15.1).

This takes me to a second possible worry. I have argued that describing producing, underlying and maintaining mechanisms in scientific explanations emphasizes causal, componential, and continuous aspects, respectively. But I have also said that explanations describing maintaining mechanisms can be understood as explanations describing either underlying or producing mechanisms. If this is so, are there not really just two different aspects that mechanistic explanations can emphasize? And if so, why bother with maintaining and continuity at all? My answer is that while the producing and underlying aspects of mechanistic explanations are rather well known and directly contrast with one another (see Sect. 15.3.1), the maintaining aspect of mechanistic explanations lies on a different dimension (see Table 15.1). It contrasts continuous with individual (more or less finite) product or process generation and captures that something is repetitive and recurrent. There thus is a clear epistemic benefit of including the typically more general explanations describing maintaining mechanisms in the triad: maintaining mechanistic explanations can capture larger-scale organization and temporal dynamics that the often more specific and typically linear producing or underlying mechanistic explanations tend to miss. As a result, maintaining mechanistic explanations are ideally suited to capture, e.g., important regulatory functions within living systems. This not only ensures that mechanistic explanations can be applied to a wider range of phenomena, but may also help defend mechanistic theory against critics from, e.g., dynamical systems theory.[14]

[14] Thanks to an anonymous discussant for pointing this out.

15.4 Applications and Payoffs: Integration, Scientific Progress, and the Lac Operon

Thus far, I have distinguished four different kinds of explanatory projects, individuated by the kinds of phenomena to be explained. Each of these projects goes hand in hand with specific discovery strategies that will lead scientists to construct differently structured mechanistic explanations. Rather than being mutually exclusive, combining the insights gained from such different explanations will typically promote understanding; just like using different measurement tools uncovers different features of, say, a physiological system (cf. Kästner 2018). However, this is *only* possible if we know where and how to fit the pieces of the puzzle together. This is the challenge of *scientific integration.*

My examination above provides a toolbox for scientific integration. I highlight how different mechanistic explanations are conceptually tied to specific kinds of explananda and how shifting the explanandum can shift the emphasis on causal, constitutive, and continuous aspects, respectively, of what is going on in the world. Being clear about what the explanandum is in any given case, and what the mechanistic explanation for it looks like, will thus help to identify potential links and relations between different explanatory and discovery projects. Some mechanistic explanations may "fill in the details" of others (Sect. 15.3.1). Provided that we are clear on what the explanandum is in each case (overall processes or end product), we can, e.g., provide an explanation in terms of underlying mechanisms for different stages in a producing mechanism. Or we can situate a producing mechanism within an underlying one, etc. For an application in pharmacy in the analysis of thyroid gland hormones' actions in the human body see Abdin, Jacob & Kästner (2020).

The same basic rationale can also be applied once we include explanations describing maintaining mechanisms into the picture. For illustration, consider lactose metabolism in E. coli. Escherichia coli are bacteria whose preferred energy source is glucose. When glucose is unavailable, E. coli will also be able to digest more complex sugars, such as lactose. But this requires enzymes that split lactose into simple sugars (glucose and galactose). Since enzyme production is costly, E. coli has evolved such that it will only produce the relevant enzymes when they are actually needed. The corresponding regulatory gene sequence is known as the lac operon (Jacob and Monod 1961). By default (in the absence of lactose), the operon is blocked by a repressor binding to the operator region. This prevents RNA polymerase to transcribe those genes coding the enzymes relevant for lactose digestion; the enzymes cannot be synthesized. If lactose is present, however, it will bind to the repressor and inactivate it. The repressor will be removed and RNA polymerase will transcribe the genes coding for the enzymes; the enzymes relevant for lactose digestion will now be synthesized and E. coli can metabolize lactose. Once all the lactose is split, the repressor becomes active again blocking the lac operon and stopping transcription enzyme genes.[15]

[15]This is of course a highly simplified description but it will do for my purposes here.

To understand lactose metabolism in E. coli, we may look for answers to a whole range of different questions. We may ask, for instance, why beta-galactosidase (one of the relevant enzymes for lactose metabolism) happens to be in the cell when lactose is present. To this question we expect an answer that tells us something about how beta-galactosidase was produced, just like in the case of protein synthesis. But we may also ask why E. coli can digest lactose at all. With this, we could be asking (similar to the case of the action potential) for a mechanistic explanation describing the mechanism underlying lactose digestion. Rather than being pointed to a unidirectional causal chain at the end of which we find a certain molecule, we would then expect a description of a complex set of entities and their interactions. This description will, once we have sufficient knowledge, include the causal story of how beta-galactosidase is produced. Thirdly, we may ask how the metabolism is regulated, i.e. why the proportion of enzymes and lactose molecules present over time stays somewhat constant. As an answer to this question we expect—like in the case of the resting potential or circadian rhythms—an explanation in terms of the relevant regulatory (maintaining) mechanisms, viz. the interplay of lactose, repressor and enzyme production. The mechanistic explanations we seek with this question will include the causal story of beta-galactosidase production but also add the stopping of its production. Both will be incorporated in the overall mechanistic explanation describing what underlies lactose metabolism (though it might not reflect repeated operation).

This example makes it quite plain that we cannot only aim for different explanatory targets while investigating a single set of "goings-on" in the world but that we can also fruitfully integrate our discoveries as we switch back and forth between different explananda. Moreover, it might only be possible to satisfactorily answer a question about what underlies a phenomenon by referring to productive or maintaining aspects. But combining different mechanistic explanations may not only serve to add details; it can also serve to add constraints, e.g. on spatio-temporal organization or the kinds of entities and activities that may feature in a related mechanistic explanation. Either way, the benefit of integration is to gain a more complete understanding of the world as we carve it up in different ways and combine the insights gained from different discovery strategies. And for this combination to be successful, we must carefully consider what exactly we are explaining and how different kinds of mechanistic explanations relate.

Let me briefly point to a second but related benefit of distinguishing different kinds of (mechanistic) explanations. Consider the lac operon again; the second question I suggested we might ask about it was why E. coli can digest lactose. As said, this question might aim for explanation in terms of an underlying mechanism. But on a different reading it might also aim for an evolutionary account of the ability to digest lactose. In this case we would seek a productive mechanism that takes the ability to digest lactose to be the end product of an evolutionary process. Being clear on the difference between these two projects helps us prevent errors and foreclose confusions. Having an explanation of how a capacity came about is quite different from an explanation of how it is implemented, after all.

15.5 Conclusions and Outlook

Different kinds of mechanistic *explanations* carve out different aspects of the causal-mechanical structure of the world as they account for different kinds of phenomena. If scientists seek to explain (i) how a final outcome or end product was generated they seek to discover a *producing* mechanism and focus on causal relations or the transition between different stages of a causal process. If scientists seek to explain (ii) a temporally extended finite overall process they seek to discover an *underlying* mechanism by decomposing the system into its working parts and examining how the components work together; the explanations they construct will thus focus on constitutive aspects. If scientists seek to explain (iii) how a property is kept stable or (iv) how a continuous process is actively upheld over time they aim to discover a *maintaining productive* or a *maintaining underlying* mechanism, respectively. While the former can be viewed as an iterative version of producing mechanisms, the latter can be viewed as a continuous version of underlying mechanisms. In both cases, the explanations scientists construct will emphasize the open-ended operation and continuous (rather than finite) character of the mechanisms responsible for the phenomenon to be explained.

In summary then, producing, underlying, and maintaining mechanistic explanations embody complementary ways of capturing the world. While producing and underlying mechanistic explanations are usually somewhat specific, maintaining mechanistic explanations exhibit a certain regularity or generality. Although the taxonomy I introduced above suggests rather clear criteria for classifying mechanistic explanations, it is important to acknowledge that *in practice* explaining complex phenomena will often require looking at different but related explananda and hence a combination of different kinds of mechanistic explanations. To combine these different explanations into an integrated mechanism mosaic, one must understand the relations between different explananda and identify points of linkage between different mechanisms (such as shared components). To achieve this, it is vital to know how to (at least partly) transform different kinds of maintaining mechanistic explanations into one another.

The above treatment of the mechanistic triad illustrates what such transformations may look like and what they tell us about the relations between producing, underlying and maintaining mechanisms. With these insights in place, we gain an understanding of mechanistic integration that might well serve as model for integration in many special science contexts, such as, e.g. evolutionary biology (see Green et al. 2015 for a concrete case).

Some questions remain, however. For instance, while transforming maintaining mechanistic explanations into producing or underlying ones is rather straightforward while the reverse is limited due to the special characteristics of maintaining mechanisms. These special characteristics warrant further investigation. For example, how exactly should detection be specified? And do mechanisms responsible for *active* forms of maintenance (e.g. by sodium-potassium pumps) and *passive* maintenance (e.g. by concentration gradients) differ systematically? But that discussion makes for a different paper.

Acknowledgments I'm indebted to Lindley Darden, Carl Craver, Ruey-Lin Chen, Jens Harbecke, Marie Kaiser, Beate Krickel, Lara Pourabdolrahim, Richard Moore, Michael Pauen, Astrid Schomäcker, Alfredo Vernazzani, Dan Burnston, and two anonymous reviewers for comments on earlier versions of this paper.

References

Abdin, A. Y., Jacob, C., & Kästner, L. (2020). Disambiguating "Mechanisms" in pharmacy: Lessons from mechanist philosophy of science. *Environmental Research and Public Health, 17*, 1833. https://doi.org/10.3390/ijerph17061833.

Baumgartner, M., & Gebharter, A. (2015). Constitutive relevance, mutual manipulability, and fat-handedness. *British Journal for the Philosophy of Science, 67*, 731–756. https://doi.org/10.1093/bjps/axv003.

Bechtel, W. (2011). Mechanism and biological explanation. *Philosophy of Science, 78*, 533–557. https://doi.org/10.1086/661513.

Bechtel, W., & Abrahamsen, A. (2005). Explanation: A mechanist alternative. *Studies in History and Philosophy of Biological and Biomedical Sciences, 36*, 421–441. https://doi.org/10.1016/j.shpsc.2005.03.010.

Bechtel, W., & Abrahamsen, A. (2009). Decomposing, recomposing, and situating circadian mechanisms: Three tasks in developing mechanistic explanations. *Manuscript.*

Bechtel, W., & Richardson, R. (2010). *Discovering complexity. Decomposition and localization as strategies in scientific research.* Cambridge: MIT Press.

Bogen, J., & Woodward, J. (1988). Saving the phenomena. *Philosophical Review, 97*, 303–352. https://doi.org/10.2307/2185445.

Casini, L., Illari, P., Russo, F., & Williamson, J. (2011). Models for prediction, explanation and control: Recursive Bayesian networks. *Theoria, 70*, 5–333.

Clarke, B., Leuridan, B., & Williamson, J. (2014). Modeling mechanisms with causal cycles. *Synthese, 191*, 1651–1681. https://doi.org/10.1007/s11229-013-0360-7.

Colaço, D. (2018). Rip it up and start again: The rejection of a characterization of a phenomenon. *Studies in History and Philosophy of Science Part A, 72*, 32–40. https://doi.org/10.1016/j.shpsa.2018.04.003.

Couch, M. B. (2011). Mechanisms and constitutive relevance. *Synthese, 83*, 375–388. https://doi.org/10.1007/s11229-011-9882-z.

Craver, C. F. (2007a). *Explaining the brain: Mechanisms and the mosaic unity of neuroscience*. New York: Oxford University Press.

Craver, C. F. (2007b). Constitutive explanatory relevance. *Journal of Philosophical Research, 32*, 3–20. https://doi.org/10.5840/jpr20073241.

Craver, C. F. (2013). Functions and Mechanisms: A perspectivalist view. In P. Huneman (Ed.), *Functions: Selection and mechanisms* (pp. 133–158). Dordrecht: Springer.

Craver, C. (2014). The ontic account of scientific explanation. In M. Kaiser, O. Scholz, D. Plenge, & A. Hüttemann (Eds.), *Explanation in the special sciences: The case of biology and history* (pp. 27–52). Dordrecht: Springer.

Craver, C.F. (2015). Levels. In T. Metzinger, & J. Windt (Eds.), *Open MIND 8. Frankfurt am Main: MIND Group.*https://doi.org/10.15502/9783958570498.

Craver, C. F., & Darden, L. (2013). *In search of mechanisms: Discoveries across the life sciences.* Chicago: University of Chicago Press.

Craver, C. F., & Kaplan, D. M. (2018). Are more details better? On the norms of completeness for mechanistic explanations. *British Journal for the Philosophy of Science*, 1–33. https://doi.org/10.1093/bjps/axy015.

Craver, C.F., & Tabery, J. 2015. Mechanisms in science. The Stanford Encyclopedia of Philosophy (Spring 2016 Edition).http://plato.stanford.edu/archives/spr2016/entries/science-mechanisms/. Accessed: June 2018.

Darden, L. (2006). *Reasoning in biological discoveries*. Oxford: Oxford University Press.

Darden, L. (2008). Thinking again about biological mechanisms. *Philosophy of Science, 75*, 958–969. https://doi.org/10.1086/594538.

Darden, L. (2016). Reductionism in biology. *eLS*, 1–7. https://doi.org/10.1002/9780470015902.a0003356.pub2.

Darden, L., Pal, L. R., Kundu, K., & Moult, J. (2018). The product guides the process: Discovering disease mechanisms. In D. Danks & E. Ippoliti (Eds.), *Building theories: Heuristics and hypotheses in sciences* (pp. 101–117). Dordrecht: Springer.

Fagan, M. B. (2012). The joint account of mechanistic explanation. *Philosophy of Science, 79*, 448–472. https://doi.org/10.1086/668006.

Fazekas, P., & Kertész, G. (2011). Causation at different levels: Tracking the commitments for mechanistic explanations. *Biology and Philosophy, 26*, 365–383. https://doi.org/10.1007/s10539-011-9247-5.

Feest, U. (2016). Phenomena and objects of research in the cognitive and behavioral sciences. *25th Biennial Meeting of the Philosophy of Science Association, Nov 3–5 2016,* Atlanta, GA, USA.

Gebharter, A., & Kaiser, M. (2014). Causal graphs and biological mechanisms. In M. Kaiser, O. Scholz, D. Plenge, & A. Hüttemann (Eds.), *Explanation in the special sciences: The case of biology and history* (pp. 55–86). Dordrecht: Springer.

Giere, R. (2006). *Scientific perspectivism*. Chicago: Chicago University Press.

Glennan, S. (1996). Mechanisms and the nature of causation. *Erkenntnis, 44*, 49–71.

Glennan, S. (2017). *The new mechanical philosophy*. Oxford: Oxford University Press.

Green, S., Fagan, M., & Jaeger, J. (2015). Explanatory integration challenges in evolutionary systems biology. *Biological Theory, 10*, 18–35. https://doi.org/10.1007/s13752-014-0185-8.

Harbecke, J. (2010). Mechanistic constitution in neurobiological explanations. *International Studies in the Philosophy of Science, 24*, 267–285. https://doi.org/10.1080/02698595.2010.522409.

Harbecke, J. (2015). Regularity constitution and the location of mechanistic levels. *Foundations of Science, 20*, 323–338. https://doi.org/10.1007/s10699-014-9371-1.

Illari, P., & Williamson, J. (2012). What is a mechanism? Thinking about mechanisms across the sciences. *European Journal of Philosophy of Science, 2*, 119–135. https://doi.org/10.1007/s13194-011-0038-2.

Jacob, F., & Monod, J. (1961). Genetic regulatory mechanisms in the synthesis of proteins. *Journal of Molecular Biology, 3*, 318–356. https://doi.org/10.1016/S0022-2836(61)80072-7.

Kaiser, M., & Krickel, B. (2016). The metaphysics of constitutive mechanistic phenomena. *British Journal for the Philosophy of Science, 68*, 745–779. https://doi.org/10.1093/bjps/axv058.

Kästner, L. (2017). *Philosophy of cognitive neuroscience: Causal explanations, mechanisms and empirical manipulations*. Berlin: Ontos/DeGruyter.

Kästner, L. (2018). Integrating mechanistic explanations through epistemic perspectives. *Studies in History and Philosophy of Science, 68*, 68–79. https://doi.org/10.1016/j.shpsa.2018.01.011.

Kästner, L., & Andersen, L. (2018). Intervening into mechanisms: Prospects and challenges. *Manuscript.*

Kästner, L., & Haueis, P. (2019). Discovering patterns: On the norms of mechanistic inquiry. *Manuscript.*

Krickel, B. (2017). Constitutive relevance – What it is and how it can be defined in terms of interventionism. *Manuscript.*

Lange, M. (2000). *Natural laws in scientific practice*. Oxford: Oxford University Press.

Leuridan, B. (2012). Three problems for the mutual manipulability account of constitutive relevance in mechanisms. *The British Journal for the Philosophy of Science, 63*, 399–427. https://doi.org/10.1093/bjps/axr036.

Machamer, P. (2004). Activities and causation: The metaphysics and epistemology of mechanisms. *International Studies in the Philosophy of Science, 18*, 27–39. https://doi.org/10.1080/02698590412331289242.

Machamer, P. K., Darden, L., & Craver, C. F. (2000). Thinking about mechanisms. *Philosophy of Science, 67*, 1–25. https://doi.org/10.1086/392759.
O'Rourke, M., Crowley, S., & Gonnerman, C. (2016). On the nature of cross-disciplinary integration: A philosophical framework. *Studies in History and Philosophy of Biological and Biomedical Sciences, 56*, 62–70. https://doi.org/10.1016/j.shpsc.2015.10.003.
Ohta, H., Yamazaki, S., & McMahon, D. G. (2005). Constant light desynchronizes mammalian clock neurons. *Nature Neuroscience, 8*, 267–269. https://doi.org/10.1038/nn1395.
Reddy, A. B., Field, M. D., Maywood, E. S., & Hastings, M. H. (2002). Differential resynchronisation of circadian clock gene expression within the suprachiasmatic nuclei of mice subjected to experimental jet lag. *Journal of Neuroscience, 22*, 7326–7330. https://doi.org/10.1523/JNEUROSCI.22-17-07326.2002.
Romero, F. (2015). Why there isn't inter-level causation in mechanisms. *Synthese, 192*, 3731–3755. https://doi.org/10.1007/s11229-015-0718-0.
Salmon, W. (1984). *Scientific explanation and the causal structure of the world*. Princeton: Princeton University Press.
Tabery, J. (2004). Synthesizing activities and interactions in the concept of a mechanism. *Philosophy of Science, 71*, 1–15. https://doi.org/10.1086/381409.
Van Fraassen, B. (1977). The pragmatics of explanation. *American Philosophical Quarterly, 14*, 143–150.
Wayne, A. (2018). Explanatory integration. *European Journal for Philosophy of Science, 8*, 347–365.

Chapter 16
Constraints on Localization and Decomposition as Explanatory Strategies in the Biological Sciences 2.0

Michael Silberstein

Abstract This paper is a follow up to Silberstein and Chemero (2013), wherein it was argued that contra the new mechanist philosophy, localization and decomposition often fail to obtain in complex biological systems. Herein it is argued that: (1) Mechanistic explanation is historically and still often defined exhaustively by the new mechanists in terms of localization and decomposition; and (2) There are several key features of most complex biological systems related to contextuality and global constraints that violate localization and decomposition, and this fact is not an artifact of network approaches or formal models. Thus, new mechanists must either concede that there are many such cases wherein complex biological systems fail to be fully explicable via mechanistic explanation or, they must reject the claim that localization and decomposition are both necessary and sufficient for mechanistic explanation. Either horn of the dilemma creates problems for the new mechanists. On the first horn, the mechanistic philosophy is often false because, localization and decomposition generally fail to obtain and definitely fail to obtain in crucial cases such as systems neuroscience and systems biology. On the second horn, giving up the claim that localization and decomposition are both necessary and sufficient for mechanistic explanation, threatens to make the new mechanist philosophy too broad, non-unique or downright trivial. The essence of mechanistic explanation, what distinguishes it from mere causal or dynamical explanation, is its compositional or constitutive character. If the new mechanists jettison this feature of mechanistic explanation, if they fully acknowledge the essentially dynamical nature of such explanations and systems, it is not clear what if anything is unique about mechanistic explanation. Indeed, it is argued that many of the more liberal

Special thanks to Carlos Zednik, Daniel Burnston and two anonymous referees for detailed comments.

M. Silberstein (✉)
Department of Philosophy, Elizabethtown College, Elizabethtown, PA, USA

Department of Philosophy, University of Maryland, College Park, MD, USA
e-mail: silbermd@etown.edu; msilberstein@umd.edu

© Springer Nature Switzerland AG 2021
F. Calzavarini, M. Viola (eds.), *Neural Mechanisms*, Studies in Brain and Mind 17,
https://doi.org/10.1007/978-3-030-54092-0_16

approaches to mechanistic explanation, suggest a picture of complex biological systems that comports more with contextual emergence than with the compositional and constitutive origins of the new mechanist philosophy. Thus, the new mechanistic philosophy is either largely false, non-unique or retreats to being just a description of scientific methodology.

16.1 Introduction

In 2013 Chemero and myself published a paper in Philosophy of Science entitled "Constraints on Localization and Decomposition as Explanatory Strategies in the Biological Sciences." In the intervening years there have been several responses to that paper in the literature, some who cite us approvingly (e.g., Venturelli 2016; Rathkopf 2018) and others who want to use us as a foil (e.g., Kaplan 2018). Kaplan for example says the following: "Silberstein and Chemero (2013) recently argue that sometimes the nature of dynamical systems prevents the application of these strategies [decomposition and localization], and that in such cases mechanistic explanation will be unavailable in principle" (2018, 275). That is indeed what we argued. Kaplan notes that, "The mechanistic approach to dynamical explanation faces challenges along several fronts", and he includes the apparent non-decomposability of many dynamical systems (2018, 275). Nonetheless Kaplan largely wants to defend localization (loc) and decomposition (decomp) from our argument. He concludes that, "dynamists have also appealed to notions of emergence or downward causation in complex dynamical systems to argue for the limitations of the mechanistic approach", and he ends by saying that, "defenders of the mechanistic approach have not satisfactorily addressed the challenge" (2018, 278). On this last point there is agreement with Kaplan.

However, unlike Kaplan, who seems to think these challenges can be easily addressed by the new mechanist, this paper will argue to the contrary. Kaplan and other new mechanists of course appreciate that mechanistic explanation is not the only game in town, but he and many others do appear to be committed to the claim that most any complex biological system that can be analyzed in terms of dynamical systems and networks, can in principle be subjected to loc and decomp. This is one of the claims this paper attempts to refute (Sect. 16.3). Since Kaplan and other new mechanists concede in principle that there are exceptions to mechanistic explanation, then they can either accept the examples given herein as instances of such exceptions, or they can broaden the definition of mechanistic explanation, such that neither loc nor decomp are either necessary or sufficient conditions for mechanistic explanations in general.

The problem with the first move is that failures of loc and decomp are textbook, such failures are not merely exceptions, indeed, it is the norm in many complex biological systems (Sect. 16.3). This means that mechanistic explanation defined essentially in terms of loc and decomp will rarely obtain and thus the old school new mechanistic view is largely false. The problem with the second move is that it threatens to make the definition of mechanistic explanation too broad or downright trivial (Sect. 16.4).

What follows, is in a nutshell, the argument this paper defends. The argument herein is essentially the 2.0 version of the argument defended in Silberstein and Chemero (2013). The argument is as follows:

P1. Mechanistic explanation is historically and still often, defined exhaustively by the new mechanists in terms of loc and decomp, i.e., loc and decomp are both necessary and sufficient for mechanistic explanation according to the new mechanists (section 2).

P2. There are several key features of most complex biological systems related to contextuality and global constraints that violate loc and decomp (sections 3).

C1/New mechanists must either concede that there are many such cases wherein complex biological systems fail to be explicable via mechanistic explanation or, reject the claim that loc and decomp are both necessary and sufficient for mechanistic explanation (from premise 1 and 2).

P3. Either horn of the dilemma creates problems for the new mechanists. On the first horn, loc and decomp generally fail to obtain and definitely fail to obtain in crucial cases such as systems neuroscience and systems biology. On the second horn, giving up the claim that loc and decomp are both necessary and sufficient for mechanistic explanation, threatens to make the new mechanist philosophy too broad or downright trivial. (4).

C2/The new mechanistic philosophy is either largely false or too trivial to be of interest.

Assuming they grant the historical claim, there are three types of responses a new mechanist might make to the first premise: (a) that loc and decomp are not necessary; (b) that loc and decomp are not sufficient; or (c) loc and decomp are neither necessary nor sufficient, but that is not a problem because mechanistic explanation has no essence. If the new mechanist claims that loc and decomp are not necessary, but they are sufficient for mechanistic explanation, then they need to respond to the challenge that loc and decomp are relatively rare and thus this sufficient condition is often not met. We also need to know, what are the necessary conditions for mechanistic explanation? If on the other hand, the new mechanist claims that loc and decomp are not sufficient, but they are necessary for mechanistic explanation, then we need to know what else *uniquely* constitutes the essence of mechanistic explanation—what are the sufficient conditions? We also need a response to the challenge that the necessary condition is rarely met. Without clear and agreeable answers to these questions, if, as argued herein, the failure of loc and decomp is the rule with complex biological systems (Sect. 16.3), if loc and decomp are largely just idealizations, then we are left again with the conclusions of the argument above.

If the new mechanist chooses option (c), then we need to know what exactly does demarcate mechanistic explanation from other types of explanation? Again, if one takes option (c), the worry is that the new mechanistic philosophy now becomes too broad or too trivial to be of interest. Keep in mind, for example, that historically what is supposed to separate mechanistic explanation from simply being just another case of causal explanation, is its constitutive and thus *reductive* nature. In Sects. 16.3

and 16.4 it will be argued that option (c) fails to be constitutive or reductive in any essential sense.

Indeed, it will be argued in Sect. 16.3 that complex biological systems are best seen as exhibiting contextual emergence. Contextual emergence is in many ways closer to the type of emergence defended by C.D. Broad than it is to the new mechanist philosophy defined in terms of loc and decomp. Contextual emergence will be compared with related views in the literature that have sprung up since our 2013 paper was published (e.g., Zednik 2014, 2015, 2019; Anderson 2016; Stinson 2016; Burnston 2017; Bechtel 2017a; Winning and Bechtel 2018; and Winning 2018). These views are mostly an attempt to defend option (c), and it will be argued that they all fail to be reductive in any deep sense and thus they fail to adhere to the spirit of the new mechanistic philosophy. Or perhaps the lesson is that the new mechanist's account is now compatible with a brand of emergence once thought antithetical to a mechanistic vision of biological systems, thus again trivializing it and robbing the new mechanistic philosophy of its reductive essence.

Before we turn to Sect. 16.2, just a word about how it sets-up Sect. 16.3. In Silberstein and Chemero (2013), the focus was on network or topological explanations in systems neuroscience. That was and still is an excellent case study for illustrating the in-principle failure of loc and decomp in neural systems (Sect. 16.3). The problem, as the next section illustrates, is that the network examples too easily conflates concerns about explanation and abstraction on the one hand, with claims about organizational features of complex biological systems that tell against loc and decomp, on the other. Obviously these two concerns are related but it is also important to disentangle them (Sect. 16.2).

The main point of our 2013 paper and this paper is that various global constraints and other kinds of context sensitivity tell against loc and decomp, not merely as explanatory strategies, but as bio-physical principles and actual causal-spatiotemporal organization. In our 2013 paper our focus was on the way global constraints and other kinds of context sensitivity given by being a certain kind of network structures (e.g., a small-world network), enable certain tasks to be performed by constraining the behavior of relatively more local (both topologically and structurally local) components. As will be discussed in Sect. 16.3, there are many other textbook examples from systems biology that make the same point but are less prone to being conflated with issues purely about abstraction or explanatory strategies. The point is, there is nothing unique about systems neuroscience and one need not worry that focusing on graphical explanations is an illicit instance of cherry picking of cases.

16.2 Defending and Clarifying Premise 1

This section has two purposes. First, to establish that new mechanists have and still often do define their position exhaustively in terms of loc and decomp. Second, to make clear that the issue here is not primarily about abstraction or idealization in systems neuroscience, but whether or not complex biological systems really do embody loc and decomp as organizational features. It is clear from the response in the literature to Silberstein and Chemero (2013) that these two concerns are easily conflated.

Many would say that the first point requires no substantiation. After all, it is well-known that mechanistic explanation is constitutive explanation, and loc and decomp plausibly follow directly from this. However, people often do question it, or they question exactly what the new mechanists mean by loc and decomp. My claim is that loc and decomp given its *constitutive* nature, is a form of reductive explanation. The sense of reduction here is not intertheoretic of course, it is not logical or mathematical reduction, but a kind of part/whole reductionism. Furthermore, this reductive aspect of mechanistic explanation is precisely what, historically at least, many new mechanists found appealing about it. Some new mechanists might retort that it is very explicit in Machamer, Darden, Craver (2001) and Craver (2007), and other places that a major point of the mechanistic approach is to get away from traditional debates about reductionism, in favour of integrated "mosaics" of multi-level explanation. That is all certainly true. However, the traditional debates about reductionism they wanted to get away from were primarily about intertheoretic reductionism and in spite of their emphasis on multiscale explanation, many new mechanists did and still do also emphasize the reductive nature of such explanations. This section will establish this claim beyond doubt.

Perhaps the new mechanist movement have now come so far, that most will happily concede and accept that mechanistic explanation, properly understood, is not reductive. Indeed, perhaps most new mechanists will now happily accept what is argued in Sect. 16.3, that what they call mechanistic explanation strongly suggests "contextual emergence" of a sort C.D. Broad would have appreciated. Assuming all that is the case, if going forward from 2013, there is no longer any disagreement of fact between people such as myself and the new mechanists, if we are now only squabbling over what to call this kind of explanation, then I can only hope that Silberstein and Chemero (2013) played some part in that. We will return to this discussion in Sect. 16.4.

16.2.1 The Place of Localization and Decomposition in the New Mechanistic Philosophy and What They Mean

The core idea of loc and decomp is to break down a mechanism as a whole, into operations of interrelated parts, organize them into modules, which when properly ordered, explain the workings of larger mechanisms or sub-mechanisms that they make up. Thus, we see how interacting and hierarchically organized parts causally produce the phenomenon in question (Bechtel and Abrahamsen, 2005; Bechtel 2011; Machamer, Darden, and Craver, 2001). One should not get the idea however that such explanations are strictly about *intra*-level causal relations.

As Craver makes clear (Craver 2007; Craver and Bechtel 2007), for the new mechanists, when it comes to biological mechanisms, causal relations are *intra*-level relations only. Whereas constitutive relations are *inter*-level non-causal synchronic relations. That is, compositional or constitutive relations, are "*non-causal determination relations* that are synchronous," and involve highly localized and hierarchically organized elements. The components of a mechanism "are spatially contained within the constituted individual, and such that the properties of the individuals in the team realize the properties of the constituted individual and the processes grounded by the individuals in the team implement the processes grounded by the constituted individual" (Gillett 2013, 317–18). What makes something a compositional constituent of an individual, is if it is a working part – i.e., if it does work "that non-causally results in the 'work' done by the relevant whole" (2013, 319). The point is that the components that compose a mechanism and their properties (the realizers)–the *intra*-level causal relations–are always *localized* at smaller spatial and temporal length scales than the entities they compose and the properties of the entities they realize, i.e., of the *intra*-level causal mechanism itself. Most importantly of all, such synchronic constitutive composers and realizers, determine the causal powers of such *intra*-level causal mechanisms. What makes such an explanatory strategy reductive is that in addition to its *intra*-level modular commitments, the causal powers of all *intra*-level causal mechanisms are discharged by "lower-level" causal mechanisms, all of which are discharged by *inter*-level non-causal synchronic relations residing at and localized at smaller spatial and temporal length scales. What could be more reductive than this? As Green, Serban, Scholl, Jones, Brigandt, and Bechtel note, this old school new mechanist explanatory strategy works well if and only if, "the functioning of such a part is due to its internal organization *and largely unaffected by its context* [emphasis added], so that parts can be investigated in isolation (via decomposition) and their joint operation is relatively easy to understand (via recomposition)." (2018, 1751).

This now relatively old school new mechanist's definition of loc and decomp implies that:

A) For every *intra*-level mechanism, the most fundamental components of mechanisms, those that instantiate the mechanism itself, are always at a smaller scale/lower-level of organization than the system as a whole or *intra*-level mechanism in question.
B) Functional explanation must ultimately be fully grounded in the most fundamental components of the mechanism and the interactions between them.
C) The interactions between the most fundamental components of a mechanism must be relatively insensitive to multiscale contextuality.

Some would argue that this old school definition of the new mechanist is a strawman that no longer needs attacking because all new mechanists have by now absorbed the lessons of systems biology. For example, perhaps many mechanists are now happy to grant that *inter*-level relationships can be causal, dynamical, and diachronic. And perhaps many mechanists now acknowledge that such integrated, interdependent and interconnected multiscale relations are essential to explain the functioning of most mechanisms. That is, perhaps most new mechanists are now willing to let go of A-C. First, if that is so, then I am glad to hear it. However, given the following very recent definitions of the new mechanistic paradigm as given by its key defenders, I am skeptical:

> All things are physical and their causal capacities must depend upon their basic physical constituents (Glennan 2017, 207).

> Mechanisms as wholes do what they do because of the activities of the parts (Glennan and Illari 2018, 1).

> The behavior of the whole contains the behaviors of the parts, and the behaviors of the parts collectively and exhaustively constitute the behavior of the whole (Povich and Craver 2018, 193).

Of course, one could read the preceding passages as just making the trivial claim that to have mechanisms one must have parts, but clearly much more is meant here, as there would be no point in asserting such a trivial claim to begin with. However, if these passages are not explicit enough, take the following:

> The effects of context, organization and constraints can all be accounted for in terms of *the causal influences of lower level entities and Activities* [emphasis added]. That is, within the mechanistic framework, the causal autonomy of higher levels cannot be established (Fazekas and Kertesz 2018, 1).

As will be discussed in Sects. 16.3 and 16.4, it is happily and readily granted that there are many new mechanists (one might call them, 'new, new mechanists'), who do reject the old school new mechanistic philosophy. As will become clear in Sects. 16.3 and 16.4, the new, new mechanists must now face two new questions that old school new mechanists had ready answers to: (1) Are loc and decomp now to be rejected? If so, what then is the essence of mechanistic explanation? If no, then how are loc and decomp going to be reconceived in such a way as to retain their fundamental explanatory role? (2) what now explains the formation, order, stability, causal capacities and functionality of complex biological mechanisms? For the old school mechanist, the answer to this question was based on a reductive account of mechanisms in terms of composition and realization. Needless to say, these two questions are not unrelated.

16.2.2 Concerns about Explanatory Strategies and Abstraction in Topological Explanation vs. Concerns about Causal and Spatiotemporal Organization of Mechanisms Such as Localization and Decomposition

The purpose of this subsection is to make it clear, that the main point being made throughout, is the failure of loc and decomp in complex biological systems (construed as causal and organizational features of complex biological mechanisms in the real world) based on global constraints and other kinds of context sensitivity. The following are the questions to be focused on herein:

1. Do loc and decomp often fail to explain key features of complex biological mechanisms because of global constraints, organizational features and other kinds of context sensitivity?
2. What best explains the formation, stability, functionality and causal capacities of complex biological systems? Where does such order come from and what maintains it?

Why is it necessary to point this out? As Zednik notes, several authors incorrectly took our original argument to be primarily about abstraction (2018, 23). Indeed, in their response to Silberstein and Chemero (2013), many people took our focus to be on, or at least our argument to be based on, abstractions or other explanatory features of network neuroscience. Furthermore, others do sometimes make such arguments based on abstractions, etc. For example, Ross claims that the reason graphical/network-based explanations fail to be mechanistic is because of "the role of abstraction in explaining universal behavior" (2015, 51). Here Ross is alluding to the "minimal model" account of Batterman and Rice (2014). Ross asserts that the failure of graphical/network models to be mechanistic has nothing to do with the failure of loc and decomp but only the aforementioned features of minimal models (2015, 51). Brigandt, Green and O'Malley assert that it is the fact that topological explanations "abstract away from structural detail in favor of 'design principles'", that makes them deviate from mechanistic explanation (2018, 367).

However, we never meant to claim that the primary reason such topological explanations fail to be mechanistic is *merely* because they are abstract, represent "design principles" or even merely because such networks are multiply realizable. Brigandt, Green, and O'Malley are right when they say, "In contrast, design explanation proceeds in the opposite direction, as the functions to be performed explain the presence of some structural organization (integral feedback control)" (2018, 370), but for us this is not merely a mode of description or design-stance. The amazing thing is that nature does this universally without a designer or engineer because of various global constraints and other kinds of context sensitivity.

Our primary claim was that topological explanations fail to be mechanistic for the simple reason that loc and decomp fail in such cases. This is because the difference-making topological properties, such as being a small-world network, act as *global constraints* on the structural elements participating in them, and that such networks are relatively insensitive to their mechanistic implementation; where "global" is not a synonym for "abstract."[1] The fact that network properties are global, as they

[1]Skepticism has started to arise about whether or not the brain truly instantiates small-world networks, and perhaps more generally about network neuroscience. This skepticism is based on various methodological considerations and concerns, as well as alternative analyses (Markov et al. 2013; Hilgetag and Goulas 2015; and Damicelli et al. 2018). There is no space here for me to address this issue at length. But briefly, of course whether or not the brain truly instantiates small-world networks in particular is an open empirical question, yet to be fully resolved. No doubt to

involve various order parameters at work over the whole system, is enough in itself to negate loc and decomp..

Furthermore, there is no reason to regard network models as merely or only "mathematical explanations." As with much of science, such explanations are given via mathematical representations but that does not make them nothing but mathematics. The only thing that matters *for such explanations* are the topological properties and there is no reason to regard such properties as fictions, abstracta or Platonic entities, merely because they can be modelled mathematically. Nor are we required to believe that topological structures exist independently of physical instantiations.

As with much of science, just because network models are "idealized" and relatively abstract, does not mean they do not refer to real features of biological system. Just as there is geometry in the world, there is topology in the world, and neither sort of property reduces to purely structural or 'atomic' properties. As Huneman notes, what else is the "organization" of a mechanism but its geometry and topology (2018). But again, just because such topological features are not Platonic entities or abstracta hovering over the spatiotemporal organization of biological processes, it does not follow that they are nothing but said spatiotemporal organization taken as a series of snap shots at various times.

The point again is that, the reason such systems have the spatiotemporal organization they do, is in part *because* of the relevant topological properties. As we stressed in the original paper (2013, 967), it is the self-maintenance and preservation of certain network structures that constrains the behavior of the structural parts. As Sporns puts it, "a reentrant system operates less as a hierarchy and more as a heterarchy, where super-and subordinate levels are indistinct, most interactions are circular, and control is decentralized" (2011, 193.).

As regards the specific case of network explanations, Huneman notes that the global topological features "explain why a set of mechanisms is constrained in specific way", and this implies the "stronger, metaphysical, claim that in some cases the reason why some systems are displaying a constant or regular behavior of some sort (e.g., with a specific steady state, a typical outcome, or inversely, an absence of some particular outcome etc.) is a mathematical—in the present context, topological—fact" (Huneman 2018, 120). We can make exactly the same point using the lingo of "global organizing principles" if you prefer. As Hooker puts it:

> But global biological organization challenges this overly 'mechanical' conception: components are often not stable but variously created and dissolved by the processes themselves

answer to this question we need more data and analysis of anatomical studies, big data imaging studies, neural simulations, etc. And we need to better triangulate between all these approaches. It must also be noted that the outcome of such research might vary depending on what scale, region, or level of analysis of the brain is being considered. Given how nascent network neuroscience is, it would not be very surprising if in the future a more sophisticated network neuroscience revealed topological profiles and other global organizing principles in the brain that deviate from standard small-world networks in important ways. However, there is no reason to doubt more generally that the brain instantiates various global organizational principles and many kinds of context sensitivity.

> and the globally coherent organization this requires for overall persistence in turn requires a conception of globally coherent mechanisms. Mechanisms are conceived as organized processes, but a serious incorporation of organization within them remains an outstanding issue (2011, 206).

The punchline here is that real-world global constraints and other contextual features are the reason we need network-based types of explanations, the former are not merely artifacts of the latter. Thus, given that loc and decomp are not the answer, we want to know why and how such global organizing features are possible and how they work.

Having defended premise 1 in the master argument above and having relatedly clarified the intent of the original argument and the argument herein, in the next section the focus will be on defending premise 2 of the master argument herein.

16.3 Defending Premise 2: The Frequent Failure of Localization and Decomposition in Complex Biological Systems

The main purpose of this section is to defend the claim that loc and decomp frequently fail as explanations when it comes to key properties of complex biological systems. Rather, instead of loc and decomp and the hierarchical structure they imply, what we see in such systems is contextual emergence. Something like this fact is now acknowledged by some new mechanists such as Winning and Bechtel.

16.3.1 A Brief Reminder of the Key Features of Network-Based Explanation from Systems Neuroscience

Network analyzes of the brain are based on the thought that brain function is not just relegated to individual regions and connections, but emerges instead from the topology of the brain's entire network, i.e., the connectome of the brain as a whole. In such graphical models of neural activity, the basic units of explanation are not neurons, cell groups, or brain regions, but multiscale networks and their large-scale, distributed, and nonlocal connections or interactions (Silberstein and Chemero 2013). The study of this integrative brain function and connectivity is mostly based in topological features or architecture of the network. Such multiply realized networks are partially insensitive to, decoupled from, and have a one-to-many relationship with respect to lower-level neurochemical and wiring details.

More specifically, a graph in this case is a mathematical representation of some actual many-bodied biological systems. The nodes in such models can represent neurons, cell populations, brain regions, etc., and the edges represent connections between the nodes. The edges can represent structural features such as synaptic pathways and other wiring-diagram-type features, or they can represent more topological features such as graphical distance. What matters in such graphical explanations is the topology or pattern of connections. Different geometries or arrangements of nodes and edges can instantiate the same topology.

When mapping the interactions (the edges) between the local neighborhood networks, then we are interested in global topological features, i.e., the topological architecture of the brain as a whole. While there are local networks within networks, it is the global connection between these that is often of greatest interest in systems neuroscience. Graph theory has many different kinds of network topologies, but one of great interest to systems neuroscience are small-world networks. This is because various regions of the brain and the brain as a whole are thought to instantiate such networks. The key topological properties of small-world networks are

- Sparseness: relatively few edges given the large number of vertices;
- Clustering: edges of the graph tend to form knots, for example, if X and Y know Z, there is a higher-than-normal chance they know each other;
- Small diameter: the length of the most direct route between the most distant vertices, for example, a complete graph, with n2/2 edges, has a diameter of 1, since you can get from any vertex to any other in a single step. Most nodes are not neighbors of one another yet can be reached through a short sequence of steps.

That is, (1) there is a much higher clustering coefficient relative to random networks with equal numbers of nodes and edges and (2) short topological path length. Small-world networks thus exhibit a high degree of topological modularity and nonlocal or long-range connectivity. There are many different types of small-world networks and other types of networks with unique topological properties that allow researchers to make predictions about the robustness, plasticity, functionality, health, etc., of brains that instantiate these networks (Sporns, 2011). One type of network of particular interest is called the "Rich-Club" network (Pedersen and Omidvarnia, 2016; van den Heuvel and Sporns 2011). Such network architectures are called "Rich-Club" based on the analogy with wealthy, well-connected people in society. "Members" of this club constitute a few "rich" brain-regions or central hubs that distribute a large number of the brain's global neural communications. The "Rich-Club" topological brain architecture is instantiated when the hubs of a network tend to be more densely connected among themselves than nodes of a lower degree.

As we argued in our original paper, the dynamical interactions in such networks are recurrent, recursive, and reentrant. Therefore, the arrow of explanation or determination in such systems is both top-down (graphical to structural) and bottom-up (structural to graphical). Global topological features of complex systems are not explicable in principle via localization and decomposition. The many-to-one relationship between the structural and the graphical features demonstrates that specific structural features are neither necessary nor sufficient for determining global topological features, i.e., topological features such as the properties of small-world networks exhibit a kind of "universality" with respect to lower-level structural details. In the case of random networks for example, power laws and other scale-invariant relations can be found. These laws, which by definition transcend scale, help to predict and explain the behavior and future time evolution of the global state of the brain, irrespective of its structural implementation. Power laws are explanatory and unifying because they show why the macroscopic dynamics and topological features exist across heterogeneous structural implementations.

However as discussed in the last section, the concern was raised that any inference to the failure of loc and decomp of cognitive functions in the brain based on network neuroscience are suspect, likely to be artifacts of the formalism, because such models are highly idealized and abstract. Thus, the point of the next subsection is to step away from the emphasis on the topological features of brain networks (mathematical models) and look at the wider evidence from across systems biology that loc and decomp often fail in principle in complex biological systems. The various key examples of such failure herein are from genetics, epigenomics, molecular biology, developmental biology and synthetic biology, are by now, textbook cases.

16.3.2 Other Examples from Systems Biology

One reason for looking at these other cases is to make the point that network explanations in neuroscience, neural reuse, neural plasticity, etc., are not unique to neural and cognitive systems, in addition to the brain, one can find networks, reuse, plasticity, robustness, autonomy and universality in many complex biological systems. That is, key failures of loc and decomp are rife across the biological sciences, so it is no surprise that neuroscience is no exception and topological explanation is not some suspect special case. As Bateson and Gluckman put it, "The central elements underlying many forms of plasticity are epigenetic processes, and plasticity operating at different levels of organization often represents different descriptions of the same process. Underlying behavioral plasticity is neural plasticity and underlying that is the molecular plasticity involving epigenetic mechanisms" (2011, 43). The point here being that brains inherit their network properties and other global organizing constraints from even more fundamental biological processes.

It goes without saying that while brains have unique biological features and functions, many of the biological processes discussed herein also happen in the brain. Perhaps then the best way to start this section off is with a quote from Michael Anderson from Brain and Behavioral Sciences (BBS) wherein he is responding to my reaction to his excellent book *After Phrenology* (2014):

> Hence, I completely agree with Silberstein that good neuroscience must also be what he calls "big picture" biology, and I suspect that part of what it will take to make substantial progress understanding the brain is a reform of graduate training in psychology and neuroscience to include more evolutionary and developmental biology, mathematical physics, and, yes, even philosophy (some of which is happening already). (2016, 34).

This indeed is the point of this subsection (Silberstein 2016).

Let us begin by reminding ourselves of a recent episode in the history of molecular and developmental biology. The Human Genome Project (HGP) grew out of the 1980s and 1990s in which genetic determinism was a common belief among biologists. HGP was motivated in large part by its own version of loc, decomp and modularist thinking, for example, in the form of the gene doctrine (one gene for one protein). Upon completion of sequencing the human genome many scientists were shocked to discover that humans only had 30,000 genes as opposed to the predicted 100,000. They were also surprised to find out that we humans have only 300 unique genes distinguishing us from a mouse. And again, surprised to learn that genes (more accurately gene networks) can give rise to many different proteins. Genes are pleiotropic and phenotypic traits are multigenic. In reaction to learning all this famed Harvard biologist Stephen Jay Gould in the New York Times said: "The collapse of the doctrine of one gene for one protein, and one direction of causal flow from basic codes to elaborate totality, marks the failure of [genetic] reductionism for the complex system we call cell biology."

In the years since this statement was made more and more evidence has accumulated that Gould was right. As Boi puts it, "There is unidirectional flow of information from one class of biological molecule to another. Genomic functions are inherently interactive, isolated DNA is 'virtually inert', Cells have many mechanisms to "complexify, modify and change" the structure and function of DNA" (2018, 196). Hence the rise of epigenomics and the choir of people claiming genetic determinism is dead. As Zimmer notes, "Yet the epigenome is not simply a rigid program for turning genes on and off in a developing embryo. It is also sensitive to the outside world" (2018, 160). Examples include methylation, diet and all kinds of stress. Other key variables include chromatin configurations, chromosome configurations, organelles, membranes, hormones, morphogenesis, etc. The very meaning of epigenetics is that the genetic 'code' by itself is not sufficient to determine what is happening in development or afterwards. It is now well known that there are environmentally induced epigenetic alterations that arise early in development, that happen throughout our lives and can be inherited via transgenerational epigenetic effects (so called "epigentic inheritance"). Such extra-genetic inheritances can outlive actual mutations and can be reversible.

As we are now all well aware, the relationship between genes and proteins is many-many, and we must now think in terms gene networks (genomics), RNA networks, protein networks (proteome), and the complex non-linear interactions between them. Furthermore, these relationships are also affected by several global constraints and multi-scale contextual features including cellular environment, the wider organismic environment and various features of the external environment in which the organism is situated. These interactions are obviously multi-scale, multi-level, inextricably interrelated and interdependent (Noble, 2006, 105). In the developmental process, what any key biological player does, such as genes and proteins and what it results in as output, is a function of multiscale contexts (Noble, 2006, 17 and 34; Francis 2011, 159; Bechtel 2019, 461 and 488).

All of this 'contextuality' goes for protein folding and protein function as well. Amino acid sequences alone determine neither three-dimensional structure or function. Other factors include many features of the cellular environment and cellular activities such as various properties of water, lipids and the interactions of "many other molecules that are not coded for by genes" (Noble, 2006, 35). Nothing in the genome determines the topology of proteins. As Boi says, "the biological information of proteins does not derive only from structural information, but also from the complex functional networks that connect specific binding sites at the molecular level to the cell's activity and to the more global organismic level of organization and functioning" (2017, 195).

Embryogenesis is likewise not determined by the genome and also exemplifies contextuality. Early on in the embryo a ball of identical pluripotent cells becomes differentiated into various cell types and organs as a result of a network of physical and chemical environmental gradients and signals. It is because such biological processes are so contextual and not driven by instructions from the genome, that many biologists call them self-assembling, self-organizing and self-maintaining regulatory networks, with interdependent interactions at all scales and "levels."

Povich and Craver (2018, 193) have expressed skepticism about any departure from modularity in complex biological systems on the grounds that modularity is a clear evolutionary advantage since such modular networks will be better able to survive contextual and environmental changes, damages, etc. However, DNA-RNA-protein networks are modular in the sense that they do retain a certain autonomy across a number of different contexts. There are two things to note here. First, the reason the modularity in such networks obtains is because the relevant context, e.g., the relevant network or sub-network itself, comes along for the ride when transplanted into a new environment. Second, even given this sort of modularity it does not follow that the modular network in question will produce the same output/effect given *any* change in its context. The very same network as defined structurally can produce different effects, operations or products in the context of different networks. The point is that modularity and "contextualism" can go hand-in-hand.

Perhaps all of this is best illustrated by the relationship between plasticity, robustness and autonomy. There are many different forms of robustness and plasticity, such as developmental, phenotypic, a variety of neural, behavioral, immunological, etc. Let's take phenotypic plasticity and robustness as an example. This is the phenomenon in which genetically identical individuals will develop different phenotypic traits in different environmental conditions (Kaplan, 2005, 2008). Because of phenotypic plasticity, a single genotype or genome can produce many different phenotypes depending on environmental and developmental contingencies (Gilbert and Epel 2009). Phenotypic plasticity is just one example of epigenomic processes in which various mechanisms create phenotypic variation without altering base-pair nucleotide gene sequences, altering the expression of genes but not the gene sequence.

In contrast, there are cases in which genetic or environmental changes have no phenotypic effect. This persistence of a particular organism's traits across environmental or genetic changes is called robustness. Robustness is illustrated by various knock-out experiments in synthetic biology whereby a particular gene (or group of genes) known to be involved in the development of some protein or phenotypic trait is disabled without disturbing the presence or production of the developmental end product in question (Jablonka and Lamb 2005).

To account for and model plasticity and robustness, developmental biologists have called upon network/dynamical explanations. The ongoing development of an organism acts as a global constraint that 'enslaves' the components necessary to maintain its dynamics. Because of this, a developing system will have highly plastic boundaries, and will be composed of different enslaved components over time. This plasticity serves the autonomy and robustness of the developing organism, making it more likely to be viable and adaptive.

Robustness is closely related to autonomy, another key concept in evolutionary developmental biology. Autonomy is the property of living systems to make use of their environments to maintain themselves. Autonomy is sometimes explained in terms of recursive self-maintenance. Some systems are plastic such that they can maintain stability not only within certain ranges of conditions, but also within certain ranges of changes of conditions: they can switch to deploying different processes depending on conditions they detect in the environment.

As Bateson and Gluckman note, robustness and plasticity are two-sides of the same coin, they are interdependent, "Indeed, plasticity is often regulated by robust mechanisms and robustness is often generated by plastic mechanisms" (2011, 46), in an interplay of evolutionary and developmental processes. The ever-growing varieties of robustness and plasticity are co-creating and co-maintaining, allowing a complex biological system to have autonomy, "Development involves both internal regulation and reciprocity with the environment. Careful analysis of what happens during development suggests that it is no longer helpful to retain a hard and fast distinction between robustness and plasticity" (2011, 62); Bateson and Gluckman (2011).

Plasticity, robustness and autonomy are universal features of complex biological systems. Such global or systemic contextual constraints are well known from a variety of different fields of biology, they have been well known for a long time and they are well confirmed without any appeal to abstract mathematical models (Koonin 2011, viii–ix; Jaeger and Calkins 2012, 27). It is also well known that such global constraints can impose the same function even across different species, using different structural components. One can demonstrate equivalence classes of networks across species, including network function. Jaeger and Calkins further infer that, "Given that function is conserved and the mode [i.e., specific structural mechanism] isn't, it suggests regulation by the organism as a whole" (2012, 27). This is what Povich and Craver are missing. Jaeger and Calkins characterize such global organizational constraints in terms of "top-down information control" via multilevel causation, wherein the working parts are constrained in their behavior in service to the larger function, and in their view the cell itself is one such unit of control. One can certainly model such processes mathematically in terms of network motifs, etc., but, with or without such formal models, these are well known facts about complex biological systems.

We have seen that, even without relying on mathematical modelling, network theory or anything unique to neural systems, there is ample evidence from textbook developmental, molecular and synthetic biology that the best explanation for why complex biological systems behave the way that they do involves various global constraints (or design principles if you prefer) and other kinds of multiscale contextuality. Noble is exactly right when he says, "systems biology is neither vitalism or reductionism in disguise" (2006, 65); and we might add, nor is it merely abstract mathematical modelling or Platonism in disguise.

It seems clear that global topological features of networks model, help explain, and are in turn explained by global organizational features of complex biological systems such as robustness, plasticity, autonomy and universality. Thus, Levy and Bechtel are exactly right when they say, "New tools such as graph theory, mathematical modeling, and dynamical systems theory provide holists with a research program that they previously lacked. Beyond merely criticizing extant mechanistic accounts for failing to take into account the context of the system in which mechanisms operate, they can represent and analyze the behavior of complex systems" (Levy and Bechtel 2016, 12). It also seems clear that global network properties, global organizational features and other contextual constraints of complex biological systems are not mathematical fictions. Moreno, Ruiz-Mirazo and Barandiaran express a similar picture about complex biological and cognitive systems in what follows:

> Everywhere in biology and cognitive science we deal with systems made of parts or elements with different functionalities acting in a selective and harmonized way, coordinating themselves at different time scales, interacting hierarchically in local networks, which form, in turn, global networks and, then, meta-networks . . . The organization of living systems consists in different nested and interconnected levels which, being somewhat self-organized in their local dynamics, depend globally one upon the others. This means that both the components and the sub-networks contribute to the existence, maintenance and propagation of the global organizations to which they belong. And, in turn, those global organizations contribute to the production, maintenance and propagation of (at least some of) their constitutive components (2011, 322).

16.3.3 Contextual Emergence

Perhaps many a new, new mechanist will happily accept all the conclusions made thus far. When it comes to complex biological systems such as the brain, perhaps there truly are no factual disagreements remaining between many who fly the flag of 'emergence' and those who call themselves 'mechanists.' Considering the history of evolutionary theory and neuroscience, this would certainly be historically interesting and newsworthy in its own right (Cobb, 2020, 374–75). In order to explore whether or not this is truly the case, let us recall that earlier in the paper it was said there are two questions the new, new mechanists must now answer:

(1) Are loc and decomp now to be rejected? If so, what then is the essence of mechanistic explanation? If no, then how are loc and decomp going to be reconceived in such a way as to retain their fundamental explanatory role?
(2) what now explains the formation, order, stability, causal capacities and functionality of complex biological mechanisms? This is what Winning and Bechtel call the "mysteriousness problem" (2018).

We shall return to these questions after we attempt to articulate the kind of emergence at work in complex biological systems. I call it "contextual emergence" (Silberstein 2018; Bishop and Silberstein 2019; Bishop et al. forthcoming). The contention is that contextual emergence has a central feature that C. D Broad himself, arguably the leader of the classical British Emergentist movement, the enemy of the "pure mechanists" of his day, would greatly appreciate and feel somewhat vindicated by. Conversations about Broad's view of "strong emergence" often focus on "transordinal" laws, i.e., brute bridge-laws connecting essentially different hierarchical and reified levels in nature, such as the physical and chemical, chemical and biological, or neural and psychological. In discussions about Broad's account of emergence, people often focus on such laws, and his emphasis on the "in principle" failure of derivability, prediction or explanation as the hallmark of emergence (1925). Emergents (those things that emerge) for Broad are brute facts.

All of this is what leads Povich and Craver to say the following, "*ontic emergence* is suspect or promising (depending on one's perspective) precisely because it involves such *discontinuity* [emphasis added]: there are higher-level properties and capacities that have no sufficient (ontic) explanation in terms of the parts, activities and organizational features of the system in the relevant conditions" (2018, 190).

As the result of these discontinuous and inexplicable jumps, Broad was typically considered an archenemy of the mechanists of his time, a compositional view of nature he called "pure mechanism." According to Broad, this is the view that the 'laws governing' the parts of a system operate in a purely context-independent fashion (Broad 1925, 58–61). Contextual emergence keeps the context-dependence feature of Broad's account of emergence but rejects the claim that emergents are brute or inexplicable.

With contextual emergence, global constraints and other kinds of context sensitivity are *fundamentally* at play. As Broad puts it, "[A]n emergent quality is roughly a quality which belongs to a complex as a whole and not to its parts" (Broad 1925, 23). According to him, if the properties of an irreducible whole are not given by the properties of the basic parts in isolation, they are emergent (see Humphreys 2016 for more details). For Broad, the global or systemic properties P of a system S are only reducible when the parts in isolation are sufficient to explain the existence of P. That is, there is reducibility when P can be derived or predicted in principle from the parts of S in isolation or when embedded in simpler systems (Stephan 1992, 55).

Contextual emergence emphasizes the ontological and explanatory fundamentality of multiscale *contextual constraints,* often operating globally over interconnected, interdependent, and interacting entities and their relations at multiple scales, e.g., topological constraints and organizational constraints in complex biological systems. Contextual emergence focuses on the fact that scientific explanation is often inherently and irreducibly multiscale. Contextual emergence is about the inherently interactive, multiscale interdependence of phenomena at all scales. Contextuality is conceived as a particular confluence of circumstances at multiple scales that produce a combination of constraints and stability conditions. These constraints and stability conditions will in turn open up and close off new modal spaces to the system, i.e., contextual emergence reduces a system's degrees of freedom and also opens up new possibility spaces that were previously closed off outside of that context, thus adding new degrees of freedom. Here are some key features of contextual emergence:

1. Contextual emergence is a type of scientific explanation that emphasizes the equal fundamentality of what are often multiscale *contextual constraints* and interdependent relations at multiple interacting scales.
2. Such constraints can include global or systemic constraints such as topological constraints, dimensional constraints, network or graphical constraints, order-parameters, etc. Contextual constraints therefore need not involve anything like direct causal-mechanical or dynamical interactions.
3. Such constraints can be causal-mechanical and dynamical, but they can also involve 'non-causal' or 'non-dynamical' difference makers, such as conservation laws, free energy principles, least action principles, symmetry breaking, etc.
4. Such constraints can include global organizing principles such as plasticity, robustness, and autonomy in complex biological systems. Contextual constraints can even be behavioral, social, normative, etc.
5. Contextual constraints can be symmetric, such that X and Y can simultaneously act as contextual constraints for one another.
6. Contextual constraints represent both the screening off and opening up of new areas of modal space, i.e., degrees of freedom, and thereby new patterns emergence and become robust.
7. Contextual emergence provides a framework to understand two things: (A) how novel properties are produced, and (B) why those novel properties matter.

Let us return to our two questions above and let us begin with the second question. According to contextual emergence, novelty, order and stability–modality of all varieties (e.g., nomological and causal)–are grounded in the fact that reality is a network of evolving, contextually sensitive extrinsic dispositions. Order comes not from any reductive compositional or realization-base. Order comes not from anything second-order or META-physical added. Order comes not from any causal or nomological glue, not from any metaphysical grounding whatsoever.

What biology shows us in case after case, is that the arrow of explanation and determination is not strictly bottom-up, not unidirectionally from smaller length and time scales to larger scales. It is for these reasons that contextual emergence is common, universal, non-spooky and does not defy scientific explanation. Nor does contextual emergence imply any kind of discontinuity or disunity in nature.

As for the first question above, given contextual emergence, it just is not very weighty. Of course, loc and decomp in some sense, will always be a useful strategy for the manipulation of and intervention upon biological mechanisms. But again, all such talk of loc and decomp will be purely pragmatic and contextual. That is, what functions components perform at any given time, will be determined by various interdependent and interconnected multiscale interrelations. Loc and decomp are now just research strategies, they no longer answer the second question above about the origin of order in mechanisms. And given contextual emergence, mechanistic explanation is certainly not fundamentally compositional, it is causal-dynamical-transformational. Thus, I would say that mechanistic explanation is just another instance of contextual emergence and not the other way around. Surely this is something of a win for C.D. Broad and company and demands a serious re-conceiving of mechanistic explanation.

Assuming contextual emergence is a reasonable characterization of what we are learning about complex biological systems, is this a characterization that a new, new mechanist can accept and still remain a mechanist? If the new, new mechanist in question is Winning or Bechtel, then perhaps so. They have recently been arguing that mechanistic explanation is best conceived in terms of constraints: "We provide a new account on which the causal powers of mechanisms are grounded by time-dependent, variable constraints" (Winning and Bechtel 2018, 288; Bechtel 2018, 574). They also note that, "The framework of constraints can be applied iteratively—a macro-scale object can be further constrained by incorporating it into a yet larger-scale object" (Winning and Bechtel 2018, 293).

All of this sounds a great deal like contextual emergence, but especially the following characterizations, "Thus, on our view, when constraints enable objects to have novel, emergent behaviors, this is tantamount to the emergence of causal powers... by means of possessing such emergent powers, mechanisms and components causally produce the effects they do" (Winning and Bechtel 2018, 294). And finally, "By restricting some degrees of freedom of its components and thereby enabling the whole mechanism to do things that would otherwise not be possible, constraints determine the causal powers of a machine or mechanism. Of particular importance are those constraints that are flexible and time-dependent. These enable machines to operate in different ways on different occasions" (Winning and Bechtel 2018, 307). Winning and Bechtel argue that mechanisms conceived as constraints solve the "mysteriousness problem", thus grounding the causal powers of mechanisms (Winning and Bechtel 2018, 292). The idea seems to be that mechanisms just are sets upon sets of constraints.

Is this just contextual emergence? I can not tell without more inquiry. I do not know, for example, if Winning and Bechtel would assent to every facet of contextual emergence enumerated herein and elsewhere. In Winning (2018) he talks about constraints as ontologically primitive modal structures (13). That is, he conceives of constraints as powers which are intrinsic dispositions, i.e., part of the intrinsic nature of its bearers, even when not manifested. As he puts it:

> I will refer to such ontologically primitive, intrinsic limitations as 'constraints'. On this view, constraints are more than mere regularities; in the words of Mumford ([2004]), constraints are 'modally loaded'. They may be thought of as modal patterns. Often, patterns are conceived in philosophy as nothing more than non-modal regularities. But constraints

> are more than just occurrent regularities; constraints in a dynamical system pertain to what might happen. They are the modal facts about a dynamical system, the truthmakers for dynamical equations and modal causal claims (2018, 14).

To bring this all back to causal mechanisms, Machamer (2004) and others claim that it is mechanisms themselves (construed as "activities") that answers the metaphysical grounding question; order and stability exists in biological systems because of causal mechanisms. On his Humean view, any appeal to anything as metaphysical as 'powers' is mysterious and unnatural. Whereas, Winning and Bechtel argue the reverse. They want to explain the causal powers of mechanisms by invoking constraints as intrinsic dispositions. This then is a dispute about which facts are the brutest facts, "activities" or "constraints." Or if you prefer, it's a dispute about what ultimately counts as explanans and explanandum.

From the perspective of contextual emergence, the Winning characterization of constraints is a little too second-order or META-physical. With contextual emergence, the notion of an intrinsic disposition or constraint is an oxymoron. However, I do agree that contextual constraints are not merely Humean regularities, as the latter view simply begs off the question of where nomic and causal order come from in biological mechanisms and elsewhere.

As will be discussed in the next section, my primary concern is not about metaphysical differences with Winning and Bechtel, however. My worry is that once any new, new mechanist takes option (c) as clearly Winning and Bechtel have done, thus giving up the claim that loc and decomp are both necessary and sufficient for mechanistic explanation—indeed, possibly giving up the claim that mechanistic explanation has any essence, threatens to make the mechanist philosophy too broad or downright trivial. That is, what then defines the essence of mechanistic explanation, let alone the mechanistic worldview? I wonder how many new mechanists will happily adopt my answer or Winning and Bechtel's answer? My biggest concern however is that for those new mechanists who are willing to go that far to the left, yet who insist on keeping it all within the mechanistic tradition and under the mechanist's banner, that they are in fact obscuring what a profound departure all of this is from the old school new mechanistic philosophy. Again, if people like myself, Winning and Bechtel are right about everything, there is much here that is a win for Broad and his emergentist movement.

16.4 Defending Premise 3

There are two ways to take option (c) and thus attempt to deny the dilemma presented in premise 3 of the master argument by disarming its second horn. The first is to deny that loc and decomp are essential for mechanistic explanation, and the second is to argue that loc and decomp are compatible with global constraints and other kinds of context sensitivity. Both options will be examined in this section.

16.4.1 Loc and Decomp Is Not the Essence of Mechanistic Explanation

One way of attacking the dilemma presented in premise 3 (and premise 1 for that matter), is to claim that we *never* had any right to define mechanistic explanation in terms of loc and decomp in the first place. Again, the claim here is that our characterization of mechanistic explanation was a strawman even in our original 2013 paper. Take the following from Craver and Tabery for example:

> Mechanisms are not necessarily *localizable* (Bechtel and Richardson 2010 [1993]). Components of mechanisms might be widely distributed (as are many brain mechanisms) and might violate our intuitive or tutored sense of the boundaries of objects (as an action potential violates the cell boundary). The assumption of localization is often an important heuristic in the search for mechanisms; however, this heuristic often must be abandoned as the mechanism's organization reveals itself (Craver and Tabery 2015).

We noted as much in our original paper, but as Weiskopf says:

> It bears emphasis that localization of function is a significant constraint for mechanists. The guiding image of mechanisms as machine-like structures strongly suggests that they are made of discrete parts, each of which carries out a dedicated function (2016, 677).

Craver and Tabery (2015) also emphasize that historically the new mechanist defines mechanistic explanation in terms of some degree of loc and decomp. As Glennan puts it:

> A ubiquitous and important aspect of mechanistic organization is its hierarchical character. The parts of mechanisms can themselves be broken down into parts, and the activities within mechanisms can be broken down into further activities... Mechanistic analysis will typically bottom out in some set of entities and activities that are taken to be basic. (Glennan 2016, 802).

Furthermore, as noted in section 2 of the paper in defense of premise 1, it is clear that many new mechanists still adhere to some version of loc and decomp.

Nonetheless, as we saw in Sect. 16.3, there are new, new mechanists who do seem to reject both loc and decomp as essential to mechanistic explanation (see also Levy and Bechtel 2016). Bechtel on his own has recently said the following, "A concern I raised earlier about Craver's and my treatment of top-down causation is that it rendered all causal relations at the lowest level... This interpretation [of network representations], however, is mistaken" (2017, 269–70). He gives several reasons for this conclusion such as the fact that, "there are no grounds for treating the nodes in a given graph as at a common level" (2017, 270), and nodes "should not be treated as representing entities at some basic level", period (2017, 270). Bechtel is clear that mechanisms and even organism-wide functions or global constraints, constrain the behavior of their parts such that:

> In biology, the constraints imposed in a mechanism are specific to the conditions in the living system. From this perspective, the physical is far from closed but rather is extremely open-ended. Wherever one finds a set of components organized into a module with sufficient interactions, one will encounter constraints that limit the behavior of the components and

> how they respond to external inputs. The phenomenon described as top-down causation is not unusual, but common (2017, 272).

Bechtel is clear that network explanations often bear out this "top-down causation" via global constraints, potentially even operating over the entire network or organism (Bechtel 2017a, b, c, d, 253). Contrast all this with what Craver says:

> *Properties of parts explain aggregate properties (and not vice versa) because the parts compose the whole* [emphasis added]. Network properties are explained in terms of nodes and edges (and not vice versa) because the nodes and edges compose and are organized into networks. Paradigm distinctively mathematical explanations arguably rely for their explanatory force on ontic commitments that determine the explanatory priority of causes to effects and parts to wholes (2016, 701).

Bechtel's point is this: contra Craver, the behavior of complex mechanisms is not just a matter of local, bottom-up 'matters of fact', sometimes it is "vice versa." The kind of top-down causation described here by Bechtel, constitutes a clear rejection of the claim that mechanistic explanation must involve loc and decomp.

What then is essential about mechanistic explanation, if not loc and decomp? Perhaps it has no essence at all. Levy and Bechtel go on to say that: "We don't see much benefit in the project of defining mechanism" and, "Any explanation that appeals to underlying parts and organization is mechanistic" (2016, 25). They claim that they are "not emptying the notion of mechanism of content" (2016, 26) because, "the contrast between mechanistic explanation and DN explanation or other formalist views of explanation is retained" (2016, 26).

Whether or not other mechanists are comfortable with such a liberal definition of mechanistic explanation as proffered by the new, new mechanists probably depends on how they conceive of their project. The Levy and Bechtel or Winning and Bechtel take on mechanistic explanation seems to jettison both the normative aspect and the guiding metaphysik of the machine metaphor. However, as Rathkopf notes regarding loc and decomp:

> In order to generate a mechanistic explanation, therefore, one must be in position to individuate the relevant components and provide evidence that associates components with specific operations. That both of these goals must be achieved is supported by the observation that they are necessarily interdependent. Part of the evidence that a particular component is mechanistically relevant is the fact that it is responsible for carrying out a particular operation. Of course, one could simply stipulate that mechanistic explanation is possible without any commitment to identifying parts and operations, *but that kind of bare stipulation threatens to take the normative bite out of the mechanistic program* [emphasis added] (2018, 74).

As Craver and Tabery put it, "one might object that there's nothing left of mechanism once it sheds these historical associations. One might suspect that it has been trivialized" (Craver and Tabery 2015). This is my worry as well. What is left of the new mechanist philosophy if it does not involve loc and decomp, if it is not constitutive or compositional? What is left of the new mechanist worldview if it does not involve a hierarchical conception of physical and biological systems?

Other than the fact that new, new mechanistic explanations are not DN type explanations and they involve no spooky vital forces, what is left to define them as mechanistic? As far as I can tell it is only this "minimal" description: "A mechanism for a phenomenon consists of entities (or parts) whose activities and interactions are organized so as to be responsible for the phenomenon" (Glennan 2016, 799). You might wonder who would ever argue with this characterization, thus making it trivial. In Glennan's own words, the new, new mechanist philosophy appears to now be just a study of "epistemology and methodology" (Glennan 2016, 799).

One supposes the new, new mechanist who sees themselves as only doing epistemology, methodology or the theory of explanation can simply just beg off these questions. They can assert that loc and decomp is just one of many possible heuristic devices and there is no reason not to seek and employ other heuristics. Such a person will argue therefore that it is a mistake to define mechanistic explanation essentially in terms of loc and decomp (Stinson 2016). As we saw, some such as Bechtel think it is a mistake to define it at all. In which case we will just let science and scientific practice decide the question. There is certainly nothing wrong with this epistemological project in the theory of explanation, but many of us want to go beyond such metaphysical quietism to make science-based inferences about the actual causal and spatiotemporal organization of complex biological systems.

I am not alone in questioning why we should continue to call such explanations mechanistic given the prevalence of global constraints and other kinds of context sensitivity:

> While classical localization assumed that distinct cognitive systems would have disjoint physical realization bases, massive redeployment and network theory seem to demonstrate that different systems may have entangled realizers: shared physical structures spread out over a large region of cortex. This suggests that not only will there not be distinct mechanisms corresponding to many of the systems depicted in otherwise well-supported cognitive models, but given that the relevant anatomical structures are multifunctional in a highly context-sensitive way, *perhaps there will be nothing much like mechanisms at all—at least as those have been conceived of in the dominant writings of contemporary mechanistic philosophers of science* [emphasis added]. And while it might be that these networks should count as mechanisms on a sufficiently liberal conception of what that involves, widespread entanglement still violates Poldrack's constraint that distinct cognitive structures should be realized in distinct neural structures (Weiskopf 2016, 679-681).

"Poldrack's constraint" is of course just the idea that mechanistic explanation to be worthy of the name, ought to be constitutive, compositional and realization-based.

16.4.2 Loc and Decomp Is Compatible with Global Constraints and Other Kinds of Context Sensitivity

There are those wants to grant much of what has been said about the nature of complex biological systems but still wants to retain loc or decomp in some form. Burnston in his paper, "Getting over Atomism: Functional Decomposition in Complex Neural Systems", calls this strategy "contextual decomposition" (see also Burnston 2016a, b). The basic idea is that rather than claim that loc and decomp fail, and rather than call alternative explanations "non-mechanistic or emergent", we can just relativize loc and decomp to contextual features such as causal, temporal and spatial relations at some specific time t or over some duration of time. Burnston is explicitly rejecting what he calls "Atomism", which holds that "for any function F of a whole system S, we should decompose F according to a uniform list of functions – namely, the list of intrinsic functions of the parts", and replaces it with "contextualism", which holds that "parts of systems should be functionally individuated according to what they do... in interaction with other parts of the system" at any given time t. He claims that contextual decomposition is compatible with all the features of complex biological systems that are highlighted above, and that it only need be the case that, "for any given F, we should be able to find some difference between what the distinct parts of S are doing in the context of F." Burnston is clear that "contextual decomposition" entails multi-scale, contextually sensitive interactions that contribute to the roles that individual parts play in the system. He is also clear that the important contextual features in question are often spatially and temporally extended, to include things like the cognitive task being performed.

There are many other related things that Burnston might have in mind here so let us enumerate some of them:

1. The relationship between structural parts and functions is at least one-many (perhaps many-many), both over time and presumably at-a-time, because *what function a part contributes to* at any given time can change with various changes in context. Parts are multifunctional.
2. Not only can the 'same part' performing the same operation (e.g., produce a particular protein) exhibit "multifunctionality" in different contexts (i.e., the very same protein will contribute to different functions in different contexts), but the 'same part' might even perform different operations in different contexts (e.g., produce an entirely different protein).
3. Not just their function or specific operation, but the structural parts themselves (both token and type) that are underlying some function might change (perhaps even rapidly over time) with changing context. This could cover various kinds of multiple-realizability.

This list is no doubt not exhaustive and I do not know for sure he would assent to all this, but it is sufficient for our purposes. We know that indeed all three of these things and more happen in complex biological systems. Burnston's idea then is simply this: even given 1–3 above, so long as, at a time t or duration, relative to some relevant context, if we can zero in on a particular location and determine exactly what function(s) a part is contributing to or what new operation it is performing in contribution to some function(s), then such an explanation constitutes a kind of (contextualized) loc and decomp.

I agree that contextual decomposition is, among others, one worthy and reasonable project for biologists and cognitive scientists to pursue. But if Burnston wants to claim that giving up "Atomism" in favor of contextual decomposition is compatible with either explanatory or ontological reductionism (i.e., the original compositional spirit of the mechanistic philosophy via the machine metaphor) in any strong sense, then here I must demure, or at least I must be puzzled as to the sense of reduction he has in mind. We have seen that the relevant constraints and contexts at work in complex biological systems are sometimes global and multiscale, up to and including the wider physical environment outside the organism. And in the case of cognitive systems such context will surely include the social environment.

In what substantive sense is this reductionism? I suppose the dogged new, new mechanist could claim that the entire relevant multiscale context is ‘the mechanism’ or ‘machine’, but this really would be an Orwellian move, as it violates the very spirit of mechanistic explanation and is the very essence of holism. If Burnston is merely claiming that an explanation involves loc and decomp if it focuses on differentiated parts, their interactions and their functions in various contexts across all scales, that seems too weak to be reductionist unless, he wants to add the caveat that “lower-level” or smaller scale parts are always the more fundamental explainers, at least in principle. Otherwise, who is going to disagree that there are parts and they do stuff, and we should try and figure out what it is they do in different contexts?

The particular “contextual decomposition” at a time t or across a duration, is going to be a function of, is going to be explained by, contexts at multiple scales, including in some cases the global organizing features we have been discussing. In such cases global constraints and multi-scale contexts determine the behavior of the parts, not primarily the other way around. My question for Burnston is if contextual decomposition and contextual emergence are just two different names for the same thing? If the answer is ‘yes’, then any disagreement we might have is purely semantic. If the answer is ‘no’, I am curious where our empirical differences reside.

Zednik (2019, 26) also wants to argue that network type explanations are mechanistic for the following reason, “Thus, explanations in network neuroscience are mechanistic... because they invoke interventions to uncover the composition and/organization of network mechanisms in the brain” (2018, 27–28). Zednik grants that, say, “the degree of small-worldness” in a topological explanation, is explanatory (i.e., counts as a difference-maker) irrespective of the ever changing structural details (which may not always be difference-makers themselves), even though such network properties “supervene on the properties of the individual components” in any given instance.

Of course, Zednik and others are right that network representations sometimes help unearth new mechanistic details (Bechtel 2017a, b, c, d; Colombo and Weinberger 2018 and Matthiessen 2017), but that is not all they do. Huneman is also absolutely right that:

> topologies may constrain mechanistic explanations, for instance in the way a network topology constrains more or less the dynamics of what takes place in the network; but more interestingly topologies and mechanisms are likely to condition the explanatory power of each other (2018, 143).

Note that all of this, is completely compatible with Rathkopf's claim that, “much of network science should be seen instead as a departure from the mechanistic approach, and one that offers a completely distinct explanatory strategy" (2018, 56).

The real question here is what is packed into the word "supervene"? If Zednik simply means that in any given instance, the existence of some specific components and their properties are necessary for the existence of network properties, then of course that is true. If on the other hand he means that in each case where network properties exist, they are completely ontologically determined somehow by specific smaller scale componential interactions, then no.

Zednik is simply selling the old line that while network properties are multiply realized and even difference-makers in their own right, in any given token case, such network properties are completely determined by the componential properties they "supervene" on. More specifically Zednik says:

> A topological feature is an organizational property of a mechanism if one can change the behavior of the mechanism as a whole by intervening to change that topological feature, and one can change the topological feature by intervening to change the behavior of the mechanism as a whole (2018, 26).

Such a principle, "can be used to determine when a particular topological feature—such as a network's degree of small-worldness—is in fact an organizational property of the mechanism for some explanandum phenomenon" (Zednik 2018, 26).

Certainly, small-worldness is an organizational feature (what else?) and certainly mechanistic explanations can advert to organizational features in principle, if by "organizational features" one means some representation (however abstract) of the spatiotemporal, causal and dynamical relationships between various components. What is denied however, is that network explanations are nothing but maps or representations (however abstract) of such componential relationships. This is true not only because the nodes and edges in network explanations need not refer to components and their relations directly, but because the behavior of the components is often determined by or *constrained by* the *global* organizational feature of in this case, small-worldness. It is because network properties (or order parameters in the dynamical case) can represent/alter the global state of the system (or some sub-set of it) that one can change the behavior of the various components by tweaking the network properties.

Why are such global network constraints so prevalent, so often multiply realized in complex biological and cognitive systems even though processes at smaller scales often happen at very different time scales than the processes at larger scales they support? There is often a rapid turnover of entities and states at the smaller scales creating real world multiple realizability that belies any simplistic account of composition and realization. For me, but not for Zednik, the answer is because those global topological features, once in place, in turn constrain the behavior of the ever-changing constituents in order to maintain the relevant efficacious topological constraints.

For any particular synchronic-frame or still-shot of a biological system at a time t with some duration d, the determining features include *diachronic* multiscale interactions (context sensitivity) and *global constraints* outside the time-slice in question that cannot even be assigned a scale or 'level.' That is, when it comes to such complex biological systems one should take the word *process* very seriously and understand that such systems are spatially, temporally, functionally and in a thin sense, teleologically extended. This is not to deny of course that there are a variety of both global-to-local and local-to-global determination relations involved in such systems.

Thus, once we see that global topological network properties do not "supervene" on structural components, there is little reason to think that topological explanations are mechanistic in the sense of loc and decomp. No doubt, as we have discussed, there are other weaker and non-reductive criteria under which we might count such explanations as mechanistic. But none of those criteria will discharge the dilemma herein for the mechanistic philosophy. Furthermore, as illustrated in Sect. 16.3, without adverting to formal topological and network models, there is ample textbook evidence from systems biology in general that loc and decomp fail for key properties of complex biological systems.

16.5 Conclusion

It has been argued that the new mechanist and new, new mechanist philosophy is likely either false or trivially true. One might ask, why not just embrace explanatory pluralism as a way out of the dilemma? Can we not agree that, "Different types of models are necessary to explain relevant features of biological systems. Depending on the data available and the research question, top-down and bottom-up approaches are employed, each of which is multilevel in its own right but involves different explanatory tactics" (O'Malley et al. 2014, 823). And can we agree that, "Deploying molecular approaches is not equivalent to embracing reductionism" (2014, 823). Yes, we can agree on both these points. But as these authors also note, "Reduction does not adequately describe the integrative impulse underlying this multilevel production of new biological knowledge" (2014, 824), if reduction means, "the process will bottom out at a preferred level" (2014, 823).

Explanatory pluralism exists in part because complex biological systems really do instantiate various global constraints whereupon loc and decomp fail, and *that fact is reflected* in many of our best biological explanations, such as topological explanations. As Love says:

> First, reciprocal interactions between genetic and physical causes does not conform to the expectations that mechanism descriptions 'bottom-out' in lower-level activities of molecular entities (Darden 2006). The interlevel nature of the causal dynamics between genetic and physical factors runs counter to this expectation and is not amenable to an interpretation in terms of nested mechanisms realizing another mechanism. Second, the reciprocal interaction between genetic and physical causes does not require stable, compositional organization, which is a key criterion for mechanisms (Craver and Darden 2013). The productive continuity of a sequence of genetic and physical difference-makers can be maintained despite changes in the number and types of elements in a mechanism. Although compositional differences can alter relationships of physical causation (fluid

> flow or tension), these relationships do not require the specificity of genetic interaction predominant in most mechanistic explanations from molecular biology. (The multiple realizability of CPM outcomes is central to this conclusion). Standard mechanistic strategies of representation and explanation appear inadequate to capture these mechanisms (Love 2018, 341; see also Love 2012,120 and Love and Hüttemann 2011).

Again, all of this begs the question, what remains as to the essence of mechanistic explanation? If there is none, then there is really nothing interesting to argue about.

Regarding explanatory/causal pluralism, network models for example can be causal explanations in a variety of ways to include: difference making, counterfactuals, Granger causation and other more topological, statistical and abstract notions of causation, formal causation, and even intervention/manipulation—networks can be tweaked. Furthermore, as they get more sophisticated, relatively static graphical models and explanations can be and increasingly are full-blooded *dynamical explanations*, as with "temporal networks" and "dynamic network neuroscience" (Feldt-Muldoon and Basset 2016).

But unless one is engaged in nothing more than a methodological exercise and is completely happy with metaphysical quietism, none of these facts about explanatory pluralism change the outcome of the argument herein. The key fact, as some former new mechanists are starting to admit, is that complex biological systems look much different than loc and decomp taken as ontological descriptions would suggest. Perhaps it is time to acknowledge this fact and spend less time attempting to indefinitely expand the definition of mechanistic explanation. Whether or not it can be shoehorned into the category of mechanistic explanation, I would say the real headline here is contextual emergence.

References

Anderson, M. (2016). Précis of after phrenology: Neural reuse and the interactive Brain. *Behavioral and Brain Sciences, 39*, 1–22.

Bateson, P., & Gluckman, P. (2011). *Plasticity, robustness, development and evolution*. Cambridge University Press.

Batterman, R. W., & Rice, C. C. (2014). Minimal model explanations. *Philosophy of Science, 81*(3), 349–376.

Bechtel, W. (2011). Mechanism and biological explanation. *Philosophy of Science, 78*(4), 533–558.

Bechtel, W. (2017a). Explicating top-down causation using networks and dynamics. *Philosophy of Science*.

Bechtel, W. (2017b). Analysing network models to make discoveries about biological mechanisms. *The British Journal for the Philosophy of Science*. https://doi.org/10.1093/bjps/axx051.

Bechtel, W. (2017c). Systems biology: Negotiating between holism and reductionism. In S. Green (Ed.), *Philosophy of systems biology: Perspectives from scientists and philosophers*. Springer.

Bechtel, W. (2017d). Top-down causation in biology and neuroscience: Control hierarchies. In M. P. Paolini & F. Orilia (Eds.), *Philosophical and scientific perspectives on downward causation*. Routledge.

Bechtel, W. (2018). The importance of constraints and control in biological mechanisms: Insights from cancer research. *Philosophy of Science, 85*(4), 573–593.

Bechtel, W. (2019). Analysing network models to make discoveries about biological mechanisms. *The British Journal for the Philosophy of Science, 70*(2), 459–484.

Bechtel, W., & Abrahamsen, A. (2005). Explanation: A mechanist alternative. *Studies in the History and Philosophy of Biological and Biomedical Science, 36*(2), 421–441.

Bechtel, W., & Richardson, R. C. (2010). *Discovering complexity: Decomposition and localization as strategies in scientific research* (2nd ed.). Cambridge, MA: MIT Press.

Bishop, R., & Silberstein, M. (2019). Complexity and feedback. In S. Gibb, R. Hendry, & T. Lancaster (Eds.), *The Routledge handbook of emergence*. New York: Routledge.

Bishop, R., Silberstein, M., & Pexton, M. (forthcoming). *Contextual emergence*. Oxford University Press.

Boi, L. (2017). The interlacing of upward and downward causation in complex living systems: On interactions, self-organization, emergence and wholeness. In M. P. Paolini & F. Orilia (Eds.), *Philosophical and scientific perspectives on downward causation*. Routledge.

Brigandt, I., Green, S., & O'Malley, M. (2018). Systems biology and mechanistic explanation. Ingo Brigandt, Sara Green & Maureen O'Malley – 2018. In S. Glennan & P. M. K. Illari (Eds.), *The Routledge handbook of mechanisms and mechanical philosophy* (pp. 362–374). New York: Routledge.

Broad, C. D. (1925). *The mind and its place in nature* (1st ed.). London: Routledge & Kegan Paul.

Burnston, D. C. (2016a). "Computational neuroscience and localized neural function." Synthese, 1–22. https://doi.org/10.1007/s11229-016-1099-8.

Burnston, D. C. (2016b). A contextualist approach to functional localization in the brain. *Biology and Philosophy*, 1–24. https://doi.org/10.1007/s10539-016-9526-2.

Burnston, D. C. (2017). Real patterns in biological explanation. *Philosophy of Science, 84*(5), 879–891.

Cobb, M. (2020). *The idea of the brain: The past and future of neuroscience*. New York: Basic Books.

Colombo, M., & Weinberger, N. (2018). Discovering brain mechanisms using network analysis and causal modeling. *Minds and Machines, 28*(2), 265–286. https://doi.org/10.1007/s11023-017-9447-0.

Craver, C. F. (2001). Role functions, mechanisms, and hierarchy. *Philosophy of science, 68*(1), 53–74.

Craver, C. F. (2007). *Explaining the brain: Mechanisms and the mosaic unity of neuroscience*. New York: Oxford University Press.

Craver, C. F. (2016). The explanatory power of network models. *Philosophy of Science* (forthcoming).

Craver, C., & Bechtel, W. (2007). Top-down causation without top-down causes. *Biology and Philosophy, 22*, 547–563.

Craver, C. F., & Darden, L. (2013). *In search of mechanisms: Discoveries across the life sciences*. University of Chicago Press.

Carl Craver & James Tabery. (2015). Mechanisms in science. http://plato.stanford.edu/entries/science-mechanisms/. Published, 10/01/2015.

Damicelli, F. Claus C. Hilgetag, M.-T. H., & Messen, A. (2018). Topological reinforcement as a principle of modularity emergence in brain networks. *bioRxiv* preprint first posted online Sep. 4, 2018. http://dx.doi.org/10.1101/408278.

Darden, L. (2006). Reasoning in Biological Discoveries: Essays on Mechanisms, Interfield Relations, and Anomaly Resolution. *Cambridge Studies in Philosophy and Biology.*

Fazekas, P., & Kertesz, G. (2018). Are higher mechanistic levels causally autonomous? In: [2018] PSA 2018: The 26th Biennial Meeting of the Philosophy of Science Association (Seattle, WA; 1–4 November 2018). http://philsci-archive.pitt.edu/view/confandvol/confandvolPSA2018.html. URL: http://philsci-archive.pitt.edu/id/eprint/15241

Feldt Muldoon, S., & Bassett, D. S. (2016). Network and Multilayer Network Approaches to Understanding Human Brain Dynamics. Sarah Feldt Muldoon & Danielle S. Bassett – 2016. *Philosophy of Science, 83*(5), 710–720.

Francis, R. C. (2011). *The ultimate mystery of inheritance: Epi-genetics*. W. W. Norton & Company.

Gilbert, S., & Epel, D. (2009). *Ecological developmental biology: Integrating epigenetics, medicine and evolution*. Sinauer Associates, Inc. Publishers.

Gillett, C. (2013). Constitution, and multiple constitution, in the sciences: Using the neuron to construct a starting framework. *Minds and Machines, 23*, 309–337.

Glennan, S. (2016). Chapter 38: Mechanisms and mechanical philosophy. The Oxford of philosophy of science, Humphreys, P (eds.).

Glennan, S. (2017). *The new mechanical philosophy*. Stuart Glennan: Oxford University Press.

Glennan, S. & Illari, P. (2018). *The Routledge handbook of the philosophy of mechanisms*. Stuart Glennan & Phyllis Illari (eds.). Routledge.

Green, S., Serban, M., Scholl, R., Jones, N., Brigandt, I., & Bechtel, W. (2018). Network analyses in systems biology: New strategies for dealing with biological complexity. Sara Green, Maria Serban, Raphael Scholl, Nicholaos Jones, Ingo Brigandt & William Bechtel – 2018. *Synthese, 195*(4), 1751–1777.

Hilgetag. C. C., & Goulas, A. (2015). Is the brain really as small-world network? BrainStructFunct. Apr 18. [Epub ahead of print].

Hooker, C. (2011). Conceptualising reduction, emergence and self-organization in complex dynamical systems. In Hooker (Ed.), *Philosophy of complex systems* (pp. 195–222). Elsevier.

Humphreys, P. (2016). *Emergence*. Oxford University Press.

Huneman, P. (2018). Diversifying the picture of explanations in biological sciences: ways of combining topology with mechanisms. *Synthese, 195*, 115–146.

Jablonka, E., & Lamb, M. (2005). *Evolution in four dimensions: Genetic, epigenetic, behavioral, and symbolic variation in the history of life*. Cambridge: MIT Press.

Jaeger, L., & Calkins, E. R. (2012). Downward causation by information control in microorganisms. *Interface Focus, 2*, 26–41.

Kaplan, J. (2008). Review of genes in development: Rereading the molecular paradigm (Neumann-Held, E. M, & Rehmann-Sutter, C, eds.). *Biological Theory, 2*, 427–429.

Kaplan, M. D. (2015). Moving Parts: the Natural Alliance Between Dynamical and Mechanistic Modeling Approaches. *Biology and Philosophy 30*(6):757–786.

Kaplan, D. M. (2018). *Mechanics and dynamical explanation* (pp. 267–280). Routledge.

Koonin, E.V. (2011). *The logic of chance: The nature and origin of biological evolution* (Koonin, E. V, ed., 528 pages). FT Press.

Levy, A., & Bechtel, W. (2016). Towards mechanism 2.0: Expanding the scope of mechanistic explanation. In: [2016] PSA 2016: The 25th Biennial Meeting of the Philosophy of Science Association (Atlanta, GA; 3–5 November 2016). http://philsci-archive.pitt.edu/view/confandvol/confandvol2016PSA.html. URL: http://philsci-archive.pitt.edu/id/eprint/12567

Love, A. C. (2012). Hierarchy, causation and explanation: Ubiquity, locality, and pluralism. *Interface Focus, 2*, 115–125.

Love, A. C. (2018). Developmental mechanisms. In S. Glennan & P. Illari (Eds.), *The Routledge handbook of the philosophy of mechanisms and mechanical philosophy* (pp. 332–347). New York: Routledge.

Love, A. C., & Hüttemann, A. (2011). Comparing part-whole explanations in biology and physics. In D. Dieks, W. J. Gonzalez, S. Hartmann, T. Uebel, & M. Weber (Eds.), *Explanation, prediction, and confirmation* (pp. 183–202). Berlin: Springer.

Machamer, P. K., Darden, L., & Craver, C. F. (2000). Thinking about mechanisms. *Philosophy of Science, 67*, 1–25.

Machamer, P. (2004). Activities and causation: The metaphysics and epistemology of mechanisms. *International Studies in the Philosophy of Science, 18*, 27–39.

Markov, N. T. Mária Ercsey-Ravasz, David C. Van Essen, Kenneth Knoblauch. (2013). "Cortical high-density counterstream architectures." 1 Nov. 2013 Vol. 342 578. Science.

Matthiessen, D. (2017). Mechanistic explanation in systems biology: Cellular networks. *The British Journal for the Philosophy of Science, 68*, 1–25.

Moreno, A., Ruiz-Mirazo, K., & Barandiaran, X. (2011). The impact of the paradigm of complexity on the foundational frameworks of biology and cognitive science. In Hooker (Ed.), *Philosophy of complex systems* (pp. 311–333). Elsevier.

Noble, D. (2006). *The music of life: Biology beyond genes*. Oxford UK: Oxford University Press.

O'Malley, M. A., Brigandt, I., Love, A. C., Crawford, J. W., Gilbert, J. A., Knight, R., Mitchell, S. D., & Rohwer, F. (2014). Multilevel research strategies and biological systems. *Philosophy of Science, 81*, 811–828.

Pedersen, M., & Omidvarnia, A. (2016). Further insight into the Brain's Rich-Club architecture. *Journal of Neuroscience, 36*(21), 5675–5676.

Povich, M., & Craver, C. F. (2018). Mechanistic levels, reduction, and emergence. Mark Povich & Carl F. Craver – forthcoming. In S. Glennan & P. M. K. Illari (Eds.), *The Routledge handbook of mechanisms and mechanical philosophy*. Routledge.

Rathkopf, C. (2018). Network representation and complex systems. *Synthese, 195*, 55–78.

Ross, L. (2015). Dynamical models and explanation in neuroscience. *Philosophy of Science, 82*, 32–54.

Silberstein, M. (2016). The implications of neural reuse for the future of cognitive neuroscience and the future of folk psychology. *Brain and Behavioral Sciences, 39*, E132.

Silberstein, M. (2018). *Contextual emergence* (Special issue of philosophica on emergence) (Vol. 91, pp. 145–192)., Carruth, A. D., & Miller, J. T. M., eds.

Silberstein, M., & Chemero, A. (2013). Constraints on localization and decomposition as explanatory strategies in the biological sciences. *Philosophy of Science, 80*(5), 958–970.

Sporns, O. (2011). *Networks of the brain*. Cambridge, MA: MIT Press.

Stephan, A. (1992). Emergence—a systematic view on its historical aspects. In Beckermann, A., et al. (eds.), pp. 25–47.

Stinson, C. (2016). Mechanisms in psychology: Ripping nature at its seams. *Synthese, 193*(5), 1585–1614. https://doi.org/10.1007/s11229-015-0871-5.

van den Heuvel, M. P., & Sporns, O. (2011). Rich-Club Organization of the Human Connectome. *Journal of Neuroscience, 31*(44), 15775–15786.

Venturelli, N. A. (2016). A cautionary contribution to the philosophy of explanation in the cognitive neurosciences A. Nicolás Venturelli. *Minds and Machines 26*(3), 259–285.

Weiskopf, D. A. (2016). Integrative modeling and the role of neural constraints. *Philosophy of Science, 83*(December 2016), 674–685.

Winning, J. (2018). Mechanistic causation and constraints: Perspectival parts and powers, non-perspectival modal patterns. *British Journal for the Philosophy of Science*.

Winning, J., & Bechtel, W. (2018). Rethinking causality in biological and neural mechanisms: Constraints and control. *Minds and Machines, 28*(2), 287–310.

Zednik, C. (2014). Are systems neuroscience explanations mechanistic? In *Preprint volume for philosophy science association 24th biennial meeting* (pp. 954–975). Chicago: Philosophy of Science Association.

Zednik, C. (2015). Heuristics, descriptions, and the scope of mechanistic explanation. In *Explanation in biology* (pp. 295–318). Springer.

Zednik, C. (2019). Models and mechanisms in network neuroscience. *Philosophical Psychology, 32*(1), 23–51.

Zimmer, C. (2018). *She has her mother's laugh: The powers, perversions and potential of heredity*. Dutton Press.

Chapter 17
Compare and Contrast: How to Assess the Completeness of Mechanistic Explanation

Matej Kohár and Beate Krickel

Abstract Opponents of the new mechanistic account of scientific explanation argue that the new mechanists are committed to a 'More Details Are Better' claim: adding details about the mechanism always improves an explanation. Due to this commitment, the mechanistic account cannot be descriptively adequate as actual scientific explanations usually leave out details about the mechanism. In reply to this objection, defenders of the new mechanistic account have highlighted that only adding *relevant* mechanistic details improves an explanation and that relevance is to be determined relative to the phenomenon-to-be-explained. Craver and Kaplan (B J Philos Sci 71:287–319, 2020) provide a thorough reply along these lines specifying that the phenomena at issue are *contrasts*. In this paper, we will discuss Craver and Kaplan's reply. We will argue that it needs to be modified in order to avoid three problems, i.e., what we will call the *Odd Ontology Problem*, the *Multiplication of Mechanisms Problem*, and the *Ontic Completeness Problem*. However, even this modification is confronted with two challenges: First, it remains unclear how explanatory relevance is to be determined for contrastive explananda within the mechanistic framework. Second, it remains to be shown as to how the new mechanistic account can avoid what we will call the 'Vertical More Details are Better' objection. We will provide answers to both challenges.

17.1 Introduction

It is widely agreed among the new mechanists that complete explanations are better than incomplete explanations, and that the closer an explanation is to being complete the better. However, there is an on-going discussion about what completeness of explanation amounts to (Baetu 2015; Miłkowski 2016; Craver and Kaplan 2020). Opponents of the mechanistic account of explanation object that the new mechanists

M. Kohár · B. Krickel (✉)
Institut für Philosophie, Literatur-, Wissenschafts- und Technikgeschichte,
Technische Universität Berlin, Berlin, Germany
e-mail: matej.kohar@gmail.com; beate.krickel@tu-berlin.de

© Springer Nature Switzerland AG 2021
F. Calzavarini, M. Viola (eds.), *Neural Mechanisms*, Studies in Brain and Mind 17,
https://doi.org/10.1007/978-3-030-54092-0_17

are committed to assuming that adding *any kind of details* about a mechanism will improve an explanation (Batterman and Rice 2014; Chirimuuta 2014; Levy 2014). As a consequence, the new mechanists are committed to the claim that, for example, mentioning quarks in the explanation of spatial memory will improve the explanation of the latter, listing all kinds of activities of all billions of neurons in the brain will improve explanations in neuroscience, and mentioning the exact location of the ion-channels or mentioning the exact diameter of the axon will improve the explanation of the action potential. This, according to the opponents, is obviously problematic, as actual scientific explanations do usually not provide all kinds of details. Real explanations are sketchy, they abstract away from details, and they do not necessarily mention quarks. And these explanations are good especially *because* they leave out details. Hence, the new mechanistic account fails as an adequate account of scientific explanation or is incomplete at best. This is what Craver and Kaplan label the *'More Details are Better' (MDB) objection* (Craver and Kaplan 2020).

Defenders of the new mechanistic approach, however, argue that they are not committed to such a 'More Details are Better' claim. They highlight that clearly only *relevant* details improve an explanation and that relevance is to be determined *relative to the phenomenon-to-be-explained* (Baetu 2015; Boone and Piccinini 2016; Miłkowski 2016; Craver and Kaplan 2020). Many authors focus on defending the view that explanations that abstract away from details can still be mechanistic (Boone and Piccinini 2016; Miłkowski 2016), or on how to empirically establish whether a given explanation is complete (Baetu 2015). In contrast, in a recent paper, Craver and Kaplan provide a detailed analysis of how the norm of completeness is to be understood in the context of the new mechanistic approach by elaborating on the *relevance relative to the phenomenon*-idea. In a nutshell, they argue that mechanistic explanation (or models) aim at explaining *contrasts*, such as the spiking of the action potential at -70 mV rather than -50 mV. Relevance has to be determined relative to these contrasts.

In this paper, we discuss Craver and Kaplan's (2020) reply to the MDB-objection. More specifically, the paper will proceed as follows: In Sect. 17.2, we present the MDB-objection and Craver and Kaplan's reply. In Sect. 17.3, we will highlight three problems for Craver and Kaplan's account that we will call the *Odd Ontology Problem*, the *Multiplication of Mechanisms Problem*, and the *Ontic Completeness Problem*. In Sect. 17.4, we will suggest modifications to Craver and Kaplan's reply that solve these problems. We will, in Sect. 17.4.1, introduce a distinction between ontic mechanisms, mechanism descriptions, and mechanistic explanatory tests that helps to solve the *Odd Ontology Problem* and the *Multiplication of Mechanisms Problem*. In Sect. 17.4.2, we will show that completeness is a predicate of mechanism descriptions and mechanistic explanatory texts rather than ontic mechanisms. We thereby solve the *Ontic Completeness Problem*. In Sect. 17.5, we will argue that even based on these modifications, the reply to the MDB-objection is confronted with two challenges: First, it remains unclear how explanatory relevance can be determined for contrastive explananda within the mechanistic framework (Sect. 17.5.1). Second, it remains to be shown how the new mechanistic account can avoid what we will call the 'Vertical More Details are Better' objection (Sect. 17.5.2). We will provide answers to both challenges that essentially hinge on the idea that mechanistic explanations aim at identifying *crucial points of intervention.*

17.2 The MDB-Objection & Craver and Kaplan's Reply

According to opponents of the new mechanistic approach, the mechanistic approach fails as it is unable to account for the fact that many explanations are good because they leave out details about the mechanism that is responsible for the phenomenon (Woodward 2013; Batterman and Rice 2014; Chirimuuta 2014; Levy 2014; Rice 2015). They hold that the mechanistic account entails or that mechanists even explicitly defend the claim that the more details an explanation mentions, the better this explanation is. This is the 'More Details Are Better' or 'MDB' objection (Craver and Kaplan 2020). More specifically, different authors seem to accuse the new mechanists of being committed to the following claims:

a) Explanations should be complete.
b) Explanations are complete if and only if they describe every detail of a mechanism.
c) Adding details about a mechanism to an explanation always improves the explanation.

However, so the argument goes, often explanations are good because *they leave out* certain details of the mechanism (Woodward 2013; Batterman and Rice 2014; Chirimuuta 2014; Levy 2014; Rice 2015). For example, when cognitive neuroscientists explain cognitive capacities, they do not mention all 86 billion neurons of the brain and what they are doing (Chirimuuta 2014, 149). Network explanations are good because they mention the generic features of the network topology rather than more specific details about the implementing mechanism (Woodward 2013, 41). Optimality models provide explanations because they do *not* accurately represent the causes of a phenomenon, and they would provide "a worse (or perhaps no) explanation" at all if they described the causes in all detail (Rice 2015, 591). Completeness of mechanistic explanations, thus, must be interpreted in a way that is less demanding than b), or a) has to be rejected. In any case, c) has to be rejected as at least sometimes adding details about the mechanism does *not* improve the explanation.

Craver and Kaplan (2020) provide a detailed reply to this objection. They argue that their opponents are wrong in assuming that the new mechanists are committed to claims a), b), and c). According to Craver and Kaplan, the new mechanists are rather committed to what we summarize as claims a'), b'), c'). These claims are indeed compatible with the fact that some explanations are good even if they leave out details about the mechanism, or even if they are not complete.

a') Completeness concerns stores of *explanatory knowledge* rather than models or explanations (Craver and Kaplan 2020, 310). The closer a store of explanatory knowledge is to being complete, the better. Although a complete model is in principle possible, actual models serve further non-explanatory purposes that such a complete model would fail to account for (such as understandability, computational tractability, unification) (Craver and Kaplan 2020, 308).

b') A mechanistic explanation/model is complete if and only if it mentions all details of a mechanism that are *relevant for explaining a contrastive phenomenon P vs. P′* (all other things being equal) (Craver and Kaplan 2020, 300). If an explanation leaves out relevant details this will decrease its explanatory power and can only serve non-explanatory purposes (Craver and Kaplan 2020, sec. 7). However, an explanation does not have to be complete in this sense in order to have explanatory force (Craver and Kaplan 2020, 301).

c') Adding details about a mechanism to a mechanistic explanation/model improves the explanation if and only if *the details are relevant for explaining the phenomenon P vs. P′* (Craver and Kaplan 2020, 303).

According to Craver and Kaplan, claims a'), b'), and c') are compatible with the fact that good explanations often do not mention all details about a mechanism. Mentioning neurons in the brain is unlikely to provide a good mechanistic explanation as it is unlikely that all neurons and their activities will be relevant to explaining a given contrast about a cognitive capacity. However, if the activities of all neurons in the brain were explanatorily relevant for a given contrast, adding more details about the neurons would improve the explanation. This does not imply that the less detailed explanation has no explanatory force at all. Similarly, many network explanations highlight relevant details about the underlying mechanism (e.g. the structural, causal, or functional connectivity within a mechanism; see Craver (2016)), and are, thus, explanatory. Adding details about the mechanism will improve the explanation if these details are relevant for a given contrast-to-be-explained. Importantly, adding details to an already explanatory powerful network explanation does not imply that the original explanation is thereby rejected. Mechanistic explanations are essentially multilevel (Craver and Kaplan 2020, 304).

For the purpose of this paper, we will focus on Craver and Kaplan's reply to claims b) and c). We agree that it is odd to read into the mechanistic account the idea that adding *any kind* of detail would improve a given explanation. Early on, the new mechanists (including Craver (2007a)) have made clear that mechanistic details that are to be mentioned in an explanation are individuated *relative to the phenomenon-to-be-explained*, and they have pointed out that mechanistic details need to be *relevant* to the phenomenon at hand. Especially Craver has provided a detailed account of what relevance amounts to in the mechanistic context—constitutive explanation being the primary focus. In his so-called *mutual manipulability account of constitutive relevance*, Craver makes use of Woodwardian interventionism to specify the notion of relevance:

(*Mutual Manipulability Account of Constitutive Relevance*)

A mechanistic detail (an acting entity, X's ϕ-ing) is constitutively relevant for a given phenomenon (an acting entity, S's ψ-ing) if and only if:

(i) X's ϕ-ing is a spatiotemporal part of S's ψ-ing,
(ii) there is an ideal intervention on X's ϕ-ing by means of which one can change S's ψ-ing, and
(iii) there is an ideal intervention on S's ψ-ing by means of which one can change X's ϕ-ing. (Craver 2007a, 153)[1]

Based on these assumptions about the relevance of mechanistic detail for a given phenomenon, Craver and Kaplan introduce the following two-step account as a reply to the MDB objection and as a strategy to argue for b') and c'): First, they define what they call 'Salmon-Completeness' or 'SC' to spell out in which sense ontic mechanisms are complete:

Salmon-Completeness (SC): The Salmon-complete constitutive mechanism for [the phenomenon] P versus P′ is the set of all and only the factors constitutively relevant for P versus P′. (Craver and Kaplan 2020, 300)

Then, based on this notion of SC, Craver and Kaplan define under which conditions adding details to an explanation (they speak of "models" instead of "explanations") improves its explanatory power:

More Relevant Details Are Better (MRDB): If [explanatory] model M contains more explanatorily relevant [i.e., constitutively relevant] details than M* about the SC mechanism for P versus P′, then M has more explanatory force than M* for P versus P′, all things equal. (Craver and Kaplan 2020, 303).[2]

As already mentioned, mechanists have always stressed the importance of *relevance* of mechanistic detail *for a given phenomenon*. In addition, Craver and Kaplan argue that phenomena are contrasts of the form P vs. P′ (such as 'Socrates died vs. remained alive', 'Socrates died quickly vs. at some other rate') (Craver and Kaplan 2020, 296). They stress that this is not an entirely new idea and not foreign to the mechanistic thinking: Craver has explained already in his Craver 2007a, b book that phenomena are "multifaceted" (Craver 2007a, 125) and that one needs to explain various contrasts in order to fully explain a phenomenon. Furthermore, the idea that mechanistic explananda are contrastive follows naturally from the application of Woodwardian interventionism to make sense of explanatory relevance. However, in the present context, this addition is still new. The reason is that the notion of constitutive relevance has, so far, only been defined for non-contrastive phenomena, i.e., only for acting entities/Ss' ψ-ings.

The general spirit of Craver and Kaplan's reply to the MDB-objection is convincing. Clearly, only adding detail that is *explanatory relevant for the phenomenon-to-be-explained* increases the power of an explanation. However, there are three

[1]For the sake of argument, we ignore the challenges for the mutual manipulability account. For a discussion of these challenges see (Romero 2015; Baumgartner and Gebharter 2016; Baumgartner and Casini 2017; Kästner 2017; Baumgartner et al. 2018; Krickel 2018b).

[2]Craver and Kaplan use the label 'MDB_r' (with an index). For our purposes, it is more convenient to use the label 'MRDB'.

problems with the specifics of Craver and Kaplan's account and two general challenges. In the next two sections, we introduce the three problems and offer a solution. This solution will allow for a modification of Craver and Kaplan's reply without departing much from their approach. However, two challenges remain, which will be discussed in Sects. 17.5 and 17.6.

17.3 Three Problems: Odd Ontology, Multiplication of Mechanisms, Ontic Completeness

We will call the first problem for Craver and Kaplan's reply to the MDB objection the *Multiplication of Mechanisms Problem.* Given that mechanisms are individuated relative to the phenomena they are supposed to explain, the introduction of the contrastive phenomenon brings with it a multiplication not only of phenomena but also of mechanisms. On the original mutual manipulability account (see previous section), there was only one mechanism for each phenomenon such as the action potential, muscle contraction, or a rat's navigating through a maze (S's ψ-ing). Based on the contrastive interpretation of phenomena, Craver and Kaplan are committed to the view that there are multiple mechanisms—one for each contrast. For example, the action potential has many different features that may figure in such a contrast. It has a certain speed, voltage, etc., its refraction period has a certain length, and presumably other features. For each of these features, we can formulate an unbounded number of contrastive phenomena. For instance, in connection with voltage, there would be the contrastive phenomenon "action potential with voltage 70mV rather than 30mV", another one "action potential ... rather than 35mV" and so on. None of these contrastive phenomena is inherently more or less worthy of explaining. Furthermore, each of these contrastive phenomena would individuate a mechanism that is responsible for it. This generalizes to each of the unbounded number of contrasts one can formulate with respect to action potential or any other phenomenon taken as an acting entity. But on the grounds of parsimony, such multiplication of phenomena and mechanisms should be avoided.

The second problem is the *Odd Ontology Problem.* It can be formulated as a dilemma: If mechanisms and phenomena in Craver and Kaplan's account are supposed to be ontic, then they either cannot be contrasts, or Craver and Kaplan are committed to an odd ontology. The original mutual manipulability account had a straightforward ontic reading: X's ϕ-ing and S's ψ-ing are both acting entities that are real things in the world (Machamer et al. 2000). The part-whole relation between the mechanistic component and the phenomenon is a mind-independent ontic relation between these two acting entities. Since ideal interventions need only be "logically possible" (Woodward 2003, 128), also the mutual manipulability condition is a mind-independent relation between the phenomenon and the mechanistic component. Craver and Kaplan commit to an *ontic conception of scientific explanation.* According to this conception, explanations are objective things in the world (Craver and Kaplan 2020, sec. 5). This implies that mechanisms, which are the explanations according to the ontic conception, as well as their explanantia, i.e., phenomena, are real, ontic things. Based on the contrastive interpretation of the phenomenon, however, it becomes unclear in which sense phenomena and the relation between a mechanism and a phenomenon can be ontic. Clearly, contrasts of the form P vs. P′ are not real entities. P′ does not actually occur, and contrasts in this context are not entities at all but comparisons that scientists make. If at all, contrastive phenomena can be *descriptions* of things that are compared. But how does an ontic mechanism cause or constitute a contrast if the latter is a description?

The third problem, we call the *Ontic Completeness Problem*. It consists in the fact that Craver and Kaplan talk about 'ontic completeness' and define Salmon-Completeness (see Sect. 17.2) in terms of ontic mechanisms. It is a category mistake to speak of ontic mechanisms as complete or incomplete. Ontic mechanisms are the way they are, and it does not make sense to say that an ontic mechanism is complete or incomplete as if we had to build mechanisms similar to an IKEA cupboard where one screw is always missing. In explanatory contexts, mechanisms are already in existence. They are neither complete, nor incomplete. They just are. This reasoning is based on a thesis that may be called the *They Just Are Principle* (inspired by Craver 2007a, b, p. 27): ontic, mind-independent things on their own do not have normative or evaluative properties, they are neither good, nor bad, neither complete, nor incomplete. It is either our *descriptions* of the things in the world that are complete or incomplete; or it is a feature of a set of ontic things *relative* to a normative description (such as in the IKEA case: There should be six screws in the box!). Based on the They Just Are Principle, another way to formulate the Ontic Completeness Problem is in terms of the following argument:

1. Craver and Kaplan take explanation to be ontic: the mechanism that explains the phenomenon and the phenomenon itself are mind-independent things in the world.
2. There is no norm about which things *should* be components of a mechanism.[3]

(K1) Hence, it does not make sense to say that a mechanism that explains a phenomenon is complete or incomplete (from 1, 2, and the They Just are Principle).

3. Salmon-completeness is defined for (constitutive) mechanisms.

(K2) Hence, according to Craver and Kaplan it is possible to say about a mechanism that it is complete or incomplete (from 3).

Thus, Craver and Kaplan run into a contradiction (between K1 and K2). However, the They Just Are Principle already suggests a way to modify Craver and Kaplan's account such that the *Ontic Completeness Problem* can be avoided: define com-

[3]Note that the mutual manipulability account is not a description about what *should* be a component of a mechanism for a given phenomenon but rather a recipe for determining what *is* a component of a mechanism for a given phenomenon.

pleteness as a feature of *descriptions* of mechanisms rather than for mechanisms themselves. This is what we will do in the next section.

17.4 Solving the Problems: A Threefold Distinction & Two Notions of Completeness

Before we can provide solutions for the problems presented in the previous section, we have to do a little bit of stage setting. For the purposes of this paper, we will make the following assumptions that most new mechanists accept (including Craver and Kaplan):

Mechanism Characterization Mechanisms are entities and activities organized such that they are responsible for a phenomenon. (Machamer et al. 2000; Craver 2007a; Illari and Williamson 2012; Glennan 2017)

Etiological vs. Constitutive Mechanisms consist of those and only those acting entities that are either causally or constitutively relevant for a phenomenon. (Craver 2007a)

Constitutive Relevance Constitutive relevance is spelled out in terms of two necessary conditions: (i) spatiotemporal parthood, (ii) mutual manipulability. (Craver 2007a, b)

Levels of Mechanisms Mechanisms and mechanistic explanations come in hierarchies that are determined by relations of constitutive relevance and that are local to the phenomenon-to-be-explained. (Craver 2007a; Craver and Bechtel 2007)

Phenomena Phenomena are acting entities and to explain a phenomenon means to explain various contrasts. (Craver 2007a; Kaiser and Krickel 2017)

Singularism/Nominalism Mechanisms, entities, and activities are concrete particulars. Types are descriptions/models summarizing details about concrete particulars. (Glennan 2017; Krickel 2018a)

Purpose of Explanation The core function of explanation is to show how a phenomenon is situated in the causal structure of the world (Craver 2007a, 200), chiefly for the purpose of *intervening* into the phenomena (Craver 2007a, 93). Abstraction, in the sense of *ignoring explanatorily relevant details*, has only non-explanatory virtues (Craver and Kaplan 2020, sec. 7).

Unique Endeavour Explaining a phenomenon is a unique scientific endeavor that is distinct from prediction and description (Craver and Kaplan 2020, sec. 3).

Most new mechanists (including Craver and Kaplan) would also accept the following commitment:

Explanatory Relevance Explanatory relevance is constitutive relevance (in the case of constitutive explanation) or causal relevance (in the case of etiological explanation).

However, the equivalence between constitutive and explanatory relevance cannot hold assuming the contrastive view of explananda and the view that constitutive relevance holds between ontic things. Explanatory relevance and constitutive relevance are identical only if explanatory relevance is a relation between ontic things as well. If explananda are contrasts, however, we run into the Odd Ontology Problem if we assume that they can stand in constitutive relevance relations. If we resolve the Odd Ontology Problem by viewing contrastive explananda as descriptions, then they cannot enter into relations of constitutive relevance. One way to solve this problem is to reject that explanatory relevance and constitutive relevance are identical and allow for explanatory relevance to relate descriptions of ontic things that stand in constitutive relevance relationships. Our account will ultimately decouple explanatory relevance from constitutive relevance in that way.

It is important to note that unlike Craver and Kaplan, not all new mechanists accept the ontic conception of explanation mentioned in the previous section. Neither will we, as already indicated in the previous paragraph. Indeed, as we will argue in the next section, it is the conflation between ontic mechanisms and explanations that leads into the two problems mentioned in the previous section. Roughly, our solution will be to introduce a threefold distinction that helps us to keep apart the ontic and the epistemic aspects of explanation (Sect. 17.4.1). This threefold distinction will illuminate the role and meaning of completeness within the mechanistic framework (Sect. 17.4.2).

17.4.1 Ontic Mechanisms, Mechanism Descriptions, Explanatory Texts

In order to be able to unambiguously talk about ontic and epistemic issues, about matters of description vs. matters of explanations, we will distinguish between three elements:

1. **Ontic phenomena and mechanisms**: Ontic phenomena and ontic mechanisms are concrete particulars (see *Singularism/Nominalism* above). Ontic phenomena are acting entities such as a neuron firing, an axon terminal releasing neurotransmitter, a muscle contracting, a mouse navigating the Morris Water maze (Kaiser and Krickel 2017). Ontic phenomena are objects of explanatory endeavors, and targets of investigations. The mechanistic ontology is committed to the view that ontic phenomena are constituted by equally real ontic mechanisms (Illari and Williamson 2011), which are composed of acting entities with a spatiotemporal organization particular to each phenomenon. However, explanatory practices are only mediately concerned with ontic phenomena and mechanisms. Instead, explanation consists in constructing two types of texts: mechanism descriptions and mechanistic explanatory texts.
2. **Mechanism descriptions:** Mechanism descriptions are texts or other knowledge items that can be found in textbooks, journals, or other scientific media. The ontic mechanism is the truthmaker of the mechanism description. Mechanism descriptions are not guided by any particular explanatory interest but aim at

neutral description of the mechanism that later (via explanatory texts — see below) can be used for various explanations. Ideally, mechanism descriptions mention *all acting entities that are constitutively relevant* for a given ontic phenomenon *with maximal detail*. For example, ideally, the description of the mechanism responsible for a neuron's firing will mention, say, how many ions and ion-channels are involved, where they are located, what size they have, etc. for every point in time of the occurrence of the mechanism. The description of a single mechanism may span a number of publications, with only a part of the whole description exhibited in one place. In this they are close to Craver and Kaplan's "stores of explanatory knowledge". It is important to note that this is to be understood as a regulative ideal (see Railton (1980) for a similar view). Much scientific work consists in refining mechanism descriptions and filling in any gaps in them, although in practice, all mechanism descriptions actually available in the scientific community are incomplete.

3. **Mechanistic explanatory texts:** Mechanistic explanatory texts are vehicles of explanation, i.e., they are the explanantia. Each mechanistic explanatory text is an answer to a particular why-question. Why-questions, in our account, following Dretske (1972) and the spirit of Craver and Kaplan (2020) require explaining a particular contrast, whether explicitly, or implicitly stated. Mechanistic explanatory texts contain information from mechanism description relevant for explaining a particular contrastive explanandum. Note that it is only in mechanistic explanatory texts that contrasts play a role. Neither ontic phenomena, nor mechanism descriptions are in any way concerned with contrasts. Although mechanistic explanatory texts depend on mechanism descriptions, in practice even incomplete mechanistic descriptions can furnish the researcher with enough information to construct numerous mechanistic explanatory texts concerning various contrasts. Additionally, research that aims at answering particular why-questions, i.e. at constructing particular mechanistic explanatory texts can lead to the discovery of hitherto unknown ontic constituents, thus enriching the overall mechanism description. The question remains, however, what information goes into the mechanistic explanatory text and whether these texts always improve with the addition of further details. This will be taken up in Sect. 17.5.1.

As we will see in Sect. 17.4.2, the distinction between mechanism descriptions and mechanistic explanatory texts allows us to formulate different norms of completeness for descriptions and explanatory texts. Kaplan and Craver's talk of mechanistic models which at the same time describe a mechanism for a phenomenon and provide the explanatorily relevant factors for a contrast precludes one from acknowledging that depending on the purpose of the model a different completeness norm is appropriate. Therefore, what Craver and Kaplan call "mechanistic models" can on a case-by-case basis be classified as either mechanism descriptions or mechanistic explanatory texts.

One advantage of this threefold distinction is that it allows us to maintain the idea that mechanistic explanations have to pick out the ontic relations between a mechanism and a phenomenon—in contrast to a strict epistemic view that assumes that explanation has purely pragmatic or psychological functions. Furthermore, it is only by drawing a distinction between mechanism descriptions and explanatory texts that one can account for the idea that to explain a phenomenon is more than merely describing the mechanism that is responsible for it. Only *specific* ways of describing the ontic mechanism are explanatory. For example, assume that a certain volume of water has a temperature of −17 °C and is therefore frozen (under otherwise normal conditions) (Craver and Kaplan 2020, 304–305). If we want to explain as to why the water is frozen in contrast to not frozen, the explanation will mention that the temperature is below 0 °C rather than simply that the temperature is −17 °C. Both answers provide correct descriptions of the ontic mechanism and are thus explanatory according to the ontic conception of scientific explanation. However, only the fact that the temperature is below 0 °C is explanatorily relevant as it does not matter whether the temperature is −17, −16, ..., or − 1 °C for its being frozen. Take another example: assume that we want to explain why neurotransmitter are released rather than not released. As a matter of fact, the release was caused by the rise in intracellular Na + concentration. As both descriptions of the neurotransmitter release mechanism in terms of 'rise in intracellular Na+ concentration' as well as in terms of 'rise in the membrane voltage' are true and causally relevant, both aspects should count as equally explanatory according to the ontic conception of scientific explanation. However, what is explanatory relevant is the rise in the membrane voltage that goes along with it and that could have been induced by any other cations (Craver 2007a, 205). This shows that not all ways of describing an ontic mechanism are explanatory. Only describing a mechanism by providing what we call a "mechanistic explanatory text" is explanatory. We will say more on how explanatory texts are generated in Sect. 17.5.1.

A further advantage of our threefold distinction is that it allows us to solve the Multiplication of Mechanisms and the Odd Ontology problem. We can uphold the idea that *explananda are contrastive* without being committed to the view that ontic phenomena are contrastive. We can thereby avoid being committed to an odd ontology without having to deny that phenomena are ontic. Explananda are Why-questions or requests for explanation of the form 'Why P rather than P'?'. Explanantia are explanatory texts, i.e., representations of ontic mechanisms that identify the elements of a mechanism description that are relevant to a given explanatory contrast P vs. P′. Furthermore, our account does not multiply mechanisms. Mechanisms are not individuated with regard to explanatory contrasts. Rather, they are individuated relative to ontic phenomena, which are acting entities, such as neurons that fire, rats that navigate mazes, or muscles that contract (Craver 2007a; Kaiser and Krickel 2017). If at all, we multiply explanatory texts. It makes sense to say that the number of contrasts that can be formulated for a given ontic phenomenon is at least possibly infinite.

Finally, our threefold distinction between ontic mechanisms, mechanism descriptions and mechanistic explanatory texts helps us to solve the Ontic Completeness Problem. Ontic mechanisms are the way they are, and it does not make sense to say that an ontic mechanism is complete or incomplete. What can be complete or incomplete, though, is knowledge about or representations of ontic mechanisms.

These representations can be either mechanism descriptions or explanatory texts. However, the norms of completeness for mechanism descriptions, on the one hand, and for explanatory texts, on the other, are different. The different norms of completeness will be developed in the next section.

17.4.2 Descriptive Vs. Explanatory Completeness

As already indicated in our characterizations of mechanism descriptions and explanatory texts, the primary difference between the norms of completeness for mechanism descriptions and mechanistic explanatory texts is that mechanism descriptions obey *norms for descriptive completeness*, while mechanistic explanatory texts obey *norms for explanatory completeness*.

As was stated in Sect. 17.4.1, mechanism descriptions should describe every detail of the ontic mechanism in order to be descriptively complete. We will, thus, formulate a notion of descriptive completeness for mechanism descriptions:

> (*Descriptive Completeness*) A mechanism description is descriptively complete if and only if it contains all and only the constitutively relevant details about the ontic mechanism for phenomenon P.

Here, 'constitutive relevance' can be straightforwardly interpreted in line with the original mutual manipulability account as presented in Sect. 17.2. Based on this notion of ontic completeness, a 'More Relevant Details Are Better'-claim for mechanism descriptions can be derived:

> **More Descriptively Relevant Details are Better (MDRDB):** If a mechanism description D contains more constitutively relevant details than D* about the ontic mechanism for phenomenon P, then D has more descriptive power than D* for P, all things being equal.

As we saw in Sect. 17.4.1, we endorse this MDRDB claim as a regulative ideal for mechanism descriptions. However, this does not imply that mechanistic *explanations* are always better with additional detail, since mechanism descriptions are not vehicles of explanation on our account.

Completeness of explanatory texts is more closely tied to the explanatory interests expressed by the request for explanation. Furthermore, as specified in the characterization of our threefold distinction, explanatory texts are formed based on mechanism descriptions rather than ontic mechanisms directly. Therefore, we will call the norm of completeness for explanatory texts 'explanatory completeness':

> (*Explanatory Completeness*) An explanatory text is explanatorily complete if and only if it mentions all and only the explanatorily relevant details for P vs. P′ contained in the mechanism descriptions for P and P′.

As we saw above, in order to make sense of explanatory completeness, the notion of explanatory relevance cannot straightforwardly be interpreted along the lines of constitutive relevance. We provide a positive account of explanatory relevance in

Sect. 17.5.1. Still, we can formulate a 'More Relevant Details Are Better'-claim for explanatory texts based on the norm of explanatory completeness:

> **More Explanatorily Relevant Details are Better (MERDB):** If an explanatory text T contains more explanatorily relevant details for P vs. P′ than T* from the mechanism descriptions for P and P′, then T has more explanatory power than T* for P vs. P′, all things being equal.

Drawing the distinction between descriptive and explanatory completeness is important for the following reason: remember the example of the freezing water; the ontic mechanism, as a matter of fact, involves a temperature of −17 °C. However, the exact value is not explanatorily relevant for why the water is frozen. Rather, the temperature's being below 0 °C is explanatorily relevant. Still, as a matter of fact, the property of having a temperature of below 0 °C is realized by the actual temperature of −17 °C. An ontic mechanism cannot have the property of involving a temperature of below 0 °C without instantiating a specific value. This indicates that the norms for evaluating the completeness of mechanism descriptions differ from the norms of completeness of explanatory texts. Take another example. Imagine a neuron's firing that is brought about by the transmission of electric current along an axon where exactly 14,560 ions are involved. Among the components of the ontic mechanism, thus, are 14,560 ions. However, for the explanation of why a specific neuron fired rather than not, the exact number of ions is irrelevant. What matters is that some number of ions above some threshold >0 is involved. The difference is captured by the contrastive formulation of the phenomenon: it only matters that *some rather than none* ions are involved in order to explain *why the neuron fired rather than not.*

Opponents of the mechanistic account might have confused descriptive and explanatory completeness, or in other words, they may have mistaken mechanism descriptions for mechanistic explanations. If one, as Craver and Kaplan do (though they use the term 'model'), insists that explanations have a contrastive explanandum, it becomes clear that descriptive completeness is not an ideal of mechanistic explanation (though it is an ideal of mechanism description as formulated above).

However, given that Craver and Kaplan define Salmon-Completeness already with respect to a contrastive phenomenon, they blur the distinction between explanatory completeness and descriptive completeness. Thereby they are committed to the view that ontic mechanisms have properties such as 'a temperature below 0°C' without any determinant of this property being among their components. The mechanism cannot be said to have a temperature of 17 °C as this property is explanatorily irrelevant for why the water is frozen. Similarly, they are committed to the claim that the neurotransmitter release mechanism is composed of just a rise of the membrane voltage without any corresponding rise in Na+. However, determinables (such as 'a temperature below 0 °C' or 'a rise in the membrane voltage') have to be realized by a determinant (such as 'a temperature of -17°C', or 'rise in intracellular Na+ concentration') in order to exist. It would be odd to assume that the ontic mechanism consists of the determinables but does not contain the determinants. By introducing the distinction between ontic mechanisms, mechanism descriptions, and explanatory texts, we can account for the reality of determinables: ontic mechanisms contain the determinants (such as a temperature of −17 °C) and these determinants are mentioned in the mechanism description; explanatory texts however may mention determinables (such as a temperature of below 0 °C) that are explanatorily relevant for a given explanatory contrast.

In a nutshell: the general spirit of Craver and Kaplan's reply to the MDB objection, i.e., that mechanistic explanations are only improved by adding details if the details are explanatorily relevant to a given contrastive phenomenon, is correct. However, in order to avoid the Multiplication of Mechanisms Problem, the Odd Ontology Problem, and the Ontic Completeness Problem, their reply has to be modified. We introduce a distinction between ontic mechanisms, their descriptions, and the explanatory texts that are generated based on the descriptions. However, two challenges remain—as we will show in the next section.

17.5 Two Challenges: Constitutive Relevance & Vertical Completeness

The two remaining challenges are not only challenges for Craver and Kaplan's reply to the MDB-objection but for the mechanistic account in general. The first challenge stems from the fact that, on the one hand, Craver and Kaplan and many other mechanists want to think of the explanantia of mechanistic explanation in terms of contrasts (i.e., what we call explanatory texts). On the other hand, they hold that explanatory relevance is constitutive relevance. However, constitutive relevance is spelled out in terms of mutual manipulability between *ontic phenomena and their spatiotemporal parts* and not in terms of a contrastive account of phenomena. There is at least a gap here: How can we determine what is to be part of a mechanistic explanatory text and how can this be combined with constitutive relevance? We will discuss and answer this question in Sect. 17.5.1.

The second challenge stems from the fact that Craver and Kaplan only address *one version* of the MDB-objection. We will show that there are two different versions of this objection—the *vertical* and the *horizontal* version. So far, there is no successful answer to the vertical version of the objection. We will discuss this objection and a possible reply in Sect. 17.5.2.

17.5.1 Explanatory Relevance, Contrasts, and Constitutive Relevance

As we saw in Sect. 17.4.1, mechanists typically equate explanatory relevance with constitutive relevance. However, this option is not open to any account of mechanistic explanation which views explananda as contrastive. Craver and Kaplan persist in viewing explanatory relevance and constitutive relevance as equivalent, but the mutual manipulability account of constitutive relevance has never been explicitly combined with a contrastive account of the phenomenon. Indeed, as presented earlier, the mutual manipulability account of constitutive relevance defines under which conditions an acting entity (X's ϕ-ing) is constitutively relevant for another acting entity (S's ψ-ing). In their paper, Craver and Kaplan do not explain how the combination of the mutual manipulability account and a contrastive phenomenon is supposed to work.

In our view, explanatory relevance is distinct from constitutive relevance, though constitutive relevance constrains explanatory relevance in the sense that all explanatorily relevant details concern mechanism constituents, i.e., entities and activities that are constitutively relevant to the explanandum phenomenon. However, not all constitutively relevant detail is explanatorily relevant. Therefore, we must provide a further criterion for identifying explanatorily relevant details.

Giving a satisfactory account of explanatory relevance requires spelling out exactly the relation between mechanism descriptions and mechanistic explanatory texts. As we saw in Sect. 17.4.1, mechanistic explanatory texts answer contrastive why-questions. The why-question, or *explanation request*, determines the class of contrast phenomena P′, with which the actual phenomenon is compared. For example, one might ask "Why is the water in the glass frozen?". This explanation request is incomplete, because it does not specify a contrast class explicitly. However, implicitly, we could read the contrast in as "Why is the water in the glass frozen, rather than thawed?". This picks out the phenomenon P: *water in the glass is frozen*, and the contrast class P′: *water in the glass is thawed*. Other why questions could be posed which pick out different contrast classes. For instance, one could ask: "Why is the water in the glass frozen rather than evaporating?". This why-question picks out the same phenomenon P, but a different contrast class P′: *water in the glass is evaporating*.

Based on the mechanistic idea that explanation is about finding crucial points of intervention (see *Purpose of Explanation* in Sect. 17.4), explanatory texts should identify these crucial points of intervention. Crucial points of intervention with respect to the contrast P vs. P′ are those which allow one to change P into P′ *with minimal effort*. A preliminary characterization of the contents of mechanistic explanatory texts, thus, is the following:

(*Contents of METs$_{priliminary}$*) A mechanistic explanatory text T explaining a contrast "P vs. P′" has the form "because C rather than C′", where C is a set of constituents of the description of the actual mechanism M_{actual} for P and C′ is a set of constituents of a description of a possible mechanism $M_{possible}$, where the following holds:

(i) $M_{possible}$ is a member of a set S of possible mechanisms each sufficient to bring about P′,
(ii) C and C′ contain all and only constituents that differ in the description of M_{actual} and the description of $M_{possible}$ and that are also differences between M_{actual} and all other members of S,
(iii) $M_{possible}$ could in principle be created by means of *intervening into* M_{actual} *with minimal effort* compared to the effort that would be necessary for the creation of each other member of S.

Condition (i) in the definition above should be self-explanatory. The mechanistic explanatory text explains why P occurred rather than P′, by exhibiting the constituents which were crucial to the occurrence of P and not P′. Therefore, we must compare the mechanism for P with some mechanism for P′, in order to isolate these crucial constituents. There are usually many ways in which a phenomenon from the contrast class P′ could be constituted. Consider the example of the action potential. If we want to explain why the neuron fired as opposed to maintaining resting membrane potential, we must contend with the fact that there are innumerable differences within the class of neurons maintaining resting potential (e.g., in the number of ion channels open, etc.). We must select one mechanism for P′ from the set of possible mechanisms to compare to the actual mechanism.

Condition (ii) further specifies that once we have selected one particular possible mechanism for comparison, we are interested only in those differences between this contrast mechanism and the actual mechanism for the phenomenon that apply to all the other mechanisms sufficient to bring about the contrast phenomenon. This is what enables us to say that the correct explanation for "Why is the water in the glass frozen rather than thawed?" is that the temperature of the water is below 0 °C, or more exactly, that the temperature of the water is −17 °C rather than *above 0* °C. The difference between −17 °C in the actual mechanism and above zero degrees holds for all instances of the contrast class (all other things being equal).

Condition (iii) identifies which of the many possible mechanisms sufficient for bringing about the contrast phenomenon should serve as the basis of the comparison. 'Minimal effort' can be defined as a function of (a) the *amount* of required interventions, and (b) the *similarity* of the required interventions (the more similar the required interventions are, the least effort). Hence, we have to know how to count and compare interventions in order to identify the mechanism $M_{possible}$ that is to figure in our comparison.

Woodward (2003, 98) defines interventions with the help of three variables {*I*, *X*, *Y*}, where *I* is the intervention variable. When I takes a particular value, it sets the value of *X*; *X* is the putative cause; and *Y* the putative effect. Since we are here concerned with constitutive rather than causal explanation, we do not take *X* and *Y* to be putative causes and effects. Instead *X* is a variable representing the mechanism constituent, and *Y* is a variable representing the phenomenon P (or temporal parts thereof, see Krickel (2018b)). According to this view, two interventions are identical, if and only if their defining *I*-variables, *X*-variables, and *Y*-variables are the same. Since we are interested in mechanism descriptions that describe mechanisms that are sufficient to bring about the contrast phenomenon P′, we can assume that *Y* is the same for all relevant interventions (*Y* represents the phenomenon and can take the values that represent P and P′). Hence, in order to determine which mechanism description of the counterfactual mechanisms that bring about P′ are most similar to the mechanism description of the actual mechanism, we have to determine which counterfactual mechanism requires the least effortful interventions individuated by their *I*s and their *X*s, giving us a set of required interventions $\{[I_1, X_1], \ldots [I_n, X_n]\}$ for each possible mechanism for P′.

In order to determine which of these sets of interventions involves the ‘least effort’, we have to know how to count *I*s and *X*s and how to determine their similarity.

There is a practical problem for the counting and comparing of the intervention variables I_1-I_n. In many cases, scientists know *what* would have to be changed in order to build a counterfactual mechanism M* from an actual mechanism M. However, they do not know *how* this change could be brought about. For example, scientists do often know what component of a pathological mechanism is responsible for the symptoms and therefore know that changing this component would lead to an improvement of the symptoms. But they do not know how to change this component. Much effort in medical research goes into inventing better drugs to be able to change mechanisms in the right way. In the present context, the consequence is that, in practice, formulating the explanatory text for a given contrastive request for explanation often is conditional on the fact that we do not know what the intervention variable represents and how similar or different it will be compared to other interventions. We therefore need to decouple the measure of minimal changes from the count of intervention variables. However, the measure we ultimately choose should respect the interventionist insight that the number of intervention variables matters. This can be achieved if we make our measure sensitive to similarities and differences between *X*s. If the targets of interventions are similar in specific ways, it is likely that they can be intervened on with just a single intervention variable *I*.

The practical problem does not arise for the counting and comparison of the *X*s. Counting and comparing *X*s means to count and compare mechanistic components. These mechanistic components, in our framework, are described in the mechanism descriptions. Hence, in the end in order to determine which interventions require the least effort, we have to count and compare the differences between the *descriptions* of all (nomologically) possible mechanisms for P′. This results in the following characterization of the contents of mechanistic explanatory texts:

(*Contents of METs*) A mechanistic explanatory text T explaining a contrast “P vs. P’” has the form “because C rather than C’”, where C is a set of constituents of the description of the actual mechanism for P M_{actual} and C′ is a set of constituents of a description of a possible mechanism $M_{possible}$, where the following holds:

(i) $M_{possible}$ is a member of a set S of possible mechanisms each sufficient to bring about P′,
(ii) C and C′ contain all and only constituents that differ in the description of M_{actual} and the description of $M_{possible}$ and that are also differences between M_{actual} and all other members of S,
(iii) the description of $M_{possible}$ is more similar to the description of M_{actual} than the description of any other member of S.

Thus, we have to be able to determine when a mechanism description D is more similar to another mechanism description D’ than some further mechanism descriptions D*, D** etc.

Let us look at two more informal examples of the kind of comparison that figures in constructing METs. For example, suppose we are trying to construct an explanatory text answering the question: “Why did the car go straight, as opposed to turning right?". The mechanism description for the ontic mechanism in which the car is going straight at speed 90 km/h with rattling bumpers includes a number of constituents describing the activity of the spark plugs. However, the mechanism description of the most similar mechanism $M_{possible}$, which would underlie the car's turning right includes the very same constituents describing the activity of spark plugs. Therefore, when answering the question regarding going straight in contrast to turning right, this information will not be included in the explanatory text. Spark plug activity is not different across the two cases. To make a car turn right, rather than go straight, one should intervene on the wheels, not on the spark plugs.

In practice, the problem of constructing the correct MET will be compounded by the fact that there may be numerous ways of exhibiting the contrast phenomenon. For instance, let's look at explaining why the car goes straight rather than standing still. Will the explanatory text mention spark-plug activity? Perhaps surprisingly, the answer is still no. Although the paradigm case in which the car stands still is one where the engine does not run, and spark plugs do not spark, there is another class of situations in which cars stand still, i.e. when they are idling with the engine running in neutral gear, or when brakes have been applied. In these cases, spark-plugs do spark in the same way as when the car goes straight. The mechanism description for the idling case, or the braking case will be closer to the description of the actual mechanism, because all the (many) engine parts will work in the same way, and thus receive the same description, as in the actual mechanism. In fact, the contrast class might be too heterogeneous to admit any set of differences satisfying point (ii) of the definition. This would suggest that the contrast must be explained piecemeal.

The matter of comparing mechanism descriptions is complicated by the fact that mechanism descriptions can be given in various forms, such as spoken word, written text, diagram, etc. and two mechanism descriptions can contain the same information about the same mechanism, even though they superficially differ. In order to resolve this issue, we stipulate that mechanism descriptions can be transformed into a canonical form:

> (*Canonical Form of MDs*): A mechanism description in its canonical form is a set of 4-tuples <E, A, S, T>, where E stands for some entity, A, for the activity this entity is performing, S for the (relative) spatial region in which this activity is performed, and T for the time during which the activity is performed. A single mechanism description will consist of many such 4-tuples stringed together.

In the rest of Sect. 17.5, we will need to distinguish between 'constituents', i.e., 4-tuples in a mechanism description and 'elements', which are any of the 4 parts which make up a constituent. Note that constituents in a mechanism description describe ontic constituents. When we refer to constituents of ontic mechanisms, this will always be specified in full. Mechanism descriptions in sentential or diagrammatic form can be, at least in principle, converted to this canonical form.

Two further questions arise with respect to mechanistic descriptions: the question of grain, and the question of sameness. The question of grain asks how detailed mechanism descriptions are. In practice, the answer varies, because different particular mechanism descriptions will be exhibited with varying detail. However, in Sects. 17.3 and 17.4 we saw that mechanism descriptions follow the norms for ontic completeness. This means that we can at least specify the conditions under which a mechanism description is better than another one characterizing the same ontic mechanism. The answer is in line with the MRDB claims formulated in the previous section: the more detailed a mechanism description the better. The best mechanism description describes all ontic constituents, and it describes all of them to maximum detail. A scientific community which has more fine-grained mechanism descriptions at its disposal is better off than a scientific community with only coarse-grained mechanism descriptions. This is because the former scientific community can explain more contrasts than the latter one. Further note that in practice scientific communities have descriptions at various levels of grain available to them, and they can construct coarser descriptions if need be by substituting less determinate denotations for entities, activities, places and times. Thus, the scientific community with the more fine-grained description will always also be in possession of the coarse-grained description.

Secondly, when are two mechanism descriptions equivalent? For mechanism descriptions in non-canonical forms, the answer is simple — two such descriptions are equivalent if and only if they can be transformed into the same canonical form without adding or leaving out any empirical content. Two mechanism descriptions in the canonical form are the same, if they contain the same constituents.

Can the comparison between mechanism descriptions be formalized such that a general recipe for how to compare mechanisms can be made available? Our proposal is that the minimal set of differences between mechanism descriptions M and M* can be computed based on adapting the concept of generalized *edit distance*. This measure is frequently used in computer science to reason about string matching and indirectly about graph matching. Adopting this framework is licensed by the fact that mechanism descriptions can be transformed into our canonical form.

In computer science edit distance of string s from s* is based on the number of steps required to transform s into s*. Each step consists of applying one of a set of permitted edit operations to one character of s. Different applications sanction different sets of permitted edit operations. Additionally, a cost, or weight, is associated with each permitted edit operation. The edit distance is the sum of these costs (Cohen et al. (2003); see also papers cited therein). The most well-known version of edit distance for strings is the so-called Levenshtein distance (Levenshtein 1966). Levenshtein distance permits the following operations: character insertion, character deletion and character replacement, all of which are of equal cost 1. The Levenshtein distance from the string 'dogged' to 'froggy' is 4 – 1 addition, 2 replacements, and 1 deletion.[4] Other related measures use a more restrictive set of edit operations (e.g. disallowing direct replacement; Wagner and Fischer (1974)), or on the contrary a more permissive set of edit operations (e.g. allowing direct transposition, etc.; Damerau (1964)). For some applications, weights different from 1 are used, so that some operations are more costly to perform than others (Monge

[4]'dogged' → 'fdogged' → 'frogged' → 'froggyd' → 'froggy'

and Elkan 1997). Although the edit-distance framework was originally devised for imprecise string-matching, similar measures have now been used for comparing graphs, such as semantic networks (Bunke and Shearer 1998).

A version of edit distance can be straightforwardly applied to mechanism descriptions in their canonical forms. Instead of performing edit operations on characters in a string, we can define edit operations on constituents in mechanism descriptions. Two mechanism descriptions M and M* are the same, if the edit distance from M to M* is 0.[5] Alternative ways of exhibiting contrast phenomena described by contrast mechanism descriptions M*, M**, M*** etc. can be ranked according to their edit distance from the actual mechanism description M. The one with the lowest edit distance from M is the appropriate contrast.

At this point we are left to specify the appropriate set of edit operations for mechanism descriptions. Firstly, there are straightforward equivalents for insertion and deletion. Constituent-addition and constituent-deletion are equivalent to character-insertion and character-deletion respectively. There is also an operation roughly equivalent to character-replacement. This is element-replacement, which consists in replacing one of the 4 elements in a constituent with a different element of the same category. Thus, one is permitted to change <Ei, Ai, Si, Ti> to any of the following: <Ei*, Ai, Si, Ti>, <Ei, Ai*, Si, Ti>, <Ei, Ai, Si*, Ti> and <Ei, Ai, Si, Ti*>.

Constituent-insertion and constituent-deletion are both weighted 1. Element-replacement, on the other hand, is weighted 0.5. This is to ensure that the cost for changing <Ei, Ai, Si, Ti> into <Ei*, Ai*, Si*, Ti*> is higher than the cost of changing fewer elements in a constituent. The distance from <Ei, Ai, Si, Ti> to <Ei*, Ai*, Si*, Ti> is 1.5, by 3 element-replacements. The distance from <Ei, Ai, Si, Ti> to <Ei*, Ai*, Si*, Ti*> is 2 — either 4 element-replacements at 0.5 each, or 1 constituent-deletion and 1 constituent-insertion at 1 each.

Apart from equivalents for the standard string edit operations, we introduce an edit operation unique to comparing mechanism descriptions, called *mass element-replacement*. Mass element-replacement is our attempt at discounting such systematic changes to multiple constituents, for which a single intervention variable is likely to be responsible. Such systematic changes should be discounted because in formulating mechanistic explanatory texts we are interested in finding crucial points of intervention, where one can intervene with minimal effort.

In mass element-replacement, applying the same change to a group of relevantly similar constituents has the same cost as a simple element-replacement: 0.5. By relevant similarity, we mean that:

a) The entity elements E of these constituents can be subsumed by the same type description. Thus, we can apply a change to, e.g., all constituents whose entity element is an electron.

[5] This criterion is equivalent to the one on p. 21 above. The edit distance from M to M* is 0 iff M and M* have the same constituents.

b) The activity element A of these constituents can be subsumed under the same type description. Thus, we can apply a change to, e.g., all constituents whose activity element is a fission.
c) The space element S of these constituents falls in a specific range, say a sphere, with a defined centre and radius.
d) The time elements T of these constituents are synchronous or fall into a determined interval.

The range of constituents targeted by a mass element-replacement at once can be narrowed down by specifying that they be similar according to two or more of these similarity criteria. For example, we can specify that we want to change constituents involving electrons, but only those within 20 centimeters of an electric coil. The change applied to a group of constituents specified in this way need not concern the element on which their similarity depends. We can specify a group of constituents by noting that they involve electrons, and systematically change their locations in space, or slow them down, for instance.

The notion of applying the *same change* to a group of constituents also requires elucidation. Mass element-replacements are element-replacements on every constituent in the specified group. However, only certain types of replacement should be discounted. Specifically, we propose that:

a) The activity elements of a group of relevantly similar constituents can be replaced at once with cost 0.5, if all the activity elements A are replaced by elements A* which can be subsumed under the same type descriptions. Replacing all fissions in the mechanism description with fusions, for example, is an edit operation with cost 0.5.
b) The space elements of a group of relevantly similar constituents can be replaced at once with cost 0.5 by scaling (changing the size of constituents by a constant ratio), translation (moving the constituents in a uniform fashion, e.g., all 30 cm to the right) and rotations (tilting the constituents over).
c) The time elements of a group of relevantly similar constituents can be replaced at once with a cost of 0.5 by scaling (changing the duration of constituents by the same ratio).

The mechanism description edit distance with these edit operations (constituent-insertion, constituent-deletion, element-replacement, and mass element-replacement) as specified here, is meant to guide judgments about the minimal differences between mechanism descriptions in a way that parallels the results one would obtain by counting interventions, without requiring us to know the intervention variables but only based on the target variables X_i-X_n. In particular, the rules for mass-element-replacement are founded on the intuition that similar things can be changed by a single intervention in a systematic way.

At the same time, however, this proposal is provisional, and subject to amendment as the framework is further developed. In this paper, we include it to demonstrate the possibility of developing sophisticated semi-formal modes of reasoning about mechanistic explanation.

17.5.2 The Vertical Version of the MDB-Objection

As explained above, the second challenge for Craver and Kaplan's reply consists in the fact that they only address the *horizontal version of the MDB-objection* but do not provide a satisfying reply to the *vertical version*. In this section, we explain the difference between these two versions and why Craver and Kaplan fail to address one of them.

Ontic mechanisms form hierarchies in such a way that the same acting entity can be a phenomenon that is constituted by a mechanism, but at the same time a constituent in another higher-level phenomenon (Craver 2007a, chap. 5). Therefore, mechanistic hierarchies can be said to have a horizontal and a vertical dimension. The *horizontal dimension* of mechanism hierarchies is the one along which constituents are related by non-constitutive causal relations and by relations of temporal precedence. It is called 'horizontal', because it corresponds to the horizontal axis of the Craver diagram (Craver 2007a, 121). The *vertical dimension* of mechanism hierarchies is the one along which constituents are related by part-whole relations. This corresponds to the vertical axis of the Craver diagram.

Based on this distinction, the MDB-objection can be read as a claim about the *horizontal* dimension of hierarchies of mechanisms or as a claim about the *vertical* dimension of hierarchies of mechanisms. For example, opponents accuse the new mechanistic account of claiming that adding more *horizontal details* to an explanation by, say, listing the exact positions of ion-channels improves an explanation. And they object that the new mechanistic account implies that adding *vertical detail*, by adding, for example, details about quarks always improves an explanation. As we will show, Craver and Kaplan's reply to the MDB-objection accounts only for horizontal completeness but fails to account for vertical completeness. To see this, more accurate definitions of horizontal and vertical completeness are required.

Horizontal completeness can be defined for mechanism descriptions as well as explanatory texts:

(*Horizontal Completeness$_{description}$*) A mechanism description is horizontally complete if and only if the description mentions at least one set of constitutively relevant acting entities of the ontic mechanism for phenomenon P that is minimally sufficient for bringing about P.

(*Horizontal Completeness$_{text}$*) An explanatory text is horizontally complete if and only if the text mentions all explanatorily relevant factors for P vs P′ from the minimally sufficient sets of acting entities mentioned in the horizontally complete mechanism description for phenomenon P and the horizontally complete mechanism description for P′.

The acting entities contained in a horizontally complete set of mechanism components will either constitute the mechanism at different points in time (i.e., they form horizontal chains), or they will occur in different spatial location. This is implied by the requirement that the set of acting entities described in horizontal completeness$_{\text{description}}$ must be *minimally sufficient*. In the given context, the term 'minimally sufficient' is supposed to imply that the set of acting entities does not contain redundant members, i.e., acting entities that, given the other members of the set, do not make a difference to the higher-level phenomenon. This implies that the acting entities that are members of this set will not spatiotemporally overlap.

The two notions of horizontal completeness are usually not co-extensional. A horizontally complete mechanism description will usually mention more acting entities and more details about them than an explanatory text based thereon. On the assumption that the physical realm is causally closed, and each physical event has a physical effect (we ignore quantum events for the sake of argument), an ontic mechanism will at each point in time of its occurrence be composed of at least one acting entity. A mechanism description, ideally, mentions all of them. However, horizontal completeness of explanatory texts is compatible with there being gaps in the text. For the explanation of, say, why the action potential peaks at +40 mV rather than +50 mV it may not be explanatorily relevant what happened one millisecond after the stimulus onset. Hence, the explanatory text will be silent about what happened one millisecond after stimulus onset and leave a gap.

Horizontal completeness in the sense defined above is a goal of mechanism description as well as of explanation. A mechanism description that is not horizontally complete misses some acting entities that are crucial for the occurrence of the ontic phenomenon; in other words, it misses some parts of what might be called the 'constitutive basis' of a phenomenon. Similarly, an explanatory text that is not horizontally complete does not fully explain why the phenomenon P occurred rather than P′. Thus, the closer a description or text is to horizontal completeness the better. Two MRDB-claims can be formulated (where 'horizontal' details are those that bring us closer to horizontal completeness):

> **Horizontal descriptive MRDB-claim:** If a mechanism description D contains more horizontal details than D* about the ontic mechanism for phenomenon P, then D has more descriptive power than D* for P, all things being equal.

> **Horizontal Explanatory MRDB-claim:** If a mechanism text T contains more horizontal details from the mechanism description that are explanatorily relevant for P vs. P′ than T*, then T has more explanatory power than T* for P vs. P′, all things being equal.

Does Craver and Kaplan's reply to the MDB objection apply to both descriptive and explanatory horizontal completeness? As Craver and Kaplan's account makes use of the contrastive formulation of the phenomenon it only captures the horizontal *explanatory* MRDB-claim (on the assumption that it is modified in line with our threefold distinction). The original mutual manipulability account, however, that defined constitutive relevance relative to an ontic phenomenon P can account for the horizontal descriptive MRDB-claim only. Hence, even though the mutual manipulability account as well as Craver's and Kaplan's solution to the MDB objection each on its own fail to account for the horizontal descriptive *and* the horizontal explanatory MRDB-claim, taken together they capture both.

However, as will become clear, this combinatory strategy does not work for vertical completeness. Again, vertical completeness norms can be defined for mechanism descriptions as well as for explanatory texts:

> (*Vertical Completeness*$_{description}$) A mechanism description is vertically complete if and only if the description is (descriptively) horizontally complete at each mechanistic level.

(*Vertical Completeness$_{text}$*) An explanatory text is vertically complete if and only if the text is (explanatorily) horizontally complete at each mechanistic level.

Vertical completeness of mechanism descriptions is a descriptive matter. A mechanism description is vertically complete iff it describes the whole mechanism. This implies that a mechanism description is vertically complete if and only if it goes down to the fundamental level (if there is one). On the assumption of physicalism, everything supervenes on the fundamental physical level. As a consequence, every mechanistic hierarchy bottoms-out at the fundamental physical level. A vertically complete mechanism description will thus mention all acting entities that are constitutively relevant for a given phenomenon P at each level until down to the fundamental level. This gives us a further MRDB-claim:

Vertical Descriptive MRDB-claim: If a mechanism description D contains more vertical details than D* about the ontic mechanism for phenomenon P, then D has more descriptive power than D* for P, all things being equal.

In contrast to mechanism descriptions, explanatory texts do not always go down to the fundamental level. As a matter of fact, mechanistic explanations that can be found in the life sciences and other special sciences do not always mention, say, fundamental particles. Therefore, mechanistic explanatory text should be considered vertically complete even if they do not go down to the fundamental level. How to account for this? Craver and Kaplan briefly discuss this question in a footnote. They explain that their account

> is consistent with the possibility that higher-level causes can 'screen off' lower-level components. Once the higher-level component is fixed, and the relevant background assumptions of the request for explanation are made explicit, differences among the lower-level components no longer make an additional difference to the explanandum phenomenon. In Woodward's [(Woodward 2018)] terms, the lower-level parts can be irrelevant conditional on the behaviour of higher-level components. For example, once we know the sodium current across the membrane, the precise locations of the individual sodium ions are irrelevant. Likewise, it does not matter which of the thousands upon thousands of sodium channels do and do not open. This is why the total current equation can explain current in terms of conductance changes, while bracketing future knowledge of precisely how these changes are brought about. These low-level differences make no relevant difference once the higher-level behaviour is fixed. One consequence of this is that a complete explanation for a properly specified explanandum phenomenon need not (and typically does not) end in quarks (for further discussion, see Craver [2007a, b]). Which of the various multilevel relevance relationships happens to be screened off depends on the contrastive specification of the explanandum. (Craver and Kaplan 2020, n. 16)

This footnote suggests that Craver and Kaplan take Woodward's 2018-account of *conditional irrelevance* to be a potential answer to the question of how to account for the fact that explanatory texts do usually not go down to the fundamental level. According to Woodward, a set of variables Y_k *is irrelevant for a variable E conditional on some additional variables* X_i iff (i) changes in the variables X_i are causally relevant to E, (ii) changes in the variables Y_k are causally relevant to E, and (iii) given the values of X_i are fixed, changes in Y_k make no difference to E (Woodward 2018). Applied to the present context, 'causal relevance' mentioned in (i) and (ii), has to be replaced by 'constitutive relevance'. On the assumption that E is the phenomenon variable at level L_0; the X_i variables represent the mechanistic components at level

L_-1, and the Y_k variables represent the mechanistic components at L_-2, we end up with the following account of *conditional irrelevance of lower mechanistic levels*:

(*Conditional Irrelevance of Lower Mechanistic Levels*) A set of variables Y_k representing mechanistic components at level L_-n (n > 1) is irrelevant for a phenomenon E at level L_0 conditional on variables X_i representing mechanistic components at level L_-n + m (n > m > 0) iff:

a) changes in the variables X_i are constitutively relevant to E,
b) changes in the variables Y_k are constitutively relevant to E, and
c) given the values of X_i are fixed, changes in Y_k make no difference to E.

Conditions a) and b) are clearly satisfied for all components at all mechanistic levels (otherwise they would not be at lower levels as 'being at a lower level' is defined in terms of 'being constitutively relevant'). The problem is that c) will be necessarily satisfied as well. The reason is that variables Y_k are also constitutively relevant for X_i. This follows from the definition of mechanistic levels: Y_k are at a lower level than X_i iff the former are components of the mechanism for the latter, which is the case iff the former are constitutively relevant for the latter (Craver 2007a, 189). As a consequence, each change in Y_k makes a difference to E only *via* a change in X_i. If a change in Y_k did not induce a change in X_i but still changed E, this would imply that the change in Y_k is *not* constitutively relevant for X_i. Thus, it would not be at a lower level than X_i. As a consequence, if Woodward's account of conditional irrelevance was applied in the present context, necessarily, all levels lower than L_-1 would turn out to be irrelevant for the phenomenon at L_0 conditional on the first lower level L_-1. In other words, all lower levels (except for the first lower level) turn out to be always explanatorily irrelevant. Note that each lower level would be explanatorily relevant to the level directly above, but never to any higher level. Entities and activities at level L_-2, for example, would be relevant to phenomena on L_-1, but never to the original explanandum at L_0. Explanations would always stop at the first lower level. However, this may be too restrictive.[6] We should allow for

[6]Note that the question we are interested in differs from the question that Woodward answers with his account of conditional irrelevance. Our question is 'When is an explanation improved by going down the mechanistic hierarchy?' Woodward's question is 'When is a higher-level explanation better than or as good as a fundamental level explanation?' Woodward's perspective differs from ours in the sense that in his context it is commonly assumed that (i) there are different explanations at different levels (whereas we assume that there is one explanation that can extend over multiple levels), and (ii) that lowest-level explanations are by default the preferred ones (due to considerations of causal closure and exclusion). Based on these considerations, the question arises whether higher-level explanations can at least sometimes be better or at least as good as lowest-level explanations. Here, Woodward provides a convincing answer: a given higher-level explanation is at least as good as the lowest-level explanation if the lowest-level explanation is irrelevant for the explanandum conditional on the higher-level explanation. In the mechanistic picture, however, explanation is a top-down matter: while the first lower-level is clearly explanatorily relevant for the phenomenon (say, the activity of the hippocampus is clearly explanatorily relevant for spatial memory), the lowest level is clearly not (say, the interactions between quarks is clearly irrelevant for the explanation of spatial memory). The question, then, is where in the mechanistic hierarchy explanatory relevance stops.

the possibility that *at least sometimes* going further down the mechanistic hierarchy improves an explanation.

The fact that Woodward's notion of conditional irrelevance makes lower levels always irrelevant shows that there is a further problem for Craver and Kaplan's reply to the MDB objection: either their account is too restrictive if they adopt Woodward's notion, or it is to permissive if they do not provide an alternative way of determining where explanatory texts bottom-out. As a consequence, they cannot account for vertical explanatory completeness, i.e., the vertical completeness norms for explanatory texts. Therefore, they are still confronted with what may be called the 'Vertical MDB-objection':

> (*Vertical MDB-objection*) According to the new mechanists, an explanation of a higher-level phenomenon is always improved by adding more constitutively relevant lower-level details. However, explanations of higher-level phenomena do not usually go down to the fundamental level. Hence, the new mechanistic account of explanation fails.

Avoiding the vertical MDB-objection requires providing a criterion based on which one can decide how much lower-level detail is necessary to fully explain a phenomenon but just enough to not add explanatory irrelevant details. We have already seen that one possible solution fails: Woodward's account of conditional irrelevance seems to be too restrictive, as it implies that adding lower-level details never improves an explanation once the details about the first lower-level are fixed.

A second potential solution can be found in Machamer et al. (2000). They argue that explanations "bottom out" at levels that "are accepted as relatively fundamental or taken to be unproblematic for the purposes of a given scientist, research group, or field" (2000, 13). Even though this may be a good pragmatic answer to what it means for an explanation to be complete *relative to what someone finds interesting*, it does not provide an objective criterion of when an explanation is vertically complete (by "objective" we mean "not dependent on the interests of any individual" and "not pragmatic"). For example, a sociologist may be happy with explaining a social phenomenon just at the macro-level, and she may therefore declare it to be complete given her explanatory interests. But that does not mean that the explanation is complete in any objective way. The sociologist may happily admit that the explanation indeed is not (objectively) complete, but that the division of labour in science allows her to leave the further bits of the explanation to other researchers.

Such a pragmatic, non-objective criterion is problematic as a reply to the vertical MDB-objection. Craver and Kaplan and other mechanists accepting the ontic conception of explanation aim at developing an account of *objective* explanation (Craver and Kaplan 2020, 300). In order to be able to stick to this aim, the vertical MDB-objection should not be addressed by adding pragmatic constraints to the account. Indeed, if it turned out that the vertical MDB-objection can only be rejected if one were to integrate pragmatic considerations, this would constitute a rejection of the ontic conception of explanation, and in the end, admitting that the opponents were right: mechanists who aim at providing objective criteria for explanation in fact would turn out to be committed to what the vertical MDB-objection ascribes to them.

Luckily, we think that an objective criterion can indeed be provided. This criterion can be inferred from the purpose of explanation of reaching "understanding of where, and sometimes how, to intervene and change the world for good or for ill" (Craver 2007a, 93) (see Sect. 17.4). Based on this, one can infer that an explanation is better than some other explanation if it identifies more crucial points of intervention, i.e., if it identifies where to intervene such that the intended phenomenon is produced in the most economic fashion (with minimal effort). On our account, mechanistic explanatory texts exhibit the differences between the description of the mechanism for the actual phenomenon and the description of the mechanism for the contrast phenomenon that is most similar to the description of the actual mechanism compared to the descriptions of all other possible mechanisms for the contrast phenomenon. The vertical completeness issue is resolved by extending this principle to the choice of the appropriate bottoming-out level. That is, in choosing the appropriate level to stop the explanation, we are attempting to find the crucial points of intervention. Crucial points of intervention are those, where we can find the most systematic and least disruptive way to transform the description of the actual mechanism to a description of the contrast phenomenon.

The choice of the appropriate bottoming-out level is assessed in an equivalent way to the choice between two or more competing ways to exhibit the contrast phenomenon from Sect. 17.5.1. The only difference is that instead of computing edit distances for complete mechanism descriptions of possible mechanisms, one finds edit distances from the mechanism description for P to the mechanism description for P′ for each competing level. The appropriate bottoming-out level is the one with the smallest edit distance.

> (*Bottoming-Out in METs*) Mechanistic explanatory texts bottom out at that level, for which the edit distance from P to P′ is minimal in comparison to other available levels. Where there is a tie, lower bottoming-out levels are preferred.

The criterion proposed here has three interesting features. Firstly, in most cases it only gives us defeasible justification for the belief that our explanation of any particular contrast is vertically complete. This is because it is always possible that on a lower, so far unexplored level of mechanism, the differences between the actual mechanism description and the mechanism for the contrast phenomenon will be more systematic, thus allowing a shorter transformation procedure from the one to the other. Even if we find a level where the transformation procedure only has one step, it is possible, though unlikely, that a lower level will be found at which the transformation procedure also has just one step. In this situation, we think it uncontroversial that one should prefer the deeper explanation. In practice, though, such situations would be exceedingly rare.

Secondly, the criterion does not intrinsically favour either lower, or higher-level explanation. Rather, the appropriate level at which the explanation is complete is contingent on the result of empirical investigations. Further, vertical completeness of explanation may differ across phenomena, and across contrasts related to the same phenomenon. This means that the criterion we propose is non-arbitrary, but intimately tied to explanatory practice.

Lastly, this criterion helps explain why mechanistic models in individual special sciences tend to bottom out at levels containing similar entities and activities specific for each discipline or sub-discipline. It can be hypothesized that the entities and activities at these levels contain crucial points of intervention for the contrastive explananda of interest to the sub-disciplines in question. For instance, even though the mechanism underlying certain depressive episodes is highly complex, it appears that serotonin mediated synapses in a number of brain circuits play a crucial role. On higher levels of mechanistic description, the contrast between a depressed episode and normal functioning must be accounted for by citing a number of disparate differences in many brain regions. But on a lower level, this contrast is accounted for by a higher number of highly systematic differences having to do with neurotransmitter concentrations. Other seemingly complex contrasts on higher levels of mechanism in the brain may turn out to be due to systematic differences in neurotransmitter concentration, secretion or inhibition. The research into these kinds of differences constitutes the discipline of psychopharmacology.

17.6 Conclusion

The aim of this paper was to find a satisfactory solution to the MDB-objection. We showed that the most promising extant account due to Craver and Kaplan (2020) introduces new problems, namely the Odd Ontology Problem, the Multiplication of Mechanisms Problem, and the Ontic Completeness Problem. Furthermore, that account is still incomplete, as it leaves open how explanatory relevance with respect to contrasts is to be determined. And even worse, it is still vulnerable to a version of the MDB-objection, i.e. the vertical MDB-Objection.

Our account builds on the foundational idea by Craver and Kaplan, and it resolves all five of these issues. The Odd Ontology and the Multiplication of Mechanisms problems are avoided because our threefold distinction between ontic phenomena, mechanism descriptions and mechanistic explanatory texts only introduces contrasts as a feature of mechanistic explanatory texts. Ontic phenomena are not contrastive, and there are no ontic mechanisms for every conceivable contrast. The Ontic Completeness problem is solved, because instead of formulating completeness norms for ontic mechanisms (Craver and Kaplan's SC) we provide completeness norms for both mechanistic descriptions and mechanistic explanatory texts.

Additionally, we provide criteria for explanatory relevance of mechanistic details relative to contrastive explananda. In our account, this means to determine the contents of mechanistic explanatory texts, which enables us to determine which constituents from the mechanism description should be cited to account for any particular contrast. Since in our account contrasts are not ontic, we can keep the original account of constitutive relevance as an account of the dependency relation between ontic mechanisms and ontic phenomena.

Finally, in contrast to Craver and Kaplan's account, our account avoids the vertical version of the MDB-objection. According to our proposal, mechanistic explanatory texts bottom out at those levels of the mechanistic hierarchy where the edit distance from P to P′ cannot be decreased by going a level down. This level will contain the crucial points of intervention for turning phenomenon P into contrast phenomenon P′.

References

Baetu, T. M. (2015). The completeness of mechanistic explanations. *Philosophy of Science, 82*, 775–786. https://doi.org/10.1086/683279.

Batterman, R. W., & Rice, C. C. (2014). Minimal model explanations. *Philosophy of Science, 81*, 349–376. https://doi.org/10.1086/676677.

Baumgartner, M., & Casini, L. (2017). An abductive theory of constitution. *Philosophy of Science, 84*, 214–233. https://doi.org/10.1086/690716.

Baumgartner, M., & Gebharter, A. (2016). Constitutive relevance, mutual manipulability, and fat-handedness. *The British Journal for the Philosophy of Science, 67*, 731–756. https://doi.org/10.1093/bjps/axv003.

Baumgartner, M., Casini, L., & Krickel, B. (2018). Horizontal Surgicality and mechanistic constitution. *Erkenntnis*. https://doi.org/10.1007/s10670-018-0033-5.

Boone, W., & Piccinini, G. (2016). Mechanistic abstraction. *Philosophy of Science, 83*, 686–697. https://doi.org/10.1086/687855.

Bunke, H., & Shearer, K. (1998). A graph distance metric based on the maximal common subgraph. *Pattern Recognition Letters, 19*, 255–259. North-Holland. https://doi.org/10.1016/S0167-8655(97)00179-7.

Chirimuuta, M. (2014). Minimal models and canonical neural computations: The distinctness of computational explanation in neuroscience. *Synthese, 191*, 127–153. Springer Netherlands. https://doi.org/10.1007/s11229-013-0369-y.

Cohen, W. W., Ravikumar, P., & Fienberg, S. E. (2003). A comparison of string metrics for matching names and records. In *ACM international conference on Knowledge Discovery and Data Mining (KDD), Workshop on data cleaning, record linkage, and object consolidation, 2003*, 3:73–78.

Craver, C. F. (2007a). *Explaining the brain: Mechanisms and the mosaic Unity of neuroscience*. New York: Oxford University Press.

Craver, C. F. (2007b). Constitutive explanatory relevance. *Journal of Philosophical Research, 32*, 1–20. https://doi.org/10.5840/jpr_2007_4.

Craver, C. F. (2016). The explanatory power of network models. *Philosophy of Science, 83*, 698–709. https://doi.org/10.1086/687856.

Craver, C. F., & Bechtel, W. (2007). Top-down causation without top-down causes. *Biology and Philosophy, 22*, 547–563. https://doi.org/10.1007/s10539-006-9028-8.

Craver, C. F., & Kaplan, D. M. (2020). Are more details better? On the norms of completeness for mechanistic explanations. *The British Journal for the Philosophy of Science, 71*, 287–319. https://doi.org/10.1093/bjps/axy015.

Damerau, F. J. (1964). A technique for computer detection and correction of spelling errors. *Communications of the ACM, 7*, 171–176. ACM. https://doi.org/10.1145/363958.363994.

Dretske, F. I. (1972). Contrastive statements. *The Philosophical Review, 81*, 411. https://doi.org/10.2307/2183886.

Glennan, S. (2017). *The new mechanical philosophy*. Oxford: Oxford University Press.

Illari, P. M. K., & Williamson, J. (2011). Mechanisms are real and local. In *In Causality in the Sciences* (pp. 818–844). Oxford: Oxford University Press. https://doi.org/10.1093/acprof:oso/9780199574131.003.0038.

Illari, P. M. K., & Williamson, J. (2012). What is a mechanism? Thinking about mechanisms across the sciences. *European Journal for Philosophy of Science, 2*, 119–135. https://doi.org/10.1007/s13194-011-0038-2.

Kaiser, M. I., & Krickel, B. (2017). The metaphysics of constitutive mechanistic phenomena. *The British Journal for the Philosophy of Science, 68*, 745–747. https://doi.org/10.1093/bjps/axv058.

Kästner, L. (2017). *Philosophy of cognitive neuroscience: Causal explanations, mechanisms and empirical manipulations*. Berlin/Boston: De Gruyter.

Krickel, B. (2018a). *The mechanical world* (Studies in brain and mind) (Vol. 13). Cham: Springer. https://doi.org/10.1007/978-3-030-03629-4.

Krickel, B. (2018b). Saving the mutual manipulability account of constitutive relevance. *Studies in History and Philosophy of Science Part A, 68*, 58–67. https://doi.org/10.1016/j.shpsa.2018.01.003.

Levenshtein, V. I. (1966). Binary codes capable of correcting deletions, insertions, and reversals. *Soviet Physics Doklady, 10*, 707–710. https://doi.org/10.1023/A:1022689900470.

Levy, A. (2014). What was Hodgkin and Huxley's achievement? *The British Journal for the Philosophy of Science, 65*, 469–492. https://doi.org/10.1093/bjps/axs043.

Machamer, P., Darden, L., & Craver, C. F. (2000). Thinking about mechanisms. *Philosophy of Science, 67*, 1–25.

Miłkowski, M. (2016). Explanatory completeness and idealization in large brain simulations: A mechanistic perspective. *Synthese, 193*, 1457–1478. https://doi.org/10.1007/s11229-015-0731-3.

Monge, A., & Elkan, C. (1997). An efficient domain-independent algorithm for detecting approximately duplicate database records. In *The proceedings of the SIGMOD 1997 workshop on data mining and knowledge discovery*.

Railton, P. A. (1980). *Explaining explanation: A realist account of scientific explanation and understanding*. Princeton University.

Rice, C. (2015). Moving beyond causes: Optimality models and scientific explanation. *Nous, 49*, 589–615. John Wiley & Sons, Ltd (10.1111). https://doi.org/10.1111/nous.12042.

Romero, F. (2015). Why there isn't inter-level causation in mechanisms. *Synthese, 192*, 3731–3755. https://doi.org/10.1007/s11229-015-0718-0.

Wagner, R. A., & Fischer, M. J. (1974). The string-to-string correction problem. *Journal of the ACM, 21*, 168–173. https://doi.org/10.1145/321796.321811.

Woodward, J. (2003). *Making things happen: A theory of causal explanation*. New York: Oxford University Press.

Woodward, J. (2013). Mechanistic explanation: Its scope and limits. *Aristotelian Society Supplementary Volume, 87*, 39–65. https://doi.org/10.1111/j.1467-8349.2013.00219.x.

Woodward, J. (2018). Explanatory autonomy: The role of proportionality, stability, and conditional irrelevance. *Synthese*. Springer Netherlands: 1–29. https://doi.org/10.1007/s11229-018-01998-6.

Part V
Computation and Representations

Chapter 18
(Mis)computation in Computational Psychiatry

Matteo Colombo

Abstract An adequate explication of *miscomputation* should do justice to relevant practices in the computational sciences. While philosophers of computation have neglected scientific practices outside computer science, here I focus on computational psychiatry. I argue that computational psychiatrists use a concept of *miscomputation* in their explanations, and that this concept should be explicated as interest-relative and perspectival, although non-arbitrary, relatively clear-cut, experimentally evaluable, and instrumentally useful. To the extent my argument is convincing, we should reconsider the general adequacy of the mechanistic view of computation for illuminating relevant methodological and explanatory practices in the computational sciences.

Keywords Miscomputation · Computational psychiatry · Aberrant prediction error · Aberrant precision · Malfunction · Representation

18.1 Introduction

Because computing systems are kinds of rule-governed systems, they can perform computations *wrong*. A computing system can return an output o_2 that deviates to a greater or a lesser extent from the output of the function f on input i, $f(i) = o_1$, which the system ought to return. When this happens, the system *miscomputes*.

Philosophers of computation have explicated the concept of *miscomputation* without paying attention to relevant scientific practices outside computer science (Fresco and Primiero 2013; Dewhurst 2014; Piccinini 2015; Tucker 2018). In this paper, I extend this line of work on *miscomputation* to computational psychiatry, and address these two questions: Does a concept of *miscomputation* have any place in computational psychiatry? If it does, how should it be explicated?

M. Colombo (✉)
Tilburg center for Logic, Ethics and Philosophy of Science, Tilburg University, LE Tilburg, The Netherlands
e-mail: m.colombo@uvt.nl

© Springer Nature Switzerland AG 2021

F. Calzavarini, M. Viola (eds.), *Neural Mechanisms*, Studies in Brain and Mind 17,
https://doi.org/10.1007/978-3-030-54092-0_18

My answer to the first question is that a concept of *miscomputation* figures at least in Bayesian and Reinforcement Learning computational modelling practices in psychiatry. Psychiatrists often use this concept for explaining impairments associated with psychiatric illnesses. These explanations involve expressions like "malfunctioning computations," "false inference," "aberrant prediction error" or "aberrant precision estimates," which are meant to indicate that a target system is *performing a computation wrong*, as opposed to performing different kinds of computations.

My answer to the second question is that this concept of *miscomputation* should be understood as interest-relative and perspectival, although non-arbitrary, relatively clear-cut, experimentally evaluable, and instrumentally useful. If any concept of *computation* entails the concept of *miscomputation*, then at least one adequate explication of *computation* should also be interest-relative and perspectival.

To be clear: my focus, here, is not on whether brains are objectively physical computing systems, or whether they must have representational properties if they actually are computers. My focus is on certain scientific practices, imputations and interpretations. My overall point is that a purely mechanistic notion of *miscomputation* does not fit some imputations, interpretations and practices central to computational psychiatry. This meta-scientific conclusion is meant to put pressure on the idea that a mechanistic explication of *miscomputation* suffices to do justice to relevant practices involved in the computational sciences.

I begin by outlining the aims and methodologies of contemporary computational psychiatrists, showing that the concept of *miscomputation* has a place in psychiatry and that *miscomputation* cannot be chalked off as indicating only a difference in computing (Sect. 18.2). After I lay out two possible explications of *miscomputation* (Sect. 18.3), I argue that a satisfactory explication of *miscomputation* in computational psychiatry should refer to psychiatrists' expectations and pragmatic concerns in relation to the (mal)functioning and representational properties of a target system modelled as a computational system. I develop this argument based on the idea that computational psychiatrists rely on *specifications* of target systems (Sect. 18.4). In a short conclusion, I summarize the contribution of this paper, and draw one implication for the mechanistic view of computation.

18.2 Miscomputation in Computational Psychiatry

Computational psychiatrists use computer simulation, computational and mathematical modelling, and computational methods for pursuing the goals of classification, diagnosis, prediction, understanding, and treatment (e.g., Ahmed et al. 2009; Huys et al. 2011; Montague et al. 2012; Deco and Kringelbach 2014; Friston et al. 2014; Adams et al. 2016; Kurth-Nelson et al. 2016; Brugger and Broome 2019).

To pursue these goals, there are theory-driven and data-driven approaches. Typical in data-driven approaches is the use of machine-learning techniques to mine large sets of genetic, neural and behavioural data from psychiatric patients and healthy controls, for patterns, clusters, and causal dependencies (Huys et al. 2016, 405–8). Theory-driven approaches generally seek to assess people's performance in experimental tasks, to evaluate the effectiveness of therapies, and to explain psychiatric phenomena by imputing mathematical functions to be computed to experimental participants or target neural systems, and by modelling the activities and components of these systems in terms of computations of these functions (e.g., Maia and Frank 2011).

Computational psychiatrists need not be committed to the idea that neural systems are actual computing systems to successfully pursue their goals. Computational psychiatrists may or may not believe that the brain is actually a computing system, or that it is in some sense an information-processing system. But this does not matter to the success of their modelling practices.

Like in other fields in the sciences of mind and brain, the emphasis is on successful computational modelling. On successfully representing target systems in terms of rule-governed transitions from mathematical inputs to mathematical outputs (Egan 2019). This requires that researchers fit computational models to various sets of data, and generate simulation data from the best fitting model to ensure the model is empirically adequate. Given a set of candidate models for a clinically relevant phenomenon, the most empirically adequate model will be the most useful to pursue the goals of classification, diagnosis, explanation, or treatment with respect to that phenomenon.

Let me describe a typical study in computational psychiatry, which illustrates this point. Schlagenhauf and collaborators (2014) wanted to explain why patients diagnosed with schizophrenia show an impairment in certain learning tasks. Using a model-based brain imaging methodology (e.g., Colombo 2014a), they collected behavioural and neural data from un-medicated patients diagnosed with schizophrenia and healthy controls. Their experimental participants performed a probabilistic reversal learning task,[1] while undergoing magnetic resonance brain imaging.

Schlagenhauf and collaborators formulated various computational models corresponding to different hypotheses about the rule-governed transitions from inputs to output, which could describe participants' behaviour in their task. They evaluated the empirical adequacy of these competing models based on individual participants' trial-by-trial choice and neural data. One model had the best fit to data from both healthy controls, and only some schizophrenia patients. For most schizophrenia patients, the best fitting model was a different one.[2]

[1]This task requires participants to learn from probabilistic feedback, where the structure of the task can change so that what used to be positive outcomes (i.e., a positive reward) are now negative outcomes (i.e., a punishment, or negative reward), and what used to be negative are now positive outcomes.

[2]Specifically, the best fitting model for healthy controls and some schizophrenia patients was a Hidden Markov Model. According to this model, participants built and updated a representation of the structure of the task, based on the past history of choices and resulting rewards. Their belief about the current state of the task would be used to make a choice. Instead, the best fitting model for the other schizophrenia patients was a Rescorla-Wagner model. According to this model,

Schlagenhauf and collaborators identified strong associations between the activity of target neural systems in individual participants and trial-by-trial variation in specific components of the best fitting models. For all participants, activity in the ventral striatum in response to the same patterns of state-reward contingencies in the learning task was most strongly associated with a component of the models called "reward prediction error"—more on this component in Sect. 18.4 below. Compared to healthy controls, all schizophrenia patients exhibited reduced activity in the ventral striatum. Schizophrenia patients, whose choice and neural data were captured by the same model as in the healthy controls, showed a level of prefrontal activity similar to that of healthy controls, but higher than that in the other patients. Overall, both reduced activity in the striatum and in the prefrontal cortex of participants correlated with higher scores of positive symptoms of schizophrenia such as delusions and hallucinations assessed with the Positive and Negative Syndrome Scale (PANSS) (Kay et al. 1987).

From these findings, Schlagenhauf et al. (2014) concluded two things. First, reduced reward prediction error signals in the ventral striatum is a general dysfunction in schizophrenia—even when the performance of both schizophrenia patients and healthy controls is captured by the same type of computational model. Second, reduced reward prediction error signals in the ventral striatum explains schizophrenia patients' impaired performance in reversal learning tasks—even when we control for differences in computational ascriptions to different sub-groups of patients.

Although Schlagenhauf et al. (2014) did not mention the term "miscomputation," they framed their conclusions in terms of a "dysfunction" consisting in the "impairment" (172) or "deficit" of "ventral striatum prediction error signaling" (178). This way of talking is plausibly associated with the idea of *performing a computation wrong*, as opposed to performing different kinds of computations, or implementing different kinds of computational architecture. After all, Schlagenhauf et al. (2014) relied on computational modelling exactly to reach "more definitive conclusions... about processes more directly related to the disease that diminishes problems of interpretation due to behavioural differences associated with adaptive disease dependent strategies" (172).

Several other studies can be cited to show that the concept of *miscomputation* has a place in computational psychiatry, and that this concept cannot be understood only in terms of differences in computations, or differences in computational architecture.

In the context of Bayesian and Reinforcement Learning modelling (cf., Colombo 2019; Montague et al. 2012) are explicit that computational psychiatrists seek "to characterize mental dysfunction in terms of *aberrant computations*" (72, emphasis added). Even more explicit are King-Casas et al. (2008), when they write that com-

participants did not build a representation of the structure of the task. For each trial, participants would choose an option based on its expected value. After a trial, the expected value of only the chosen option would be updated on the basis of a prediction error (Schlagenhauf et al. 2014, 172–3).

putational modelling "offers the opportunity to understand some of the components of [psychiatric] disorders in terms of *malfunctioning computations*" (806, emphasis added). Huys et al. (2015b) distinguish three classes of "failure modes" that computational modelling uncovers in mental illnesses, namely: performing the right computations to solve the wrong problem, performing *poor* or *wrong computations* to solve the right problem, and performing the right computations to solve the right problem but in an unfortunate environment (for example, an environment that makes generalization from experience more difficult or maladaptive).

Two prominent examples of Bayesian and Reinforcement Learning miscomputations concern *prediction error* and *precision estimates*. A prediction error is a component of many Bayesian and Reinforcement Learning models, which quantifies the difference between an expected outcome and the actual outcome—for example, the difference between the expected monetary value of making a choice and the actual amount of money received in making that choice. Precision estimates of an outcome are components of many Bayesian models, and quantify the inverse variance of the outcome—for example, they are estimates of how far a set of monetary gains is spread out from the mean monetary gain in the set and from one another.

Schlagenhauf et al. (2014) refer to prediction error computations in a Reinforcement Learning model when they conclude that schizophrenia patients have a dysfunction in ventral striatum reward prediction error signalling. Fletcher and Frith (2009) also refer to prediction error computations when they suggest that psychotic symptoms of schizophrenia, such as hallucinatory experiences and delusional beliefs, can usefully be explained "in terms of a disturbed hierarchical Bayesian framework," specifically in terms of a disruption in prediction error signalling (48). Examining autonomic arousal and cortical activity in patients with autism spectrum disorder (ASD), Gu et al. (2015) refer to precision estimates to interpret their findings. They say: "the current findings provide direct support for recent proposals suggesting that failures in Bayesian inference, and particularly aberrant precision (i.e., inverse variance) of the information encoded at various levels of sensorimotor hierarchies, may contribute to socioemotional deficits in ASD" (3335). Lawson et al. (2017) also talk about precision estimates in their study of learning in autistic patients. Testing the computational prediction that aberrant precision estimates explain autistic patients' reduced behavioural surprise to atypical events, they conclude that their "findings provide preliminary empirical evidence for neurobiologically informed Bayesian accounts of autism that emphasize... *inappropriate* setting of gain (precision) on cortical responses (prediction errors) under conditions of uncertainty" (1298).

This overview highlights two aspects of contemporary practice in computational psychiatry. First, computational psychiatrists use terms like "aberrant prediction error" and "aberrant precision estimates" for explaining psychiatric phenomena. For example, aberrant prediction error computations would explain schizophrenia patients' impairment in reversal learning, as well as their hallucinatory experiences and delusional beliefs. Aberrant computations of precision estimates would explain autistic patients' socioemotional deficits, as well as their impaired responses to environmental volatility. Second, when computational psychiatrists say that prediction error signaling is aberrant, what they plausibly mean is not that the target of their best-fitting computational model is functioning a-typically, or in a statistically abnormal way. What they mean is that it malfunctions, or presents some dysfunction.

Given these two aspects of existing practice in computational psychiatry, one may ask the following question: how should we exactly understand terms like "aberrant prediction error" or "aberrant precision estimates"? What is a good thing to mean by these terms in the specific context of Bayesian and Reinforcement Learning modelling for the purpose of explaining clinically relevant phenomena? (on the idea of an explication as a "good thing to mean" see Gupta 2015 Sec. 1.5).

18.3 Two Explications of *Miscomputation*

Here are two possible answers to these questions:

[*m-miscomputation*] A target system miscomputes just in case the best-fitting computational model of the system captures some malfunction in the system, where (i) the system's malfunctioning is determined in relation to the system's objective goals or selective history, and where (ii) the ascription that the system (mis)computes a certain function in a task does not presuppose any representational ascription to the system.

[*p-miscomputation*] A target system miscomputes just in case the best-fitting computational model of the system captures some malfunction in the system, where (i') the system's malfunctioning is determined in relation to certain expectations, interests and conventions of a relevant scientific community, and where (ii') the ascription that the system (mis)computes a certain function in a task presupposes representational ascriptions to the system.

Both explications are committed to the idea that the condition that is both necessary and sufficient to apply the concept of *miscomputation* in psychiatry is that a computational model successfully represents some malfunction of a target system in a task. As explained in Sect. 18.2, the criterion of success here is fitting the experimental data into computational models that predict the data itself. So, for example, we can say that ventral striatum prediction error signalling counts as a miscomputation if and only if the best-fitting model of striatal activity in a task posits computations of prediction errors, these posits predict relevant neural and choice data generated by striatal activity sufficiently well, and the striatum is somehow malfunctioning.

The two explications differ in their commitment as to whether or not a system's malfunctioning is determined objectively just on the basis of mind-independent properties of the system, and in their commitment as to whether or not ascribing that the system (mis)computes a certain function in a task presupposes representational ascriptions to the system.

The first explication has much in common with prominent accounts of concrete computing systems in the mechanistic tradition, such as Piccinini's (2015). I call it *m-miscomputation*. The second explication is the conjunction of a perspectival view about the function of performing computations in a task (e.g., Dewhurst 2018b) and a pragmatist view about representation (e.g., Egan 2010, 2014; Coelho Mollo 2020). I call it *p-miscomputation*.

To unpack the commitments of *m-miscomputation* and *p-miscomputation*, it will help to rehearse relevant ideas from the literature on concrete computation. I start from the mechanistic account of concrete computation, and focus on various treatments of (mal)function. Then, I briefly review three popular accounts of how representational content is determined.

18.3.1 On Malfunction

According to the mechanistic account, concrete computing systems are mechanisms that perform computations, that is: systems of spatially and temporally organized, causally related components with functions to perform. At least one function of computing mechanisms is that of performing computations (Miłkowski 2013; Fresco 2014; Piccinini 2015; Coelho Mollo 2019).

There are at least three options about what determines the function to compute of a mechanism. According to the first option, the function to compute of a mechanism is determined by the stable causal contributions that performing this function makes in relation to some objective goal, where the objective goals of an organism are its survival and inclusive fitness (cf., Maley and Piccinini 2017).

The second option is that the function to compute of a mechanism is determined by the stable causal contributions that performing this function made, in the past, to processes of differential reproduction and differential retention (e.g., processes of evolution, development, and learning) involving organisms with that type of mechanism in a population (cf., Neander 1991; Garson 2019).

While the second option says that a mechanism's function to compute depends on the selective history of the mechanism, the first option does not appeal to any historical process, but only to how a mechanism's performing computations contributes, now, to the survival and inclusive fitness of organisms with that kind of mechanism.

However, the first and second option are analogous because they share the idea that what fixes the function to compute of a mechanism are objective, mind-independent properties of the mechanism. An explication of *miscomputation* committed to this idea will recommend that the ascription that the brain's computational function in a given task is, say, to compute posterior probabilities, or to map situations to actions so as to maximize some measure of reward, should be understood independently from any human interest or expectation. These computational functions would amount to biological functions. Their ascriptions to human brains would be warranted to the extent we have warranted beliefs that computing posterior probabilities (or maximising some specific measure of reward), now, furthers the objective goals of humans; or that computing posterior probabilities (or maximizing some specific measure of reward), in the past, causally contributed to the differential retention of a brain with certain features in humans within a population.

According to a third option, the function to compute of a mechanism is partly determined by certain expectations, interests and conventions of a relevant scientific community. In particular, Dewhurst (2018b, 581) argues that it is determined by certain interpretations of the physical structure of the mechanism, where these interpretations are grounded in an "explanatory perspective." An explication of *miscomputation* committed to this idea will say that the meaning of the ascription that the brain's computational function is to compute posterior probabilities is dependent on certain expectations, explanatory interests, and conventions of some relevant community. Computational functions would not just amount to biological functions. Ascriptions of certain computational functions to target systems in a task would be warranted to the extent relevant, perspectival interpretations of structural and causal features of the target system are warranted.

Now that we have a better idea of how a mechanism's function to compute can be determined, let's consider *mal*function. In the context of artificial computers, Piccinini (2015, 149–50) claims that miscomputation is a "failure of a hardware component to perform its function." This failure can be caused by some (non-essential)[3] component of the system being missing, or by the alteration of the spatial, temporal or causal organization of the hardware. Regardless of how it is caused, a hardware component's failure to perform its computational function consists in a deviation between the function the component should compute and what the component actually computes.[4] That is, the system "M is computing function f on input i, $f(i) = o_1$, M outputs o_2, and $o_2 \neq o_1$" (Piccinini 2015, 13).[5] Fresco and Primiero (2013) call this deviation "operational error," and Turing (1950, 449) calls it "error of functioning."[6] Depending on the right option about what determines a

[3]If an essential component of a computing system is missing, altered or broken, then the system may not compute anymore. If a system does not compute at all, then it cannot miscompute.

[4]There's no consensus among proponents of the mechanistic view about how we should individuate what a computing system actually computes at a time. For example, unlike Piccinini (2015), Tucker (2018, 8) argues that a system's computational structure is individuated without any reference to factors external to the system; what the system is actually computing at a time is determined by the actual inputs to the system at that time, in addition to its computational structure.

[5]In Sect. 18.2, I referred to Huys et al. (2015), who distinguished three classes of "failure modes" that computational modelling highlights in mental illnesses. One failure mode, *viz.* performing the right computations to solve the wrong problem, arises when the system M returns o_2, while computing a function $g(i)$, which differs altogether from the $f(i)$ it ought to compute. In this case, o_2 may be the right output to solve the wrong problem, $g(i)$.

[6]Writes Turing: "We may call [... these two types of errors] 'errors of functioning' and 'errors of conclusion'. Errors of functioning are due to some mechanical or electrical fault which causes the machine to behave otherwise than it was designed to do. In philosophical discussions one likes to ignore the possibility of such errors; one is therefore discussing 'abstract machines'. These abstract

mechanism's function to compute, there are three ways to articulate the nature of this deviation, and, thereby, the nature of computational malfunction.

First option: when a system computes function f on input i, the system returns output o_2; o_2 deviates from the output $f(i) = o_1$; and o_1 would make, now, a causal contribution to some objective goal of the system.

Second option: when a system computes function f on input i, the system returns output o_2; o_2 deviates from the output $f(i) = o_1$; and o_1 made a causal contribution, in the past, to processes of differential reproduction and differential retention for some trait.

Third option: when a system computes function f on input i, the system returns output o_2; o_2 deviates from the output $f(i) = o_1$; and o_1 is the output a relevant community expects for systems of that type, given a certain "explanatory perspective," interests, and conventions.

The first and second way to articulate computational malfunction are reflected in *m-miscomputation*. If an adequate explication of *miscomputation* reflects either of these two options, then warranted ascriptions that a brain is malfunctioning in a given task when it computes, say, posterior probabilities depends on warranted beliefs that its output o_2 deviates from that output o_1, which either furthers the objective goal of the organism, or causally contributed to the differential retention of brains in a certain population of organisms. Instead, if an adequate explication of *miscomputation* reflects the third way of articulating the idea of computational malfunction, then warranted ascriptions that the brain is malfunctioning when it computes posterior probabilities would depend on warranted, communal expectations about outputs o_2 and o_1, given certain pragmatic interests and conventions.

18.3.2 On Representation

Unlike *m-miscomputation*, *p-miscomputation* is committed to the idea that (ii') (mis)computation in a task should presuppose representational ascriptions. This idea is reflected in the semantic view of concrete computation, according to which a system cannot compute unless it possesses representational properties (e.g., Fodor 1975; Churchland and Sejnowski 1992; Sprevak 2010; Rescorla 2014; Shagrir 2018). According to this view, computing systems differ from non-computing systems because computing systems can manipulate representations, while non-computing systems cannot.

machines are mathematical fictions rather than physical objects. By definition they are incapable of errors of functioning. In this sense we can truly say that 'machines can never make mistakes'. Errors of conclusion can only arise when some meaning is attached to the output signals from the machine. [. . .] When a false proposition is typed we say that the machine has committed an error of conclusion. There is clearly no reason at all for saying that a machine cannot make this kind of mistake." (Turing 1950, 449).

It is plausible that the *individuation* of systems that compute does not involve any representation. After all, a machine can systematically manipulate strings of digits, following a rule defined over the appropriate degrees of freedom of its possible input strings, outputs and internal states, even if the strings have no representational property (see, e.g., Dewhurst 2018a).[7]

Yet, in the computational sciences, representation plays several fruitful roles. For example, some computer scientists and engineers design and build certain machines to execute appropriate mathematical computations. They, and anybody else, describe these machines as doing maths. But it is only by presupposing that the states of these machines represent numbers that these descriptions and practices make sense. So, even if the semantic view of concrete computation is false, it remains an interesting question what practices and ascriptions in the computational sciences presuppose the ascription of representational properties to a system, and what purposes these ascriptions could serve.

To evaluate the role of representational ascriptions in relation to *miscomputation* in computational psychiatry, it will help to briefly rehearse different proposals about how the content of a representation gets fixed—that is, how the condition for a representation's being right (or wrong) about a subject matter is determined.

Three proposals are prominent in the existing literature. According to the first proposal, the contents of a system's representations are determined, narrowly, by the system's intrinsic properties. The idea is that the content of a subject's representation does not require the subject to stand in any relation to anything in the environment. The contents of our thoughts would depend only on the causal goings-on inside our heads (cf., Fodor 1987). The condition for a representation's being right about a subject matter would be an intrinsic property of our brains. If this condition is fulfilled, that representation is accurate (or true).

If content is determined narrowly, then computing systems with the same intrinsic properties must have the same representations. In the context of computational modelling in psychiatry, this proposal invites the prediction that modellers ascribe representations to target systems without appealing to features of the systems' environment, focusing only on features intrinsic to the systems.

According to the second proposal, the contents of a system's representations are determined, widely, by relevant extrinsic properties of the system. The idea is that the content of a subject's representation depends on the way the subject is embedded in the environment. Thus, the contents of our thoughts would depend both on the internal interactions between various states of our brain, as well as their relations to external circumstances. A brain state would represent the presence of a green tree in the environment because of some causal, information, historical or biological relation with green trees in the outside world (cf., e.g., Dretske 1981; Millikan 1984). The condition for a representation's being right about a subject

[7] By 'degrees of freedom', I mean one of two things: either certain formal syntactic differences, or certain concrete physical differences between inputs and outputs and states of a system along some dimension of variation (e.g., voltage levels, rate of activation, or timing of activation).

matter would be an extrinsic property of our brains; it would involve the external condition required for the behavioural effects, which the representation prompts, to achieve certain ends. If this condition is fulfilled, that representation is accurate (or true).

If content is determined widely, then computing systems with the same intrinsic properties, but embedded in different social or physical environments, need not have the same representations. In the context of computational modelling in psychiatry, this proposal invites the prediction that modellers ascribe representations to target systems by appealing to features of the systems' environment, focusing on stable relations between features intrinsic to the systems and conditions in the world.

According to the third proposal, the content of a representation is fixed in a perspective-dependent fashion, or as Shagrir (2018) puts it "interpretatively." The idea is that the contents of a subject's representations are not objective properties, either narrow or wide. Although statements involving representations aim to state certain facts, they do not aim at truth. Because they aim at serving pragmatic purposes of a certain community—such as classification, prediction, explanation and intervention—these statements should be accepted if they actually serve these purposes (cf., Dennett 1987; Egan 2014; Sprevak 2013).

If content is determined interpretatively and pragmatically in this way, then computing systems with the same intrinsic properties and embedded in the same social and physical environments need not have the same representations. In the context of computational modelling in psychiatry, this proposal invites the prediction that modellers ascribe representations to target systems pragmatically and interpretatively, based on the extent to which these ascriptions serve their purposes.

18.4 Explicating *(Mis)computation* in Computational Psychiatry

Piccinini (2015) claims that "miscomputation finds an adequate explication within the mechanistic account" (275). In this section, I examine whether this claim is true in the context of Bayesian and Reinforcement Learning approaches in computational psychiatry. I use Schlagenhauf et al.'s (2014) study introduced above as a case study, and address these questions: When researchers say that a system's performing aberrant prediction error computations explains a certain psychiatric phenomenon, what is it that warrants their ascriptions of aberrant prediction error computations in a given task? What is it that warrants the idea that the system is malfunctioning? Is it some of the researchers' pragmatic interests, conventions and warranted "perspective"? Or, is it their warranted beliefs about the selective history or objective goals of the system? And should the ascription that the system (mis)computes prediction errors in a given task presuppose any representational ascription to the system?

18.4.1 Perspectival Malfunction

Let's start from malfunction. Schlagenhauf et al. (2014) wanted to better understand why schizophrenia patients show an impairment in reversal learning tasks. The most successful behaviour in these tasks can be defined as the behaviour that maximises rewards, where rewards may consist in money, food, water, or some other good participants would find rewarding. Accordingly, one's behaviour is successful in this task to the extent it brings about specific rewarding outcomes.

Maximising rewards (and minimising losses) in reversal learning tasks depends on various capacities. One is the capacity to learn the state-reward contingencies in the task from experience. Another is the capacity of converting beliefs about the reward values into choices. Yet another one is the capacity to inhibit actions that are learned in response to certain cues when they no longer result in reward. These capacities can work more or less well. For example, learning can be more or less quick, the motivation to pursue subjectively rewarding outcomes can be more or less strong, or the inhibition of learned actions can be more or less effective. Where these capacities are impaired, participants in a reversal learning task will be less likely to flexibly change their behaviour in response to changes in the structure of the task, and so, less likely to maximise rewards and minimize losses in the task.

From behavioural, neural, and computational modelling results, Schlagenhauf et al. (2014) concluded that a dysfunction in prediction error computations in the ventral striatum could explain schizophrenia patients' impaired reversal learning. This dysfunction would explain why schizophrenia patients' behaviour is less successful in this task compared to healthy participants.

According to *m-miscomputation*, the ascription of a dysfunction in prediction error signalling in the ventral striatum means that, in schizophrenia patients, either dopamine-dependent activity in the striatum does not return the outputs it was selected to return in reversal learning tasks, or it does not return those outputs that would promote schizophrenic patients' objective goals of survival and reproduction when they face these tasks.

This explication does not do justice to relevant practices. For two reasons. Call the first reason "the critical range problem." The problem is that an adequate explication of *miscomputation* should make sense of how and why computational psychiatrists often conclude that *reduced* or *increased* prediction error signalling in the ventral striatum is a dysfunction.

To illustrate the problem, suppose that some particular response activity in the ventral striatum is widespread among the participants in reversal learning tasks, but some smaller groups of participants exhibit reduced (or increased) activation.

If we accept *m-miscomputation*, then we need three premises to license the conclusion that ventral striatal prediction error computing is dysfunctional in the subgroups of participants. First, one has to map features of the task faced by the participants onto features of some real-world environment, with which humans recurrently interacted, or interact now. Second, one has to map participants' ventral striatal activations in this task onto ventral striatal activations in response to some matching real-world environment, with which humans recurrently interacted, or interact now. And finally, one has to show that given these mappings, a specific range of ventral striatal activation in response to reversal learning tasks was adaptive, or is adaptive now, and activations outside that range were likely, or are likely, to impede one's chance of survival and reproduction. If either of these premises is unwarranted, then the ascription that reduced (or increased) ventral prediction error computing is dysfunctional is unwarranted too.

Although researchers could rely on various types of evidence—e.g., ecological data, genetic data, phylogenetic data, comparative data—to warrant those premises, we have so far very little knowledge about *a critical range* of dopamine turnover in the ventral striatum for adaptive reversal learning (cf., Alcaro et al. 2007; O'Connell and Hofmann 2011; Howes and Kapur 2009). So, *m-computation* is currently of little help to explicate in what sense *reduced* or *increased* prediction error signalling in the ventral striatum counts as a dysfunction for computational psychiatrists.

The second reason why *m-miscomputation* is not a good thing to mean by expressions like "dysfunction of prediction error signalling" concerns the "mismatch problem." The problem is that an adequate explication should capture normal psychiatric usage of the term "dysfunction" in the face of possible *mismatches* between the computational function ascribed to a system and the environment with which the system would now compute that function. Let me explain.

Suppose that certain patterns of activations in the human dopamine system in response to certain physiological or environmental conditions are selected effects—one possible example might be the pattern of activation underlying the formation of certain beliefs in response to very surprising perceptual experiences. Based on *m-miscomputation*, we would consider those patterns to be a biological function of the dopamine system. Suppose that prediction error signals within a certain range in certain computational models in a given task show a good degree of fit with those patterns exhibited, now, by patients with delusions diagnosed with schizophrenia. We would then be warranted to say that computing certain prediction errors is (probably) a biological function of those dopamine responses. Suppose finally that there is an evolutionary mismatch between the way the dopamine system is designed to respond to surprising perceptual experiences, and the response that would be adaptive with respect to the perceptual experiences in the current environment (Pani 2000). On *m-miscomputation*, one would not be warranted to say that those patterns of dopamine activity are dysfunctional, though they are statistically abnormal and are now associated with delusions exhibited by patients with schizophrenia. They would be functional responses of the dopamine system, which may produce delusions associated with schizophrenia given the current (mismatched) perceptual environment (cf., Garson 2019, 180–1).

Let's grant that existing evidence warrants this kind mismatch, and that the pattern of dopamine activation underlying the formation of certain beliefs in response to surprising perceptual experiences is a selected effect. One problem with *m-miscomputation* is that its recommendations go against normal psychiatric judgement. If the patterns of activation exhibited by schizophrenia patients are both mismatched and functional, then conclusions like the one drawn by Schlagenhauf et al. (2014) that reduced prediction error signals in the ventral striatum is a "signature dysfunction" of schizophrenia are false; we should not take them seriously. It would also be wrong to say that "that *dysfunction* of the mesocorticolimbic dopamine system causes delusion formation via *disrupted* prediction-error signalling" (Corlett et al. 2007, 2387–8, emphases added; see also Feeney et al. 2017).

If these conclusions are false, then one practical consequence is that interventions targeting changes in dopamine activity in schizophrenia would be misguided and potentially bad for patients. Because these interventions are often effective and have contributed to elucidate common characteristics of the pathophysiology of schizophrenia patients (Tsou 2012), understanding expressions like "dysfunctional striatal prediction errors" in terms of *m-miscomputation* would be practically unfruitful too.

P-miscomputation provides us with a better explication, which can make good sense of both the critical range problem and the mismatch problem. Both problems can be addressed if we understand ascriptions of computational (mal)function in a task as dependent on pragmatically useful representational ascriptions and a relevant explanatory perspective.

Let's start from the idea of an explanatory perspective. In the context of computational psychiatry, this idea can helpfully be understood by analogy with *specifications* in computer science (Turner 2011; Fresco and Primiero 2013).

Specifications of a computational system are sets of documented, explicit requirements at various levels of abstraction, which a computer should satisfy. Specifications stipulatively define the vehicles of computing of a system (e.g., voltages, electric currents) and their rules of transformation, given the relevant degrees of freedom of a concrete physical system. Since specifications could be used to fabricate computers, and to evaluate their performance in a given task along various dimensions (e.g., processing power, energy consumption, memory, scalability, sturdiness), they function as blueprints and reference documents for computer scientists, engineers, programmers, computer manufacturers and users. They also enable consistent, transparent communication about a certain type of system.

Most importantly, they provide us with *stipulative definitions* of when and to what extent computing machines malfunction. As Turner puts it: "it is the act of taking a definition to have normative force over the construction of an artefact that turns a mere definition into a specification... Whether a [computational system] malfunctions is then not a property of the [system] itself but is determined by its specification" (Turner 2011, 140–1). Or, in the words of Schweizer (2019, 41): "[i]t is only at a *non-intrinsic* prescriptive level of description that 'breakdowns' can occur, and we characterize these phenomena as malfunctions only because our extrinsic ascription has been violated."

Computational psychiatrists' explanatory perspectives can helpfully be understood by analogy with computer scientists' specifications. Such perspectives warrant "extrinsic ascriptions" that the range of activity exhibited by a certain neural system modelled as a computing system in a task is (dys)functional, or that the activity exhibited by that system in certain populations in a certain environment is plausibly dysfunctional, even though it may be an adaptation.

A computational framework like Reinforcement Learning is an example of a specification, which provides researchers with an explanatory perspective, or explanatory template, for studying and understanding the behaviour of certain biological and artificial systems, and of psychiatric phenomena too (Sutton and Barto 2018; Niv 2009; Maia and Frank 2011).

P-miscomputation handles the critical range problem by saying it is computational psychiatrists' explanatory perspective or specification that can warrant their ascription that a certain range of prediction error computation counts as a malfunction. When psychiatrists model ventral striatal activity in terms of prediction error signals, warranted claims about what range of the mathematical function returning prediction errors is dysfunctional and what range indexes well-functioning computing in various experimental participants depend on three sources of information belonging to their explanatory perspective. First, on optimality results in mathematics and computer science; second, on known associations between various profiles of prediction error signalling exhibited by different groups participants in different experimental tasks; third, on diagnostic information about participants' general levels of suffering and "adaptive functioning" outside the lab (e.g, participants' PANSS scores).

Claims of (sub)optimality depend on mathematical results and on computer simulations. These results demonstrate under what conditions (e.g., under what parametrizations, in problems with what statistical or topological structure) a given Reinforcement Learning model quickly, and with little energy expenditure, can converge to a global (or local) maximum (or minimum) value of a function to be computed. These results set a normative standard, a yardstick, against which the learning performance of biological or artificial systems can be evaluated (Sutton and Barto 2018; Niv 2009).

Apart from results about optimality and computational complexity, the kind of specifications shared by computational psychiatrists can be related to individual and group differences in general levels of adaptive functioning and symptom severity. Computational psychiatrists form warranted expectations about these differences based on their clinical experience, calibrated scales like PANSS, and on widely shared diagnostic manuals like the DSM-5 and ICD-10, which define adaptive functioning in terms of "how well a person meets community standards of personal independence and social responsibility, in comparison to others of similar age and sociocultural background" (DSM-5, 31).

Now, computational psychiatrists sometimes find that some neural systems of some groups of psychiatric patients can adequately be modelled as performing optimal computations, or computations that are more efficient, or more accurate than the computations ascribed to healthy individuals to solve the same task—patients with depression, for example, show an absence of unrealistic optimism, which may captured with optimal computations in some tasks (cf., Huys et al. 2015a). And yet, psychiatrists understand these optimal computations as *miscomputation*, either because, based on results from computer science and mathematics, these optimal operations are known to involve trade-offs in efficiency, reliability and timeliness with other computations in other tasks, or because, based on clinical experience and diagnostic information, these optimal computations are known to be associated with low levels of adaptive functioning or with some debilitating symptom.

Recall that Schlagenhauf et al. (2014) found that the range of magnitudes of prediction error signalling in healthy controls was larger than in schizophrenia patients, who displayed reduced prediction error signals in the ventral striatum. Computer simulations show that reduced prediction error signals lead to blunted updates of the expected values of outcomes in a given state for future trials, which means that learning becomes slower and worse than learning driven by relatively higher prediction errors. So, the outputs returned by the dopamine system of schizophrenia patients diverged from the outputs a dopamine system ought to return, where this "ought" is grounded in a communal specification (or explanatory perspective) of the dopamine system as a reinforcement learning computing system, and on warranted expectations based on clinical experience and shared tools for diagnosis.

In summary, the "right" (or "wrong") range for the values of prediction error signals is based on "extrinsic ascriptions," on a communal specification. Such ascriptions are non-arbitrary, because they are based on reproducible and transparent optimality results and on communal expectations about certain illnesses. They are relatively clear-cut, because they give us determinate answers for many profiles of prediction error signalling. They are experimentally evaluable and revisable in the light of new optimality results, accumulating clinical experience, and revisions of widely shared diagnostic tools. They are instrumentally useful too, since psychiatrists can use these perspectival ascriptions of dysfunction for classification and devising targeted therapies (cf. Colombo and Heinz 2019 on classifications based on computational phenotypes).

18.4.2 Pragmatist Representation

Let's finally consider the role of representation in ascriptions of miscomputation in a task. Unlike *m-miscomputation*, *p-miscomputation* invites us to understand these ascriptions by positing representations. But what epistemic or practical role could representations play here exactly?

As I noted at the beginning of this section, Schlagenhauf et al. (2014) started with a cognitive task, *viz.* with a reversal learning task, where schizophrenia patients show an impairment. Successful performance in this task can be defined in terms of the relationships between participants and the environment, *viz.* as participants' interactions with the environment that maximize their rewards. Defining the task and the performance to be explained in this way involves representational ascriptions to participants. For example, it involves the ascription that participants have beliefs and expectations about reward contingencies in the task, the desire to obtain as much reward as possible, or the ability of using their beliefs and desires to make choices.

Schlagenhauf et al. (2014) adopted the explanatory perspective of Reinforcement Learning to explain participants' performance in this task. The Reinforcement Learning models they formulated need not involve any representational posit. Prediction error signals in these models quantify the difference between the learned predictive value of some current state and the sum of the current reward and the value of the next state. Specifically, a reward prediction error signal $\delta(\mathrm{t})$ computed at time t is equal to $r(\mathrm{t}) + V(\mathrm{t}+1) - V(\mathrm{t})$, where $V(\mathrm{t})$ is the predicted value of some option at time t, and $r(\mathrm{t})$ is the reward outcome obtained at time t. Because *any* distal state could in principle bear predictive value, Reinforcement Learning models compute prediction errors *regardless* of the environment they would find themselves in. In this specific sense, they are environment-neutral (or domain general). This means that the models Schlagenhauf et al. (2014) formulated would compute the same mathematical functions $V(\mathrm{t})$ and $\delta(\mathrm{t})$ over certain inputs, had the input states been strings of sounds instead of the geometrical shapes Schlagenhauf et al. (2014) actually used to distinguish different states in their learning task.

Representational ascriptions played the role of connecting ascriptions of Reinforcement Learning (mis)computations with the behaviours exhibited by experimental participants in a given task. In order to clarify how their computational modelling results explained performance in reversal learning, and, in particular, how aberrant prediction error signals explained impaired performance in patients with schizophrenia, Schlagenhauf et al. (2014) interpreted operations and components of computational models in terms of representations of specific states and reward outcomes in their task. And these interpretations allowed them to ascribe representational content to neural signals too—for example, to say that phasic dopamine firing represents errors of reward prediction. Thus, representational ascriptions enable researchers to connect computational modelling and neural systems with participants' performance in a given cognitive task (cf. Egan 2010, 2014; Coelho Mollo 2020). This connection affords researchers with an "explanatory gloss" (Egan 2014), which allows them to say *what* task participants are trying to solve, and *how*, on the basis of what inferences and reasoning steps, they are trying to solve it.

This way of connecting computational model and neural activity with behaviour in a given task also dissolves the mismatch problem, which, recall, is the problem of accounting for normal psychiatric usage of the term "dysfunction" in the face of possible mismatches between the computational function ascribed to a system and the environment in which the system operates now. If Reinforcement Learning models are environment-neutral (or domain general), and computational psychiatrists using these models ascribe content to their target systems pragmatically, then the mismatch problem does not arise. Experimental tasks just are the environment in which participants operate now. And the specific computational functions ascribed to experimental participants cannot be mismatched, since these ascriptions depend on the degree of empirical adequacy of alternative computational models in capturing participants' data in the task.

Psychiatrists choose experimental tasks that are relevant to evaluate different dimensions of psychiatric illnesses—for example, they use reversal learning tasks to assay belief updating and cognitive control. Based on the task of interest and on the computational functions ascribed to participants, it may turn out that computational psychiatrists ascribe different representations to participants with similar neurophysiological profiles and embedded in similar social environments—Schlagenhauf et al. (2014), for example, ascribed beliefs about the (hidden) state of their reversal learning task only to some of their patients.

These representational ascriptions enable them to clarify in what sense observed performance in a task is impaired, connecting (mis)computation, neural activity and behaviour. Thus, for example, because delusions are species of rigid beliefs, one might expect that schizophrenia patients with delusions would be less likely to flexibly switch their behaviour in a reversal learning task after reversals in reward contingencies in the task. Yet, Schlagenhauf et al.'s (2014) patients exhibited too much switching, and this behavioural profile correlated with reduced ventral striatal activity and higher levels of the severity of their delusions as measured with the PANSS scale. If one appeals to representational ascriptions to make sense of how miscomputations of prediction errors explain these results, then one could hypothesise that delusions, hallucinations and other symptoms of schizophrenia are all "expressions of the same core pathology: namely, an aberrant *encoding* of the precision" of prediction errors. Many symptoms of schizophrenia, that is, would amount to dysfunctions in neural computations involving representations of uncertainty (Adams et al. 2013, 1).

Though perspectival, these representational ascriptions need not be arbitrary or untestable. The content of dopamine activity is generally understood as a reward prediction error (Schultz et al. 1997). But this ascription is now contested (Colombo 2014b), and will be probably revised, as recent computational and neuroscientific results indicate that dopamine activity encodes dimensions of an error in prediction unrelated to reward (Langdon et al. 2018). While other researchers believe that dopamine activity represents the precision of a prediction error (Adams et al. 2013), different representational ascriptions motivate further testing of alternative computational models of a given task formulated in different modelling frameworks. Results of these tests will help researchers find more adequate explanations of psychiatric phenomena and targets for more effective treatment.

In summary, the mismatch problem does not arise if we understand miscomputation as *p-miscomputation*, and representational ascriptions pragmatically. Representational ascriptions enable computational psychiatrists to link computational and neural results, with the behaviour to be explained in a given task. While representational ascriptions are pragmatic, they are not arbitrary. They are based on a natural, common, pre-formal understanding of a given task, and of the computational models for that task. While revisable, computational psychiatrists' representational ascriptions remain warranted to the extent they contribute to further explanatory and practical purposes psychiatrists care about.

18.5 Conclusion

One of the aims of existing accounts of physical computation is to do justice to actual practices in the computational sciences. In this paper, I focused on central modelling practices in computational psychiatry. I considered a perspectival and pragmatist explication, which I called *p-miscomputation*, and a purely mechanistic explication I called *m-miscomputation*. I argued that, compared to *m-miscomputation*, *p-miscomputation* is a better thing to mean by terms like "aberrant prediction error" or "aberrant precision" in the specific research context of Reinforcement Learning and Bayesian modelling for the purpose of explaining clinically relevant phenomena like impaired learning in schizophrenia.

Perhaps, mechanistic accounts as encapsulated in *m-miscomputation* better comport with successful practices in psychiatry grounded in connectionist (e.g., Cohen and Servan-Schreiber 1992), dynamicist (Globus and Arpaia 1994; Durstewitz et al. 2020), or network approaches to computational modelling (Wang and Krystal 2014), of which I said nothing here. But, if the point I made in this paper is right, then ideas from some prominent mechanistic accounts of computation, ideas about how to determine computational functions and what role representation should play in computation, are detrimental to achieve the aim of doing justice to actual practices in the computational sciences. An explication of *(mis)computation* grounded in a perspectival pragmatism will be more descriptively adequate and practically fruitful.

Acknowledgements I am grateful to Andreas Heinz, J. Brendan Ritchie, Corey J. Maley, Dimitri Coelho Mollo, Joe Dewhurst, Nir Fresco, and an anonymous reviewer for their generous comments on previous versions of this paper. This work was supported by the Alexander von Humboldt Foundation through a Humboldt Research Fellowship for Experienced Researchers at the Department of Psychiatry and Psychotherapy, at the Charité University Clinic in Berlin.

References

Adams, R. A., Stephan, K. E., Brown, H. R., Frith, C. D., & Friston, K. J. (2013). The computational anatomy of psychosis. *Frontiers in Psychiatry, 4*, 47. https://doi.org/10.3389/fpsyt.2013.00047.

Adams, R. A., Huys, Q. J., & Roiser, J. P. (2016). Computational psychiatry: Towards a mathematically informed understanding of mental illness. *Journal of Neurology, Neurosurgery & Psychiatry, 87*(1), 53–63.

Ahmed, S. H., Graupner, M., & Gutkin, B. (2009). Computational approaches to the neurobiology of drug addiction. *Pharmacopsychiatry, 42*(Suppl 1), S144–S152.

Alcaro, A., Huber, R., & Panksepp, J. (2007). Behavioral functions of the mesolimbic dopaminergic system: An affective neuroethological perspective. *Brain Research Reviews, 56*(2), 283–321.

Brugger, S., & Broome, M. (2019). Computational psychiatry. In M. Sprevak & M. Colombo (Eds.), *Routledge handbook of the computational mind* (pp. 468–484). New York: Routledge.

Churchland, P. S., & Sejnowski, T. J. (1992). *The computational brain*. Cambridge: MIT Press.

Coelho Mollo, D. (2019). Are there teleological functions to compute? *Philosophy of Science, 86*, 431–452.

Coelho Mollo, D. (2020). Content pragmatism defended. *Topoi, 39*, 103–113.
Cohen, J. D., & Servan-Schreiber, D. (1992). Context, cortex and dopamine: A connectionist approach to behavior and biology in schizophrenia. *Psychological Review, 99*, 45–77.
Colombo, M. (2014a). For a few neurons more: Tractability and Neurally informed economic modelling. *The British Journal for the Philosophy of Science, 66*(4), 713–736.
Colombo, M. (2014b). Deep and beautiful. The reward prediction error hypothesis of dopamine. *Studies in History and Philosophy of Science Part C: Studies in History and Philosophy of Biological and Biomedical Sciences, 45*, 57–67.
Colombo, M. (2019). Learning and reasoning. In M. Sprevak & M. Colombo (Eds.), *The Routledge handbook of the computational mind* (pp. 381–396). New York: Routledge.
Colombo, M., & Heinz, A. (2019). Explanatory integration, computational phenotypes and dimensional psychiatry. The case of alcohol use disorder. *Theory and Psychology*.https://doi.org/10.1177/0959354319867392.
Corlett, P. R., Murray, G. K., Honey, G. D., Aitken, M. R., Shanks, D. R., Robbins, T. W., et al. (2007). Disrupted prediction-error signal in psychosis: Evidence for an associative account of delusions. *Brain, 130*(9), 2387–2400.
Deco, G., & Kringelbach, M. L. (2014). Great expectations: Using whole-brain computational connectomics for understanding neuropsychiatric disorders. *Neuron, 84*(5), 892–905.
Dennett, D. C. (1987). *The intentional stance*. Cambridge: MIT Press.
Dewhurst, J. (2014). Mechanistic miscomputation: A reply to Fresco and Primiero. *Philosophy & Technology, 27*(3), 495–498.
Dewhurst, J. (2018a). Individuation without representation. *The British Journal for the Philosophy of Science, 69*, 103–116.
Dewhurst, J. (2018b). Computing mechanisms without proper functions. *Minds & Machines, 28*, 569–588.
Dretske, F. I. (1981). *Knowledge and the flow of information*. MIT Press.
Durstewitz, D., Huys, Q. J., & Koppe, G. (2020). Psychiatric illnesses as disorders of network dynamics. *Biological Psychiatry: Cognitive Neuroscience and Neuroimaging*.
Egan, F. (2010). Computational models: A modest role for content. *Studies in History and Philosophy of Science Part A, 41*(3), 253–259.
Egan, F. (2014). How to think about mental content. *Philosophical Studies, 170*, 115–135.
Egan, F. (2019). The nature and function of content in computational models. In M. Sprevak & M. Colombo (Eds.), *The Routledge handbook of the computational mind* (pp. 247–258). New York: Routledge.
Feeney, E. J., Groman, S. M., Taylor, J. R., & Corlett, P. R. (2017). Explaining delusions: Reducing uncertainty through basic and computational neuroscience. *Schizophrenia Bulletin, 43*(2), 263–272.
Fletcher, P. C., & Frith, C. D. (2009). Perceiving is believing: A Bayesian approach to explaining the positive symptoms of schizophrenia. *Nature Reviews Neuroscience, 10*(1), 48–58.
Fodor, J. A. (1975). *The language of thought*. Cambridge: Harvard University Press.
Fodor, J. A. (1987). *Psychosemantics*. Cambridge, MA: MIT Press.
Fresco, N. (2014). *Physical computation and cognitive science*. Heidelberg: Springer.
Fresco, N., & Primiero, G. (2013). Miscomputation. *Philosophy & Technology, 26*(3), 253–272.
Friston, K. J., Stephan, K. E., Montague, R., & Dolan, R. J. (2014). Computational psychiatry: The brain as a phantastic organ. *The Lancet Psychiatry, 1*(2), 148–158.
Garson, J. (2019). *What biological functions are and why they matter*. Cambridge: Cambridge University Press.
Globus, G. G., & Arpaia, J. P. (1994). Psychiatry and the new dynamics. *Biological Psychiatry, 35*(5), 352–364.
Gu, X., Eilam-Stock, T., Zhou, T., Anagnostou, E., Kolevzon, A., Soorya, L., et al. (2015). Autonomic and brain responses associated with empathy deficits in autism spectrum disorder. *Human Brain Mapping, 36*, 3323–3338.
Gupta, A. (2015). Definitions. In E. N. Zalta (Ed.), *The Stanford encyclopedia of philosophy*. https://plato.stanford.edu/archives/win2019/entries/definitions.

Howes, O. D., & Kapur, S. (2009). The dopamine hypothesis of schizophrenia: Version III—The final common pathway. *Schizophrenia Bulletin, 35*(3), 549–562.
Huys, Q. J., Moutoussis, M., & Williams, J. (2011). Are computational models of any use to psychiatry? *Neural Networks, 24*(6), 544–551.
Huys, Q. J., Daw, N. D., & Dayan, P. (2015a). Depression: A decision-theoretic analysis. *Annual Review of Neuroscience, 38*, 1–23.
Huys, Q. J., Guitart-Masip, M., Dolan, R. J., & Dayan, P. (2015b). Decision-theoretic psychiatry. *Clinical Psychological Science, 3*(3), 400–421.
Huys, Q. J., Maia, T. V., & Frank, M. J. (2016). Computational psychiatry as a bridge from neuroscience to clinical applications. *Nature Neuroscience, 19*(3), 404–413.
Kay, S. R., Fiszbein, A., & Opler, L. A. (1987). The positive and negative syndrome scale (PANSS) for schizophrenia. *Schizophrenia Bulletin, 13*(2), 261–276.
King-Casas, B., Sharp, C., Lomax-Bream, L., Lohrenz, T., Fonagy, P., & Montague, P. R. (2008). The rupture and repair of cooperation in borderline personality disorder. *Science, 321*(5890), 806–810.
Kurth-Nelson, Z., O'Doherty, J., Barch, D., Deneve, S., Durstewitz, D., Frank, M., & Tost, H. (2016). Computational approaches for studying mechanisms of psychiatric disorders. In A. D. Redish & J. A. Gordon (Eds.), *Computational psychiatry: New perspectives on mental illness* (pp. 77–99). Cambridge, MA: MIT Press.
Langdon, A. J., Sharpe, M. J., Schoenbaum, G., & Niv, Y. (2018). Model-based predictions for dopamine. *Current Opinion in Neurobiology, 49*, 1–7.
Lawson, R. P., Mathys, C., & Rees, G. (2017). Adults with autism overestimate the volatility of the sensory environment. *Nature Neuroscience, 20*(9), 1293–1299.
Maia, T. V., & Frank, M. J. (2011). From reinforcement learning models to psychiatric and neurological disorders. *Nature Neuroscience, 14*(2), 154–162.
Maley, C., & Piccinini, G. (2017). A unified mechanistic account of teleological functions for psychology and neuroscience. In D. M. Kaplan (Ed.), *Explanation and integration in mind and brain science*. Oxford: OUP.
Miłkowski, M. (2013). *Explaining the computational mind*. Cambridge, MA: MIT Press.
Millikan, R. G. (1984). *Language, thought, and other biological categories*. Cambridge: MIT Press.
Montague, P. R., Dolan, R. J., Friston, K. J., & Dayan, P. (2012). Computational psychiatry. *Trends in Cognitive Sciences, 16*(1), 72–80.
Neander, K. (1991). Functions as selected effects: The conceptual analyst's defense. *Philosophy of Science, 58*(2), 168–184.
Niv, Y. (2009). Reinforcement learning in the brain. *Journal of Mathematical Psychology, 53*(3), 139–154.
O'Connell, L. A., & Hofmann, H. A. (2011). The vertebrate mesolimbic reward system and social behavior network: A comparative synthesis. *Journal of Comparative Neurology, 519*(18), 3599–3639.
Pani, L. (2000). Is there an evolutionary mismatch between the normal physiology of the human dopaminergic system and current environmental conditions in industrialized countries? *Molecular Psychiatry, 5*(5), 467–475.
Piccinini, G. (2015). *Physical computation: A mechanistic account*. Oxford: Oxford University Press.
Rescorla, M. (2014). A theory of computational implementation. *Synthese, 191*, 1277–1307.
Schlagenhauf, F., Huys, Q. J., Deserno, L., Rapp, M. A., Beck, A., Heinze, H. J., & Heinz, A. (2014). Striatal dysfunction during reversal learning in unmedicated schizophrenia patients. *NeuroImage, 89*, 171–180.
Schultz, W., Dayan, P., & Montague, P. R. (1997). A neural substrate of prediction and reward. *Science, 275*(5306), 1593–1599.
Schweizer, P. (2019). Computation in physical systems: A normative mapping account. In *On the cognitive, ethical, and scientific dimensions of artificial intelligence* (pp. 27–47). Cham: Springer.

Shagrir, O. (2018). In defense of the semantic view of computation. *Synthese*. https://doi.org/10.1007/s11229-018-01921-z.

Sprevak, M. (2010). Computation, individuation, and the received view on representation. *Studies in History and Philosophy of Science Part A, 41*(3), 260–270.

Sprevak, M. (2013). Fictionalism about neural representations. *The Monist, 96*(4), 539–560.

Sutton, R. S., & Barto, A. G. (2018). *Reinforcement learning: An introduction*. Cambridge, MA: MIT Press.

Tsou, J. Y. (2012). Intervention, causal reasoning, and the neurobiology of mental disorders: Pharmacological drugs as experimental instruments. *Studies in History and Philosophy of Science Part C: Studies in History and Philosophy of Biological and Biomedical Sciences, 43*(2), 542–551.

Tucker, C. (2018). How to explain miscomputation. *Philosophers' Imprint, 18*(24), 1–17.

Turing, A. (1950). Computing machinery and intelligence. *Mind, 59*(236), 433–460.

Turner, R. (2011). Specification. *Minds and Machines, 21*(2), 135–152.

Wang, X. J., & Krystal, J. H. (2014). Computational psychiatry. *Neuron, 84*(3), 638–654.

Chapter 19
What Is the Job of the Job Description Challenge? A Study in Esoteric and Exoteric Semantics

Colin Klein and Peter Clutton

Abstract Ramsey's *Job Description Challenge* has received substantial attention in debates over mental representation. We distinguish two senses of representational semantics: *esoteric* semantics, which concern the relationship between a representation and the world, and *exoteric* semantics, which concern the semantics of representations within a computational framework. Given that pair, we argue that there are three ways in which you could look to cognitive science to answer the job description challenge. Two of those ways make the job description challenge trivial—that is, answerable but uninteresting. We think that the recent literature has focused on readings to the challenge that possess this failing. We argue that Ramsey's challenge is best understood in terms of the interaction between esoteric and exoteric semantics. This third reading is more complicated to address but more interesting to answer. Understood in this way, the answers to the challenge will be local and case-by-case. On some of these cases, as we review, the challenge is met.

19.1 Introduction

In an influential passage, William Ramsey sets out what he calls the *Job Description Challenge* for representationalist theories in cognitive science. If we want to understand certain cognitive processes as representational, then

> ... we need to be told, in presumably computational, mechanical or causal/physical terms, just how the system employs representational structures. Principally, there needs to be some sort of account of just how the structure's possession of intentional content is (in some way) relevant to what it does in the cognitive system. After all, to be a representation, a state or structure must not only have content, but it must also be the case that this content is in some way pertinent to how it is used. We need, in other words, an account of how it actually serves as a representation in a physical system; of how it functions as a representation (2007, p. 27)

C. Klein (✉) · P. Clutton
The Australian National University, Canberra, ACT, Australia
e-mail: colin.klein@anu.edu.au

© Springer Nature Switzerland AG 2021

F. Calzavarini, M. Viola (eds.), *Neural Mechanisms*, Studies in Brain and Mind 17,
https://doi.org/10.1007/978-3-030-54092-0_19

Positing representations, in other words, must do some useful work for cognitive scientists; conversely, if nothing empirical turns on whether something is a representation or not, then naturalistically inclined philosophy of mind ought to avoid the term.

One way to read the job description challenge is as asking whether explanations in terms of representation involve something over and above mere "intentional glosses" (Egan 2014, p. 128). The question is then whether that intentional explanation is "part of the essential characterization of the device" or simply "ascribed to facilitate the explanation of the relevant cognitive capacity", that is, simply to make clear "how the computational/mathematical theory addresses the intentionally characterised phenomena with which we began and which it is the job of the theory to explain" (Egan 2014, p. 128).

This debate still animates philosophy of cognitive science. We will focus on debates where specific sorts of representations are invoked. To take a recent example, consider Gadsby and Williams (2018)'s argument that the cognitive neuropsychiatric theories of body representation and misrepresentation provide a good case against the standard anti-representationalism of Hutto and Myin (2012). They argue that "a robustly representational concept—the body schema—is explanatorily central within this research" and that these representations have "satisfaction conditions of some kind", allowing them to "misrepresent" (p. 5298). Neander (2017), in another recent example makes a similar case regarding cognitive neuropsychological explanations of certain visual deficits. Again what is at issue is that "what is posited is intentional, insofar as the relevant mental content permits the possibility of error and hence is not mere (i.e., natural-factive as opposed to intentional) informational content" (Neander 2017, p. 27).

Both Neander and Gadbsy & Williams explicitly address Ramsey (2007) in this line of argumentation. Ramsey argues that naturalism entails that in some sense, representational explanations are not completely necessary, and that the use of representational explanations in cognitive science would be cause for representational realism only when those explanations provide non-trivial explanatory purchase. Neander and Gadsby & Williams in their arguments aim to show that indeed this is true of their respective case examples—the representational explanations do provide non-trivial explanatory purchase—and draw representational realism conclusions from that. If these fields are on the right track, then we have good reason for representational realism.

Note that in each case, the Job Description Challenge is typically accepted as legitimate, and responses try to show how the challenge is met by the practice of ordinary cognitive scientists.

We think that there is a problem with the Job Description Challenge itself, at least on some readings of it (and these readings have received a lot attention). In what follows, we will distinguish three different readings of Ramsey's challenge. Two of those readings, we will argue, make the Job Description Challenge relatively uninteresting. The final way makes the challenge more interesting, but also requires local and case-by-case responses. On some of these cases, we argue, the challenge will be met. More broadly, we suggest, the delineation of different readings of the

Job Description Challenge shows how distinct readings have been confused, and generated philosophical puzzles beyond what is necessary.

19.2 Two Types of Semantics for Representations

The attraction of representationalism in cognitive science arguably derives from the attraction of computationalism. Conversely, many of the same issues about representation occur in the computational literature (Piccinini 2008). So to begin, let's start with a relatively concrete computational example. Suppose I'm writing an app to help tourists navigate Canberra's sprawling light rail system. I will need to represent the stops and their relationship to one another. *What* I need to represent is thus fixed by the world plus what I need to do. I still have considerable latitude in *how* I represent that information. I might use a list of stops, or a directed graph, or a set of tuples, or a matrix, or any of an indefinite variety of other data structures.

The choice of data structure should be driven, in part, by what I want to *do* with the information I represent. As Marr puts it:[1]

> ...even though one is not restricted to using just one representation system for a given type of information, the choice of which to use is important and cannot be taken lightly. It determines what information is made explicit, and hence what is pushed into the background, and it has a far-reaching effect on the ease and difficulty with which operations may be subsequently carried out on that information. (1982, pp. 21–22)

Different data structures have different properties. These determine which algorithms can be run on them, and how efficiently they can be run. Philosophers of psychology often talk about tasks being performed by algorithms *tout court*, but strictly speaking an algorithm can only be specified relative to a (possibly abstract) data type (Wirth 1976). The algorithm for finding the next stop given an index might be simple if we represent stops with a list, but require an exhaustive search if we use a set of tuples.

This is not merely a point about programming. As Pylyshyn (1984) perceptively notes, cognitive science cares not just about weak, input-output equivalence but also a strong equivalence in terms of algorithms. In practice, this can often be determined by things like relative timing data or resource usage.[2] The algorithms available to solve a computational process, and the efficiency with which they run, are dependent on the datatype used.

---

[1]Ritchie (2019) notes that Marr's classic presentations of computational analysis repeatedly recognized a duality between algorithmic process on the one hand and representation on the other, and that this duality is present at each of the classic levels of analysis.

[2]Or in the most general sense, by the computational complexity of different algorithms. Aaronson (2015) provides a nice introduction of complexity pitched at philosophers. Complexity profiles show how various parameters scale with input size, which is why relative scaling is often useful evidence in cognitive science.

Marr's point about what is 'made explicit' and what is 'pushed into the background' is also important. A list emphasizes the ordering of stops, as well as embodying a certain pessimism about the likelihood that the system will expand beyond a single line. A graph emphasizes connectivity, and builds in optimism and future flexibility at the price of added complexity. A large map-like matrix might emphasize spatial layout: this makes connectivity difficult to extract but it's easier to link up to other maps. And so on. Each may carry the same information about the target domain, but are more or less difficult to use for different purposes. All of these are utterly familiar sorts of tradeoffs to programmers.

Suppose we settle on using a list. Following Cantwell Smith (1996, 33ff), we note that there will be two kinds of questions one can ask about that list. *Exoteric* questions about representations concern whether and how the representation we chose hooks up to the world. Does it make sense to treat our list as a list of *stops*? Or is it really a list of *shops*? Does it have the right sort of systematic relationship, or use by internal consumers, or selection history, or whatever to really *represent* the light rail line? Note that we might think that a mere list falls short of whatever my brain does, and so my app does not reach full-fledged representation. But it still makes sense to ask exoteric questions.

Esoteric questions, by contrast, are those internal to the logic of programs and the computations they give rise to. Esoteric semantics determine what it means to say that I am using a *list*, rather than some other data structure: that is, what it means to *be* a list. Esoteric semantics are thus tied up with what expectations I can have about lists when I manipulate them. Is a successor always defined? Is my representation the sort of thing that you can sort, concatenate, and duplicate? If I access it twice in a row in the same way, am I guaranteed to get the same result? The answers to esoteric questions are determined by the semantics of my programming language, not about the world.

Note, following Smith's (1996) critique of Fodor, that esoteric questions are still questions about *semantics* of programming languages (and so ultimately about the computational objects that they designate) rather than the syntax of expressions in a programming language. Syntax affects semantics, of course, but important aspects of esoteric semantics outstrip syntax. So, for example, the syntax of Python tells you that `string_one+string_two` is a well-formed expression, but nothing at all about the operation of string concatenation that is indicated by that expression. We are concerned with the latter: that is, with data structures and operations on them, rather than the syntax of expressions which refer to data structures and operations.

Esoteric semantics thus detail a set of *guarantees* about what operations I can perform on instances of that datatype. As a standard textbook puts it, datatypes are simply "...defined by some collection of selectors and constructors, together with specific conditions that those procedures must fulfill in order to be a valid representation" (Abelson et al. 1996, p. 91). These 'specific conditions' are relative to operations on the datatype itself, not the relationship between data and the world.

Guarantees might include not just what can be done to a datatype, but how efficiently those operations can be performed and what sort of resources different operations might require. Datatypes are defined and distinguished by their guarantees, which is why details of algorithms depend on the datatypes upon which they operate.[3]

Esoteric and exoteric conditions clearly come apart. Consider maps. Rescorla (2009) makes a useful distinction between representing geometric structure and *replicating* geometric structure. Concrete maps represent geometric structure by replicating it, but there are numerous possible ways to represent the same geometric structure. Hence there are esoterically different datatypes which can play the same exoteric role. Conversely, there are things which meet the esoteric conditions for maps without representing anything at all. A map of Middle-Earth is still a map, precisely because it has a spatial structure of the right sort. Hence "replicating" geometric structure cannot mean simply mirroring a geometric structure which actually exists: to have a geometric structure is instead to meet certain esoteric guarantees (density, the triangle inequality, etc.) that maps must have to be maps.

Both esoteric and exoteric semantics come equipped with their own variety of normativity. On the exoteric side, my list *misrepresents* if it fails to line up with the world in the right way: if it misses stops, or gets the order wrong. A list which misrepresents might still be perfectly good as a *list*, though. Conversely, on the esoteric side, there is a straightforward sense in which my list can misfire as a representation by failing to live up to its guarantees. My list might become corrupted, or fail to return the right things in the right order. Any of these would count as a form of 'miscomputation.' Note, however, that not all exoteric failures need be esoteric failures: my list might be deficient *qua* list while still being usable for the purposes required. Conversely (and this surely happens), my list be slightly unreliable in fulfilling its guarantees while still being reliable enough to satisfy the exoteric conditions on representation.

19.3 Structure and What This Is Not

We speak of a representations, considered as instances of a datatype, as having a *structure*. But it is important not to confuse the esoteric structure of a list with either the structure of an implementation or the structure of the thing it is used to represent.

On the one hand, the thing which *implements* a list need not have the structure of a list (except in the completely derivative sense that it is an implementation of a list). My list of station names has a linear ordering. But what implements it need not, and often will not, have anything like a linear ordering. Different entries on a

[3]"We must decide in each case how much structure to represent in our tables, and how accessible to make each piece of information. To make such decisions, we need to know what operations are to be performed on the data. For each problem considered in this chapter, therefore, we consider not only the data structure but also the class of operations to be done on the data; the design of computer representations depends on the desired function of the data as well as on its intrinsic properties. Indeed, an emphasis on function as well as form is basic to design problems in general" (Knuth 2011, Volume 1, p. 238).

list can reside in arbitrarily dispersed parts of memory. If it's a very long list, some might be in memory and some swapped out to disk. What makes this a list, in the esoteric sense, is precisely the operations which I can perform on it.

A list is thus an odd sort of object, metaphysically speaking: what it takes to count as a list is entirely defined by the esoteric semantics, and those merely pick out a collection of things that you can *do* to a list. This is true of datatypes in general. Indeed, although there is a sense in which persistent data must be stored as something which persists, the structure of data itself need not be implemented by anything conceptually static.

Consider, for example, Abelson, Sussman, and Sussman's demonstration of different ways to implement an ordered pair. They note that one could store each item in an ordinary variable. One could also implement a pair purely functionally, by using a pair of functions which return specific values when invoked along with a setting function which creates new functions as needed. This procedural representation, they note, "... is a perfectly adequate way to represent pairs, since it fulfills the only conditions that pairs need to fulfill" (1996, p. 92). Thus there is an important sense in which a 'representation' like a list need not involve anything 'object'-like at all, because the relevant guarantees might be met purely functionally. That functionality must ultimately be cashed out in physical stuff, but the structure of the stuff needn't bear any straightforward relationship to the esoterically defined structure.

There is also an important distinction between our sense of esoteric structure and what has come to be known as *Structural* or *S-representations.* Ramsey-style S-representational accounts understand "cognitive representations as internal, structure-preserving models or map-like mechanisms." (Lee Forthcoming; see also O'Brien and Opie 2004; Ramsey 2007 3.2ff). A good S-representation preserves structure which can be exploited to solve particular tasks (Gładziejewski and Miłkowski 2017). The success or failure of a representation depends on the degree to which a representation resembles its target in exploitable ways (Lee Forthcoming). As Ramsey puts it, structural representations can function as models, and are useful when there is "a type of isomorphism between the sketch and the target that can be exploited to learn certain facts about the target" (Ramsey 2007, 82). There has been considerable debate about whether all neural representations are of the structural sort, and whether proposed conditions are too strict (Shagrir 2012; Morgan 2014). We think that S-representationalism has important insights.

Yet we again emphasize that the esoteric sense of 'structure' is fundamentally a matter of the constraints placed on the datatype, not the particular instances of data represented. The structure of a list is determined by the esoteric guarantees on lists. One can concatenate lists, or create functions which return a new list by copying the original and removing the first element. You can't do that with train lines. Just as the format of a representation can diverge arbitrarily from the format of the thing represented, so too can the (esoteric) structural demands on a representation diverge from the (exoterically relevant) structures that are represented by that item.[4]

Of course, it is not a mystery that the structure of a list is particularly handy for representing a linearly ordered set of stops: the easy operations on a list are ones that make exploiting it as an S-representation particularly handy. As a different way of making the point, we note that the structure of a datatype is relevant to other, counterfactual uses of the same data. Computational libraries, for example, often make available particular kinds of structured data. That structure is relevant insofar as it permits and constrains other, as-yet-unconceived, uses of the same bits of data. Hence the relevant notion of esoteric structure is one that *must* be able to vary independently of the exoteric uses to which data is put.

19.4 Two Uninspiring Versions of the Job Description Challenge

Having made the distinction between esoteric and exoteric aspects of representations, we return to the job description challenge. Recall that the job description challenge required a story about how representations are useful to cognitive science. We use the esoteric/exoteric distinction to suggest that this question can be disambiguated three ways. Two of those ways, we will argue, provide a response to the job description challenge, but a fairly trivial and unsatisfactory response. We clear them away to focus on a more interesting case in the following section.

First, one could read the Job Description Challenge as focusing on esoteric aspects of representations *alone*, in the absence of any further exoteric questions. That is, one could read it as asking whether thinking about vector operations or syntactic tree structures or maps could be explanatorily useful. Since the irrelevance of exoteric links to the world is presupposed here, the usefulness would have to come down to the usefulness of particular data structures in supporting various kinds of computations.

This is an unusual way to think about the Job Description Challenge, in part because it makes the challenge too easy. Arguments about the computational usefulness and necessity of particular kinds of data structures abound in cognitive science. This is, arguably, simply because these things are so useful in computer science, and that usefulness carries over to computational explorations of the mind. Inverting the question, a putative data structure earns its keep precisely *because* of the explanatory work it promises to do. On this reading, then, the Job Description Challenge is satisfied, though that success is unlikely to surprise anyone.

[4]For a neural example, see Goddard et al. (2018), who examine this issue in the case of dimensionality reduction of single-cell recordings. They note an important distinction between the structure of feature spaces and the structure of representational spaces, and suggest that apparently conflicting results about population coding can be reconciled by careful distinction of the two.

Second, one can read the challenge as focusing on exoteric conditions *alone*. So suppose that some exoteric conditions are wholly distinct from esoteric conditions. That is, it takes something for a cognitive item to count as a real, full-fledged representation, and that something doesn't have anything to do with the esoteric conditions that make it the computational sort of representation that it is. Whatever these additional exoteric conditions are, we can pose the job description challenge with respect to them: what does this additional stuff do, and do cognitive scientists care?

We suspect that this is how a lot of philosophers think of the Job Description Challenge. It is what generates staple puzzles about Twin Earth and Swamp Man. Those are cases where all the esoteric conditions are met, if they are discussed at all; it's only some aspect of exoteric conditions that are missing.

Many philosophers seem to read the challenge in this way. They look to cognitive science practice to find out whether scientists use the notion of exoteric error to answer the challenge. Where they find evidence for it, they claim that the challenge has been met and that representational realism follows. Thus there is a kind of *methodological* argument meant to settle a question about the existence of representations. Indeed, in one of the examples mentioned earlier, Neander explicitly frames her overall argument as a methodological argument: from the practice of cognitive scientists, we can draw conclusions about representations. That is, when the practice of cognitive science includes the use of representational explanations that provide 'non-trivial explanatory purchase' (2017, p. 85), we have real cause for accepting representational realism.

Yet we have also come to find this way of putting the challenge a bit puzzling. The interesting questions cannot simply be about whether cognitive science provides explanations that use explicitly exoteric terms. Everyone agrees that cognitive scientists themselves talk about representation in this sense all the time. Nearly all the answers to 'how does an organism do thus-and-such' will be cashed out in terms of error-supporting representations. Given such a question, there is an explanation in exoteric intentional terms that involves the possibility of error relative to the task at hand. This is because the task description itself is typically given in intentional terms: something like 'how does the organisms distinguish (these) EDGES from (those) SURFACES?', and that only makes sense if failure is possible.

This is especially true of cognitive neuropsychology, which was drawn on by both the Neander and Gadbsy & Williams papers we cited at the outset. Cognitive neuropsychology is explicitly built around the assumption that a particular task of interest can succeed or fail to be performed (see, for example, Coltheart 2001). A typical starting point is to ask *how do we represent this thing and how can we go wrong* (see, for example, Striem-Amit et al. 2018).

If this is your picture of how the science works, then it shouldn't be a condition on the naturalist picture that it sort out the exoteric semantics. Because, in an important sense, it's a background *presupposition* of doing cognitive science in the first place that there's some useful way to do so. We take this attitude to be exemplified by Chomsky; consider, for example, his remarks that:

> Thus I understand "mental" to be on a par with "chemical", "optical", or "electrical". Certain phenomena, events, processes and states are informally called "chemical" etc., but no metaphysical divide is suggested thereby. The terms are used to select certain aspects of the world as a focus of inquiry. We do not seek to determine the true *criterion of the chemical*, or the *mark of the electrical*, or the *boundaries of the optical*. I will use "mental" the same way, with something like ordinary coverage, but no deeper implications. By "mind" I just mean the mental aspects of the world, with no more interest in sharpening the boundaries or finding a criterion than in other cases. (1995, p. 1)

In other words, cognitive science *begins* by assuming that there is a useful notion of representation to be had, and investigating the conditions on it. Giving exoteric semantics is useful, but it is fleshing out a presupposition that is already there.

Indeed, Ritchie (2019) suggests that a similar view is present even in Marr. On Ritchie's account, representational content is actually part and parcel of Marr's computational level. As he notes, such an inclusion can "be motivated on more principled grounds by considering that representation and process are core to the very idea of an information-processing task, and Marr's levels are supposed to explain different aspects of how a system carries out such tasks" (2019, p. 1087). Hence even in classical presentations, some sort of exoteric validity is a foundational assumption.

Indeed, we think that a purely exoteric reading might end up making the Job Description Challenge seem needlessly difficult for the naturalist. The challenge appears to involve methodological deference to the natural sciences on the one hand, combined with a claim that scientists themselves might be systematically wrong about the core explanatory concepts they presuppose. It thus invites the naturalistic philosopher to take up a stance external to cognitive science and decide whether certain practices live up to an additional, extra-scientific set of criteria. Understood that way, the naturalist ought to simply reject the challenge.

19.5 Combining Esoteric and Exoteric Conditions Together

The first two ways of reading the challenge made a relatively sharp distinction between *whether* and *how* something represents the world—that is, between exoteric and esoteric. We argued that if the job description challenge is addressed to one of these questions in isolation, it fails to be compelling.

A sharp line between the two questions, however, is a philosophical artefact. We suggest that in cognitive science, the questions are usually treated as interacting. That is, it is a general assumption that one can't figure out what a particular bit is representing without *also* knowing some things about the nature and structure of the representation itself.

On this reading, the job description challenge may be read as demanding to know whether the particular combination of esoteric and exoteric criteria employed by cognitive scientists are actually useful enough to continue employing them. Note that this is no longer a general question: it depends on particular uses of particular representational posits.

We think that this version of the Job Description Challenge is both non-trivial and a naturalistically respectable question. We first review some examples of the approach in cognitive science, and then step back to think about how these explanations might answer a more robust and interesting version of the job description challenge.

For example, consider recent debates over the representation of the space near the body. The brain treats the space near the body in special ways. Graziano and colleagues (1994) showed that the precentral gyrus of macaques contained neurons with bimodal visual and tactile receptive fields which densely overlap near the body (Graziano and Cooke 2006, fig. 4). Stimulation of the same neurons can evoke complex defensive behaviors directed towards near space (Graziano et al. 2002a, b; Graziano 2006). Reviewing these findings, Graziano and Cooke (2006, p. 846) suggested that "a major function of these cortical areas is to maintain a margin of safety around the body and to coordinate actions that defend the body surface." This margin has come to be known as *peripersonal space* (PPS).

There is an active and vigorous debate about the particular properties of PPS representation. A parallel debate concerns whether we really represent *peripersonal* space, or if we merely represent (e.g.) potential actions in space, and the space near our body happens to be a place where many of our interests coincide. Grush, for example considers neurons in these areas to underly the "capacity to represent an environment of actionable objects" (2007, pp. 342–43), rather than an action-neutral space.

What is useful about this debate is that it shows how exoteric and esoteric questions are deeply intertwined. One cannot separate the questions of what PPS representations refer to from the question of how many there are, and that in turn depends on questions of what format that they might be in.

De Vignemont and Iannetti (2015), for example, argue that there must be two (and only two) distinct representations of PPS. They argue this on the basis that certain tasks distort PPS representations, and that the distortions for defense and for tool use affect PPS in different ways. Klein (forthcoming) responds with a model which accounts for the data using a single representation with a more complicated coordinate basis. The debate between the two, as Klein notes, depends on whether the format of PPS representations is a low-dimensional cartographic representation or a higher-dimensional space with a nonlinear basis.

In a recent review, Bufacchi and Iannetti (2018) similarly suggest that PPS is represented by a series of 'fields'. These fields, like physical fields, take a scalar quantity at every point in space. Noel and Serino (2019) respond that a representation that has continuous coverage of an indefinite region of space is biologically unrealistic. In response, Bufacchi and Iannetti (2019) argue that the representation is functional: what is important is that PPS representations allow the brain to systematically recover and transform action values given a specific point in space. Thus while a dense matrix would be unrealistic, there are other lightweight ways to meet the necessary guarantees.

This strikes us as paradigmatic cognitive science. In each of these debates, note, there is a crucial interplay between esoteric and exoteric semantics: whether something represents, and what is representing, is taken to depend crucially on *how* it represents. Esoteric semantics demands stories about plausible formats and plausible operations that can take place on those formats, while the exoteric part constrains those stories by the external role that the representation plays. Klein, for example, points out that if one is restricted to linear readout functions, then only a nonlinear basis will allow for a unified representation of PPS in the domains that De Vignemont and Iannetti care about.

This reading of the job description challenge takes seriously Ramsey's demand to know "just how the system employs representational structure". If esoteric and exoteric semantics partially interact, then whether and how a representation is fit for purpose will depend both on the details of that representation's structure and how that structure is used by the cognitive system.

Thus the job description challenge is not, first and foremost, something that can be solved once and for all for cognitive science, at least in any substantial way. (This is another parallel with computer science, where there is relatively little discussion of representation as such, as compared to endless discussion of different representational structures and their merits.) Instead, the job description challenge is fought case-by-case.

Of course, on this reading, there remains the possibility of interesting debates about exoteric semantics. Indeed, it is at least possible that we might end up denying that some esoterically respectable representation really counts as a representation.

So for example, Barbara Webb (2006; Webb and Wystrach 2016) suggests that many of the computations in insect navigation should be counted as mere transformations, rather than representations of space. Her argument turns on fine-grained details of the properties of the underlying computations, and is done against a background against which some neural computations *are* supported by representations. Barron and Klein (2016), by contrast, suggest that the integrative action of the central complex is sufficiently complex to elevate these same procedures to representational status.

Whichever side you fall on in the insect navigation debate, we suggest, this debate shows that the job description challenge is met. "Insects don't represent space because they use a specific kind of landmark-based pattern matching combined with gradient descent" is a substantive and scientifically fruitful hypothesis; pushing back against it has required detailed discussion of the neural details.

Indeed, we note that among the anti-representationalists, there is considerable variation on this very question. When Van Gelder (1995) introduces a dynamic, anti-representationalist approach to cognition, he is fairly clear that he is presenting an empirical *alternative* to traditional computational representations. In our terms, he posits entities with radically different esoteric conditions, and suggests that these can be conjoined with exoteric conditions which meet this challenge. Chemero (2009) similarly takes the interesting fight to be wherever he and traditional computationalists agree on exoteric questions, and argues instead about whether alternative dynamic notions might be explanatorily fruitful.[5] We are (by and large) representationalists, but we think this form of anti-representationalism is playing on the naturalistic grounds that the Job Description Challenge demands.

We have said that there is a background presupposition regarding the use of representations in many areas of cognitive science. For that reason, we discouraged the use of easy arguments from scientific realism to realism about representations. Explanations that advert to representations come cheaply in many areas of cognitive science. But that does not mean we endorse the kind of instrumentalist reading of the scientific practice of representation that has sometimes been offered (Chemero 2009; see also Lee 2018). We encourage instead a way of looking at these practices that asks questions with substantive empirical force, as in our example regarding the nature of the computations used in insect navigation.

Our position also does not choose sides in the standard breakdown between ontological and methodological naturalism (Caiani 2018). Ontological naturalists about representation looks for certain types of physical objects and properties to play the role that representation plays; methodological naturalists look to the explanatory utility of representation in our best scientific practices.

Our position doesn't stake out a claim on this familiar territory in any straightforward way. We have said that on certain ways of looking at this question, the use of representation in explanation, the methodological side, is more of a presupposition than the kind of practice that ought to be used to verify the presence of representation in any way. And further, that on the ontological side, there is again a type of trivial answer in the area that of course there are naturalistic structures that play an interesting role in cognition when performing various tasks. Where there are interesting questions, we have said, they will require examining particular combinations of esoteric and exoteric criteria employed by cognitive scientists and deciding whether those particular combinations are useful enough to continue using.

This is the most interesting reading of the job description challenge. It poses a fruitful, substantive question, the answer to which has both scientific and philosophical import. And, at least sometimes, this more substantive version of the job description challenge can be met. That is an interesting result.

19.6 Conclusion

Naturalistic challenges in philosophy always walk a fine line: they must balance what philosophers think scientists should care about with what scientists actually

[5]"I take it that using the newer, more restrictive definition to try to argue in favor of nonrepresentational cognitive science would be problematic. 'Using my new definition of representations, none of these systems has representations' is a near neighbor of the Hegelian arguments deplored [earlier]. That is, it allows radical embodied cognitive scientists or their opponents to win arguments by re-defining terms. For purposes here, then, the traditional views are more appropriate . . . " (Chemero 2009, 66).

care about. In the case of representation, much of the philosophical interest comes from the putative power of representational theories to solve old problems about intentionality and the nature of the mental. Yet solving those problems is not the reason why representations appear in empirical explanations, and the surrounding apparatus of cognitive science was built to tackle very different questions.

The Job Description Challenge is posed in a naturalistic spirit. We suggested that the reading which combines esoteric and exoteric conditions on representations is faithful to that spirit, and is interesting enough that it is the grounds for meaty fights between representationalists and anti-representationalists.

Why all the heat and noise, then? We conclude with a tentative diagnosis. We noted that focusing on either esoteric or exoteric questions in isolation was relatively uninteresting: indeed, they make the Job Description Challenge itself seem like a mistake. There is an understandable philosophical tendency to break difficult problems down into their component parts. Furthermore, keeping one part of a problem fixed while investigating another is, for many philosophical problems, the best way to get general solutions.

So, for example, much of the description of 'pure' exoteric problems ends up bracketing questions of esoteric structure: internal representations might as well be blinking lights. But then all that one can say is that of course cognitive science cares about representation: look how often they talk about it! Bracketing exoteric questions leads to a similarly unproductive sort of stalemate. So the Job Description Challenge seems like it ought to have bite—yet many ways of actually trying to approach it end up solving a much less interesting problem.

That is perhaps what should be expected. The interesting questions about representation, if the above is correct, are primarily *local* questions: we can ask about whether this or that way of dealing with peripersonal space is a representation of PPS, and in what sense it is. This depends in intimate ways on both the structure of the representation and the domain of the representation, however, and there is comparatively little that carries over to a discussion about how insects navigate the world.

Splitting esoteric and exoteric questions thus creates confusions without delivering generality. We take it that the main contribution of our paper is distinguishing two types of question that have often been considered separately. We have done so, however, in order to warn against pursuing them separately.

References

Aaronson, S. (2015). Why philosophers should care about computational complexity. In B. J. Copeland, C. Posy, & O. Shagrir (Eds.), *Computability: Gödel, turing, church, and beyond*. Cambridge: The MIT Press.

Abelson, H., Sussman, G. J., & Sussman, J. (1996). *Structure and interpretation of computer programs*. Cambridge: MIT Press.

Barron, A. B., & Klein, C. (2016). What insects can tell us about the origins of consciousness. *Proceedings of the National Academy of Sciences, 113*(18), 4900–4908.

Bufacchi, R. J., & Iannetti, G. D. (2018). An action field theory of peripersonal space. *Trends in Cognitive Sciences, 22*(12), 1076–1090.
Bufacchi, R. J., & Iannetti, G. D. (2019). The value of actions, in time and space. *Trends in Cognitive Sciences, 23*(4), 270–271.
Caiani, S. Z. (2018). Intensional biases in affordance perception: An explanatory issue for radical enactivism. *Synthese*.https://doi.org/10.1007/s11229-018-02049-w.
Cantwell Smith, B. (1996). *On the origin of objects*. Cambridge: The MIT Press.
Chemero, A. (2009). *Radical embodied cognitive science*. Cambridge: The MIT Press.
Chomsky, N. (1995). Language and nature. *Mind, 104*(413), 1–61.
Coltheart, M. (2001). Assumptions and methods in cognitive neuropsychology. In *The handbook of cognitive neuropsychology: What deficits reveal about the human mind* (pp. 3–21). Philadelphia: Psychology Press.
De Vignemont, F., & Iannetti, G. (2015). How many peripersonal spaces? *Neuropsychologia, 70*, 327–334.
Egan, F. (2014). How to think about mental content. *Philosophical Studies, 170*(1), 115–135.
Gadsby, S., & Williams, D. (2018). Action, affordances, and anorexia: Body representation and basic cognition. *Synthese, 195*(12), 5297–5317.
Gładziejewski, P., & Miłkowski, M. (2017). Structural representations: Causally relevant and different from detectors. *Biology and Philosophy, 32*(3), 337–355.
Goddard, E., Klein, C., Solomon, S. G., Hogendoorn, H., & Carlson, T. A. (2018). Interpreting the dimensions of neural feature representations revealed by dimensionality reduction. *NeuroImage, 180*, 41–67.
Graziano, M. (2006). The organization of behavioral repertoire in motor cortex. *Annual Review of Neuroscience, 29*, 105–134.
Graziano, M. S., & Cooke, D. F. (2006). Parieto-frontal interactions, personal space, and defensive behavior. *Neuropsychologia, 44*(6), 845–859.
Graziano, M. S., Yapp, G. S., & Gross, C. G. (1994). Coding of visual space by premotor neurons. Science, *266*(5187), 1054–1057
Graziano, M. S., Taylor, C. S., & Moore, T. (2002a). Complex movements evoked by microstimulation of precentral cortex. *Neuron, 34*(5), 841–851.
Graziano, M. S., Taylor, C. S., Moore, T., & Cooke, D. F. (2002b). The cortical control of movement revisited. *Neuron, 36*(3), 349–362.
Grush, R. (2007). Skill theory v2.0: Dispositions, emulation, and spatial perception. *Synthese, 159*, 389–416.
Hutto, D. D., & Myin, E. (2012). *Radicalizing enactivism: Basic minds without content*. Cambridge: The MIT press.
Klein, C. (Forthcoming). Do we represent peripersonal space? In F. de Vignemont, A. Serino, H. Y. Wong, & A. Farné (Eds.), *The world at our fingertips: Exploration in peripersonal space*. Oxford: Oxford University Press.
Knuth, D. E. (2011). *Art of computer programming, volumes 1-4A boxed set*. Reading: Addison-Wesley Professional.
Lee, J. (2018). Mental representation and two kinds of eliminativism. *Philosophical Psychology, 31*(1), 1–24.
Lee, J. (Forthcoming). Structural representation and the two problems of content. *Mind & Language*. https://doi.org/10.1111/mila.12224.
Marr, D. (1982). *Vision: A computational investigation into the human representation and processing of visual information*. New York: WH Freeman.
Morgan, A. (2014). Representations gone mental. *Synthese, 191*, 213–244.
Neander, K. (2017). *A mark of the mental: In defense of informational teleosemantics*. Cambridge: The MIT Press.
Noel, J.-P., & Serino, A. (2019). High action values occur near our body. *Trends in Cognitive Sciences, 23*(4), 269–270.
O'Brien, G., & Opie, J. (2004). Notes toward a structuralist theory of mental representation. In H. Clapin, P. Staines, & P. Slezak (Eds.), *Representation in mind* (pp. 1–20). Oxford: Elsevier.

Piccinini, G. (2008). Computation without representation. *Philosophical Studies, 137*(2), 205–241.
Pylyshyn, Z. W. (1984). *Computation and cognition*. Cambridge: Cambridge University Press.
Ramsey, W. M. (2007). *Representation reconsidered*. Cambridge: Cambridge University Press.
Rescorla, M. (2009). Cognitive maps and the language of thought. *The British Journal for the Philosophy of Science, 60*(2), 377–407.
Ritchie, B. (2019). The content of Marr's information-processing framework. *Philosophical Psychology, 32*(7), 1078–1099.
Shagrir, O. (2012). Structural representations and the brain. *British Journal for Philosophy of Science, 63*, 519–545.
Striem-Amit, E., Wang, X., Bi, Y., & Caramazza, A. (2018). Neural representation of visual concepts in people born blind. *Nature Communications, 9*(1), 5250.
Van Gelder, T. (1995). What might cognition be, if not computation? *The Journal of Philosophy, 92*(7), 345–381.
Webb, B. (2006). Transformation, encoding and representation. *Current Biology, 16*(6), R184–R185.
Webb, B., & Wystrach, A. (2016). Neural mechanisms of insect navigation. *Current Opinion in Insect Science, 15*, 27–39.
Wirth, N. (1976). *Algorithms + data structures = programs*. Englewood Cliffs: Prentice Hall.

Chapter 20
Categorically Perceiving Motor Actions

Chiara Brozzo

Abstract In this chapter, I will present an empirical conjecture to the effect that some bodily actions are categorically perceived. These are bodily actions such as grasping or reaching for something, which I am going to call *motor actions*. My conjecture builds on one recently put forward about how the categorical perception of facial expressions of some emotions works. I shall motivate my own conjecture on the basis of both theoretical and empirical considerations describe how it could be operationalised and what explanatory gain could be obtained from it.

20.1 Introduction

In this chapter, I am going to present an empirical conjecture about the way in which some bodily actions are perceptually processed. These are bodily actions such as grasping or reaching for something, which I am going to call *motor actions* (more on their characterisation in Sect. 20.2). My conjecture has it that humans *categorically perceive* motor actions (the notion of categorical perception will be explained in Sect. 20.3). This conjecture both builds on and complements one recently put forward by Stephen Butterfill (2015), which is based on evidence to the effect that humans categorically perceive facial expressions of some emotions (e.g., Etcoff and Magee 1992; Calder et al. 1996; see also Kotsoni et al. 2001). This evidence and the related conjecture will be presented in Sect. 20.4.

In Sect. 20.5, I shall present my own conjecture, to the effect that motor actions are categorically perceived. I will motivate this conjecture on the following grounds: a significant structural analogy exists across motor actions, the expressions of some emotions and the articulations of phonemes, insofar as all of them are actions directed to motorically represented outcomes (a notion that will be explained in due course), and may be categorically perceived as such.

C. Brozzo (✉)
Philosophy Department, Durham University, Durham, UK

© Springer Nature Switzerland AG 2021
F. Calzavarini, M. Viola (eds.), *Neural Mechanisms*, Studies in Brain and Mind 17,
https://doi.org/10.1007/978-3-030-54092-0_20

Why should you be interested in a conjecture based on another conjecture? Because I believe these to be two sides of the same coin, and considering them jointly provides a useful unifying theoretical framework for interpretation and for further testing. In Sect. 20.6, I will describe the explanatory gain that can be obtained from my proposed conjecture in terms of the interpretation of data about the neural mechanisms involved in the processing of motor actions.

20.2 What Are Motor Actions?

I am now going to introduce the notion of motor action.[1] Consider a situation in which I pick up a plum. What are the conditions under which it is correct to say that the plum has been *grasped* (as opposed to, e.g., *pushed away*)? In the most ordinary cases, these conditions will include that I end up with the plum held between my palm and my fingers. But it is also possible to imagine a situation in which my hands are occupied, or are somehow blocked, or do not exist at all, and I have to grasp the plum with my mouth or with my feet. Not *any* body part will do (I cannot grasp anything with my nose, due to anatomical constraints), but *some* body part has to be employed. The essential feature, then, for the characterisation of a motor action such as grasping something is that it involves one of a *circumscribed set of body parts*, which has to undergo a certain change—one, for instance, that brings it in a specific relation to a given object. In the light of the above, I shall call *motor actions* actions such as grasping, whose characterisation unavoidably involves mention of body parts and their configurations and an object in relation to which these configurations unfold. To clarify, compare a motor action—e.g., grasping—with the action of, e.g., designing an experiment. No specific body part is involved in the accomplishment of the latter, nor is any circumscribed set of bodily movements.

Notice that motor actions cannot be specified exclusively in terms of the corresponding final bodily configurations. Consider as an example the action of catching something. One may believe it sufficient to characterise the action of catching a ball in terms of the ball being held between one's hands. But this is not sufficient, insofar as it cannot be said that a ball has been *caught* if the ball being held between one's hands is the result of someone carefully placing it there: without the specification that one has performed certain bodily movements, it is false that she has caught the ball (see Pacherie 2008). Because of this, a motor action should be specified by reference to a circumscribed set of *sequences* of bodily configurations,

[1]The term *motor action* has been widely used in the study of action production, both in neuroscience (e.g., Gallese et al. 1996; Hamilton and Grafton 2007; Jeannerod 1994, 2006; Rizzolatti et al. 1996) and philosophy (e.g., Butterfill and Sinigaglia 2014; Ferretti and Zipoli Caiani 2019; Mylopoulos and Pacherie 2017; Nanay 2013; Pavese 2015). I am, however, introducing this term with a specific meaning, to be illustrated shortly, that does not straightforwardly coincide with how this term has been employed in the aforementioned literatures, although it is likely consistent with it.

and not only the final one. So, motor actions are actions whose characterisation unavoidably involves mention of sequences of bodily configurations. This is a characterisation in purely behavioural terms. In the following sections, I shall propose a conjecture to the effect that humans categorically perceive motor actions. I will begin by characterising the notion of categorical perception.

20.3 The Premise: Humans Categorically Perceive Facial Expressions of Some Emotions

When human subjects are asked to discriminate between pairs of faces that express a certain emotion, their responses exhibit a very specific pattern of discrimination. Presented with several pairs of faces that differ by a fixed physical amount, created by morphing a face expressing a certain emotion (happiness) into a face expressing another emotion (fear), subjects are better at discriminating pairs where each member expresses a different emotion, rather than pairs where both members express the same emotion (Etcoff and Magee 1992; see also Calder et al. 1996; Kotsoni et al. 2001). In a separate task, subjects are also asked to identify the stimuli that they are presented with as expressing one of two possible emotions (happiness or fear). Subjects are consistent in identifying stimuli that do not fall too close to the category boundary, whereas they are at chance (that is, they identify the face stimulus as happy or sad with equal frequency) when it comes to identifying stimuli that fall too close to the category boundary (Calder et al. 1996). When such patterns of discrimination are exhibited in relation to a certain domain—i.e., some pairs of stimuli are easier to discriminate than others, and, moreover, what explains this is that those pairs of stimuli fall in different categories recognised by the subjects—it is said that *categorical perception* of that domain occurs (Repp 1984; Harnad 1987; McKone et al. 2001; Harnad 2003).[2] According to Harnad (2003), specifically,

[2]Some important clarifications are in order. Hereafter I will discuss conjectures concerning the *recognition* of emotions on the part of an observer, but I shall not make any claims about the *nature* of emotions. As to the latter, there is a controversy (on which I do not mean to adjudicate) concerning between the nature of emotions is *categorical* or *basic* rather than *dimensional*. According to the *basic* view of emotions (Ekman 1992; Izard 1971; Tomkins 1962), emotions fall into discrete categories, which are reflected in the information provided by cues such as facial expressions and body postures. According to the *dimensional* view of emotions, by contrast, rather than falling into discrete categories, emotions arise from combinations of degrees of arousal and valence, two distinct dimensions whose values vary in a continuous way, without giving rise to clear-cut category boundaries (Russell 1980). It is crucial to notice that some authors have taken the aforementioned results supporting the view that the *recognition* of emotions takes place by means of categorical perception as support for a categorical view of the *nature* of emotions. Fugate (2013) points out that this is a mistake: she refers to evidence provided by Young et al. (1997) and Fujimura and colleagues (2012) suggesting that both categorical and dimensional information might be drawn on in categorical perception. I am grateful to the editors of this volume for pointing out this potential source of misunderstanding.

perceived differences between stimuli within a category are smaller than the actual physical differences between those stimuli, and/or perceived differences between stimuli across a category boundary are larger than the actual physical differences between those stimuli (Harnad 2003).[3]

Evidence suggests that humans show categorical perception for a number of other domains in addition to facial expressions of emotion, including speech (Liberman et al. 1957; Eimas et al. 1971), colour (Bornstein and Korda 1984), orientation (Wolfe et al. 1992) and face identity (Beale and Keil 1995; Kikutani et al. 2008).

The colour case is illustrative of the sort of phenomena that the notion of categorical perception is supposed to explain. For example, I mentioned that pairs of stimuli falling in different categories are easier to discriminate. Bornstein and Korda (1984) showed that pairs of hues can be told apart comparably quickly, even though they may be more or less different in purely physical terms, as long as each belongs to a different colour category. Conversely, pairs of hues that belong each to a different colour category can be told apart more quickly than any pair of hues that belong to the same colour category. This is so in spite of the fact that the pair of hues belonging to the same colour category might be more different from each other in physical terms than the pair of hues that belong each to a different colour category. Thus, ease of discrimination is shown not to depend straightforwardly on physical differences, but, rather, is a function of the categories to which the stimuli belong.[4]

Another phenomenon that the notion of categorical perception is supposed to explain is the occurrence of pop out effects. Daoutis et al. (2006) have shown that, given an array of coloured dots, all of the same colour except for one, the time it takes to find the odd one out does not increase as a function of the number of dots when the odd one out is of a different colour category with respect to the other dots. That is, hues falling into different colour categories pop out.

The evidence reported earlier in this section (Etcoff and Magee 1992; Calder et al. 1996) gives us reasons for thinking that humans show categorical perception of facial expressions of some emotions. The qualification *some* is justified by the fact that the stimuli employed in the earlier reported experiments typically involve happy and fearful faces. It might therefore be safer to claim that it is only for the expression of *some* emotions that humans show categorical perception.

There might be a principled reason behind this, namely that some emotions lend themselves to a more straightforward connection with their bodily expression than

[3]Harnad (2003) also offers a version of this definition to accommodate the case of learned categorical perception. In this version, the term of comparison is not actual physical differences between stimuli but, rather, perceived similarity between the stimuli within and across category boundaries before learning. I am grateful to a reviewer of this volume for inviting me to report Harnad's definition.

[4]Studies that hinge on physical differences are subject to a potential objection: couldn't it be that same physical differences are treated differently by the retina and therefore end up being perceived differently by the subjects? A study by Witzel and Gegenfurtner (2014) counters this objection by using just-noticeable differences instead of physical differences. A just-noticeable difference (JND) is the smallest difference between two stimuli that a subject can perceive.

others (such as *Schadenfraude*).[5] That is, it is plausible that some emotions may have more easily identifiable characteristic expressions associated with them.

So far, I have introduced the notion of categorical perception and have reported evidence that the facial expressions of some emotions are categorically perceived. On the basis of this evidence, Butterfill (2015) puts forward the following conjecture: facial expressions of emotions could be categorically perceived insofar as they are actions directed to motorically represented outcomes (a notion that will be defined in the next section). I shall now present this conjecture, along with how it is supported by current evidence. This will provide the springboard for my own conjecture, which I will introduce in Sect. 20.5.

20.4 What Are Facial Expressions of Emotions? The AMROs Conjecture

In this section, I am going to present Butterfill's conjecture that facial expressions of emotions are actions directed to a motorically represented outcome (AMROs), and are processed as such within the context of categorical perception (Butterfill 2015). I shall explain the notion of AMRO first by reference to the case of speech, and will then show how this notion can be applied to the case of facial expressions of emotions.

20.4.1 Phoneme Articulations as AMROs

Speech is one of the most extensively studied cases of categorical perception (e.g., Liberman et al. 1957; Eimas et al. 1971; Harnad 1987; Nygaard and Pisoni 1995; Harnad 2003). Here is evidence that we categorically perceive speech. It is possible to create a series of test stimuli consisting in sounds that spread across the phonemes *ba* and *pa*. These are designed in such a way that each two neighbouring test sounds differ from one another by the same amount (in terms of frequency) as any other pair of neighbouring sounds (the test stimuli consisting in facial configurations described in Sect. 20.3 were created on the basis of an analogous principle). Subjects find it hard to discriminate neighbouring pairs of test sounds, except when two neighbouring pairs fall on two different sides of a category boundary—i.e., one is perceived as *ba* and the other as *pa*. Within the same category, on the other hand, subjects will hear the same phoneme, e.g., *ba* (Liberman et al. 1957).

So, humans categorically perceive speech, and the categories consist in phonemes. But what are phonemes? An interpretation that has been put forward

[5]This leaves it open that the connection in question could be mediated by factors such as conceptual knowledge (Brooks and Freeman 2018) or culture (see Caruana and Viola 2018).

(e.g., by Liberman and Whalen 2000) is that a phoneme is an *outcome*, i.e. a state of affairs, to which an action is directed.[6] What distinguishes outcomes (in the case of speech, consisting in phonemes) from mere acoustic signals? The distinction is twofold. First, different acoustic signals could be employed to articulate the same phoneme. This is shown by the fact that we have categorical perception of speech: as mentioned earlier, a number of different acoustic signals will be treated as the same phoneme (e.g., *pa*) by a perceiver. In addition to this (and this goes beyond the idea that we categorically perceive speech), single acoustic signals by themselves may not be diagnostic of what phoneme is being articulated: the same single acoustic signal, depending on contextual factors such as speed of articulation or dialect, could result from the articulation of different phonemes (see, e.g., Repp and Liberman 1987).

So far, I have presented reasons in support of the idea that phonemes should be considered outcomes, and how this differs from considering them merely acoustic signals. The idea, in short, is that the same phoneme could be articulated through different acoustic signals, and the same acoustic signal could result from different phonemes being articulated. Building on this, Butterfill (2015) hypothesises that articulations of phonemes may be characterised as *actions directed to motorically represented outcomes*—henceforth, AMROs for short.

But what is a *motorically represented* outcome? It is an outcome represented by motor areas of the brain. The best evidence that an outcome is represented motorically is that a marker of motor processing, e.g. neuronal discharge that is recorded in motor areas of the brain, or motor evoked potentials, can be found in correlation with an outcome being brought about (Butterfill and Sinigaglia 2014, p. 122).[7] Butterfill suggests that phonemes could be motorically represented outcomes

[6]This closely resembles the idea that phonemes are *intended gestures* of a speaker, which is the heart of the Motor Theory of Speech Perception (Liberman et al. 1957; Liberman and Mattingly 1985). The Motor Theory of Speech Perception has a complex history, and its evaluation is made difficult by the fact that it encompasses several different claims, whose fate has proved very different. Galantucci et al. (2006) helpfully break the Motor Theory of Speech Perception down into different claims: "(1) speech processing is special, (2) perceiving speech is perceiving gestures, and (3) the motor system is recruited for perceiving speech." Galantucci and colleagues argue that (1) is likely false, but that (2) and (3) still find support. Claim (3) has recently been vindicated by Whalen (2019). In this chapter, I am exploiting precisely claims (2) and (3) of the Motor Theory of Speech Perception, but not (1).

[7]These markers of motor processing are often discussed under the heading of *motor representations*. The idea that motor representations might represent outcomes rather than just fine-grained bodily movements has given rise to what Butterfill and Sinigaglia (2014) call the Interface Problem: how do the outcomes represented by intentions and the outcomes represented by motor representations non-accidentally match? Answers to this problem have been discussed, e.g., by Butterfill and Sinigaglia themselves (2014), as well as by Mylopoulos and Pacherie (2017), Burnston (2017), Ferretti and Zipoli Caiani (2019) and Shepherd (2019). There are also motor representations representing an action in greater detail—for example, that represent grasping with a specific body part (e.g., one's hand) and with a specific kind of grip (e.g., a precision grip—the one you would typically adopt to grasp a peanut; Rizzolatti et al. 1988). For a more extended discussion of what motor representations represent, see Ferretti (2016).

insofar as articulating a phoneme requires coordinated movements of the vocal organs—lips, tongue tip, tongue body, tongue root, velum and larynx (see, e.g., Goldstein and Fowler 2003).

To this, I would like to add that precisely this sort of rationale—that bringing about a certain outcome requires a series of movements coordinated around the outcome—has led neuroscientists to posit that actions should be represented in the brain, in a way that abstracts away from the details of the bodily movements but is also sensitive to a certain outcome being achieved (Jeannerod 1994, 2006; Rizzolatti and Sinigaglia 2008). So, it seems plausible to suppose that not only are phonemes outcomes, but they are *motorically represented* outcomes.

On the basis of this idea, it is possible to hypothesise that the categories into which the categorical perception of speech sorts acoustic signals are motorically represented outcomes, and that articulations of phonemes are processed as AMROs in the context of cateogorical perception. This is an interpretation of what goes on in the categorical perception of speech that suggested by, e.g., Liberman and Whalen (2000).

20.4.2 Expressing Emotion as an AMRO

Why is this relevant to the categorical perception of facial expressions of emotions? Because Butterfill (2015) conjectures that facial expressions of emotions, just like phonemes according to the interpretation reviewed in the previous subsection, could be categorically perceived as AMROs, rather than merely as facial configurations. This is based on the idea that an emotion being facially expressed (henceforth *an emotion being expressed*) is a motorically represented outcome. Let me show how this idea is justified, before moving on to the conjecture in the next subsection.

A contrast between facial configurations and emotions being expressed can be set up, analogously with that between acoustic signals and phonemes. As with speech, the evidence to the effect that we categorically perceive facial expressions of emotions indicates that multiple facial configurations (e.g., more or less wide smiles) can be involved in the same emotion (e.g., happiness) being expressed. But, to support the idea that emotions being expressed are outcomes rather than just facial configurations, we also need evidence that single facial configurations are not necessarily diagnostic of emotions, i.e. may be taken to express different emotions depending on contextual factors.

Aviezer et al. (2008) provide just this sort of evidence. They show that the very same facial configuration can be taken to express different emotions depending on the context into which it is inserted—specifically, the overall bodily configuration of the individual exhibiting that facial configuration. For instance, the very same facial configuration on an individual's face can be verbally classified by an observer as either disgusted or proud depending on the overall bodily configuration of the observed individual—i.e., whether the individual with this facial configuration is holding a disgusting object or is engaged in a power pose.

So, both conditions for emotions being expressed to be outcomes are fulfilled: the same emotion can be expressed through different facial configurations, and the same facial configuration can express different emotions depending on contextual factors.

But why think that emotions being expressed should be *motorically represented* outcomes? In response to this, Butterfill presents a line of reasoning analogous to the one provided in relation to the case of speech:

> expressing an emotion by, say, smiling or frowning [...] involves making coordinated movements of multiple muscles [...]. That such an expression of emotion is a goal-directed action follows just from its involving motor expertise and being coordinated around an outcome [...]. (Butterfill 2015, p. 446)

That expressing emotions relies on motor expertise is further supported by evidence that motor programmes seem to have a fundamental role in the production of emotions, so that tampering with motor programmes imposes limits on one's own emotional experience (see Davis et al. 2010).[8]

20.4.3 Butterfill's Conjecture: Facial Expressions of Emotions Are Processed as AMROs in Categorical Perception

To sum up, so far I have presented reasons for thinking that emotions being expressed are outcomes, not reducible to facial configurations. Now on to Butterfill's conjecture. This has it that, when facial expressions of emotions are categorically perceived, these are processed as actions—specifically, AMROs—rather than merely as facial configurations. In other words, the stimuli consisting in facial configurations would trigger a hypothesis about which motorically represented outcome is being pursued—e.g., happiness being expressed—and, consequently, about which action is being performed in order to achieve that motorically represented outcome—e.g., expressing happiness.

Butterfill's conjecture about the categorical perception of facial expressions of emotions, by his own admission, requires that "the things categorised in in categorical perception of expressions of emotions are events rather than configurations or anything static" (2015, p. 446). While the idea that acoustic signals are processed as actions may have seemed reasonable given that acoustic signals are dynamic stimuli, the idea that facial configurations (which are static stimuli) are processed as actions might seem surprising. In response to this concern, Butterfill observes that his conjecture is not in principle incompatible with the fact that the categorical perception of expressions of emotions may be triggered by static stimuli, such as the facial configurations described in Sect. 20.3. In support of this idea, he cites

[8] I am grateful to the editors of this volume for bringing this evidence to my attention.

evidence to the effect that static stimuli are sufficient to trigger motor programmes in an observer (Borghi et al. 2007).

In the light of this conjecture, the data about the categorical perception of facial expressions of emotions reviewed in Sect. 20.3 could be explained in the following way: pairs of stimuli that fall in the same category are treated in the same way because they can be interpreted as part of actions directed to the same motorically represented outcome: that happiness (or fear) is expressed. Interpreting the data in this way makes room for the fact that, if the stimuli were made more complex so as to include wider bodily configurations, contextual factors affecting their categorisation could be taken into account, just as contextual factors may affect the categorical perception of speech.

Butterfill supports his conjecture on the basis of a few considerations. Among these, there is the idea that facial expressions of emotions and phonemes are analogous in a number of ways—e.g., facial configurations alone might not be diagnostic of emotions, in the same way in which isolated acoustic signals might not be diagnostic of phonemes, and both are open to the influence of contextual factors in determining which emotion or phoneme is detected by an observer. Moreover, Butterfill points out that when stimuli are chosen in order to test the categorical perception of facial expressions of emotions, the guiding principle is not which facial configuration is more likely to be associated with a given emotion, but rather which facial configuration is more likely to express a given emotion. Therefore, his conjecture is in line with how the stimuli are categorised in the first place, and makes sense of plausible analogies between facial expressions of emotions and phonemes. So, Butterfill's conjecture seems worth exploring.

20.5 A Complementary Conjecture: Humans Categorically Perceive Motor Actions

I would now like to go back to motor actions, introduced in Sect. 20.2, and present a conjecture that builds on and complements Butterfill's one. According to my conjecture, humans would categorically perceive motor actions. This is based on the idea that motor actions are AMROs. Let me provide reasons in support of the latter idea first, and then explain why this motivates considering the possibility that motor actions could be categorically perceived. In the next section, I will show the explanatory gain to be obtained from this conjecture.

20.5.1 Motor Actions Are AMROs

In order to show why it is reasonable to consider motor actions AMROs, i.e. actions directed to motorically represented outcomes, let me start by showing why motor actions should be thought of as directed to outcomes.

This is easily done. Recall from Sect. 20.2 that motor actions were defined in behavioural terms as actions whose characterisation unavoidably involves mention of sequences of bodily configurations. Grasping is a paradigm example of a motor action.

Now, something being grasped should be considered an outcome, as opposed to merely a bodily configuration (or series of bodily configurations), for the following reasons. First, multiple different bodily configurations may be employed to achieve the outcome of something being grasped. The latter could be achieved by using thumb and index finger in different configurations (e.g., with a smaller or greater distance between the fingertips), or using all of the fingers on one's hand, or even using a different effector (e.g., the mouth as opposed to the hand). On the other hand, the same series of bodily configuration (e.g., one's fingers closing around the handles of a pair of pliers) may achieve different outcomes (e.g., something being grasped, or something being released) depending on contextual factors (in this case, the shape of the pliers).[9] Therefore, motor actions such as grasping are directed to outcomes (something being grasped), which are interestingly different from bodily configurations: the same outcome can be achieved by different sequences of bodily configurations, and the same sequence of bodily configurations can lead to different outcomes.

Now, why think that these outcomes are motorically represented? As mentioned in Sect. 20.4.1, the ideal evidence for an outcome being motorically represented is that a given marker of motor processing should be found in correlation with an outcome being brought about. For an outcome (as opposed to a mere sequence of bodily configurations) to be represented, two conditions need to be fulfilled (as suggested most recently by Butterfill and Sinigaglia 2014, and earlier, e.g., by Sinigaglia 2010). First, the same marker of motor processing (e.g., the same rate of neural discharge) should be found by holding the outcome fixed, but varying sequences of bodily configurations. Secondly, different markers of motor processes (e.g., markedly different rates of neural discharge) should be found by holding a sequence of bodily configurations fixed, but altering the outcome, e.g. by changing contextual factors.

In the case of motor actions such as grasping, we have precisely this sort of ideal evidence, under both conditions required to say that an outcome, as opposed to a sequence of bodily configurations, is represented motorically (as has been observed, e.g., by Rizzolatti and Sinigaglia 2008, as well as by Butterfill and Sinigaglia 2014). For example, there is evidence that in the premotor cortex of the macaque monkey brain—specifically, in the area F5—there are populations of neurons that activate in correlation with a grasping act regardless of whether grasping is executed with the

[9]This clever manipulation was used in an experiment by Umiltà et al. (2008): two different pairs of pliers were constructed, such that, with one pair of pliers, closing one's fingers around the handles would result in an object being grasped, and, with the other pair of pliers, exactly the same sequence of bodily configurations would result in an object being released.

hand as opposed to with the mouth (Rizzolatti et al. 1988),[10] thus indicating that the same outcome is represented while varying sequences of bodily configurations.

But there is also evidence to the effect that there are neurons—also in the area F5—that, in correlation with the same sequence of bodily configurations—e.g., that involved in grasping an object—fire differentially depending on the context in which grasping is performed. The different contexts consisted in the presence or absence of an object to be grasped (Umiltà et al. 2001; see also Villiger et al. 2011). Therefore, motor actions are AMROs, in virtue of their outcomes being represented motorically.

20.5.2 Motor Actions Could Be Categorically Perceived

Let me take stock. In Sect. 20.4, I reported Butterfill's (2015) observation that articulating phonemes and expressing emotions are AMROs, as well as his conjecture that facial expressions of emotions could be processed as AMROs within the context of categorical perception, and, relatedly, sorted into categories consisting in motorically represented outcomes (e.g., happiness being expressed). In the previous subsection, I pointed out that motor actions are AMROs, too. On the basis of this observation and of Butterfill's conjecture, it becomes plausible that articulating phonemes, expressing emotions and motor actions should be species of the same genus—namely, AMROs. Given that both speech and facial expressions of emotions are categorically perceived, I put forward the conjecture that motor actions could be categorically perceived, too. By analogy with the case of speech and (according to Butterfill's conjecture) facial expressions of emotions, my conjecture has it that the categories into which categorical perception would subdivide motor actions is the motorically represented outcomes around which motor actions are coordinated—e.g., something being grasped.

Considering the possibility that motor actions could be categorically perceived might sound surprising, given that many instances of categorical perception that I have discussed in this chapter involve static stimuli, such as facial configurations or colour hues. Even though in Sect. 20.4.3 I mentioned the possibility that static stimuli could trigger the perception of events in relation to Butterfill's conjecture, the fact remains that motor actions themselves *are* events. How could the idea that motor actions are categorically perceived be operationalised?[11]

Let me now clarify that the notion of categorical perception is perfectly compatible with the idea that the stimuli to be categorised are events rather than objects. This is clearest if you think of the case of the categorical perception of speech. The

[10] Here I am appealing to single-cell recordings in the macaque monkey brain based on the idea, supported by Rizzolatti et al. (2002), that there is a sufficient analogy between this particular region of the macaque monkey brain and the Brodmann area 44 of the human brain.

[11] I am grateful to a reviewer of this volume for inviting me to discuss this important issue.

stimuli employed to test this phenomenon, as said in Sect. 20.3, are acoustic sounds that constitute phonemes. These are events. But how might this work in practice in the case of motor actions?

That motor actions are categorically perceived means that the following should in principle be possible. A pair of distinct motor actions should be identified—one could be grasping with the hand, since it is a widely studied case, and another could be pushing away an object with the back of one's fingers. The two different motor actions should be performed with the same hand. On the basis of these two different motor actions, a number of stimuli—either static, such as snapshots, or dynamic, such as short clips—should be obtained, such that pairs of neighbouring stimuli involve bodily configurations (or sequences of bodily configurations, if the stimuli are dynamic), that differ by the same amount in terms of their kinematic features (e.g., distance between fingertips). If it is true that humans categorically perceive motor actions, then pairs of neighbouring stimuli should be hard to tell apart when they fall within the same category (e.g., something being grasped), but easy to distinguish when each belongs to a different category (something being grasped vs. something being pushed away), despite the fact that, by design, all the neighbouring pairs of stimuli differ by the same amount.

As to what the categories could be beyond something being grasped and (maybe) something being pushed away, there is evidence that specific neural populations in the premotor cortex become active in correlation with different action types, such as grasping (Rizzolatti et al. 1988). As mentioned in Sect. 20.5.1, the activation of these populations of neurons correlates with outcomes, such as something being grasped. The fact that the organising principle is outcomes rather than sequences of bodily configurations has been already expounded in Sect. 20.5.1: the same neural activation can be observed in correlation with different sequences of bodily configurations bringing about the same outcome, and the same sequences of bodily configurations are treated differently in terms of neural discharge depending on how the context shapes the overall outcome.

Taken together, these action types constitute what has been referred to as a *motor vocabulary*, or a *vocabulary of motor acts* (Rizzolatti et al. 1988; see Jeannerod 2006; Rizzolatti and Sinigaglia 2008). The outcomes to which the actions forming this motor vocabulary are directed are therefore plausible candidates for the categories in which humans subdivide motor actions, but it is again an empirical question whether they really provide the categories that humans are sensitive to.[12] More generally, of course, whether my conjecture holds is an empirical question, to be settled by means of experimental evidence.

[12]Support for the aspect of the conjecture concerning the categories into which motor actions are sorted is given by evidence that the organisation of actions in the brain in terms of outcomes influences our processing of action-related language (e.g., Marino et al. 2017). I am grateful to a reviewer of this chapter for bringing this to my attention.

20.6 How the Conjecture Would Explain Neural Mechanisms Involved in the Processing of AMROs

In this last section, I am going to show that, if the conjecture I am proposing turned out to be true, this would provide a good explanation of data we currently have about the involvement of certain neural mechanisms in the processing of motor actions, as well as a unifying explanation for the involvement of certain neural mechanisms in the processing of other AMROs.

First of all, we need a bit more detail about how categorical perception occurs. A reasonable model of how this could occur, which has been put forward in the case of the categorical perception of speech (Liberman and Mattingly 1985), is that, in the course of categorically perceiving a certain auditory stimulus, a hypothesis is made as to what phoneme is being articulated, and the hypothesis is checked against the available evidence. If the evidence is compatible with the hypothesis, the hypothesis is reinforced. If the evidence is incompatible with the hypothesis, the hypothesis is revised.

This model has been further supplemented in the following way: hypothesising which phoneme is being articulated would involve the activation of motor processes in the observer's brain that would normally be recruited in the production of one's own speech. The Motor Theory of Speech Perception (Liberman and Mattingly 1985) makes precisely this suggestion. Thus could motor processes be involved in the categorical perception of speech.[13]

As part of his proposed conjecture, Butterfill (2015) suggests that the categorical perception of expressions of emotions could work in an analogous way: a hypothesis could be made about what emotion is being expressed, and the hypothesis would be checked against the available evidence. Specifically, Butterfill suggests that a hypothesis as to which emotion is being expressed could involve the activation of processes in an observer that would be recruited were the observer to have that emotion herself (2015, p. 448). In particular, this would result in outcomes being motorically represented in an observer (he notes this has already been proposed by Adolphs 2001).

After putting forward this aspect of the conjecture, Butterfill points out that there is evidence suggesting that this is precisely what could occur in the case of the processing of expressions of emotions. He reports evidence that, on the one hand, processes that would occur when one is having a certain emotion also occur while observing other individuals' emotions (Bastiaansen et al. 2009; Gallese et al. 2004; Rizzolatti and Sinigaglia 2008; van der Gaag et al. 2007; Wicker et al. 2003). Moreover, there is evidence that *disrupting* the occurrence of these processes in an observer interferes with the recognition of others' emotions (Niedenthal et al. 2001; Oberman et al. 2007; Pitcher et al. 2008).

[13] As I mentioned in footnote 7, this aspect of the Motor Theory of Speech Perception still stands.

Let us therefore see whether an analogous model of the processing of motor actions is viable. Indeed, various sources have proposed a model according to which motor actions are recognised through a process of making hypotheses and checking them against the available evidence that involves the activation of motor processes in an observer (see, e.g., Kilner et al. 2007).[14]

An especially pertinent source of evidence for motor process involvement in the categorical perception of motor actions is given by an experiment carried out by Cattaneo et al. (2010).[15] In this experiment, a sensory-motor adaptation paradigm was employed: participants were trained to perform either a *push away* or a *pull towards* movement with their hand while blindfolded. This motor training was shown to have an impact on the subsequent visual recognition of analogous hand actions, and resulted in ambiguous stimuli being classified as *push away* movements following a training involving *pull towards* movements, and vice versa. The fact that activating a certain motor process by means of a repeatedly performed action impacts on the labelling of a subsequently observed action suggests that motor processes are involved in the identification of outcomes: if the recognition of a motor action was a purely visual phenomenon, it is unclear why motor training in the absence of visual stimuli should impact on it.

Moreover, transcranial magnetic stimulation (TMS) over the ventral premotor cortex suppressed the adaptation aftereffect in the recognition of an action. That is, a temporary disruption of areas involved in the production of motor actions impacted on the recognition of those actions (Cattaneo et al. 2010). This result dovetails nicely with that reported previously, to the effect that disrupting the occurrence of the processes involved in expressing an emotion in an observer interferes with the recognition of others' emotions.

Now, if motor actions turned out to be categorically perceived, along with speech and facial expressions of emotions, a unifying explanation could be given as to why areas involved in the production of speech, expressions of emotions and motor actions seem to have a role in the perceptual processing of these stimuli. This would be explained in the following way. Articulations of phonemes, expressions of emotions and motor actions are all AMROs. Categorical perception, if Butterfill's conjecture and my proposed one turn out to be correct, sorts AMROs into categories corresponding to motorically represented outcomes (AMROs). This categorisation relies on a process of hypothesis testing that draws on the very processes involved in the production of these stimuli. This is how the conjecture that motor actions are categorically perceived complements Butterfill's conjecture that expressions of emotions are categorically perceived as AMROs, giving rise to a unifying theoretical

[14]The notion of *understanding from the inside* has been put forward to indicate cases in which an observer motorically represents an outcome that an observed individual is trying to fulfil (e.g., Rizzolatti and Sinigaglia 2010; see also Gallese and Sinigaglia 2011; Rizzolatti and Sinigaglia 2016). An interesting topic of investigation, which is best left to another occasion, is the relationship between the motor processes hypothesized to be involved in the processing of motor actions and mindreading.

[15]I am grateful to Corrado Sinigaglia for bringing this evidence to my attention.

framework. This same framework would accommodate the idea that categorically perceiving speech sorts stimuli into phonemes conceived as motorically represented outcomes.

20.7 Conclusion

In the foregoing, I have presented my conjecture that humans categorically perceive motor actions. I have done so by drawing an analogy with expressions of emotions and articulations of phonemes, which can be plausibly thought of as actions directed to motorically represented outcomes (AMROs), as are motor actions. My conjecture builds on an interpretation of the categorical perception of speech whereby acoustic signals are processed as actions directed to a motorically represented outcome, and the categories consist in motorically represented outcomes. My conjecture also builds on the one recently put forward by Butterfill (2015), which interprets the categorical perception of facial expressions of emotions in terms of facial expressions of emotions being processed as AMROs. The two conjectures and the interpretation of the categorical perception of speech in terms of the processing of actions naturally complement each other, and together give rise to a unifying theoretical framework, which could explain data showing the involvement of motor processes in an observer's brain when processing others' actions directed to motorically represented outcomes. While this does not show the conjecture to be true, it makes it worthy of consideration.

Acknowledgments I would like to thank Corrado Sinigaglia and Hong Yu Wong, the editors of this volume and the referees for this chapter for detailed comments on previous versions of this work, which greatly helped improve it. I would also like to thank the members and friends of the Philosophy of Neuroscience research group led by Hong Yu Wong at the University of Tübingen (especially Gregor Hochstetter, Roberta Locatelli, Alex Morgan, Jean-Moritz Müller, Krisztina Orbàn, Katia Samoilova), the members of Bence Nanay's research group at the University of Antwerp (especially Dan Cavedon-Taylor, Laura Gow, Margot Strohminger), the audiences of the Neural Mechanisms Online Conference (especially Dan Burnston and Louise Röska-Hardy), of the "The Neuroscientific Turn in the Philosophy of Mind" workshop at the University of Urbino (especially Mario Alai, Enzo Fano, Gabriele Ferretti, Pierre Jacob), of the Philosophy Colloquium at the University of Bochum (especially Tobias Schlicht and Joulia Smortchkova), of the European Society for Philosophy and Psychology conference at the University of St Andrews, of the Aegina Summer School (especially Laura Crucianelli, Elisabeth Pacherie, Laura Silva, Barry C. Smith), of the "Practical Reasoning and Motor Representation" workshop at the University of Warwick (especially Josh Shepherd), of the Corcoran Department of Philosophy at the University of Virginia, Stephen Butterfill, Matthew Longo and Wayne Wu for inspiration and feedback.

References

Adolphs, R. (2001). The neurobiology of social cognition. *Current Opinion in Neurobiology, 11*(2), 231–239.

Aviezer, H., Hassin, R. R., Ryan, J., Grady, C., Susskind, J., Anderson, A., Moscovitch, M., & Bentin, S. (2008). Angry, disgusted, or afraid? Studies on the malleability of emotion perception. *Psychological Science, 19*(7), 724–732.

Bastiaansen, J. A., Thioux, M., & Keysers, C. (2009). Evidence for mirror systems in emotions. *Philosophical Transactions of the Royal Society B: Biological Sciences, 364*(1528), 2391–2404.

Beale, J. M., & Keil, F. C. (1995). Categorical effects in the perception of faces. *Cognition, 57*(3), 217–239.

Borghi, A. M., Bonfiglioli, C., Lugli, L., Ricciardelli, P., Rubichi, S., & Nicoletti, R. (2007). Are visual stimuli sufficient to evoke motor information?: Studies with hand primes. *Neuroscience Letters, 411*(1), 17–21.

Bornstein, M. H., & Korda, N. O. (1984). Discrimination and matching within and between hues measured by reaction times: Some implications for categorical perception and levels of information processing. *Psychological Research, 46*(3), 207–222.

Brooks, J. A., & Freeman, J. B. (2018). Conceptual knowledge predicts the representational structure of facial emotion perception. *Nature Human Behaviour, 2*(8), 581–591.

Burnston, D. C. (2017). Interface problems in the explanation of action. *Philosophical Explorations, 20*(2), 242–258.

Butterfill, S. A. (2015). Perceiving expressions of emotions: What evidence could bear on questions about perceptual experience of mental states? *Consciousness and Cognition, 36*, 438–451.

Butterfill, S. A., & Sinigaglia, C. (2014). Intention and motor representation in purposive action. *Philosophy and Phenomenological Research, 88*(1), 119–145.

Calder, A. J., Young, A. W., Perrett, D. I., Etcoff, N. L., & Rowland, D. (1996). Categorical perception of morphed facial expressions. *Visual Cognition, 3*(2), 81–118.

Caruana, F., & Viola, M. (2018). *Come funzionano le emozioni: da Darwin alle neuroscienze*. Bologna: Il Mulino.

Cattaneo, L., Barchiesi, G., Tabarelli, D., Arfeller, C., Sato, M., & Glenberg, A. M. (2010). One's motor performance predictably modulates the understanding of others' actions through adaptation of premotor visuo-motor neurons. *Social Cognitive and Affective Neuroscience*, nsq099.

Daoutis, C. A., Pilling, M., & Davies, I. R. (2006). Categorical effects in visual search for colour. *Visual Cognition, 14*(2), 217–240.

Davis, J. I., Senghas, A., Brandt, F., & Ochsner, K. N. (2010). The effects of BOTOX injections on emotional experience. *Emotion, 10*(3), 433–440.

Eimas, P. D., Siqueland, E. R., Jusczyk, P., & Vigorito, J. (1971). Speech perception in infants. *Science, 171*(3968), 303–306.

Ekman, P. (1992). Are there basic emotions? *Psychological Review, 99*, 550–553.

Etcoff, N. L., & Magee, J. J. (1992). Categorical perception of facial expressions. *Cognition, 44*(3), 227–240.

Ferretti, G. (2016). Through the forest of motor representations. *Consciousness and Cognition, 43*, 177–196.

Ferretti, G., & Caiani, S. Z. (2019). Solving the interface problem without translation: The same format thesis. *Pacific Philosophical Quarterly, 100*(1), 301–333.

Fugate, J. M. (2013). Categorical perception for emotional faces. *Emotion Review, 5*(1), 84–89.

Fujimura, T., Matsuda, Y. T., Katahira, K., Okada, M., & Okanoya, K. (2012). Categorical and dimensional perceptions in decoding emotional facial expressions. *Cognition & Emotion, 26*(4), 587–601.

Galantucci, B., Fowler, C. A., & Turvey, M. T. (2006). The motor theory of speech perception reviewed. *Psychonomic Bulletin & Review, 13*(3), 361–377.

Gallese, V., & Sinigaglia, C. (2011). What is so special about embodied simulation? *Trends in Cognitive Sciences, 15*(11), 512–519.
Gallese, V., Fadiga, L., Fogassi, L., & Rizzolatti, G. (1996). Action recognition in the premotor cortex. *Brain, 119*(2), 593–609.
Gallese, V., Keysers, C., & Rizzolatti, G. (2004). A unifying view of the basis of social cognition. *Trends in Cognitive Sciences, 8*(9), 396–403.
Goldstein, L., & Fowler, C. A. (2003). Articulatory phonology: A phonology for public language use. In *Phonetics and phonology in language comprehension and production: Differences and similarities* (pp. 159–207).
Hamilton, A. F., & Grafton, S. T. (2007). Action outcomes are represented in human inferior frontoparietal cortex. *Cerebral Cortex, 18*(5), 1160–1168.
Harnad, S. (1987). Psychophysical and cognitive aspects of categorical perception: A critical overview. In S. Harnad (Ed.), *Categorical perception: The groundwork of cognition* (pp. 1–52). Cambridge: Cambridge University Press.
Harnad, S. (2003). Categorical perception. In *Encyclopedia of cognitive science*. Nature Publishing Group/Macmillan.
Izard, C. E. (1971). *The face of emotion*. New York: Appleton-Century-Crofts.
Jeannerod, M. (1994). The representing brain: Neural correlates of motor intention and imagery. *Behavioral and Brain Sciences, 17*(2), 187–202.
Jeannerod, M. (2006). *Motor cognition: What actions tell the self*. Oxford University Press.
Kikutani, M., Roberson, D., & Hanley, J. R. (2008). What's in the name? Categorical perception for unfamiliar faces can occur through labeling. *Psychonomic Bulletin & Review, 15*(4), 787–794.
Kilner, J. M., Friston, K. J., & Frith, C. D. (2007). The mirror-neuron system: A Bayesian perspective. *Neuroreport, 18*(6), 619–623.
Kotsoni, E., de Haan, M., & Johnson, M. H. (2001). Categorical perception of facial expressions by 7-month-old infants. *Perception, 30*(9), 1115–1125.
Liberman, A. M., & Mattingly, I. G. (1985). The motor theory of speech perception revised. *Cognition, 21*(1), 1–36.
Liberman, A. M., & Whalen, D. H. (2000). On the relation of speech to language. *Trends in Cognitive Sciences, 4*(5), 187–196.
Liberman, A. M., Harris, K. S., Hoffman, H. S., & Griffith, B. C. (1957). The discrimination of speech sounds within and across phoneme boundaries. *Journal of Experimental Psychology, 54*(5), 358–368.
Marino, B. F., Borghi, A. M., Buccino, G., & Riggio, L. (2017). Chained activation of the motor system during language understanding. *Frontiers in Psychology, 8*, 199.
McKone, E., Martini, P., & Nakayama, K. (2001). Categorical perception of face identity in noise isolates configural processing. *Journal of Experimental Psychology: Human Perception and Performance, 27*(3), 573–599.
Mylopoulos, M., & Pacherie, E. (2017). Intentions and motor representations: The interface challenge. *Review of Philosophy and Psychology, 8*(2), 317–336.
Nanay, B. (2013). *Between perception and action*. Oxford University Press.
Niedenthal, P. M., Brauer, M., Halberstadt, J. B., & Innes-Ker, Å. H. (2001). When did her smile drop? Facial mimicry and the influences of emotional state on the detection of change in emotional expression. *Cognition & Emotion, 15*(6), 853–864.
Nygaard, L. C., & Pisoni, D. B. (1995). Speech perception: New directions in research and theory. In J. Miller et al. (Eds.), *Speech, language and communication* (pp. 72–75). London: Academic.
Oberman, L. M., Winkielman, P., & Ramachandran, V. S. (2007). Face to face: Blocking facial mimicry can selectively impair recognition of emotional expressions. *Social Neuroscience, 2*(3–4), 167–178.
Pacherie, E. (2008). The phenomenology of action: A conceptual framework. *Cognition, 107*(1), 179–217.
Pavese, C. (2015). Practical Senses. *Philosophers' Imprint, 15*(29), 1–25.

Pitcher, D., Garrido, L., Walsh, V., & Duchaine, B. C. (2008). Transcranial magnetic stimulation disrupts the perception and embodiment of facial expressions. *Journal of Neuroscience, 28*(36), 8929–8933.

Repp, B. H. (1984). Categorical perception: Issues, methods, findings. *Speech and language: Advances in basic research and practice, 10*, 243–335.

Repp, B. H., & Liberman, A. M. (1987). Phonetic category boundaries are flexible. In S. Harnad (Ed.), *Categorical perception: The groundwork of cognition* (pp. 89–112).

Rizzolatti, G., & Sinigaglia, C. (2008). *Mirrors in the brain: How our minds share actions and emotions*. Oxford: Oxford University Press.

Rizzolatti, G., & Sinigaglia, C. (2010). The functional role of the parieto-frontal mirror circuit: Interpretations and misinterpretations. *Nature Reviews Neuroscience, 11*(4), 264–274.

Rizzolatti, G., & Sinigaglia, C. (2016). The mirror mechanism: A basic principle of brain function. *Nature Reviews Neuroscience, 17*(12), 757.

Rizzolatti, G., Camarda, R., Fogassi, L., Gentilucci, M., Luppino, G., & Matelli, M. (1988). Functional organization of inferior area 6 in the macaque monkey: II. Area F5 and the control of distal movements. *Experimental Brain Research, 71*(3), 491–507.

Rizzolatti, G., Fadiga, L., Gallese, V., & Fogassi, L. (1996). Premotor cortex and the recognition of motor actions. *Cognitive Brain Research, 3*(2), 131–141.

Rizzolatti, G., Fogassi, L., & Gallese, V. (2002). Motor and cognitive functions of the ventral premotor cortex. *Current Opinion in Neurobiology, 12*(2), 149–154.

Russell, J. A. (1980). A circumplex model of affect. *Journal of Personality and Social Psychology, 39*(6), 1161–1178.

Shepherd, J. (2019). Skilled action and the double life of intention. *Philosophy and Phenomenological Research, 98*(2), 286–305.

Sinigaglia, C. (2010). Mirroring and understanding action. In *EPSA philosophical issues in the sciences* (pp. 227–238). Dordrecht: Springer.

Tomkins, S. S. (1962). *Affect, imagery, consciousness. Vol. 1: The positive affects*. New York: Springer.

Umiltà, M. A., Kohler, E., Gallese, V., Fogassi, L., Fadiga, L., Keysers, C., & Rizzolatti, G. (2001). I know what you are doing: A neurophysiological study. *Neuron, 31*(1), 155–165.

Umiltà, M. A., Intskirveli, I., Grammont, F., Rochat, M., Caruana, F., Jezzini, A., Gallese, V., & Rizzolatti, G. (2008). When pliers become fingers in the monkey motor system. *Proceedings of the National Academy of Sciences, 105*(6), 2209–2213.

Van der Gaag, C., Minderaa, R. B., & Keysers, C. (2007). Facial expressions: What the mirror neuron system can and cannot tell us. *Social Neuroscience, 2*(3–4), 179–222.

Villiger, M., Chandrasekharan, S., & Welsh, T. N. (2011). Activity of human motor system during action observation is modulated by object presence. *Experimental Brain Research, 209*(1), 85–93.

Whalen, D. H. (2019). The motor theory of speech perception. In *Oxford research Encyclopedia of linguistics*. https://doi.org/10.1093/acrefore/9780199384655.013.404.

Wicker, B., Keysers, C., Plailly, J., Royet, J. P., Gallese, V., & Rizzolatti, G. (2003). Both of us disgusted in My insula: The common neural basis of seeing and feeling disgust. *Neuron, 40*(3), 655–664.

Witzel, C., & Gegenfurtner, K. R. (2014). Category effects on colour discrimination. In W. Anderson, C. P. Biggam, C. Hough, & C. Kay (Eds.), *Colour studies: A broad spectrum* (pp. 200–211). John Benjamins Publishing Company.

Wolfe, J. M., Friedman-Hill, S. R., Stewart, M. I., & O'Connell, K. M. (1992). The role of categorization in visual search for orientation. *Journal of Experimental Psychology: Human Perception and Performance, 18*(1), 34–49.

Young, A. W., Rowland, D., Calder, A. J., Etcoff, N. L., Seth, A., & Perrett, D. I. (1997). Facial expression megamix: Tests of dimensional and category accounts of emotion recognition. *Cognition, 63*(3), 271–313.

Chapter 21
On the Possibility of Multimodal Bodily Immunity to Error Through Misidentification

Krisztina Orbán and Hong Yu Wong

Abstract On the classical internal account of bodily immunity to error through misidentification (IEM), bodily self-ascriptions are immune when they are based solely on perception of one's own bodily properties 'from the inside'. De Vignemont (Bodily immunity to error. In: Prosser S, Recanati R (eds) Immunity to error through misidentification. CUP, Cambridge, 2012) has criticised this account on the basis of the multimodal character of internal perception and the marginality of the cases covered by the internal account. She proposes a multimodal account of bodily IEM. We argue that de Vignemont's account of multimodal bodily IEM is open to counterexamples and thus fails to explain bodily IEM. We catalogue different kinds of multimodal body experience and the self-ascriptions they can support so as to find the difference between the self-ascriptions which are immune and those which are not. We suggest that the difference is due to whether the self-ascription is based on *externally perceiving oneself*. Using this insight, we propose a revised version of the internal account which allows for multimodal bodily IEM. To address the marginality challenge, we also offer a new account of bodily IEM in terms of tracking-freedom, which draws on the style of explanation for the IEM of perceptual demonstratives.

We are grateful to Chiara Brozzo, Herman Cappelen, Malte Hendrickx, Ville Paukkonen, Wesley Sauret, Lucas Thorpe, and especially François Recanati, Matthew Nudds and Alfredo Vernazzani. Earlier versions of this paper were delivered in Freiburg, Istanbul, Rijeka, and Tübingen. We thank the audiences on all these occasions for their reactions. This publication was made possible partly through the support of a grant from the John Templeton Foundation to Hong Yu Wong. The opinions expressed in this publication are those of the authors and do not necessarily reflect the views of the John Templeton Foundation.

K. Orbán (✉) · H. Y. Wong
University of Tübingen, Tübingen, Germany

© Springer Nature Switzerland AG 2021

F. Calzavarini, M. Viola (eds.), *Neural Mechanisms*, Studies in Brain and Mind 17,
https://doi.org/10.1007/978-3-030-54092-0_21

21.1 What Is Immunity to Error Through Misidentification?

Some self-ascriptions have the distinctive property of being immune to error through misidentification relative to 'I' (for short: immune). When I self-ascribe *my legs are crossed* based on proprioception, then I cannot be wrong about whose legs are crossed. My self-ascription *my legs are crossed*, based on proprioception, is immune, because I cannot misidentify whose legs are crossed. For an immune self-ascription made on a basis (such as proprioception), the subject cannot misidentify who is F.

Immunity to error through misidentification (IEM) for a self-ascription (made on a specific basis) excludes one kind of mistake: it guarantees that a self-ascription, F*i* (B), on this basis, B, is such that the subject cannot misidentify who has the property (F). What are the information channels on which self-ascriptions can be based? Some bases are internal informational channels, such as proprioception, whilst others are external information channels, like vision.[1]

Our aim in this paper is to consider whether it is possible to make multimodal bodily self-ascriptions that are immune to error through misidentification relative to 'I' – and if so, to understand what explains their immunity. Frederique de Vignemont (2012) has powerfully criticised the classical internal account of bodily IEM, which draws on distinctive features of bodily awareness. After introducing the internal account of IEM, we examine de Vignemont's arguments that the internal account cannot accommodate the multimodal nature of bodily awareness. We will consider a wide range of different cases of self-ascriptions and reflect on whether they are immune to error through misidentification relative to 'I', using this range of cases as a data set that any successful theory of IEM must be able to classify correctly and explain. We argue that while the challenges she issues to the internal account are powerful, de Vignemont's accounts of multimodal bodily IEM fail. We respond to the challenges by proposing two new accounts of multimodal bodily IEM.

21.2 The Classical Internal Account of Bodily IEM

Certain bodily self-ascriptions can be made on the basis of internal perception. Internal perceptual channels include proprioception (which registers body posture), kinaesthesia (which registers body movement), interoception (which registers hunger, thirst, change of heart beat), nociception (which registers pain), the sense of balance, and the sense of effort, among others. Unlike the exteroceptive (external) senses, such as vision and audition, these internal channels track the properties of

[1]Certain demonstrative judgements, and judgements concerning 'here' and 'now' are immune as well (Evans 1982; Campbell 2002; Peacocke 2008; Prosser and Recanati 2012). However, our discussion is restricted to the IEM of self-ascriptions.

a single object: one's body (Martin 1995). A judgement or thought[2] is based on an internal channel just if that judgement or thought expresses information that is acquired through an internal channel. On the classical internal account of bodily IEM, bodily self-ascriptions are immune when they are based *solely* on perception of one's own bodily properties 'from the inside' (Evans 1982). Call this the *internal mode account of bodily IEM* – the *internal account* for short. Examples of such cases of bodily self-ascriptions are ascriptions of the position or movement of body parts, posture, balance, temperature, and pressure on the basis of internal perception. Typical expressions of the relevant judgements, *based only on internal perception*, would include, for example: 'My arm is bent', 'My wrist is twisting', 'I am standing', 'I am out of balance', 'My tongue is burning', and 'My toe is squished'. These judgements are immune just in case they are based only on internal perception. The internal account (or something very similar) is held by Shoemaker (1968), Evans (1982), and Recanati (2007), among others.[3]

The internal account is motivated by the difference between the statuses of judgements with respect to errors of identification when they are made on the basis of external perception as opposed to internal perception. I can judge 'My legs are crossed' because I perceive them to be crossed from the inside (internal perception) or because I see a pair of legs below to be crossed which look to be mine (external perception). In the former case, my judgement is immune relative to 'I'. It is not in the latter case; I may have mistaken someone else for me. This difference, in turn, is underpinned by the contrast between the possible objects of perception in the external and internal cases. External perception permits a field which may contain multiple objects whilst internal perception allows one to access nothing but one single object, oneself. The internal account relies on the fact that internal perception is underpinned by internal information channels which only transduce information about the system which they are situated within. Such channels are self-reflexive because their functional architecture ensures that only x gains information only about x.[4] To elaborate, (a) there is an identity requirement guaranteeing reflexivity: the one who gains the information, s, and the one who the information is about, x, has to be identical (sGx, *where* $x = s$); (b) there is a requirement that it must be *the subject* who comes to know: only the subject, s, employing the channel, is able to

[2]We will use 'judgement' and 'thought' interchangeably.

[3]Morgan (2015) holds that perceptual experience via a demonstrative to an object grounds immune judgement about that object. The experience enables the subject to know which property to attribute to which object – the object being fixed via a demonstrative. On his demonstrative model of 'I' the reference of 'I' is fixed by a demonstrative (or 'I' is a demonstrative). If such a demonstrative is an internal demonstrative, which relies on internal experience (via internal information channels), then Morgan's account is an elaboration of the internal account.

[4]This does not mean that when information is gained which is in fact about the subject, the subject will take it to be about herself. Some delusional subjects may not take the information to concern herself (e.g. in the case of somatoparaphrenia, when a limb is felt 'from the inside' yet judged not to belong to the subject, but to someone else). (Note also the opposite case of patients who misidentify other people's limbs as their own under experimental conditions. See footnote 12 below.)

gain information through the channel; and (c) the information gained through the channel can only be *about* the subject (Orbán 2014, 2018). Only I can feel that my legs are crossed through proprioception [i.e. (b)] and the information has to be about me [i.e. (c)]. In contrast, consider vision. Vision is not such an internal informational channel, since it is not the case that only the subject can see that her legs are crossed. Not only the subject can gain information about the subject via vision [i.e. not (b)] and the object one sees need not be the subject [i.e. not (c)]. If a channel is not self-reflexive then it is not an internal channel. This is guaranteed by its architecture. The proper functioning of an internal channel requires that the content it delivers is self-specific (i.e. about oneself). Thus, the internal account rests on architectural constraints on internal perception (Orbán 2014, 2018).[5]

The basic idea of the internal mode is not without philosophical precedent. Frege famously claimed that "… everyone is presented to himself in a particular and primitive way, in which he is presented to no-one else" (Frege 1956: 17). In claiming that a subject has special, private access to himself, Frege can be read as claiming that only the subject can gain information through internal channels and only information about himself ($x\mathrm{G}x$ & $x = s$). In this vein, Shoemaker (1968, 1996) emphasises the distinctive first-person access one has to oneself: "Now there is a perfectly good sense in which my *self* is accessible to me in a way in which it is not to others. There are predicates which I apply to others, and which others apply to me, on the basis of observations of behaviour, but which I do not ascribe to myself on this basis, and these predicates are precisely those the self-ascription of which is immune to error through misidentification." (Shoemaker 1968: 562). Shoemaker emphasises that the information channel through which I know that I am in pain is special and is only available to me 'from the inside'; this is not how I know that someone else is in pain.[6]

21.3 De Vignemont's Criticism of the Internal Account

This internal account of bodily IEM is de Vignemont's (2012) critical target.[7] Her starting point is that the distinction between internal and external perception is not so clear cut. The internal account implicitly assumes that internal sensory channels are separate and independent from external sensory channels that open the possibility of errors of identification.

[5]A neurosurgeon may rewire your proprioceptive system so that your brain is connected to another individual's body. Yet an internal channel that is constitutively self-specific would no longer qualify as internal post-rewiring (Orbán 2014). We return to consider rewiring briefly in Sect. 21.5.2.

[6]Martin's (1995) sole-object view of bodily awareness is another important point of reference, though he does not discuss IEM explicitly.

[7]All page references are to this article unless otherwise stated.

Elsewhere de Vignemont (2014) has argued powerfully for the need to understand bodily awareness as constitutively multimodal (see also Wong 2017).[8] We do not have space to survey all the sources of evidence here. First, central cases of body perception, such as the sense of body ownership, appear to be multimodal – as the rubber hand illusion and full body illusions suggest (Botvinick and Cohen 1998; Ehrsson 2012). Second, there is widespread multisensory processing in the brain, including early interaction of visual and somatosensory processing (Calvert et al. 1998). Third, vision shapes both the long-term and short-term body images of sighted people, as can be seen from differences in somatosensory perception compared with congenitally blind subjects (Röder et al. 2004). If IEM is important for understanding the first person and its use, then it has to apply to typical uses of 'I'. Typical bodily self-ascriptions are based on multimodal bases, involving both external and internal information channels. Consequently, if the internal account cannot accommodate multimodal bodily IEM then it cannot be the correct account of IEM. Call this *the multimodality challenge*.

The internal account centres on what she calls the *exclusive thesis*. According to this thesis, "bodily self-ascriptions are immune to error if and only if (i) they are based on somatic [i.e. internal] perception and (ii) there is no further ground" (236). If internal perception is multimodal, then a further challenge arises. Since central cases of bodily awareness are typically multisensory, the internal account only covers marginal cases of self-awareness. Call this the *marginality challenge*. On the internal account, either bodily IEM is so rare as to be marginal or we need to articulate a kind of bodily IEM that isn't dependent only on the internal mode.

Thus, if bodily IEM is not to be a marginal phenomenon limited to isolated unimodal exercises of internal perception, then we must provide an account of multimodal bodily IEM that answers the two challenges.

21.4 De Vignemont's Account of Multimodal Bodily IEM

How does de Vignemont address these challenges? According to de Vignemont, "bodily self-knowledge is primarily multimodal" (235) and there are cases of multimodal IEM based on multisensory integration of external and internal information (which we will discuss below). To capture this, de Vignemont's account rests on two ideas: (i) there are invariant body structures in external perception and (ii) the processes underlying multisensory integration in multimodal body perception are identification-free. They are said to be 'complementary' but it is unclear whether

[8]Following De Vignemont, we understand 'multimodal' for the purposes of this paper as the integration of vision and somatosensation (235, fn. 6). Some of the cases discussed concern bodily IEM based on vision; there the emphasis is on how bodily IEM can also be based on external perception and less on multimodality.

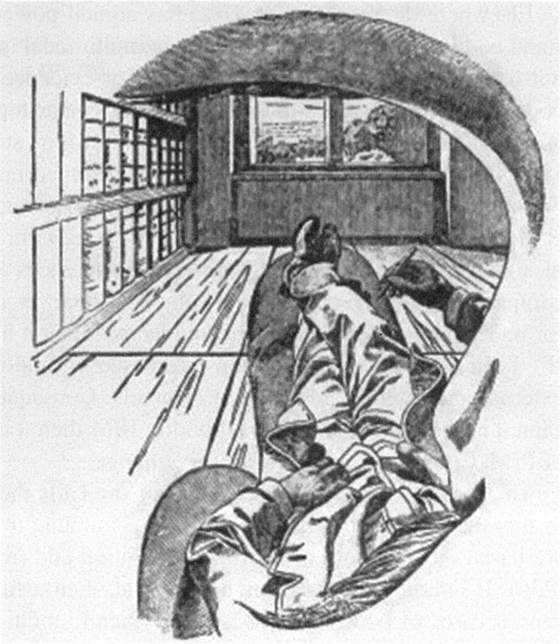

Fig. 21.1 *First-person visuo-spatial perspective of invariant body structures.* Mach's monocular view of himself. (From Mach 1922)

they are parts of one account or two accounts of bodily IEM. They are actually different accounts, as we will show.

De Vignemont's first account exploits the distinctive position of one's nose. Her reasoning is that given one's anatomy, bodily judgements based on experiences of invariant elements of one's own body – such as one's nose – in first-person visuo-spatial perspective are IEM (see Fig. 21.1). This is analogous to how the internal account draws on the fact of our anatomy (the internal loop of the information processing architecture) to argue that proprioceptive judgements are immune. Call this the *nasal account of bodily IEM.*

De Vignemont further argues that the relevant kind of multimodal basis (for an immune self-ascription) is identification-free. This move nicely connects her account to the orthodox characterisation of IEM in terms of identification-freedom (Evans 1982). When a self-ascription made on a basis, like proprioception, is immune then it cannot be that the knowledge of the self-ascription, Fa, is dependent on an identification-component with the logical form $a = b$ – in a way that it relies on presupposing Fb, $a = b$.

The cases of multimodal perception of interest for bodily IEM involve multi-sensory integration (Ernst and Bülthoff 2004). Redundant signals from multiple sensory sources concerning the same property of one object are integrated into a single robust multimodal percept with less variance. Multisensory integration proceeds on the 'unity assumption': only collateral sources of information assigned to the same source are bound together (Welch and Warren 1980). De Vignemont correctly notes that this 'unity assumption' is not akin to an identification postulate at the personal level, but rather a sub-personal process of 'assignment'. Because the relevant cases of multisensory integration involve no identification and, in particular, no self-identification, she concludes that they are identification-free (244–245). Call this the *unity account of bodily IEM*.[9]

21.4.1 Different Kinds of Experiences of Bodily Properties

There are compelling grounds for the multimodality of internal perception, as de Vignemont points out. Accordingly, it is important that we consider a range of different classes of multimodal bodily perception and examples of self-ascriptions based on these classes. We will do this presently. This will allow us to assess whether de Vignemont's nasal account and the unity account can do justice to them.

A. **Visual Proprioceptive Experience**: *Visual proprioception, optic flow and haptic flow*

Gibson (1966) suggested that a classification of sense modalities could be done according to function, as opposed to the kind of sense receptors and ambient energy transduced. He introduced the notion of visual proprioception; he observed that changes in optic flow uniquely specify the subject's own body posture and movement (see Fig. 21.2). This is self-specific exteroception; we can extract self-specific information from a sense modality that is not dedicated to perception of one's own body. Based on experiences of visual proprioception, I can judge 'I am veering to the right' which is immune. This, we suggest, is a counterexample to the internal account.

Another example which has not been considered is haptic flow in haptic perception (Harris et al. 2017; Bicchi et al. 2003). Haptic perception involves active tactile-kinaesthetic exploration of objects. Haptic features, like softness, are detected based on the relative motion of a subject's fingers or skin, which is in contact with an object. Haptic information employed to discriminate softness of an object relies on "how fast the contact area grows when the probing force is increased", which is similar to optic flow where the "distribution of apparent velocities of movement of patterns in an image, [arises] from relative motion between an object and a viewer" (Bicchi et al. 2003).

Similarly to optic flow, haptic flow provides for self-location of the relevant effector (one or more fingers, typically) relative to the explored object. Haptic

[9]In stating her unity account, De Vignemont writes that "the assignment to a common source results from a subpersonal comparative process that does not depend on self-identification [i.e. an identification component [a = me]]" (245). This statement is infelicitous. IEM is a property defined for *judgements* (based on some grounds) and not processes. Thus, we shall read her as claiming that the judgements based on multimodal experiences of looking down are not dependent on self-identification.

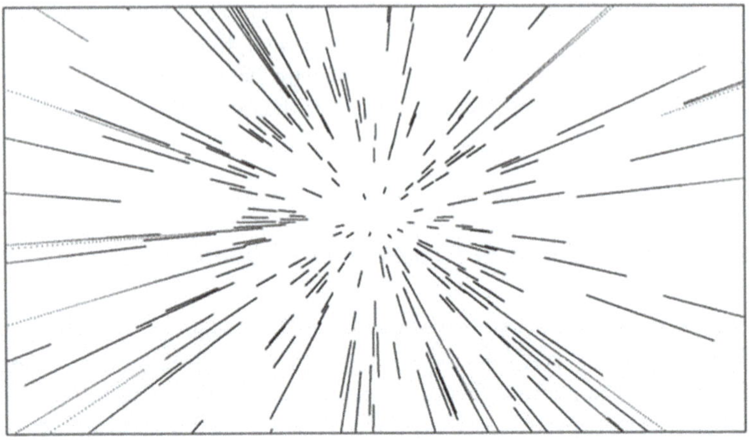

Fig. 21.2 *Visual proprioception*. Optic flow specifying egomotion. (From Kim 2015; used under CC BY)

perception is multisensory because it involves integrating *internal* information from proprioception (including kinaesthesia) and *external* information from tactile perception. In particular, in a case involving haptic exploration with my fingers, there is integration of information from the perception of the movement of the fingers, alongside information from the perception of the spatial relation of the different fingers to each other, and information from tactile perception of the object explored. A judgement 'I feel the object moving in a certain direction (relative to my hand)' based on haptic perception is immune relative to 'I'. This is another counterexample to the internal account.

B. **Self-locating Experiences**: *Self-locating visual and haptic experiences (without perceptual props)*

Typically, when you see some object, you are in a position to locate yourself relative to the object you see. Visual experiences that "represent the environment within an egocentric frame of reference" (240) can support immune judgements. Consider a judgement 'I am standing in front of the Louvre', made on the basis of normal human vision.[10] This is immune. Judgements of this sort and more mundane ones such as 'I am in front of a desk' or 'I am below the chandelier' are immune when they are made on the basis of vision (without perceptual props, such as mirrors). In contrast, self-locating judgements made on the basis of vision involving the use of a mirror are not immune, since one can mistake the person reflected in the mirror for oneself.

[10]By 'normal human vision', we mean to exclude video systems, VR, brain computer interfaces, chips built into the brain and other technical sensory enhancements.

C. **Nose Experiences**: *First-person visuo-spatial perspective on invariant body structures*

Each one of us has a first-person visuo-spatial perspective on our face (the ridge of the eye, one's nose, one's cheeks, and one's mouth in some conditions). Mach depicted the monocular first-person view he enjoyed of himself: "In a frame formed by the ridge of my eyebrow, by my nose, and by my moustache, appears a part of my body, so far as visible, with its environment. My body differs from other human bodies ... by the circumstance, that it is only seen piecemeal, and, especially, is seen without a head." (Mach 1922: 18–19; see Fig. 21.1.)

There are questions about how accurate Mach's self-portrait is of his (and our) everyday experience, since visual experience is typically binocular. In the binocular case, one's nose becomes a blurry object in the centre, with the binocular field framed by the arches of one's eye sockets. But notice that there are important invariants in these experiences: one's nose, and the ridges of one's eye sockets, etc.

Given how our anatomy is, visual experiences with this kind of first-person perspective on these invariant body structures (e.g. one's nose) are self-specific. It would appear that one only sees one's own nose in this way. If this claim is correct, then when I judge 'My nose is red' on the basis of a first-person visual experience of my nose, this would seem to be immune, as de Vignemont claims. This is the basis for her nasal account of bodily IEM (which we will critically examine in the next section). Though we do not dispute that we have a first-person visuo-spatial perspective on invariant body structures, we disagree that this can support immune judgements (see the NOSE case in Sect. 21.4.2.1).

D. **1PP Vision of Body**: *Vision of one's body from the first-person perspective*

Looking down and seeing one's trunk is probably the most familiar instance of this category (see Fig. 21.3). One can often see one's limbs coming into view during locomotion and other activities from the first-person point of view (without perceptual props, such as mirrors, virtual reality systems, video feedback, etc.). The key difference between this and the previous kind of case is that here we remove all the uniquely self-specific features of the face, such as one's nose. De Vignemont calls these cases of 'unspecific first-person perspective'. There are no uniquely

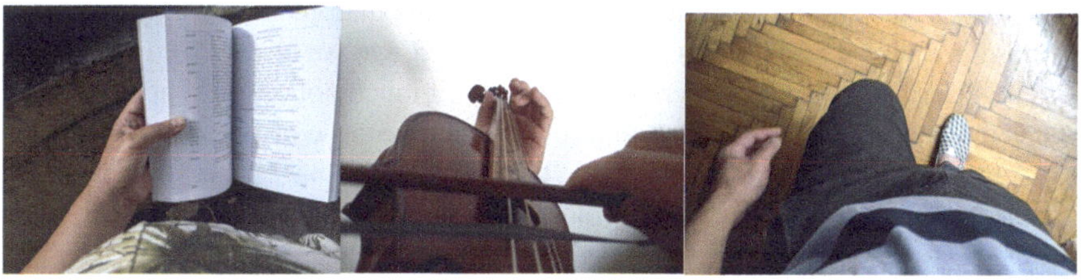

Fig. 21.3 *Vision of body (under normal conditions)*. Left: Looking down at one's hand. Middle: One's hands in action. Right: Looking down at one's trunk. (Photographs by the authors)

self-specific features, but under normal conditions, usually these visual experiences are of one's body. One sees one's fingers busy typing or one's hands dicing the vegetables. For de Vignemont, these cases of unspecific first-person perspective can ground immune judgements. We will challenge this based on the simple fact that looking down, you might misidentify whose hand or leg you perceive, even when this is a case of multimodal integration involving a unity assumption.

E. **Mirror Experiences**: *Specular perspective on myself*

Self-ascriptions based on a specular perspective (involving a mirror) are never immune because the possibility of misidentification is always open in such cases. The judgement 'I have a bump on my forehead' based on looking in a mirror leaves the possibility that I know about someone else (not me) who has a bump on her forehead open.

21.4.2 Examining De Vignemont's Two Accounts of Multimodal Bodily IEM

De Vignemont's account relies on two ideas which are independent, so we will discuss them separately. First we will discuss the invariant anatomical features of the body and its relation to IEM and then we will discuss the account relying on the unity assumption.

21.4.2.1 Examining the Nasal Account

Let's begin with our noses… The anatomical structures which are invariantly situated, depending on one's direction of gaze, include one's nose, the ridges of one's eye sockets, one's cheeks, and (occasionally) one's upper lip. Recall, on de Vignemont's nasal account, the anatomically invariant features provide a secure basis for self-ascriptions which are immune. One cannot be mistaken whose nose one sees or whose nose is red when one sees what is supposedly her own nose from the first-person perspective.

One question is whether the invariant anatomical features are always in one's visual experience. Visual experience is typically binocular. In binocular experience, the nose is a blur if one attends to one's visual field. Do we always (or even often) see our noses? In everyday experience, one doesn't much notice one's nose. What's true is that we normally 'look through' our noses – because of stereoscopy and the proximity of the nose – our noses are 'transparent'. In that sense, it is true that there is no question whose nose it is when one looks through a nose. However, what matters is attribution of properties – and it is hard to attribute anything to a nose one looks through. But let us set this aside. There are two key problems. One, the nasal account would be too narrow even if it worked. Two, there are counterexamples to the account.

Why is the nasal account too narrow? Though we agree there is a first-person visuo-spatial perspective on invariant body structures, (i) it doesn't seem to be a pervasive feature of everyday experience and (ii) the standard cases of bodily IEM judgements are not usually about the invariant body structures in first-person visuo-spatial experience. Thus, even if the nasal account could be made to work, it would be too narrow, and would fail to meet de Vignemont's marginality challenge.

But let us examine the case with more care to see if it supports bodily IEM at all.

NOSE: I see that my nose is red and I judge 'My nose is red' based on what I see.

According to De Vignemont, NOSE is a case of bodily IEM. We disagree. Looking ahead and slightly downwards, I see a nose and it is red. In all probability, my nose is red. However, nothing excludes the possibility that it is someone else's nose even though this is utterly unlikely. Remember that the question is whether misidentification is possible – not whether it is likely.

To make this vivid consider the following three scenarios.

(I) Suppose Gogol has a reconstructed nose because he was in a duel. The nose is removable and it clicks onto his face with a sophisticated mechanism. The bus we are on brakes suddenly and his nose jumps off and clicks onto my face. I say 'My nose is red' but it is Gogol's nose. The point is that it is utterly unlikely that we misidentify whose nose we see, yet it is not excluded as a possibility.

(II) Two lovers are kissing and their noses are in the way. The two noses are in close proximity and touching. One can see the other's nose where her nose typically is. A fleeting glance at her lover results in Heloise seeing a red nose, which she takes to be hers. She judges in haste, 'My nose is red' based on what she saw. But it is Abelard's nose which is red. Again, this is unlikely but not ruled out.

(III) In the Amazon, your friend swats a nasty bug on your nose. You look and see the ridge of your nose and think that you are bleeding. But actually, the bug is bleeding. When you judge 'I am bleeding', based on what you see, you can misidentify who is bleeding on this basis.

These cases are unlikely, but not impossible. So we cannot exclude the possibility of misidentification. Thus, the nasal account of bodily IEM fails.

21.4.2.2 Examining the Unity Account

De Vignemont is sensitive to the fact that the nasal account is too narrow, which is why she also puts forward the unity account.[11] The problem is that the unity account is too broad. It applies to cases where errors of identification are certainly possible. Thus, it, too, fails as an account of multimodal bodily IEM, as we will argue. The unity account predicts that all cases of multisensory bodily awareness based self-ascriptions are immune; this is clearly not the case. Looking down, you might misidentify whose hand or leg you perceive, even when this is a case of multimodal integration involving a unity assumption. This is the class of cases (1PP Vision of Body) which de Vignemont directs her unity account at. We will argue that this class of cases does not support bodily IEM. Moreover, we will argue that the unity account overgeneralises and thus fails as an account of bodily IEM.

We agree with de Vignemont that the most interesting extensions of IEM would be to the range of cases under vision of one's body (1PP Vision of Body). This would allow for an optimal trade-off between explanatory reach and epistemic security of bodily IEM. This appears to be what De Vignemont suggests in talking of cases of looking down; a similar approach is also reflected in the explanatory ambitions of related accounts such as Peacocke's (2012). De Vignemont only mentions that such cases could support immune judgement, but Peacocke provides a concrete example of this. So let us consider his example.

Peacocke's characterisation of first person IEM is of a judgement with a first-person content being immune "when the judgement is reached in a certain way W and in normal circumstances". Examples of immune judgements are (Peacocke 2014: 107): 'I'm in front of a desk' based on a "perceptual experience of being in front of a desk" or 'My arm is broken' based on a "visual experience of your own broken arm, seen as part of your own body". A case like Peacocke's first example will be discussed later (as the MONT BLANC case). The latter example is key for our discussion; this is a case of vision of one's body (1PP Vision of Body).

If you sit adjacent to me, your hand could well be in a position where mine could or even ought to be. If I judge 'My arm is broken' based on a visual experience of an arm, *seen as* part of my body, it could be someone else's arm. Nothing excludes the mistake that I see someone else's arm as mine and as if it were attached to me. IEM would require that this mistake is impossible.

Why should we think that the judgement is not immune? In this case, there is an arm I see and I take it to be mine. The judgement is based on the presupposition that *this arm is mine*. The truth of the judgement ('My arm is broken') is dependent on the truth of the presupposition: the object I see is myself (a=i). And this presupposition can be erroneous. Thus misidentification of whose arm is broken is possible. Moreover, there are actual cases when subjects see someone else's arm

[11] "Bodily self-knowledge most probably derives also from visual experiences that do not guarantee bodily IEM, such as visual experiences of the body from an unspecific first-person perspective" (243).

as their own arm (cf. the pantomime experiment in Wegner 2002).[12] A possible move is for Peacocke to consider this example as an instance of invariant body structures (Nose Experiences), although this is unlikely. There is no good reason for this treatment. Hands may move so they are not invariant in that sense.

For all cases when I look at myself from a visual first-person perspective, there is a possibility that the one who I see is not me. Why is this so? Vision is a multi-object faculty, just like all external perceptual faculties. I may perceive the wrong object as myself. Whenever we found immune self-ascriptions, F*i*, the basis ensured that it has to be the correct object, myself, to which I have grounds to assign the relevant property, F. Nothing ensures that when I think I see *o* then I cannot be mistaken about whether it is *o* which I see. I think I see *o*. But this presupposition is fragile; the object I see may not be *o*. From the fact that I think I see myself, nothing guarantees that I, in fact, see myself. Consequently, when I self-ascribe a property based on seeing myself the self-ascriptions will *never* be immune but they will involve the unity assumption when they are based on multimodal perception.

Accordingly, De Vignemont's and Peacocke's position that visual experience of my own body from the first-person perspective could be immune is precarious. This is because it is difficult to insulate judgements based on vision of one's body from errors of misidentification (as we saw even with invariant structures).

The upshot of our discussion so far is that de Vignemont is deprived of the master example that is illustrative of what her unity account can capture. Now we will argue that the unity account fails on its own terms.

Multisensory integration requires a unity assumption. One feels her body from the inside and sees a body from the outside. The brain computes these sources of information as deriving from the same body (under certain conditions). This is the unity assumption. Bodily self-ascriptions based on vision of one's body allows the possibility that the body is not the subject's body – that is, the unity assumption can be wrong. IEM would require that the unity assumption cannot be wrong, but this is not the case.

All hands agree that the judgement 'My hand is bleeding' based on visual experience of one's hand is such that misidentification is possible. But this is precisely a case to which the unity account applies: it relies on a unity assumption linking the visually perceived and internally perceived object. According to de Vignemont, the unity assumption is not an identification. This is correct. But a self-ascription such as, *I am bleeding* (B*i*), relies on the presupposition that *the object I see is myself* ($a = i$) and *the object I see is bleeding* (B*a*). This has exactly the logical structure of a judgement that is based on an identification component: B*a*, $a = i$ and so B*i* (Evans 1982). As you can see, this case involves both the unity assumption and an identification component. This is because the unity assumption does not

[12]There are also cases of patients who, while they do not explicitly deny that their limbs as belonging to themselves, misidentify other people's limbs as their own in experimental settings (Garbarini et al. 2013; Garbarini and Pia 2013). This gives rise to judgements which are not immune. Note that Peacocke would rule these cases out as not part of 'normal conditions'.

imply that the basis on which the self-ascription was made involves an identification component, but it also does not rule it out. Therefore, having a unity assumption is consistent with there being an identification component. The unity assumption is irrelevant for IEM. This explains why the unity account is too broad; it does not rule out that the ground of the judgement involves an identification component.

All cases of multisensory bodily awareness would have the unity assumption, but not all self-ascriptions involving a unity assumption in their basis are immune. Therefore, the unity account assigns IEM to self-ascriptions which are not immune. It fails as an account of bodily IEM. The unity account fails on its own terms because it overgeneralises to cases which are clearly not immune. This overgeneralisation is due to the fact that the unity assumption does not exclude the presence of an identification component.

To sum up: De Vignemont's account is trailblazing, but it fails to deliver what we need from a theory of bodily IEM. Bodily judgements based on visual experiences of the first two sorts – visual proprioception and self-locating visual experiences – are fine. But once we stray beyond these secure cases, errors of identification are possible.

21.5 Two Responses to the Multimodality and Marginality Challenges

We agree with de Vignemont that a good theory of IEM has to answer both the multimodality and the marginality challenges. It has to explain the IEM of judgments based on multisensory integration for a wide range of cases. In this final section, we develop two ways of responding to the challenges. The first draws on a key insight from the old internal model. The second draws on the way perceptual demonstratives are immune.

21.5.1 The New Internal Mode Account[13]

There is a way to answer the challenges by developing insights from the old internal model. The key is the thought that external perception is what typically opens the possibility of misidentification, in contrast to internal perception. Yet it is also correct that typical cases of perception of our own body are multisensory and some of the self-ascriptions based on such sources are immune. So, to answer the multimodality challenge, we have to explain how multimodal IEM is possible.

I have a visual experience as of my hand being blue and on this basis I judge 'my hand is blue'. The hand I attribute being blue to is part of the external perceptual

[13]The authors disagree on the preferred response to the challenges. The NIM is Orbán's view.

content. There is an object I attribute the property to, which is supposedly my hand, the relevant object for the ascription. Call such content *external relevant-object dependent content*. There is a difference between cases when the external perceptual content contains the relevant object and when it does not contain it. When I look at the Acropolis without any parts of my body coming into view, then I am not part of the external content. Call such content *external relevant-object free content*.

We suggest that what matters is whether the self-ascription is based on *externally perceiving oneself*. On the *new internal mode account of bodily IEM* (NIM, for short), a self-ascription, F*i* (B), is immune when it is not based on externally perceiving the relevant object, *i*. The relevant object for F*x* is the one to which F is attributed: i.e. *x*. In the case of self-ascription, the relevant object will be the subject, *i*. A self-ascription, F*i* (B), is not immune if F*i* (B) is based on externally perceiving (e.g. seeing) the relevant object *i* instantiating the property F. In this case, the content is external relevant-object dependent.

Whenever a self-ascription, F*i* (B), is immune, the external content (if there be such) on which the self-ascription is based has to be relevant-object free. External perception is relevant-object free when its content either does not contain the relevant object at all or the judgement is not based on external observation of the object *i* instantiating the relevant property (F). The second clause is required because the self-ascription cannot be based on externally observing myself to be a certain way (e.g. bleeding). External perception may present an object which is not me to be a certain way when I falsely assume that I am that way. For example, if I am looking at (what is supposedly) my bleeding hand and, based on this, I self-ascribe 'I am bleeding', this judgement will be not immune; this is because I base my judgment on observing (what is supposedly) myself bleeding. The relevant object instantiating the relevant feature is part of the visual content. The one bleeding can be someone else.

Why is external relevant-object free content important for IEM? When the object to which I attribute the property is given through an external channel, then I could be mistaken about whether this object is me. When an attribution of a property is made to an object based on external perception, then two conditions typically have to be satisfied: (1) I have to know of (/be acquainted with) that object the property is attributed to and (2) I have to know of (/be acquainted with) the property which I attribute to that object. Condition (1) is what can open the possibility of misidentification. However, when one knows of the object which she is, one need not think of an externally perceived object and use 'I' for it. That means if we can satisfy condition (2) based on external perception, without the need for satisfying condition (1), then my self-ascription can be based on external content without opening the possibility of misidentification.

What is the difference between knowing about myself externally or otherwise? Knowing about myself externally requires taking an object to be myself: there is an object and its features known externally and I think I am that object which has those features. This can go wrong. In contrast, this is not the case when one knows about herself from the inside e.g. through proprioception. So, the crucial point is that the object for which I use 'I' cannot be part of the external content. This is because if it were part of the external content, it would require an identification (I am that object). Without such identification misidentification is impossible.

External perception allows for the perception of multiple objects while internal perception is perception of a single object. So only external content requires identification of the object as me, because only external content allows more than one candidate to be the object. Internal perception only allows a single object, the subject herself, to be the object which one gains information about. So, whenever the object to which I attribute the property is not part of the external content, then my self-ascription based on this will be immune. This is true even if the property attributed to that object is based on external perception (condition 2 above). In this case, though the self-ascription is based on external perception, it is based on external object-free content, so misidentification cannot happen.

The reason why the object cannot be misidentified is because it need not be singled out and identified from a multiplicity of objects, like in external perception. There are at least two way this can be accomplished: either (i) the object is part of a structural presupposition of the experience which cannot be questioned (the subject of experience) or (ii) it is an internally known object, where the only object known in this way is the subject. The subject of experience or the internally known object need not be identified or singled out – in a same way that you do not need to first search for your hand in order to reach with it. NIM explains the IEM of multimodal (external and internal content based) self-ascription.

To test NIM, we will try out whether NIM can deliver an explanation of multisensory IEM and can account for the difference between cases which are immune and those which are not.

Let's return first to the NOSE case, which we met with above in discussing de Vignemont's nasal account. I judge: 'My nose is red' based on what I see. We argued that NOSE is not immune. Nothing excludes that it is someone else's nose even though this is utterly unlikely. What explains this? The basis for the judgement is relevant-object dependent external content. So, according to NIM, NOSE should be not immune relative to 'my nose' on this basis. This is what we have seen above.

To test whether our account in terms of external relevant-object free bodily ascriptions provides the correct result, consider the following cases:

> HAND: I see that my hand is blue and I judge 'My hand is frozen', based on what I see and my feeling my hand to be very cold.

When I see that my hand is blue and I think it is frozen, then my visual content contains the object which I attribute being frozen and being blue to: my hand. It could be someone else's hand which I see to be blue and infer to be frozen, even if I feel that my hand is very cold 'from the inside'. Therefore, I can still misidentify whose hand is frozen. In this case, seeing my hand is externally relevant-object dependent content. Only judgements based on externally relevant-object free content are immune. HAND is relevant object-dependent and thus is not immune. Consequently, NIM explains how misidentification can happen.

In contrast, in other cases the object to which I attribute a property is not part of the content of external experience:

MONT BLANC: When I see that the summit of Mont Blanc is just in front of me and I can attack it, then I form the thought 'I am facing the summit of Mont Blanc' based on vision.

In 'I am facing the summit of Mont Blanc' the object to which I attribute the property of facing the summit of Mont Blanc is not part of the visual content. I am not looking at any of my body parts. Thus the relevant object to which I attribute the property is not part of the externally gained content. I do not attribute a property to myself because I perceive an object, take it to be myself, and I presuppose that 'that object is facing the summit of Mont Blanc'. I cannot misidentify who is facing the Mont Blanc. The explanation, once again, is that the basis of the judgement is externally relevant-object free; I am not part of the external content.

In the case of multimodal IEM, the quality which I attribute could be such that I need external sources to know about that quality but not about its instantiation in an object. The quality in some way could be part of the external content but the object to which I attribute it cannot be part of the external content. Consider the following case for an illustration.

BALANCE 1: I judge 'I am out of balance', based on visual proprioception integrated to the sense of balance.

In BALANCE 1 it is not the case that I observe myself by external means where I am part of the content. I do not see an object being out of balance, think that this object is me and, based on this, attribute being out of balance to myself. I cannot misidentify who is out of balance on the basis of visual proprioception. Thus, this judgement, based on visual proprioception integrated to the sense of balance, is immune. This can be explained by the fact that the basis is relevant-object free. The same strategy can cover classical cases from the internal model:

PAIN: I judge 'I am in pain' based on nociception.

PAIN is the paradigmatic example of the immune self-ascription based on internal perception. This judgement is external relevant-object free because it is not based on externally observing an object to be in pain and thinking that this object is me. Thus it is immune because it is externally relevant-object free.

In short, when the object to which one attributes the property is not part of the externally acquired content, then the relevant object is not given through external perception and it is not possible to make a mistake about whether that object is me. This excludes the possibility of misidentification. When one knows about herself internally one cannot be mistaken about which object she knows about.

Recall that external perception is relevant-object free when its content either does not contain the relevant object at all or the content is not based on external observation of the object i instantiating the relevant property (F). Let us consider the second disjunct.

Intuitively, the idea is that if the attribution of the relevant property is based on an identification then misidentification is possible. But if you can have attribution of the property based on external perception, but not based on observing the relevant object instantiating the property, then the attribution is not based on identification. Do we have cases like this?

BALANCE 2: I judge 'I am out of balance' based on visual proprioception, which is integrated with my sense of balance, but where my hands and legs come into view (visually).

Here even though my hands and legs come into view (visually), the judgement remains immune. It is because there cannot be another candidate about whom I know – on this basis – that she is out of balance. So I cannot misidentify who is out of balance on this basis. Thus the self-ascription on this basis is immune; misidentification of who is out of balance is not possible.

I am not attributing being out of balance because I see an object out of balance and I think that object is me. What matters is that the attribution of the property, F, cannot be based on external observation of the relevant object being F. Cases like this show that deciding whether a self-ascription is dependent on an identification component (or not) is a delicate matter. External content could be relevant-object free, yet include a part of the body which is *irrelevant* for the attributed property. When I judge 'I am out of balance' on the basis of vision – exploiting visual proprioception – my judgement is immune (BALANCE 1). The only difference between BALANCE 1 and BALANCE 2 is that I see my hands and legs in the latter case.

Would seeing my hand mean that the judgement loses its immunity? No, in this case I am *not* attributing being out of balance *because* I see myself being out of balance. This is because seeing hands in one's visual field – even if they are in a weird position – does not license me to judge 'That person is out of balance'. Thus, this self-ascription will be immune because seeing my hand does not ground the self-ascription. So, for a self-ascription, F*i* (B), to be immune, the relevant object required for the attribution should *not* be based on perceiving *i* as instantiating the property (F) in the external content. (Consider for contrast the HAND case. Looking at a hand and *seeing that it is blue grounds* the self-ascription of thinking that it is frozen. Therefore, the judgement in HAND is not immune.)

Let us introduce a new tactile case to test the theory properly:

TOUCH: 'I feel that this object has a rough texture', based on haptic perception including proprioception.

In this case, the judgement is immune because it cannot be misidentified who feels the object to be such and such. This is because the object who is doing the touching is not part of the external content. Thus, TOUCH is external relevant object free in this case. What is acquired through external perception is only the texture of the object, but not the knowledge of who feels the texture. For this reason, the judgement is immune relative to 'I' because it is based on external object-free content.

When do we have immune self-ascriptions based on external content? It is correct that self-ascriptions based on the first two kinds of experiences in our list (visual proprioception and self-locating visual experiences) are immune, but only on the condition that they are based on external relevant-object free content. According to the NIM, the bodily self-attribution, F*i* (B), will be open to misidentification when the self-attribution is based on externally observing an object being F (externally relevant-object dependent content). The reason for this is simple. The object which is externally perceived as being F, *a*, might not be the subject. Such cases necessitate an identification that the perceived object is the subject ($a = i$). In such cases the object, *a*, to which the subject attributes the relevant property (F), is part of the content gained through an external information channel. An object is observed to be F and thought to be the subject ($a = i$). 'I am that object' is an identification component which opens the possibility of misidentification, if the attribution of the relevant property is based on it. NIM is drawing on Evans's identification-freedom characterisation of IEM to develop an account of multimodal bodily IEM. According to NIM, a bodily self-attribution, F*i* (B), is immune iff the self-attribution is externally relevant-object free (not based on externally observing the object to be F).

21.5.2 A Test: Rewiring

If NIM is correct, it should have a clear answer to a classical challenge to bodily IEM: rewiring (Wittgenstein 1958: 54, Cappelen and Dever 2013). Suppose my proprioception is rewired in a way that I not only receive information of my body posture but I can receive information from Mary's body posture as well. Whatever I have is *not* internal perception because the information channels providing information about Mary are not self-reflexive once they are rewired. Internal channels of information per definition are those which are self-reflexive. Internal content (gained through an internal channel) cannot presuppose an identification of which object is perceived precisely because only one object can be perceived. There is neither a need nor a possibility to identify one candidate from among many candidates who might have the property ascribed on this basis, precisely because there is no more than one candidate. In contrast, for rewired proprioception, I have to identify whether the object I perceive is Mary or me if I am in a position to know about Mary or me at all. Thus, rewired proprioception is not an internal information channel providing internal content. If an information channel must either be classified as internal or external, then 'my legs are crossed' based on rewired proprioception will be external relevant-object dependent. If rewired proprioception is considered neither internal nor external, the judgement will still be based on perception of multiple objects where the relevant object has to be identified.

If the perceptual content has multiple objects, as in the case of rewired proprioception, then one *has to identify the relevant object.* This suffices to open the possibility of misidentification; the relevant object cannot come from internal perception where there cannot be more than one candidate. So, rewired proprioception is relevant-object dependent because it relies on the identification of the relevant object. In such a case I acquire knowledge about myself through a non-self-reflexive and thereby external basis. Thus, judgements based on rewired proprioception are never immune.

Trouble only comes when all of the internal ways (including introspection) are rewired without exception.[14] In this case there is no longer the possibility of IEM, whether bodily or mental. A creature like that is very different from us and it is not clear that such creatures would be able to use 'I'. Their use of 'I', if it were possible, would be relevantly different from our use.[15] Such a creature cannot be sure that she thinks about her own body when she receives information of a body which is supposed to be hers. The kind of security which bodily IEM provides is only available to creatures with internal information channels – and (*de se*) self-representations based on internal channels cannot fail their self-representational function. This suggests that the functional architecture of internal perception may be crucial for understanding our use of 'I'. We may only be able to use 'I' because we have reliable internal information channels.

21.5.3 The Ecological Model[16]

We have seen that the NIM can deal with a range of multimodal cases and deliver an attractive account of multimodal bodily immunity. But one might say that the NIM does not fully meet the marginality challenge, since any case where the attribution is based partly on externally observing the object to be F would not support an IEM judgement and one might claim that the bulk of ordinary cases of self-attribution are of this sort. It is an open empirical question what the natural statistics of the range of cases of multimodal bodily IEM is. But if the bulk of ordinary cases of self-attribution are indeed based partly on external observation, then even though we have expanded the range of cases which would support IEM judgement, it could still be said with some justification that IEM remains somewhat marginal. This would not immediately prevent those judgements which are IEM or those situations which could ground IEM judgements from having a special significance, since it might be claimed that these are necessary for having self-attribution at all as Shoemaker (1968) famously claimed. Though we are sympathetic to this strategy, we will not attempt to argue for this claim here. Instead, we will propose a second model in the

[14]In schizophrenic patients, it sometimes happens that they think they know of someone else through internal information channels. In these cases, they ascribe content gained through such channels to external subjects. But IEM is only about self-ascriptions. It does not rule out the possibility that the subject in a delusional condition can ascribe the relevant property to the wrong person.

[15]How about introspection? Either it can be rewired or not. If it cannot be rewired then there is always an internal information channel available to the subject: introspection. If introspection can be rewired then our prediction is that the subject might not be able to use 'I', only 'I*', a different kind of self-referring expression.

[16]This is Wong's preferred response to the challenges. We wish to thank Matthew Nudds for discussion.

spirit of the NIM, but which is more permissive. Call this the *ecological model of bodily IEM* (for short: ecological model).

At the heart of the ecological model is the idea of multimodal tracking of individuals. The thought is that when the multimodal tracking of individuals is correct, then we are in a position to make multimodal judgements that are IEM on the basis of the multimodal perception that is tracking the individual as the same individual across different sensory modalities. It is easy to see how this account works in the case of multimodal perceptual demonstrative judgements. 'This object is round', made on the basis of sight and touch with the object sitting in one's hand in the case where the individual is correctly tracked, is immune relative to 'this object'. This is an extension of the standard account of the IEM of perceptual demonstrative judgements to the case of multimodal perceptual demonstrative judgements. The underlying thought is the same: on the very basis for which the reference of the perceptual demonstrative is fixed, misidentification is impossible because nothing else is a candidate for the predication (Evans 1982; Campbell 2002; Peacocke 2008). For multimodal perceptual demonstrative judgements, what is required is that we have correct multimodal (or crossmodal) tracking of the individual the judgements concern. In effect, if the unity assumption is correct – that is, tracking is successful – then the perceptual demonstrative judgement is immune. This follows not because of de Vignemont's reasoning that the unity assumption is not an identification component and hence the judgement remains identification-free, but because the tracking apparatus is locking on to one individual across sensory modalities. The thought is that tracking error opens errors of identification. Thus, when we have instances of multimodal perception of individuals which is multimodal tracking error-free then we don't have the possibility of an error of misidentification.

So far we have a model of immune multimodal perceptual demonstrative judgements. How can we develop an ecological model of immune multimodal self-ascriptions? If one would accept a demonstrative model of 'I' (Campbell 1994; Morgan 2015), then this would be straightforward. But we reject the model partly because of the reasons discussed by Campbell (1994). So we would have to develop a model based on the same strategy, which is not a demonstrative model of self-ascription. Let us observe that in cases of bodily self-ascriptions with a multimodal perceptual basis (i.e. they are based on both external and internal perception), the only thing which opens the possibility of misidentification is a mistake of crossmodal tracking. For example, on a crowded bus, I might think that a gloved hand in my peripersonal space in a position which is anatomically plausible for my hand to be at and which is roughly consistent with my proprioceptive awareness of hand position to be my hand when it is not. In this case I have made a mistake of cross-modal tracking. That hand is not mine. Conversely, when multimodal perceptual tracking is accurately locking on to the hand that is mine in vision, haptics, action, and proprioception, then there is no possibility of misidentification. Only cross-modal tracking mistakes open the possibility of misidentification for multimodal self-ascriptions. Thus, when there is no cross-modal tracking error for a self-ascription based on multimodal perception, then this self-ascription will be immune. This view is an ecological view because it would appear to cover the overwhelming majority of cases where we have perception of our own body. (We can see this model as achieving what the projects from de Vignemont and Peacocke sought to achieve with the unity assumption and with normal conditions, respectively.) On this view, IEM is not marginal and multimodal self-ascriptions are immune, except in cases where there are cross-modal tracking mistakes.

We suggest that theorists may pick between the two theories depending on whether they have more internalist or externalist epistemological proclivities. Note that even if a self-ascription is immune based on multimodal perception which is tracking error-free, the subject may think that there was a tracking mistake. (For example, he may have felt that he didn't pay sufficient attention to his leg as he made a self-ascription.) On the ecological model, the explanation of the IEM of self-ascription is dependent solely on the success of the multimodal tracking. It is based on the fact of the psychological mechanism successfully underpinning singular thought. On the NIM, the subject has some access to whether his judgements are IEM, something that is not the case on the ecological model, since you may not know whether you made a tracking error. The two models are compatible, but can be held independently.

21.6 Conclusion

We have argued that de Vignemont's account of multimodal bodily IEM fails. However, through her criticism of the internal account, she has raised new and important challenges for any new account of bodily IEM. According to the internal mode account whenever a self-ascription is immune it has to be based solely on internal basis (like proprioception). De Vignemont criticises the internal account of IEM on two fronts. First, the internal account only covers atypical marginal cases. This is the marginality challenge. Second, the internal account cannot answer for the multisensory character of bodily awareness and IEM based on a multimodal basis. This is the multimodality challenge. We agree with de Vignemont that bodily awareness is multisensory and judgements based on visual proprioception and self-locating visual judgements can be immune. However, we need a principled way to decide which judgements are immune and explain how multisensory IEM is possible for a wide range of cases. In particular, we need to be able to answer both the multimodality and marginality challenges.

To do this, we proposed two models: the new internal model (NIM) and the ecological model. According to NIM, a bodily self-attribution, F*i* (B), is immune relative to *i* iff the basis of the self-attribution is externally relevant-object free (not based on externally perceiving the relevant object, *i*, to be *F*). The basis of a self-ascription, F*i* (B), is externally relevant-object free iff *either* (i) the object to which the self-ascription attributes the property, *F*, is not part of the external content *or* (ii) the self-attribution is not based on externally observing the object to be *F*. When I am experiencing optic flow, I judge 'I am out of balance', based on visual proprioception and the sense of balance, then this visual content is external relevant-object free [condition (i)]. And the ground is external relevant-object free even if I see my hand [condition (ii)]. This is because I do not base my self-ascription of being out of balance on seeing my body being out of balance. In contrast, my judgement 'I am out of balance' would *not* be immune based on seeing myself in the mirror or in a live video or seeing my shadow as being out of balance. In this case, I would see someone out of balance and my self-ascription would be based on this (condition ii). The NIM explains why not all, but only some multisensory cases are included. Consequently, the NIM provides a simple explanation for why on certain external bases bodily self-ascriptions are immune while on relevant-object dependent external bases they are not immune. We also sketched another account in terms of multimodal tracking error-freedom: ecological IEM. The key idea is that when the multimodal tracking of individuals is correct, then we are in a position to make multimodal judgements that are IEM on the basis of the multimodal perception that is tracking the individual as the same individual across different sensory modalities. This view is an ecological view of IEM because it would appear to cover the overwhelming majority of cases where we have perception of our own body. On this view, IEM is not marginal and multimodal self-ascriptions are immune, except in cases where there are cross-modal tracking mistakes. The two models are compatible but can be held independently. Both NIM and the ecological model provide for the possibility of multimodal bodily IEM.

References

Bicchi, A., Dente, D., & Scilingo, E. P. (2003). Haptic illusions induced by tactile flow. In *Proceedings of eurohaptics*, pp. 314–329.

Botvinick, M., & Cohen, J. (1998). Rubber hands "feel" touch that eyes see. *Nature, 391*, 756.

Calvert, G. A., Brammer, M. J., & Iversen, S. D. (1998). Crossmodal identification. *Trends in Cognitive Science, 2*, 247–253.

Campbell, J. (1994). *Past, space, and self.* Cambridge, MA: MIT Press.

Campbell, J. (2002). *Reference and consciousness*. Oxford: OUP.

Cappelen, H., & Dever, J. (2013). *The inessential indexicals*. Oxford: OUP.

de Vignemont, F. (2012). Bodily immunity to error. In S. Prosser & R. Recanati (Eds.), *Immunity to error through misidentification* (pp. 224–246). Cambridge: CUP.

de Vignemont, F. (2014). A multimodal conception of bodily awareness. *Mind, 123*, 989–1020.

Ehrsson, H. H. (2012). The concept of body ownership and its relation to multisensory integration. In B. E. Stein (Ed.), *The new handbook of multisensory processes* (pp. 775–792). Cambridge, MA: MIT Press.

Ernst, M., & Bülthoff, H. H. (2004). Merging the senses into a robust percept. *Trends in Cognitive Sciences, 8*, 162–169.

Evans, G. (1982). In J. McDowell (Ed.), *The varieties of reference*. Oxford: Clarendon.

Frege, G. (1956). The thought: A logical inquiry, reprinted in Ludlow *Readings in the philosophy of language.* Cambridge, MA: MIT Press.

Garbarini, E., & Pia, L. (2013). Bimanual coupling paradigm as an effective tool to investigate productive behaviors in motor and body awareness impairments. *Frontiers of Human Neuroscience, 7*, 1–5.

Garbarini, F., Pia, L., Piedimonte, A., Rabuffetti, M., Gindri, P., & Berti, A. (2013). Embodiment of an alien hand interferes with intact-hand movements. *Current Biology, 23*(2), R57–R58.

Gibson, J. J. (1966). *The senses considered as perceptual systems*. Boston: Houghton Mifflin.
Harris, L. R., Sakurai, K., & Beaudot, W. H. (2017). Tactile flow overrides other cues to self motion. *Scientific Reports, 7*(1), 1–8.
Kim, N.-G. (2015). Perceiving collision impacts in Alzheimer's disease: The effect of retinal eccentricity on optic flow deficits. *Frontiers in Aging Neuroscience, 7*(218). https://doi.org/10.3389/fnagi.2015.00218.
Mach, E. (1922). *Die Analyse der Empfindungen*. Jena: Gustave Fischer Verlag.
Martin, M. G. F. (1995). Bodily awareness: A sense of ownership. In Bermúdez et al. (Eds.), *The body and the self*. Cambridge, MA: MIT Press.
Morgan, D. (2015). The demonstrative model of first-person thought. *Philosophical Studies, 172*, 1795–1811.
Orbán, K. (2014) *Fixing the reference of 'I': Immunity to error through misidentification as a guide*. PhD, Birkbeck, University of London.
Orbán, K. (2018). The view from nowhere: The zero perspective view of internal perception. *Teorema, XXXVII*(3), 39–63.
Peacocke, C. A. B. (2008). *Truly understood*. Oxford: OUP.
Peacocke, C. A. B. (2012). Explaining de se phenomena. In S. Prosser & R. Recanati (Eds.), *Immunity to error through misidentification* (pp. 144–157). Cambridge: CUP.
Peacocke, C. A. B. (2014). *The mirror of the world: Subjects, consciousness, and self-consciousness*. Oxford: OUP.
Prosser, S., & Recanati, R. (2012). *Immunity to error through misidentification*. Cambridge: CUP.
Recanati, F. (2007). *Perspectival thought*. Oxford: OUP.
Röder, B., Rösler, F., & Spence, C. (2004). Early vision impairs tactile perception in the blind. *Current Biology, 14*, 121–124.
Shoemaker, S. (1968). Self-reference and self-awareness. *The Journal of Philosophy, 65*, 555–567.
Shoemaker, S. (1996). Self-knowledge and "inner sense". In Shoemaker (Ed.), *The first person perspective and other essays*. Cambridge: CUP.
Wegner, D. (2002). *The illusion of conscious will*. Cambridge, MA: MIT Press.
Welch, R. B., & Warren, D. H. (1980). Immediate perceptual response to intersensory discrepancy. *Psychological Bulletin, 88*, 638–667.
Wittgenstein, L. (1958). *The blue and the brown books*. Oxford: Blackwell.
Wong, H. Y. (2017). On proprioception in action: Multimodality versus deafferentation. *Mind & Language, 32*(3), 259–282.

The manufacturer's authorised representative in the EU is Springer Nature Customer Service Centre GmbH, Europaplatz 3, 69115 Heidelberg, Germany. If you have any concerns regarding our products, please contact ProductSafety@springernature.com

Printed and bound by CPI Group (UK) Ltd, Croydon, CR0 4YY
30/09/2026
02225452-0017